OCEAN YEARBOOK 9

OCEAN
YEARBOOK 9

Pacem in Maribus

Sponsored by the
International
Ocean Institute

Edited by
Elisabeth Mann Borgese,
Norton Ginsburg, and
Joseph R. Morgan

Assistant Editor: Glenys Owen Miller
Editorial Assistant: Susan L. Biggs

Supported by the East-West Center

The University of Chicago Press

Chicago and London

The University of Chicago Press, Chicago 60637
The University of Chicago Press, Ltd., London

©1991 by The University of Chicago
All rights reserved. Published 1991
Printed 1991 in the United States of America

International Standard Book Number: 0-226-06612-6
Library of Congress Catalog Card Number: 79:642855

Contents

Coastal Management

Military Activities

Regional Developments

Appendices

Acknowledgment to the East-West Center

The International Ocean Institute is indebted to the East-West Center for its valuable support of and contributions to the *Ocean Yearbook*. The center houses its editorial offices and provides essential infrastructural support as well as the salary of an assistant editor. Even more important are the substantive contributions of the research staff of the Oceans and Atmosphere Program of the center's Environment and Policy Institute (EAPI), with which the *Yearbook* is closely integrated. Two of the coeditors of the *Yearbook* and its assistant editor are members of the EAPI research staff.

The East-West Center is a public nonprofit educational institution with an international board of governors. Its objectives are research and training concerning important issues in Asia, the Pacific, and the United States, in collaboration with scholars and policymakers in those areas. The center was established in 1960 by the Congress of the United States, which provides principal funding, although support also comes from more than 20 Asian and Pacific governments as well as private institutions. The EAPI of the center was established in October 1977, to further understanding of the relationships between resource endowments and developmental policies within Asia and the Pacific and to convert that understanding into practical programs designed to enhance human welfare within the region.

THE EDITORS

Acknowledgments

Many individuals and organizations have contributed to this issue of the *Ocean Yearbook*, but only a few can be cited here. We are grateful to the Intergovernmental Oceanographic Commission of Unesco, particularly for assistance in distributing copies of the *Yearbook* to institutions in developing countries. Warm thanks also go to the board of editors for numerous formal and informal contributions and to the University of Chicago Press for its editorial and infrastructural support. Special thanks go to the University of Chicago Press for its competent coordination of *Ocean Yearbook* matters at the press.

Many of the maps and illustrations in *Ocean Yearbook 9* are the creation of Lynn Indalecio. Nicholas Dunning and David Berner updated the tables, and David Berner compiled the index. Dalhousie University's Department of Political Science and the Lester Pearson Institute for International Development there provided valuable material support.

THE EDITORS

Perestroika and the Law of the Sea

Elisabeth Mann Borgese
Dalhousie University

INTRODUCTION

Throughout the 20 years of the genesis of the new Law of the Sea, we have considered the oceans as the great laboratory for the making of a new world order. The convention that emerged in 1982 has officially been characterized as a "Constitution for the Oceans," which means, potentially, a constitution for the world. Although not yet officially in force, the United Nations Convention on the Law of the Sea (LOS Convention) has already shaken the existing order and driven the engines of change and innovation.[1]

A great deal of work has been accomplished to adumbrate the main requirements of a comprehensive new world order. The official documents of the United Nations—the Declaration of Economic Rights and Duties of States and the Declaration on a New International Economic Order and the Plan of Action—are, retrospectively, conservative. Had they been implemented, they would have served to prop up a dying economy, rather than build a new one. Fairer terms of trade, better prices for commodities, debt relief, sovereign rights over natural resources, and a code of conduct for multinationals would not have been attainable in the present structural context, which the official programs did not attempt to change. Various proposals—the Brandt Commission, the Palme Commission, the Brundtland Commission, the South Commission, and the Human Rights Commission—pointed farther in the direction of change, but have remained on the drawing board.[2] One, the most recent, Gorbachev's *perestroika,* is changing the face of the earth.

1. *Law of the Sea: Official Text of the United Nations Convention on the Law of the Sea with Annexes and Index* (New York: United Nations, 1983), sales no. E.83.V.5, hereafter cited as LOS Convention. For lists of signatories to the LOS Convention, see App. G in this volume.
2. The report of the Brandt Commission focuses on the financing needed to narrow the development gap to acceptable dimensions. It envisages that the needed funds should come primarily from vastly increased official development aid (Willy Brandt et al., *North-South: A Programme for Survival: The Report of the Independent Commission on International Development Issues* [London: Pan Books, 1980]).

The report of the Palme Commission focuses on the military aspects of security.

Where does it come from, and why is its effect so dramatic, if not traumatic?

REALISM AND VISION

Perestroika, which means "restructuring," appears to have been generated by realism and vision. *Realism* is understood here as a pragmatic and realistic assessment of the situation surrounding us here and now. This situation had become untenable and explosive. Economic growth in the Soviet Union and its allies was declining to a level close to stagnation. Technological innovation was restricted to the military sector, while there was an obvious lack of efficiency in using scientific achievements for economic needs. Consequently, for all "gross output," there was a shortage of goods. The Soviet Union spent, in fact is still spending, far more on raw materials, energy, and other resources per unit of output than other developed nations.

Economic stagnation went hand in hand with intellectual stagnation. Creative thinking was driven out from the social sciences. Intellectual barrenness was breeding corruption; alcoholism, drug addiction, and crime were growing; and society was becoming increasingly unmanageable.

The combination of this situation with underlying racial, religious, or national tensions was a recipe for explosion, and the pressure was gathering within the republics of the Soviet Union and those surrounding it. In Gorbachev's words, "this society is ripe for change. It has long been yearning for it. Any delay in beginning perestroika could have led to an exacerbated international situation in the near future which, to put it bluntly, would have been fraught with serious social, economic, and political crises."[3]

Gorbachev did not create this situation. He inherited it, but he had the courage to recognize it and to act. He tried to release the pressure gently by raising the lid which would have been blown with violence in the near future. At the same time he tried to provide direction to the released energies. This

It puts forward the concept of "common security" and proposes the most advanced institutional framework for its implementation (Olof Palme et al., *Common Security: A Programme for Disarmament: The Report of the Independent Commission on Disarmament and Security Issues* [London: Pan Books, 1982]).

The report of the Brundtland Commission focuses on environmental security. It puts forward the concept of sustainable development that is to reconcile developmental and environmental concerns (G. H. Brundtland et al., *Our Common Future: The Report of the World Commission on Environment and Development* [Oxford: Oxford University Press, 1987]).

The report of the South Commission focuses on the needed changes in the south itself (South Commission, *The Challenge to the South: The Report of the South Commission* [Geneva: South Commission, 1990]).

3. Mikhail Gorbachev, *Perestroika: New Thinking for Our Country and the World* (New York: Harper and Row, 1987), p. 3.

required vision: the vision of a better future, of a genuinely new order, which, in his thought, appears to rest on certain pillars.

BASIC PRINCIPLES: PILLARS OF *PERESTROIKA*

This is the situation from which *perestroika* takes off within the USSR. The pressure of the arms race, on the one hand, and rigidity and isolation of socioeconomic structures on the other, have paralyzed the country. The arms race must be stopped; rigidity and isolation must be broken. The required changes are dramatic. They also are mutually dependent. Perhaps there is a common denominator: a new relationship between individual initiative and common cause.

The new, integrative relation between the individual and the collective has two further implications: It implies an integrated concept of what is "inside" and what is "outside": that is, the recognition of inseparable linkages between domestic and foreign issues and policies. It equally implies an integrated concept of the relationships between *continuity* (humankind extends in time as it does in space, comprising present as well as past and future generations) and *change* (which is episodic, the part that makes up the whole, but also depends on that whole, without which it cannot take place), and between *long-term* and *short-term*, neither of which can be conceived without the other.

Individual Initiative and Common Cause

Gorbachev emphasizes the need for blending private and public values and interests. "What is needed to achieve this aim is finding the most effective and modern forms of blending public ownership and the personal interest that is the ground work for all our quests, for our entire concept of radically transforming economic management."[4] And again: "We believe that combining personal interests with socialism has still remained the fundamental problem. . . . Then we will combine the advantages of a large collective economy with the individual's interests."[5]

Socialism and Market

Obviously this integration of public and private values and interests can never entail a repudiation of socialism and an embrace of capitalism. "There was an opinion, for instance, that we ought to give up planned economy and

4. Ibid., p. 19.
5. Ibid., pp. 82–83.

sanction unemployment. We cannot permit this, however, since we aim to strengthen socialism, not replace it with a different system. What is offered to us from the West, from a different economy, is unacceptable to us."[6]

Gorbachev calls for decentralization of government responsibilities, which must be devolved on the enterprises themselves, including the transfer of cost accounting, a radical transformation of the centralized management of the economy, fundamental changes in planning, and a reform of the price formation system and of the financial and crediting mechanism. Planning must start at the grass roots, at the enterprise level. This, however, does not mean the abandonment of planning. It makes planning more complex and brings it closer to people, that is, to the demand side. The enterprise itself must be democratized. Workers must be fully involved in the decision-making process, and they must have the right to elect their own managers.

The Local and the Universal

Internal progress largely depends on international conditions. Coping with the internal consequences of the arms race, of technological change, and of environmental degradation requires changes in the international system. "Now the whole world needs restructuring, i.e., progressive development, a fundamental change."[7]

Revolution from Above and from Below

To achieve change, a revolution must be both "from above" and "from below." Or, to change the metaphor, there must be "push" as well as "pull." Organized ideas and concepts are more likely to come from individuals or leadership groups; but if there is no "pull," if the masses of people are not ready for the change, it will not occur, or it will not last.

> Perestroika would not have been a truly revolutionary undertaking, it would not have acquired its present scope, nor would it have had any firm chance of success if it had not merged the initiative from "above" with the grass-roots movement; if it had not expressed the fundamental, long-term interests of all the working people; if the masses had not regarded it as their program, a response to their own thoughts and a recognition of their own demands; and if the people had not supported it so vehemently and effectively.[8]

6. Ibid., p. 72.
7. Ibid., p. 240.
8. Ibid., p. 42.

Continuity and Change

Perestroika is pervaded by a yearning to find legitimacy in the teachings of the past, especially the later writings of Lenin and his emphasis on "socialist democracy." While this undoubtedly is also politically expedient, an armor against attacks from the guardians of orthodoxy, it is more, and deeper than that: the need for an anchor in the sea of change; an attitude of piety, familial religiosity; "religion" in the sense of that which binds the past with the future, the familiar with the novel.

Common Issues and Common Ownership

If it is possible, indeed necessary, at the local and national level to combine the driving forces of individual freedom and initiative with the stabilizing power of the common interest, in a mutually reinforcing synthesis of democracy and socialism, the same applies to the international level and the relations between socialist and free-market–based states in a global system that transcends both. There are in fact a number of systems-transforming developments on the global scene that, in terms of the Palme Commission—fully endorsed by Gorbachev—are "ideology bridging": conservation of the environment; the economic cost of, and environmental dangers inherent in, the arms race, in particular; the impact of modern technology, in general.

> Another no less obvious reality of our time is the emergence and aggravation of the so-called global issues which have also become vital to the destinies of civilization. I mean nature conservation, the critical condition of the environment, of the atmosphere and the oceans, and of our planet's traditional resources which have turned out not to be limitless. I mean old and new awful diseases and mankind's common concern: how are we to put an end to starvation and poverty in vast areas of the Earth? I mean the intelligent joint work in exploring outer space and the world ocean and the use of the knowledge obtained to the benefit of humanity.[9]

If it is these issues that have challenged the socialist system, forced it to transcend itself and incorporate elements of the market system, they are equally challenging the market system, forcing it to transcend itself and to incorporate certain elements that used to be associated with socialism. The Brundtland report, also fully endorsed by Gorbachev, makes it amply clear that these are issues the "market" cannot resolve. The all-pervasive and dramatically urgent problems of the conservation of the environment require

9. Ibid., p. 123.

planning and regulation, whatever the economic system and the ideology it is based on. Poverty, like consumerism, is incompatible with the conservation of the environment and must be abolished, a goal that undoubtedly introduces a strong dose of what used to be called socialism into our market system. It pushes both sides toward the development of a new economic system, based on a new economic theory, which could be called the "economics of the common heritage." "The necessity of effective and fair international procedures and mechanisms which would ensure rational utilization of our planet's resources as the property of all mankind, becomes ever more pressing."[10]

Common and Comprehensive Security

The Palme report stressed the concept of common security, meaning that, in the nuclear age, security cannot be acquired by any one nation at the expense of the security of another nation; that security must be common security; that only the security of all is the security of each. Gorbachev fully endorses this concept: "The idea of 'security for all,' which was put forward by him [Olof Palme] and further elaborated by the International Palme Commission, has many points of similarity with our concept of comprehensive security."[11]

Comprehensive security, including economic and environmental security, together with military security, is an enlargement of the Palme concept.

Defining for itself the main principles of the concept of ecological security, the Soviet Union considers disarmament, the economy, and ecology as an integral whole.[12]

The search for military security through arms control and disarmament negotiations has thus far dominated the East-West agenda. The search for economic security through development cooperation and, eventually, the building of a new economic order, has been the theme of North-South dialogue.

The joining of the two issues in the concept of common and comprehensive security means the joining of the East-West and the North-South dialogues. It offers the best guarantee against the marginalization of the South.

Comprehensive security's scope is global, and the third component of the concept, environmental security, is perhaps the one that ties the whole concept together, as the major problems of the environment are tangibly global and do not distinguish between East, West, South, or North.

10. Ibid.
11. Ibid., p. 192.
12. Eduard Shevardnadze, *Foreign Policy and Perestroika* (Moscow: Novosti Press Agency, 1989).

Land

Gorbachev's vision is continent-centered. His historical linkages extend both to the "European homeland" and to the ancient cultures of the Far East. The Soviet Union's two faces pose a unique challenge and opportunity. "Efforts in this direction by countries of the two continents—Europe and Asia—could be pooled together to become a common Euro-Asian process which would give a powerful impulse to an all-embracing system of international security."[13]

Sea

While this worldview is continent-centered, Gorbachev is fully cognizant of the enormous importance of the oceans surrounding this landmass: the Arctic, the Baltic, the Pacific, the Mediterranean, the Indian Ocean, and the world ocean as a whole, for communication within the system as a whole, for peace and security, for the protection of the environment, for scientific research, and for development—East, West, North, and South.

The oceans, covering more than 70% of the surface of the planet, represent common heritage, communality, continuity; the continents may symbolize differentiation, "individuality." Clearly, the basic principles of *perestroika* apply to the new order in the oceans as well as to that on land.

Nowhere on the planet are local, regional, and global issues and regimes more closely interwoven than in the marine environment, necessitating new forms of interaction and cooperation between national governments and regional and global organizations.

The marine revolution, no less than *perestroika*, must be a revolution from above and from below. There must be pull as well as push. It is the lack of a constituency, putting pressures on governments, that is slowing down the ratification process and the coming into force of the LOS Convention. If the marine revolution is to be successful and lasting, mass movements in favor of the conservation of the marine environment must be mobilized, based on a better understanding of the importance of the oceans in determining the global climate, as well as of the potential contributions the oceans can make in providing food and fiber, energy and minerals; and there should be mass movements to save the great whales, based on awareness of the importance of species diversity on our planet.[14]

Continuity and change express the eternal rhythm of the world ocean itself, and reflect the progression of life from the sea to land and air and

13. Gorbachev (n. 3 above), p. 170.
14. Editors' note.—For more on this subject, see the articles in the "Living Resources" section of this volume.

of human activities, from the oldest, fishing and navigating, to the newest, high-technology–based. The Law of the Sea is the oldest of international laws, and it is also the most advanced.

Where everything flows, no rigid concept of individuality or ownership (and that the two are linked was known already to the Buddha 2,500 years ago) can resist. Rigidity and flow are contradictions in terms. It is in the Law of the Sea that the concept of the "Common Heritage of Mankind" has been developed and given legal content; the emerging economics of the common heritage promises a synthesis not only of economics and ecology, but also of individual initiative and common cause, of freedom, and of planning.

Marine resources include the water that environs them. Thus the destruction of the environment destroys development. Since the destruction of the aquatic environment results equally from underdevelopment (sewage, erosion, and urban wastes) and overdevelopment (industrial wastes); and since industrial wastes are generated as much by the industrial/military complex as for peaceful purposes, it follows that development, environment, and a stop to the arms race are closely interlinked in the ocean world. Environmental security, which is basic for the development of marine resources, is unattainable without military as well as economic security; that is, security must be comprehensive or it will not be at all.

Humans are concentrated on the land portion of Planet Earth. Nuclear holocaust is the most concentrated threat to our survival. The central element of comprehensive security on land is military security which, however, is unattainable without economic and environmental security, including the aquatic environment.

In the vast oceans, uninhabited by humans, the central element of comprehensive security is environmental. This, however, is unattainable without economic and military security on land.

The management of environmentally focused security has different geographic parameters than that of militarily focused security. Militarily focused security is land-based, terrestrial, and politically organized; Europe, Asia, and the Americas are terrestrial regions. Environmentally focused security is ocean-based and transcends political boundaries. The Mediterranean basin, the wider Caribbean, the Indian Ocean, and the circumpolar Arctic are ocean-centered regions dominated by the requirements of environmentally focused security.

Ocean-centered and land-based regions obviously overlap. All Mediterranean countries, north, south, east, and west, belong to the Mediterranean environmental security system articulated in the Mediterranean Regional Seas Programme of the United Nations Environment Programme and its action plan.[15] At the same time some of them are part of the European Community

15. Editors' note.—For more information on this subject, see Peter M. Haas, "Save the Seas: UNEP's Regional Seas Programme and the Coordination of Regional Pollution Control Efforts," and App. G, "Status of Participation in UNEP's Legal Instruments for Marine Environmental Protection," in this volume.

or of NATO; others, of the Arab League, or the Organization of African Unity: systems focused on economic or military security. In building a system of common and comprehensive security, these systems complement and depend on and reinforce one another. Ocean-centered, environmentally focused, and boundary-transcending security may be the element that holds the whole system together, makes it global and stable.

This study is organized around the convergent concepts of common heritage and comprehensive security, each with its threefold military, environmental, and economic connotations. The first part covers the military dimension and consists of three sections: (1) dealing with the Law of the Sea and the Treaty on the Prohibition of the Emplacement of Nuclear Weapons and Other Weapons of Mass Destruction on the Seabed and the Ocean Floor and in the Subsoil Thereof (1972 Treaty), (2) covering the denuclearization of regional seas, and (3) treating collective security measures, such as United Nations or regional naval units. The second part examines the environmental dimension. Starting from the UNEP-initiated regional seas project, it draws the functional and institutional consequences of the unitary concept of comprehensive security. The final part explores the economic dimension and examines, in particular, the potential for some *perestroika* proposals to develop marine industrial technology, at both global and regional levels.

MILITARY ASPECTS

Reservation for Peaceful Purposes

The LOS Convention reserves the High Seas for peaceful purposes (Article 88). According to Articles 58 and 86, this applies as well to the 200-mile exclusive economic zone. The seabed beyond the limits of national jurisdiction, furthermore, is reserved "exclusively for peaceful purposes" (Article 141).[16] According to Article 240, finally, "marine scientific research shall be conducted exclusively for peaceful purposes."

While potentially revolutionary, the concept of reservation for peaceful purposes lacks legal content and definition in the LOS Convention. UNCLOS III felt it had no mandate to deal with the military uses of the sea, the regulation of which was left to the Disarmament Committee in Geneva. Thus the concept of "Common Heritage of Mankind," which postulates both management of peaceful uses and reservation for peaceful purposes, was split in two, a consequence not only of political expediency (many of the diplomats dealing with UNCLOS III were of the opinion that the job of drafting the

16. Editors' note.—See article by Stanley D. Brunn and Gerald L. Ingalls, "Voting Patterns in the UN General Assembly on Uses of the Seas," in *Ocean Yearbook 7*, ed. Elisabeth Mann Borgese, Norton Ginsburg, and Joseph R. Morgan (Chicago: University of Chicago Press, 1988), pp. 42–64.

convention would have become unmanageably complicated had it also to
cover the regulation of military uses, arms control, and disarmament) but
also of the fragmented nature of the UN system. Thus two UN bodies, the
Disarmament Committee and UNCLOS III, worked independently and produced two independent conventions, at different points in time (1971 and
1982). Once the 1982 LOS Convention comes into force, it will be necessary
to harmonize the two instruments. In a paper written in 1984, I identified
five levels at which such harmonization should take place: geographic scope,
functional scope, the problem of verification, the considerations of technology, and dispute settlement.[17]

The spatial organization of the world ocean in the LOS Convention is
different from the one in the 1972 Treaty: the 12-mile Territorial Sea, the
exclusive economic zone, the High Seas, the Continental Shelf, the archipelagic waters, and the international seabed area each call for different
treatment.

With regard to the functional scope, prohibition in the 1972 Treaty is
restricted, adopting the U.S. formula, for "the implanting and emplacing of
nuclear weapons or other weapons of mass destruction." This, however, was
to be considered merely as a first step toward the total demilitarization of the
seabed, as advocated by the USSR. The obligation in the Preamble, "to continue negotiations concerning further measures leading to this end," introduced a time dimension, a dynamic aspect into the treaty, which clearly indicates that a process is involved with demilitarization of the seabed as the
ultimate objective.

To harmonize the two instruments, demilitarization (the more advanced
concept) should apply to the international seabed area that now is reserved
for exclusively peaceful purposes, while denuclearization could apply, for the
time being, to the seabed up to the 12-mile limit of the Territorial Sea of the
LOS Convention (the term "Contiguous Zone" in the 1972 Treaty has to be
amended), as well as to the water column above it: the High Seas, reserved,
according to the LOS Convention, "for peaceful purposes."

Verification, in the 1972 Treaty, is entirely the responsibility of the states
parties, even though a number of delegations wanted to go much further
and establish some form of international verification mechanism; in the LOS
Convention, verification with regard to seabed activities is entrusted to the
International Sea-Bed Authority, which has to establish and direct an inspectorate for this purpose. "Activities in the Area," however, are to be construed
as activities directly related to the exploration and exploitation of manganese
nodules (surveillance with regard to economic and environmental aspects).
Military activities are not in the purview of this provision as it now stands.

17. Elisabeth Mann Borgese, "The Sea-Bed Treaty and the Law of the Sea:
Prospects for Harmonisation," in *The Denuclearisation of the Oceans,* ed. R. B. Byers
(London: Croom Helm, 1986), pp. 88–103.

To harmonize the two instruments, surveillance by the inspectorate should be multipurpose, pertaining to military as well as environmental and economic aspects. The monitoring technologies, in any case, are the same. This would be in line with the contemporary thinking and would be perfectly legitimate, considering the environmental hazards inherent in the deployment of nuclear weapons on the seabed. The idea is not new, incidentally, as the delegation of Canada proposed it in the early days of the UN Sea-Bed Committee (prior to UNCLOS III). It would also be in line with the proposals of *perestroika:* the responsibilities of the World Space Organisation proposed by the Soviet Union are multipurpose. Its satellites are to monitor compliance with the provisions of arms control and disarmament agreements as well as any changes in the environment of the biosphere. They also are to serve the progress of science and economic development for the benefit of all people.

The jurisdiction of the Authority extends only to the outer edge of the continental margin, up to 350 nm, or even more in some cases, from the coast. Monitoring and surveillance of the shelf area, between 350 and 12 nm from the coast, should be entrusted to self-management through regional security arrangements, which might be perceived as less intrusive and offensive and would also be less costly.

The 1972 Treaty is subject to review and revision every 5 years. In order to be able to review the proper functioning of the treaty, delegations need information on the state of the art of seabed technology. The review of the treaty, according to Article VII, "shall take into account any relevant technological developments." As it turned out, this information was hard to come by. Time and again states would simply report that no relevant technological development had taken place. In the face of the vast sums spent on R&D in deep-sea technology such statements are not very convincing.

Meanwhile, technology transfer for the peaceful uses of the seabed has become a burning issue; developing states demanded information on the state of the art, which was crucial as a basis for their decision making with regard to the LOS Convention. The Office for Ocean Affairs and the Law of the Sea of the United Nations Secretariat has established a data base. Although this effort is as yet quite modest and in need of more funding and better cooperation, it is a step in the right direction.

To harmonize the two treaties, a technology bank that would provide information to the parties of both the 1972 and 1982 treaties is needed, as the technology relevant to both treaties is the same.

The LOS Convention of 1982, as is well known, contains the most comprehensive and the most binding system for the peaceful settlement of disputes ever devised by the international community. The 1972 Treaty contains no provisions on dispute settlement, although in a 1969 working paper, the delegation of Brazil had proposed it, stressing the importance of a credible system of dispute settlement for the acceptance of verification measures. The Brazilian proposal was ahead of its time. It was hardly even discussed.

To harmonize the two treaties, the settlement of any dispute that might arise under the 1972 Treaty could be entrusted to the International Tribunal for the Law of the Sea, established under the LOS Convention. This could easily be done through a protocol added to the 1972 Treaty at the next revision conference.

One might ask, Why all this fuss about the 1972 Treaty? There has been no trouble with it. There has been no violation of it. "Don't fix it if it ain't broke." Furthermore, the treaty has lost interest in the public eye. It prohibits what states did not intend to do anyway; it has done nothing to prevent the nuclear arms race in the seas and oceans.

One might answer that the treaty would be irrelevant, were it not for one provision: it establishes an obligation to continue negotiations concerning further measures leading to the complete exclusion of the nuclear arms race from the seas and oceans.

It is precisely this provision that has been violated during the almost two decades since the adoption of the treaty, because no such negotiations have taken place, due to the Cold War and the stubborn resistance of the United States to including naval armaments in general disarmament discussions.

Now, however, there has been a drastic change. With the propositions of *perestroika* on the negotiating table, the continuation of the dialogue on the 1972 Treaty and the widening of its geographic and functional scope become mandatory.

To widen the geographic scope would mean, first of all, to extend the jurisdiction of the treaty from the seabed, which is far less relevant for the installation of nuclear weapons, to the superjacent waters, where submarine and naval activities take place.

To widen the functional scope would mean to pass from the U.S. formula of 1972, limiting the treaty to the prohibition of nuclear weapons and other weapons of mass destruction, to the USSR formula of 1972, extending it eventually to all military activities, a formula to which *perestroika* has remained faithful.

The Denuclearization of the Oceans

The nuclearization of the oceans is incompatible with environmental security. In particular, there are three aspects that should be kept in mind:

1. At present, almost 30% of the world nuclear arsenal is seaborne.[18] The five nuclear powers together possess more than 7,200 submarine-launched ballistic missile warheads plus 5,900 tactical nuclear weapons. About 13,100 nuclear weapons in total are earmarked for naval use, that is, almost one-third

---

18. Editors' note.—For more information on this subject, see the "Military Activities" section in this volume.

of the world's stockpile of nuclear weapons. This constitutes serious danger, especially in enclosed and semienclosed seas where close encounters are possible.

2. Nuclear testing contaminates marine flora and fauna and endangers the health of human populations over wide areas.

3. The dumping of nuclear waste in canisters whose corrosion resistance is certainly shorter than the half-life of the material they contain constitutes a threat to the environmental security of future generations.

Existing Arrangements

The inhabitants of affected zones have been painfully aware of these dangers for some time, and a number of international agreements have been put into place to cope with the situation. In most cases, however, these have not yet been effectively enforced.

One should mention, in particular, the Antarctic Treaty, the Treaty of Tlatelolco, the Treaty of Rarotonga, and the United Nations Resolutions on the Indian Ocean as a Zone of Peace.

The Antarctic Treaty (1959)[19] is increasingly vulnerable from other points of view but has been the most successful in keeping a continent completely demilitarized. Article I states that "Antarctica shall be used for peaceful purposes only. There shall be prohibited, *inter alia,* any measures of a military nature, such as the establishment of military bases and fortifications, the carrying out of military maneuvers, as well as the testing of any type of weapons." Military personnel and equipment, however, may be used for scientific research or any other peaceful purpose. Article V prohibits "any nuclear explosions in Antarctica and the disposal there of radioactive waste material." It should be noted, however, that these provisions do not apply to the High Seas within the area south of 60° south latitude, which thus is neither denuclearized nor demilitarized nor reserved for peaceful purposes.

The Treaty of Tlatelolco (1967),[20] which also provides for an elaborate regional institutional infrastructure, stipulates in Article 1 that

> 1. The Contracting Parties hereby undertake to use exclusively for peaceful purposes the nuclear material and facilities which are under their jurisdiction, and to prohibit and prevent in their respective territories:
>   (a) The testing, use, manufacture, production or acquisition by any

19. Antarctic Treaty, signed 1 December 1959. The text of the treaty is found as an appendix to an article by G. L. Kesteven, "The Southern Ocean," in *Ocean Yearbook 1,* ed. Elisabeth Mann Borgese and Norton Ginsburg (Chicago: University of Chicago Press, 1978), pp. 493–99. See also the article in this volume by Francis Auburn, "Conservation and the Antarctic Minerals Regime."

20. Treaty for the Prohibition of Nuclear Weapons in Latin America (Treaty of Tlatelolco), signed 14 February 1967.

means whatsoever of any nuclear weapons, by the Parties themselves, directly or indirectly, on behalf of anyone else or in any other way, and

(b) The receipt, storage, installation, deployment and any form of possession of any nuclear weapons, directly or indirectly, by the Parties themselves, by anyone on their behalf or in any other way.

2. The Contracting Parties also undertake to refrain from engaging in, encouraging or authorizing, directly or indirectly, or in any way participating in the testing, use, manufacture, production, possession or control of any nuclear weapon.

According to the Preamble, the parties considered this as a step toward general and complete disarmament under effective international control, which is the final goal at a later stage. The provisions of the Treaty of Tlatelolco apply to the land territories of the states parties as well as to their Territorial Seas, which at the time of signing extended, in many cases, to 200 nm from the baselines. Perhaps the time has come to review and revise this pioneering treaty, which in the present circumstances has not been able to prevent, for example, the nuclearization of the Caribbean.

The Treaty of Rarotonga (1985),[21] like that of Tlatelolco, prohibits to states parties the manufacture, acquisition, control, or possession of nuclear explosive devices. They may not allow the stationing or testing of such devices or the dumping of radioactive wastes within their territory. The boundaries of the nuclear-free zone are defined in Annex I to the treaty. Whereas the Treaty of Tlatelolco applies basically to land territory, including the Territorial Sea, the Treaty of Rarotonga applies primarily to the sea. Land accounts for only 2% of the region's total area. The Treaty of Rarotonga has been signed by nine South Pacific states. Three protocols, binding third states, in particular the nuclear powers, have not been signed by France and the United States.

The Indian Ocean Resolutions, finally, are not enforceable.[22]

The LOS Convention provides a new legal framework for the denuclearization of ocean regions. Jens Evensen stated,

Part IX of the Convention contains certain provisions concerning enclosed or semi-enclosed seas. The main provisions contained in Article 123 indicate that "States bordering an enclosed or semi-enclosed sea should co-operate with each other in the exercise of their rights and in

21. South Pacific Nuclear-Free Zone Treaty (Treaty of Rarotonga), signed 6 August 1985. It was signed by 8 South Pacific states but endorsed earlier the same day by all 13 member states of the South Pacific Forum. The treaty and its appendices and protocols are republished in App. B, *Ocean Yearbook 6,* ed. Elisabeth Mann Borgese and Norton Ginsburg (Chicago: University of Chicago Press, 1986), pp. 594–605.

22. See Brunn and Ingalls (n. 16 above), pp. 53–55, 61–62.

the performance of their duties under this Convention." This suggests that such states have special rights and obligations to formulate policies with regard to the peaceful development in their enclosed or semi-enclosed seas. The establishment of zones of peace, nuclear-weapons free zones and other peace-related activities in the area, such as safe havens for home fleets in certain maritime areas, special procedures for the commencement of naval manoeuvres and the like, would be examples. Such initiatives have been proposed for the Baltic and the Caribbean. These possibilities should be further explored.

The Law of the Sea Convention does not directly address the denuclearisation of the oceans or related arms limitation issues. However, the Convention further codifies the principles which underlie the peaceful uses of ocean space. As such the Convention could serve as a legal basis for more directly addressing the issue of nuclear weapons at sea.[23]

It is on this new legal basis that one should promote the implementation of the more recent and more specific *perestroika* proposals. They draw legitimization from the convention. At the same time they contribute to the implementation and progressive development of its provisions.

Perestroika *Proposals*
A. P. Movchan, in his paper "The Law of the Sea in Light of the New Political Thinking,"[24] states,

> The essence and central arrangement or basic and principal legal requirement of the principle of the use of the World Ocean for peaceful purposes is to exclude the use or threat of force in the maritime activities of States and, consequently, to ultimately prohibit military activities of States on the seas and oceans and ensure in fact that they are used only for peaceful purposes. However, the practical realisation of this main requirement is possible only through specific prohibitions of military activities in certain specific expanses of the World Ocean or specific types of such activities. The 1982 Convention was drafted with this in view with regard to the principle of using the seas for peaceful purposes.

In his Vladivostok address in 1986, Gorbachev made specific proposals for the denuclearization of the Asian Pacific oceanic region.[25] He suggested

23. Jens Evensen, "The Law of the Sea Regime," in Byers (n. 17 above), pp. 86–87.
24. A. P. Movchan, "The Law of the Sea in the Light of the New Political Thinking," in *Perestroika and International Law*, ed. William E. Butler (Dordrecht, Netherlands: Martinus Nijhoff Publishers, 1990), p. 129.
25. For the Pacific, Gorbachev made the following suggestions, from *Time for Action, Time for Practical Work* (Moscow: Novosti Press Agency, 1988):

"that there be in the foreseeable future a Pacific conference attended by all countries gravitating towards the ocean." In a statement in Delhi the same year, he advocated a new international conference for the implementation of the UN Resolutions on the Indian Ocean as a Zone of Peace. His Murmansk statement, in 1987, contained proposals for the demilitarization of the Arctic Ocean as well as international cooperation for the protection of the Arctic environment, for scientific research and economic development.[26] In 1988,

---

First: Aware of the Asian and Pacific countries' concern, the Soviet Union will not increase the amount of any nuclear weapons in the region; it has already been practising this for some time—and is calling upon the United States and other nuclear powers not to deploy them additionally in the region.

Second: The Soviet Union is inviting the main naval powers of the region to hold consultations on not increasing naval forces in the region.

Third: The USSR suggests that the question of lowering military confrontation in the areas where the coasts of the USSR, the PRC, Japan, the DPRK and South Korea converge be discussed on a multilateral basis with a view to freezing and commensurately lowering the levels of naval and air forces and limiting their activity.

Fourth: If the United States agrees to eliminate military bases in the Philippines, the Soviet Union will be ready, in agreement with the government of the Socialist Republic of Vietnam, to give up the fleet's material and technical supply station in Camranh Bay.

Fifth: In the interests of the safety of sea lanes and communications in the region, the USSR suggests that measures be jointly elaborated to prevent incidents in the open sea and air space over it. The experience of the already existing bilateral Soviet-American and Soviet-British accords as well as the USA, USSR-Japan trilateral accord could be used in the elaboration of these measures.

26. For the Arctic, from Mikhail Gorbachev, *The Speech in Murmansk* (Moscow: Novosti Press Agency, 1987):

Firstly, a nuclear-free zone in Northern Europe. If such a decision were adopted, the Soviet Union, as has already been declared, would be prepared to act as a guarantor.

Secondly, we welcome the initiative of Finland's President Mauno Koivisto on restricting naval activity in the seas washing the shores of Northern Europe. For its part, the Soviet Union proposes consultations between the Warsaw Treaty Organization and NATO on restricting military activity and scaling down naval and air force activities in the Baltic, Northern, Norwegian and Greenland Seas, and on the extension of confidence-building measures to these areas.

Thirdly, the Soviet Union attaches much importance to peaceful cooperation in developing the resources of the North, the Arctic. Here an exchange of experience and knowledge is extremely important. Through joint efforts it could be possible to work out an overall concept of rational developments of northern areas. We propose, for instance, reaching agreement on drafting an integral energy programme for the north of Europe. According to existing data, the reserves there of such energy sources as oil and gas are truly boundless. But their extraction entails immense difficulties and the need to create unique technical installations capable of withstanding the Polar elements. It would be more reasonable to pool efforts in this endeavour, which would cut both material and other

at Belgrade, he proposed convoking a special conference to discuss questions of limiting the activities of naval forces, of reducing them in the Mediterranean, and of declaring that sea a zone of peace.

Perestroika also makes provision for the prohibition of naval activities in agreed zones of international straits and areas of intensive international navigation and fishing; for limiting the number of large-scale naval exercises in each ocean and sea theater of military operations; and for limiting the navigation of warships carrying nuclear weapons.

Two conferences in Moscow, Pacem in Maribus XVIII (1989) and Mir na moriach (Peace to the oceans, 1990), spelled out these proposals in great detail.[27]

In spite of past efforts to keep the naval arms race separate from the rest of the arms race and to exclude it from general disarmament or arms control discussions, it obviously is part of it. Naval disarmament by itself cannot solve the wider arms race problem, whereas the abandonment of the arms race in general is likely to overtake the partial solutions proposed with

---

outlays. We have an interest in inviting, for instance, Canada and Norway to form mixed firms and enterprises for developing oil and gas deposits of the shelf of our northern seas. We are prepared for relevant talks with other states as well.

Fourthly, the scientific exploration of the Arctic is of immense importance for the whole of mankind. We have a wealth of experience here and are prepared to share it. In turn, we are interested in the studies conducted in other sub-Arctic and northern countries. We already have a programme of scientific exchanges with Canada.

Fifthly, we attach special importance to the cooperation of the northern countries in environmental protection. The urgency of this is obvious. It would be well to extend joint measures for protecting the marine environment of the Baltic, now being carried out by a commission of seven maritime states, to the entire oceanic and sea surface of the globe's North.

Sixthly, the shortest sea route from Europe to the Far East and the Pacific Ocean passes through the Arctic. I think that depending on progress in the normalization of international relations we could open the North Sea Route to foreign ships, with ourselves providing the services of ice-breakers.

27. In its declaration, Pacem in Maribus XVII noted, "The new order for the seas and oceans emerging from the 1982 United Nations Convention on the Law of the Sea should be considered as a first step in international Perestroika: a restructuring, a fundamental change in an area covering over two thirds of the surface of the Earth." The conference urged "the inclusion of the naval arms race in ongoing negotiations on arms control and disarmament and the establishment of nuclear weapons free zones in the Baltic, the Sea of Japan, and the East China Sea, as well as zones of peace in the Indian Ocean and in the South Atlantic, in implementation of the Declarations by the United Nations General Assembly." It also urged the establishment of such zones in the South Pacific, the Mediterranean, and the Arctic (*Proceedings of Pacem in Maribus XVII* [Oxford: Pergamon Press, forthcoming]).

The recommendations of Pacem in Maribus XVII were based on the concept of comprehensive security, including military, economic, and environmental security, in accordance with *perestroika*.

regard to the denuclearization of the oceans. This, however, is no reason for abandoning the proposals at this point or for detracting from their long-term usefulness—likely to last beyond the end of the arms race, for the reasons pointed out next.

Perestroika looks at denuclearization of the oceans in the context of comprehensive security; that is, the proposals are linked with proposals for environmental, scientific, and developmental cooperation in the denuclearized areas. This requires a progressive development of the institutional framework of the existing regional seas program. The emerging institutional framework may be one of the essential elements of the new paradigm.

Institutional Restructuring

With regard to institutional restructuring, the Palme report, fully endorsed by Gorbachev, is more advanced than any of the reports preceding or following it. Palme's institutional framework for "common security" has three major components: changes in the political decision-making mechanism (Security Council), strengthening of the technical instruments of collective security (United Nations Peacekeeping Forces), and regional collective security arrangements.

Palme's proposals were far ahead of their time. They exemplify the kind of utopianism of yesterday that becomes the realism of today. Today, they provide the institutional framework that is absolutely essential in the light of *perestroika* and the ongoing dismantling of the arms race. Disarmament without an institutional framework for collective security (including verification) either does not happen or becomes chaotic. Palme and Gorbachev reinforce each other and must be considered together.

Changes in the Decision-Making Mechanism

The Palme Commission suggests that the permanent members of the Security Council should solemnly agree to a kind of "political concordat" to support collective security operations or, at least, not to vote against them, whenever disputes arise that are likely to cause, or actually result in, a breach of peace. Such a "concordat" obviously is intended to offset the paralyzing effect of the veto power and to restore the collective security role originally envisaged for the United Nations. The Palme report cautiously—perhaps overcautiously—limits this concordat, to start with, to disputes that might arise among Third World countries. Such a limitation would not appear to be justified today. Disputes of the kind envisaged might develop anywhere, and they must be prevented from erupting and escalating into global catastrophes. They must be brought to peaceful settlement.

The Palme report itself suggests, "The question may be asked, Why limit collective security measures to Third World disputes? In theory, there can be no objection to a global approach. Practicality, however, dictates otherwise. Disputes beyond the Third World invariably involve NATO or Warsaw Pact countries. The East-West conflict has prevented the development of international collective security in the past. It retains the potential to frustrate its evolution still."[28]

It no longer does. *Perestroika* has cleared the way for a global approach to the kind of concordat proposed by the Palme report.

United Nations Peacekeeping Forces

The decisions of the Security Council should be backed up by the strengthening of the operational structure for UN standby forces as envisaged in Article 43 of the UN Charter. The Military Staff Committee should be reactivated and strengthened for this purpose. What is of special interest in the perspective of the present study is that, according to the recommendations of the Palme report, these standby forces should include a UN naval unit. "The UN also must be prepared to respond to new kinds of challenges to international peace and security. For example, the emergence of extensive piracy in the areas off South East Asia might suggest the creation of a small UN naval patrol force based on the voluntary assignment of naval vessels and crews to UN duty by member states, and the consent of the littoral states."[29]

Regional Collective Security Arrangements

Regional approaches to security are conceived as a supplement, not as an alternative, to the global collective security structure needed. They are in fact to be closely interlinked. The measures suggested include regional arrangements to promote units for peacekeeping duties on a standby basis and the establishment of regional conferences on security and cooperation. "The [Palme] Commission recommends that the countries making up the various regions, and in some instances sub-regions, of the Third World consider the convocation of periodic or ad hoc Regional Conferences on Security and Cooperation similar to the one launched in Helsinki for Europe in 1975. Regional Conferences on Security and Cooperation could add new substance to the concept of common security."[30] "It is envisaged that the regional conferences could provide an overall framework for cooperation not only on matters directly relating to security, but in the economic, social, and cultural spheres as well."[31]

28. Palme et al. (n. 2 above), p. 133.
29. Ibid., p. 167.
30. Ibid., p. 168.
31. Ibid., p. 169. Comprehensive security, according to *perestroika*, has a cultural dimension as well: "Pooling efforts in the sphere of culture, medicine and humanitarian rights is yet another integral part of the system of comprehensive security" (Mikhail Gorbachev, *Realities and Guarantees for a Secure World* [Moscow: Novosti Press Agency, 1987]).

This is where the Palme report comes closest to *perestroika*'s concept of comprehensive security. In fact, it provides an institutional framework for its implementation and should be read in conjunction with *perestroika*'s proposals for denuclearization and cooperation in, for example, the Arctic, the Indian Ocean, the Pacific, and the Mediterranean. It might well be that comprehensive security could best be implemented at the regional level—provided there are the proper linkages backward, as it were, to the national level and forward to the global level.

The comprehensive mandate of these regional conferences would include matters such as the adoption of codes of conduct and confidence-building measures, establishment of zones of peace and nuclear weapon–free zones, and agreements on arms limitations and reductions. They might also establish a boundary commission to investigate and make recommendations on solutions for border disputes arising from the limits of Territorial Seas and exclusive economic zones. They also could establish regional research institutes to analyze security issues of relevance to the particular region and formulate recommendations for the consideration of the conference. The Palme report stresses that "in our opinion, the concept of regional security will be unlikely to take root unless it is sustained by programmes for economic cooperation to encourage countries to see themselves as having a national stake in actively working to achieve regional harmony. An important focus of the Regional Conferences must therefore be the establishment of joint projects that are designed to benefit all participating states."[32]

Examples of such joint projects are schemes for regional cooperation on the peaceful exploitation of nuclear energy in a manner that would strengthen an equitable nonproliferation regime. This again is in line with *perestroika*, which goes one step further, however, and recommends cooperative projects on nuclear fusion research.

ENVIRONMENT AND DEVELOPMENT

It should be noted that the environment was omitted from the spheres of action to be dealt with by the regional conferences. It would appear, however, that the omission is accidental rather than intentional. It is inconceivable today to consider programs in the economic, social, and cultural spheres without including environmental aspects, which are an integral part of comprehensive security.

The linkage to environmental issues—inevitable in the 1990s and in line with *perestroika*—might lead to a further suggestion. Why not utilize the existing institutional framework of the UNEP-initiated Regional Seas Programme?[33] The Conference of States Parties to the Regional Conventions

32. Palme et al. (n. 2 above), p. 169.
33. Editors' note.—For more information, see Haas (n. 15 above).

could take the place of the regional conferences. Secretarial infrastructure already exists. During the 1990s it is incumbent on the Regional Seas Programme to widen its responsibilities by finding institutional ways and means to integrate environment and development, in accordance with the recommendations of the Brundtland report.

One might suggest that pilot experiments should be undertaken in an area where development is just beginning, such as the Arctic, and that it might start with a conference of all circumpolar states, with the participation of the competent international organizations. Another pilot experiment might be initiated in the Mediterranean, where the Regional Seas Programme is most advanced and ready for the next phase of its evolution.

The widening of functions will require corresponding structural changes. Thus far, limitation to environmental concerns has resulted in a single linkage between the regional conference of states and the ministries of the environment at the national level. If the mandate of the regional conference is comprehensive security, with its military, economic, and environmental components, linkages must be more complex and cover a number of ministries at the national level. Besides environment, these will include defense, science and technology, agriculture and fisheries, energy and mines, shipping, ports and harbors, and tourism.

As the Brundtland report points out, the interdisciplinary nature of environmental issues and the unbreakable linkage between environment and development have begun to make the walls separating different ministries porous and permeable, just as the interaction of local and regional (transboundary) environmental and economic issues are breaking down the boundaries between national, regional, and global levels of management and policy.

In the wake of adoption of the LOS Convention and the establishment of exclusive economic zones, many states have established, or are in the process of establishing, interministerial coordinating mechanisms under the leadership of the prime minister or of a newly established ministry for ocean development, to enable them to formulate an integrated and comprehensive policy for the management of their economic zones. It is on these interministerial coordinating mechanisms that the regional conferences must be based. *Perestroika* suggests that the opinions of other entities, such as nongovernmental organizations, or even individuals representing a form of "citizens' diplomacy" should be considered in decision making. Such opinions might be articulated in advisory councils of some sort.

In *The Future of the Oceans: A Report to the Club of Rome*,[34] I proposed the establishment of regional councils composed of (a) plenipotentiary representatives of the interministerial ocean councils of all states of the region and (b) the regional representatives of the "competent international organisations"

34. Elisabeth Mann Borgese, *The Future of the Oceans: A Report to the Club of Rome* (Montreal: Harvest House, 1986).

(FAO, IMO, UNEP, Unesco/IOC, UNIDO, ILO, IAEA) dealing with ocean affairs in the region. These regional councils would be competent to deal with ocean affairs in an integrated manner, focused on the concept of comprehensive security, including its military, economic, and environmental dimensions. These bodies would replace the conferences of states parties to the Regional Seas Programmes. They would be neither more complex nor more costly than these well-established meetings, except that their linkages with national government would be interdepartmental rather than restricted to departments of the environment and/or foreign affairs. Group (a) in these regional councils would ensure coordination of national policies; group (b) would link regional policies with the global competent international institutions. All three levels—national, regional, and global—are essential for policy making in ocean affairs, and they must be linked properly. Here again is an area where the emerging new order in the seas and oceans and *perestroika* can reinforce and mutually advance each other.

DEVELOPMENT AND ENVIRONMENT

Perestroika abounds in suggestions for new forms of scientific industrial international cooperation for codevelopment. The USSR and the United States, *perestroika* suggests, could come up with large joint programs, pooling resources and scientific and intellectual potentials in order to solve the most diverse problems for the benefit of humankind.

With regard to cooperation in utilizing thermonuclear energy in particular, Gorbachev states that a scientific base has been created by scientists from a number of countries working on ideas suggested by their Soviet colleagues. American scientists could join in this research. Also possible are joint exploration and use of outer space and of planets of the Solar System, and research in the field of superconductivity and biotechnology.

Joint work in exploring outer space and the world ocean and the use of the knowledge obtained to the benefit of humanity would be another promising field, according to *perestroika*. Scientific/industrial codevelopment should, of course, also be implemented in Eastern Europe. "We hope to accelerate the process of integration in the forthcoming few years. To this end, the CMEA [COMECON] should increasingly focus on two major issues: First, it will coordinate economic policies . . . and promote major joint research and engineering programs and projects. In doing so it is possible and expedient to cooperate with non-socialist countries and their organizations."[35]

The proposal retains its validity in spite of recent developments in Eastern Europe. A couple of years ago, the Eastern European socialist states established a joint undertaking, under the name of Interoceanmetal, Inc., for

35. Gorbachev (n. 3 above), p. 153.

the joint exploration of a mine site in the deep seabed and for the development of the requisite technology. Far from being overtaken by centrifugal trends, this venture is now at the point of applying to the Preparatory Commission for the International Sea-Bed Authority and for the International Tribunal for the Law of the Sea for registration as the fifth pioneer investor. Here is another example for the convergence of, and mutual reinforcement between, *perestroika* and the Law of the Sea.

Joint undertakings of the same type, according to *perestroika*, should also be established between the USSR and its Asian Pacific neighbors as well as with "the European home." "The building of the 'European home' requires a material foundation. . . . We, in the Soviet Union are prepared for this, including the need to search for new forms of cooperation such as the launching of joint ventures, the implementation of joint projects in third countries, etc. We are raising the question of broad scientific and technological cooperation."[36]

The LOS Convention provides the most advanced framework for the realization of these principles of broad scientific and technological cooperation.

Proposals for technology codevelopment in the context of registered pioneer activities have been put forward at the Preparatory Commission by the delegations of Austria (1984) and Colombia (1988), and now by the Asian African Legal Consultative Committee.[37] An international joint undertaking in R&D in seabed-mining–related high technologies would be the most efficient, if not the only, way of dealing with the environmental issues stressed so eloquently by the Soviet delegation at the Jamaica session in March 1990.

A proposal to organize the regional centers for the advancement of marine science and technology, mandated in Articles 276 and 277 of the LOS

36. Ibid., p. 190.
37. United Nations, *JEFERAD,* three working papers submitted by the delegation of Austria to the Preparatory Commission for the International Sea-Bed Authority and for the International Tribunal for the Law of the Sea, 1984 and 1985; United Nations, *The International Enterprise,* three working papers submitted by the delegation of Colombia to the Preparatory Commission for the International Sea-Bed Authority and for the International Tribunal for the Law of the Sea, 1987 and 1988; and International Ocean Institute and Asian African Legal Consultative Committee, *Alternative Cost-Effective Models for Pioneer Cooperation in Exploration, Technology Development, and Training* (Malta: International Ocean Institute, 1990). Editors' note.—See earlier articles in *Ocean Yearbook* by Professor Borgese on the Preparatory Commission, for example, "Implementing the Convention: Developments in the Preparatory Commission," in *Ocean Yearbook 7,* ed. Elisabeth Mann Borgese, Norton Ginsburg, and Joseph R. Morgan (Chicago: University of Chicago Press, 1988), pp. 1–7. For a summary of the Colombian proposal, see "The International Venture: Study Submitted by the Republic of Colombia to the Preparatory Commission for the International Sea-Bed Authority and for the International Tribunal for the Law of the Sea," Special Commission 2, 5th Session, Kingston, Jamaica, 30 March–16 April 1987, also in *Ocean Yearbook* 7, pp. 469–79.

Convention, on the principle of technology codevelopment, was put forward by the government of Malta and elaborated in cooperation with the United Nations Industrial Development Organization (UNIDO) and UNEP. A first such center, the Mediterranean Centre for Research and Development in Marine Industrial Technology, is being established in Malta.[38] A feasibility study for the Caribbean, based on the Mediterranean experience but adapting it to the specific situation in the Caribbean, is in the making. Other oceanic regions will follow.

All these proposals are based on the simple concept of (a) utilizing and developing the legal framework provided by the LOS Convention; (b) filling that frame with the most advanced concepts of technology development and management as applied, for example, in the European Community (European Research Coordination Agency [EUREKA], etc.); and (c) opening them up to participation by Eastern European as well as developing countries and industries, their participation to be financed by public granting or lending institutions such as the World Bank, the United Nations Development Programme (UNDP), regional development banks, and so forth.

Thus the basic structure of the regional centers to be established under Articles 276 and 277 of the LOS Convention, and of the joint enterprise to be established by the pioneer investors under Resolution II (which would become "the Enterprise" when the convention comes into force), would be similar, unbureaucratic, and cost-effective.[39] They would be small coordinating centers, administering an R&D program in marine industrial industry. Projects would be selected by a network of national coordinators (following the EUREKA pattern) and approved by a conference of ministers in the case of the regional centers, or by the Preparatory Commission in the case of the pioneer joint venture (by the Council of the International Sea-Bed Authority, when the convention is in force). Projects would be self-financing, as the scheme would generate investments rather than "costs." Half of these investments would come from the companies that propose the project; the other half would come from their governments. The participation of developing countries (which would have to be included in all projects) would be paid for by international (or national) development cooperation institutions. This would imply a redirection of development strategy toward science, research, and development in developing countries. Such a redirection is indeed overdue.

Marine technology involves and is dependent upon the whole range of technologies constituting the new phase of the industrial revolution: micro-

38. International Ocean Institute, *The Mediterranean Centre for Research and Development in Marine Industrial Technology: Proposal and Feasibility Study* (Malta: International Ocean Institute, 1988).

39. Editors' note.—For more information see earlier *Ocean Yearbook* articles and reports on the Preparatory Commission, pioneer investors, and Resolution II (n. 37 above).

electronics and information technology, genetic engineering and bioindustrial processes, new materials, laser and space technology. With this in mind, it may safely be assumed that a breakthrough in international cooperation to "technology codevelopment" between North, South, East, and West might have far-reaching implications for bridging gaps and enhancing confidence and economic security. Projects selected for codevelopment would have a strong environmental component; that is, technologies to be developed must be "environmentally safe and socially relevant." It is hard to imagine a more direct application of the more general proposals put forward in *perestroika*.

CONCLUSION

The two basic concepts "comprehensive security" (*perestroika*) and "Common Heritage of Mankind" (Law of the Sea) are complementary. The "Common Heritage of Mankind," incorporated in the Moon Treaty (which is in force) and in the LOS Convention (soon to enter into force) is already a principle of international law. Unlike the principle of the "global commons," which assumes a free-for-all system within which resources may be exploited and depleted on a first-come–first-serve basis, the common heritage regime postulates a system of management in which all users share, and within which resources are rationally used for the benefit of humankind as a whole. This provides the basis for "economic security" in the *perestroika* concept.

"Mankind" in the common heritage concept includes future generations which also have a right to share in the common heritage. Intragenerational equity, as the Brundtland report puts it, must be complemented with intergenerational equity. This implies conservation of resources and environment and harmonization between long-term and short-term policies. It provides the basis for "environment security."

The Common Heritage of Mankind, finally, is reserved exclusively for peaceful purposes. Activities undertaken for military or strategic purposes are excluded from the area that has been declared the Common Heritage of Mankind. This provides the basis for military security.

Together, common heritage and comprehensive security provide the basis for sustainable development, that is, a development that itself has an economic as well as an environmental and military dimension (incompatibility with the arms race).

The emerging conceptual framework transcends the boundaries of the traditional concepts of sovereignty (porousness of the walls separating disciplines, governmental departments, and national, regional, and global levels of governance) and of ownership (the common heritage cannot be appropriated: it is a concept of nonownership; the common heritage can be managed, but not owned). Thus it transcends the tenets both of the market and of the centrally planned systems.

Some of the institutional implications have been dealt with in these pages. There is one point that should be added in conclusion.

To be effective, the new institutions must be funded in a different way. The traditional idea of establishing a "fund" to be nourished by quotas assigned to states or by voluntary contributions is totally inadequate. It leaves these funds at the mercy of the few economically strongest states and exposes even the best programs to the danger of never being implemented. Development economists (from Jan Tinbergen to Willy Brandt to the World Bank) have long advocated greater "automaticity" in funding, such as through a system of international taxation. The conservation of the environment implies certain costs (even though in the long term it is extremely economical), and UNEP has completed studies on, for example, a system of international taxation to pay for desertification programs. Peacekeeping activities also cost money that presently is not available, and the Palme report recommends the establishment of an appropriate funding mechanism with built-in automaticity. "We believe that collective security operations and, for other purposes, peacekeeping ones as well, need to be financed through an independent source of revenue."[40]

An international levy on international arms transfers, amounting to over US$30 billion annually, has been suggested at various times.[41] This would be based on a register of weapons sales and transfers, as a first step. "No one in the world can yet bid farewell to arms, but we can abandon, once and for all—and we can do it now—the practice of unconstrained and uncontrolled international weapons transfers. To that end the principles of glasnost and openness should be asserted here as well. The USSR reaffirms its willingness to participate in the establishment of a United Nations register of weapons sales and transfers, including work on parameters."[42] And there are other international services that will have to be paid for.

In the emerging ocean regime, there are starting points for the development of new systems of generating international revenue. One is a system of international taxation. The other is the type of international, public/private enterprises described in this article.

The legal basis for international taxation is given in Parts VI and XI of the LOS Convention. Part VI (Article 82) provides for the payment of royalties on mineral production on the Continental Shelf under national jurisdiction, but beyond the 200-mile limit of the exclusive economic zone. It should be stressed that this tax applies to resources over which the coastal state has sovereign rights. It also should be stressed that this provision (unlike Part XI) was adopted by consensus.

Part XI, and particularly Annex III, Article 13, provides for an elaborate system of royalties and production charges to be imposed on contractors/

40. Palme et al. (n. 2 above), p. 167.
41. For example, by the delegation of Malta in the Disarmament Committee, 1967.
42. Shevardnadze (n. 12 above), p. 12.

miners. This may or may not be workable. The important point is that the principle of an international tax to be imposed on companies and to be paid to an international authority has been recognized and embodied in international law. On this precedent, other systems may now be devised as necessary.

To be effective, a tax system needs (a) an institutional infrastructure and (b) a publicly accepted purpose. During the next stage of historic evolution it is likely that both will be best defined at the regional level. The regional institutions described in this study might be best qualified for the levying of such taxes for determined regional community purposes and services. For example, the International Ocean Institute is presently conducting a feasibility study on the establishment of a small levy on the 100–300 million tourists visiting the Mediterranean annually.[43] Such a levy would constitute a crucial contribution to the financing of the Mediterranean Action Plan for the conservation of the environment, and it might also pay for other determined purposes. Regional multipurpose monitoring and surveillance as well as peacekeeping services might also be financed through schemes of regional taxation, although a tax on international arms sales would be a feasible global tax.

Economic development projects should increasingly become self-financing as they are becoming self-managed, not through "privatization," which is a temporary aberration, but through new forms of public/private cooperation, generating investments rather than costs. A possible institutional framework for this kind of cooperation is described in this study. It should, of course, be expanded from the R&D sector.

There is an intimate and inseparable linkage between the marine sector and the rest of the global system—ecological, economical, strategic, and technological. As science and technology advance, this linkage becomes even stronger.

A striking example is the global transport system. Until World War II and the advent of high technology, sea transport and land transport constituted two fairly separate systems. Then came containerization and unitization, giving rise to a unitary multimodal system including the seas, railways, roads, rivers, and airways. This is now being perfected through satellite-borne global positioning systems and electronic charting, pinpointing and guiding vessels or vehicles on land, on sea, or in the air, and harmonizing their traffic.

If it is one system and we change part of it (the ocean part), we obviously are changing the whole system. *Perestroika* is on the move. If, in the terrestrial part of the system, we are struck first of all by its unsettling, occasionally chaotic and threatening effects, then it is in the wide spaces of the oceanic part of the system—due to historic circumstances as well as to the nature of the aquatic medium—that we see the restructuring taking shape. The great concepts of *perestroika* and the Law of the Sea mingle in institutions and processes, and reinforce each other. The rest, necessarily, will follow.

43. International Ocean Institute, *Alternative Ways of Financing the Mediterranean Trust Fund* (Malta: International Ocean Institute, 1990).

The Postwar Japan-Soviet Fisheries Regime and Future Prospects*

Tsuneo Akaha
Monterey Institute of International Studies, California

INTRODUCTION

This paper examines the postwar fisheries relations between the Soviet Union and Japan and explores the possibility of bilateral cooperation in the management of fisheries resources in the Seas of Japan and Okhotsk. The central questions to be addressed are, What have the two countries learned from their past experiences in dealing with each other in the exploitation and management of fisheries resources? Have they institutionalized their cooperative efforts to the point where established arrangements may guide and even define their national interests? And have they developed sufficient confidence in each other as a reliable partner? The article concludes with an exploration of interests where mutually beneficial cooperation can be expected, as well as issues that may stand in the way of optimal cooperation.

The present analysis employs the concept of "regime," which has been developed by students of international relations concerned with international cooperation in such issue areas as trade, money, ocean, energy, resources, environment, and security. According to Stephen Krasner, a "regime" is a set of "implicit or explicit principles, norms, rules, and decision-making procedures around which actors' expectations converge in a given area of international relations." "Principles" are "beliefs of fact, causation, and rectitude"; "norms are standards of behavior defined in terms of rights and obligations"; "rules" are "specific prescriptions or proscriptions for action"; and "decision-making procedures" refer to "prevailing practices for making and implementing collective choice."[1] The absence of mutually acceptable principles, norms,

*EDITORS' NOTE.—This article was developed from an earlier paper delivered by the author at the International Conference on the Seas of Japan and Okhotsk, Nakhodka, USSR, 17–22 September 1989. The conference was hosted by the Far Eastern Branch of the Soviet Academy of Sciences and cosponsored by the East-West Center's Environment and Policy Institute, Hawaii. The article was also presented at the annual Western Slavic Association meeting, University of Arizona, Tucson, Arizona, 29–31 March 1990.

1. Stephen Krasner, "Structural Causes and Regime Consequences: Regimes as

rules, and decision-making procedures normally spells discord, or the failure to attain even limited mutual benefits. As a regime develops between parties, mutually beneficial results can be expected and cooperation enhanced. The stability of cooperation between parties depends on the perceived fairness and effectiveness of the principles, norms, rules, and decision-making procedures in collective problem solving.

Cooperation, the central focus of the present analysis, does not necessarily mean a harmony of interests between parties. Indeed, as the concept is used here, cooperation implies or presupposes a manifest or potential conflict of interests between parties and the need to coordinate and adjust their policies.[2] That is, if there is no conflict of interest between parties, there is no need to "cooperate." The amount of cooperation is determined by how far the parties go in reducing the distance between their preferred policies. In other words, cooperation is a cost that the parties are willing to bear in order to achieve a mutually acceptable outcome.

POSTWAR JAPAN-SOVIET FISHERIES REGIME

Throughout most of the Second World War, provisional short-term fisheries agreements between Tokyo and Moscow allowed the Japanese to continue their salmon and crab fishing off Kamchatka. This arrangement came to an abrupt end in August 1945, when the Soviet Union entered the war against Japan. In the waning days of the war, the Soviets occupied all of the Kurile Islands and Sakhalin and took over the Japanese facilities used in the salmon, crab, and herring fisheries of these islands as well as those of the Kamchatka peninsula.[3] Moscow did not sign the San Francisco Peace Treaty in 1952, leaving the Soviet Union and Japan legally in a state of war. The central controversy had to do with the dispute between Moscow and Tokyo over the sovereignty of four of the southernmost Kurile Islands, which the Japanese claimed (and continue to claim today) and which the Soviets occupied (and continue to occupy today).

Shortly after the San Francisco Peace Treaty went into force on 28 April 1952, the Japanese resumed their North Pacific fisheries, quickly expanding the scale of fishing operations.[4] Tokyo and Moscow tentatively agreed on

Intervening Variables," in *International Regimes* (Ithaca: Cornell University Press, 1983), p. 2.

2. Robert Keohane, *After Hegemony: Cooperation and Discord in the World Political Economy* (Princeton, New Jersey: Princeton University Press, 1984), p. 12.

3. Roy I. Jackson and William F. Royce, *Ocean Forum: An Interpretive History of the International North Pacific Fisheries Commission* (Farnham, Surrey, England: Fishing News Books, 1986), p. 25.

4. Kenzo Kawakami, *Sengo no Kokusai Gyogyoseido* (The postwar international fisheries regime) (Tokyo: Dainihon Suisankai, 1972), pp. 412–13; Norinsuisansho

13 March 1956 to negotiate a fisheries agreement to regulate fishing in the Northwest Pacific after a peace treaty was concluded. On 20 March, preliminary talks in London went into recess because of the territorial dispute noted above. The next day Moscow announced the decision by the Council of Ministers to require Soviet permission for foreign salmon fishing on the High Seas adjacent to the Soviet-claimed territories in the Sea of Okhotsk and the Bering Sea.[5] Moscow unilaterally ordered a halt on Japanese salmon fishing on the High Seas and in the following year closed Peter the Great Bay to all foreign vessels and aircraft.[6]

Japan decided to separate fisheries talks from the negotiation for the resumption of diplomatic relations. The fisheries negotiations began in Moscow on 29 April and quickly (by 15 May) produced an agreement—the Convention concerning the High Seas Fisheries of the Northwest Pacific Ocean. The accord covered salmon, trout, herring, and crab fisheries in waters outside the Soviet 12-mile territorial limits in the Sea of Japan, the Sea of Okhotsk, the North Pacific Ocean, and the Bering Sea.[7] The treaty also set up a commission responsible for assessing the status of fisheries resources subject to the bilateral treaty and for setting binding quotas on the resources to be taken by each party in the High Seas of the Northwest Pacific. Tokyo and Moscow readily agreed on the shared right to seize and arrest suspected treaty violators and on the principle of flag-state court jurisdiction.[8]

Initially the Soviets proposed that the agreement become effective on the day the peace treaty would go into force. Japan objected, and the two sides eventually reached the compromise that the fisheries accord would become effective either when the peace treaty went into force or when the bilateral diplomatic relations were restored. The peace treaty was never to be. Therefore, the fisheries agreement went into effect on 12 December 1956, when diplomatic relations were restored between Moscow and Tokyo. Moscow also showed a flexible attitude by agreeing to Tokyo's request that Japanese fishing in the area be allowed on a provisional basis prior to the effective date of the fisheries pact.[9]

There is no denying that the overall political relations between Tokyo and Moscow had an impact on their fisheries negotiations. Conversely, the successful fisheries talks contributed to the normalization of the bilateral rela-

Hyakunenshi Hensan-iinkai, ed., *Norinsuisansho Hyakunenshi* (The 100-year history of the Ministry of Agriculture, Forestry, and Fisheries), vol. 2 (Tokyo: Norinsuisansho Hyakunenshi Kankokai, 1981), p. 673.

5. Kawakami (n. 4 above), pp. 411–12.
6. Jackson and Royce (n. 3 above), p. 65.
7. Douglas M. Johnston, *The International Law of Fisheries: A Framework for Policy-Oriented Inquires* (New Haven: Yale University Press, 1965), pp. 391–96.
8. Kawakami (n. 4 above), pp. 422, 440.
9. Ibid., pp. 424–26.

tions.[10] Tokyo's early decision to separate its fisheries from diplomatic talks with Moscow—as part of its policy of *seikei bunri*, or separation of politics and economics—should also be considered an important, constructive element in the Japanese-Soviet fisheries regime in the immediate postwar period.

The 1956 fisheries agreement was to be in effect for 10 years, with each party thereafter being required to give a 1-year notice of its termination. With both sides wishing to maintain the bilateral arrangement, the treaty was renewed each year after 1966. The treaty thenceforth provided the foundation for a stable bilateral fisheries regime between Japan and the Soviet Union.

Not everything went as expected, however. Nor could every twist and turn in the development of the bilateral fisheries relations have been anticipated. In fact, there were plenty of surprises, at least to the Japanese. Unexpected Soviet actions included the February 1968 Decree of the Presidium of the Supreme Soviet concerning the Continental Shelf, the December 1976 Decree of the Presidium of the Supreme Soviet for the Provisional Measures concerning the Conservation of Living Resources and Fishery Regulation in the Areas of the Ocean Contiguous to the Coasts of the Soviet Union (the Soviet decision to establish a 200-mile fishery zone), and the February 1977 Council of Ministers' decision to implement the 1976 decision. Before long, in effect, "surprises" became part of the bilateral fisheries regime.

The February 1968 Decree of the Presidium of the Supreme Soviet made it necessary for Tokyo to enter into negotiations with Moscow. It is beyond the scope of this paper to provide a full account of this case.[11] Suffice it to say that the bilateral talks from February to April 1969 produced an agreement concerning Japanese crab fisheries in the Northwest Pacific. The negotiation could not resolve the two governments' disagreement over the legal status of Continental Shelf resources, however. The Soviets argued those resources were subject to their sovereign jurisdiction, and the Japanese asserted that the sedentary species in question were High Seas resources and not subject to exclusive jurisdiction of the coastal state. The final agreement simply noted the two conflicting positions. Since the effective duration of the agreement was 1 year, a new agreement had to be negotiated each year thereafter. This continued until 1976. Each year, limits on the size of permitted Japanese effort and catch were made more stringent than the year before. For example, Japanese effort in the area west of Kamchatka started at four mother boats in 1969 and ended at two in 1976.[12]

10. See ibid., p. 436; Norinsuisansho Hyakunenshi Hensan-iinkai (n. 4 above), p. 679.

11. See Kawakami (n. 4 above), pp. 552–67.

12. Iwao Matsumoto, *Nihon Kindai Gyogyo Nenpyo (Sengo-hen)* (The chronology of modern Japanese fisheries [postwar]) (Tokyo: Suisansho, 1980), p. 97.

Japanese *tsubu* (sea snail) fishery in the Northwest Pacific met a similar fate. Subject to bilateral agreement beginning in 1972, this small but important Japanese fishery experienced a gradual, continuing decline due to increasing bilateral restrictions. The Japanese were allowed 25 boats in the area east of Sakhalin and 25 in the northern Okhotsk Sea in 1972, but by 1976 the numbers had been slashed to 15 and 22, respectively. Their catch was similarly reduced, from 2,500 mt for each area in 1972 to 1,500 mt for the first area and 1,125 mt for the second in 1976.[13]

Painful as these reductions may have been, the bilateral agreements concerning salmon, crab, and *tsubu* fisheries in the Northwest Pacific constituted the Japanese-Soviet fisheries regime, providing legal and institutional frameworks for Japanese fisheries in the area from the 1960s through the first half of the 1970s. The regime provided a degree of stability, continuity, and even predictability.

The situation had changed drastically by the mid-1970s. In December 1976 the Soviet Union announced that it planned to establish a 200-mile fishery zone. The Soviet move was not altogether unpredictable. In October 1975, Iceland had set up its 200-mile fishery zone. The United States gave notice that it was going to extend its fishery jurisdiction out to 200 miles from its coast. Washington's warning became a reality with the enactment of the Fishery Conservation and Management Act (FCMA) of 1976, effective 1 March 1977. Accordingly, Washington and Moscow concluded a new fisheries agreement on 26 November, recognizing the U.S. 200-mile jurisdiction.[14]

Against the backdrop of the Third UN Conference on the Law of the Sea (UNCLOS III), the U.S. move was followed by Canada (in January 1977) and by the European Economic Community (in January 1977). "Decentralized ocean management"[15] had become the predominant trend in the global ocean political economy. The Soviet decision to establish a 200-mile fishery zone was part of the global trend. Moscow's December declaration was followed by the announcement on 24 February 1977 that the extension of the Soviet fishery jurisdiction would go into effect on 1 March, the day the U.S. FCMA was scheduled to become effective.[16]

The Soviet interest in the development of its coastal fisheries resources did not begin with these international developments, however. In fact, the Soviet Union had shown considerable interest in the rational development

13. Ibid., p. 99.

14. For a detailed study of Japan's response to the U.S. decision, see Tsuneo Akaha, *Japan in Global Ocean Politics* (Honolulu: University of Hawaii Press and Law of the Sea Institute, 1985), pp. 110–16.

15. Robert L. Friedheim, "The Political, Economic, and Legal Ocean," in *Managing Ocean Resources: A Primer,* ed. Friedheim (Boulder, Colorado: Westview Press, 1979), pp. 26–42.

16. For a detailed study of Japan's response to the Soviet 200-mile decision, see Akaha (n. 14 above), pp. 122–49.

and management of coastal fisheries in the 1940s and 50s, when the Soviet government placed a heavy emphasis on industrialization in its economic plans.[17] Four reasons may be postulated for the relatively early Soviet concern for the rational use and management of coastal fisheries resources: (1) the heavy wartime use of fishing vessels for war efforts, which, following the end of the war, left very few ships available for distant-water fisheries, (2) the forbidding cost of developing distant-water fisheries in the context of national fiscal shortages in the early postwar years, (3) the urgent need to expand the nation's food supply in its postwar economic reconstruction, and (4) the political-ideological justification for the centralization of economic planning, including expansion of central control over coastal fisheries. In the 1960s, as industrialization began to take its toll on the environment of the nation's lakes, rivers, and coastal areas, Moscow paid even more serious attention to the management of fisheries resources in the inland, coastal, and nearshore waters of the country.[18]

The Soviet concern with coastal fisheries resources grew further in response to the proliferation of national ocean enclosures in the 1960s and 70s. The extension of national jurisdiction over coastal waters and resources in Europe and North America threatened Soviet distant-water fishing in foreign waters—an experience that Japanese blue-water fishing was also beginning to undergo in the early 1970s.

Japan vehemently protested the Soviet 200-mile decision, asserting that the global discussion on the future order of the sea was still continuing in UNCLOS III and that unilateral actions outside this international forum could not be recognized as legitimate. Moreover, Japan lodged a diplomatic protest over the Soviet decision to include in its fishery zone areas of the sea surrounding the "Northern Territories," or the islands of the Habomai group, Shikotan, Kunashiri (Kunashir in Russian), and Etorofu (Iturup), which the Soviets have controlled and the Japanese have claimed since the end of the Second World War.

The Soviets would not heed the Japanese protests and insisted that Japan recognize the new Soviet fishery zone. The Japan-Soviet Commission's meeting in Tokyo concerning salmon fisheries in the Northwest Pacific started on 15 March 1977 but was broken off by the end of the month.

As further attempts at breaking the deadlock failed, Japanese fishermen had to give up their operations in Soviet waters during the spring fishing season. To make the situation worse, Moscow served notice on 29 April 1977 that it would terminate its 1956 fisheries agreement with Tokyo effective 1

---

17. For a useful Japanese study of the evolution of postwar Soviet fishery policy, see Hajime Imanishi and Fujio Iida, "Sengo Soren no Gyogyo Seisaku: Seifu Kettei Bunsho o chushin ni shite" (The postwar Soviet fishery policy: With a focus on documents of government decisions), *Suisankai*, no. 1227 (April 1987): 33–46.

18. Ibid., pp. 38–41.

May 1978. Unless a new agreement was negotiated, Japan could lose its Northwest Pacific salmon and herring fisheries.

In the meantime, after several months of intense domestic debate, Japan decided to extend its Territorial Sea limit from 3 to 12 miles and to set up its own 200-mile fishery zone. The extension of the Territorial Sea limit had become necessary to protect Japanese coastal fishing grounds from foreign competitors, namely South Korean and Soviet fishermen. Japan's 200-mile fishery zone was established more for a political purpose than for an economic one. To counter Soviet claims to what the Japanese call the Northern Territories, Japan found it necessary to reassert its claims to the islands by including sea areas around them within its extended fishery jurisdiction.

The twin legislation for the 12-mile Territorial Sea and for the 200-mile fishery zone quickly made it through the Diet, and on 1 July the Law on the Territorial Sea and the Law on Provisional Measures relating to the Fishing Zone went into force.

With the new legislation in hand, the Japanese went back to the negotiating table on 7 May 1977. Moscow opposed Tokyo's fishery zone delimitation. Since the two sides could not settle the territorial and jurisdictional dispute, they eventually concurred on a treaty text that each side could use to support its own claims. Namely, Article 1 of the provisional agreement—the Agreement between the Government of Japan and the Government of the Union of Soviet Socialist Republics concerning Fisheries off the Coasts of the Union of Soviet Socialist Republics in 1977[19]—stated that Japan recognized the Soviet fishery zone based on the decision of the Soviet Council of Ministers, which specified the Soviet delimitation, but Article 2 acknowledged the Soviets' reciprocal recognition of Japan's fishery-zone delimitation. Article 8 provided, moreover, that nothing contained in this agreement should be deemed to affect or prejudice in any manner the positions or views of either government with respect to the questions relating to the mutual relations between the two countries.

The provisional agreement was signed on 27 May and went into force on 10 June. The agreement, scheduled to expire at the end of 1977, was renewed each year thereafter until 1984, when a new agreement was concluded to consolidate and replace that provisional agreement and another provisional agreement, concluded in August 1977, regulating Soviet fishing within Japan's newly established 200-mile fishery zone.

The Agreement between the Government of Japan and the Government of the Union of Soviet Socialist Republics concerning Fisheries off the Coasts of Japan in 1977[20] provided for procedures and conditions for Soviet fishing

19. My translation of the Japanese title of the agreement. The full Japanese text of the agreement can be found in Gaimusho Johobunkakyoku, *Nisso Gyogyo Zantei Kyotei* (The provisional fisheries agreement between Japan and the Soviet Union) (Tokyo: Gaimusho Johobunkakyoku, 1977), pp. 30–35.

20. My translation of the Japanese title of the agreement. The full Japanese text of the agreement, annexes, and exchanged notes can be found in Takao Morizane,

within Japan's 200-mile zone and included regulatory measures with respect to all fisheries resources except the highly migratory species specified in the enforcement order for the 200-mile fishery zone law. Tokyo maintained that highly migratory species should be managed not by unilateral coastal state action but through regional and international fisheries organizations.

The binational commission's talks on Japanese fishing in the Northwest Pacific in 1977 also produced results—results that were hard for the Japanese to accept but for which there was no alternative. The agreement totally banned Japanese herring fishing in the Northwest Pacific and Japanese salmon fishing within the Soviet fishery zone. The quota for Japanese High Seas salmon fishing outside the Soviet zone was reduced to 62,000 mt in 1977, a drop of 18,000 mt from the previous year.

Negotiation for the conclusion of a new, permanent agreement on Japanese salmon fisheries in the Northwest Pacific began in September 1977 and lasted until April of the following year. The talks produced an agreement on fisheries cooperation between the two countries and a protocol providing for the bilateral commission's authority to set the salmon quota for Japan. Accordingly, Japan's total quota for 1978 was set at 42,500 mt. Most importantly, the agreement indicated Japan's explicit acceptance of the principle that the state—in this case, the Soviet Union—in whose waters anadromous stocks of fish originate has a sovereign right to those resources. Moreover, Japan was for the first time required to pay "cooperation fees" (176,000 yen in 1978) to defray part of the cost of Soviet salmon reproduction effort, a feature that has since become a permanent element of the Japanese-Soviet salmon fisheries relations.[21]

Thus, the two provisional fisheries agreements and the new Northwest Pacific salmon fisheries agreement restored a modicum of stability to the bilateral fisheries relations that had been shaken by the developments in the first half of the 1970s. The new agreements served as the foundation of the Japanese-Soviet fisheries regime until 1984.

The new regime initially reflected the previous fishing efforts of the Japanese and Soviet fishermen in each other's coastal waters, giving the Japanese a larger share of the fisheries resources in Soviet waters. The total quotas for Japanese fishing in the Soviet 200-mile fishery zone in 1978, 1979, and 1980 were 850,000 mt, 750,000 mt, and 750,000 mt, respectively. The Soviet quota within the Japanese fishery zone remained at 650,000 mt for each of the 3 years.[22] This imbalance was gradually corrected until it was totally

---

Shin Kaiyoho Chitsujo to Nihon Gyogyo (The new order of the law of the sea and Japanese fisheries) (Tokyo: Sozo Shobo, 1977), pp. 252–75.

21. *Suisan Nenkan* (Fisheries yearbook) (Tokyo: Suisansha, 1988), p. 26; Hiroshi Murabayashi, "Nisso Gyogyo Kyoryoku Kyotei no Shisa suru Mono" (What the Japan-Soviet fisheries cooperation agreement suggests), *Juristo*, no. 843 (September 1985): 63.

22. Gaimusho Johobunkakyoku, ed., *200-kairi Jidai no Gyogyo: Saikin no Taigai Gyogyo Mondai to Kokusai Kyoryoku* (Fisheries in the 200-mile age: Recent external

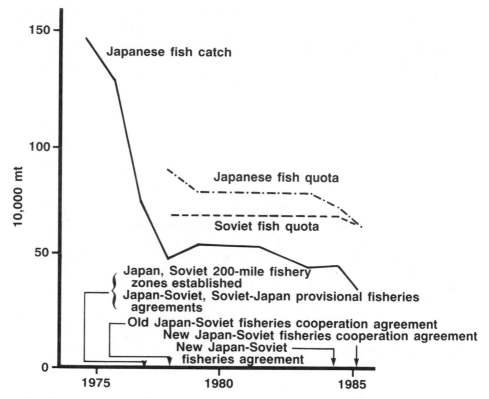

FIG. 1.—Japanese and Soviet fishing in each other's 200-mile zone. (*Gyogyo Hakusho* [Fisheries white paper] [Tokyo: Norin Tokei Kyokai, 1985], p. 22.)

eliminated in 1985, when the two countries' total quota in each other's 200-mile zone was equalized at 600,000 mt.

The bilateral fisheries regime between 1979 and 1983 was characterized by relative stability and predictability. Figure 1 shows the changes in Japanese catch and quota as well as the Soviet quota within 200 miles of each other's coasts between 1975 and 1985.

No sooner had the two provisional agreements been signed in 1977 than the Japanese began to demand that the accords be replaced by a more permanent agreement. Negotiating the renewal of those agreements each year with the fear that they might be terminated at any time did not sit well with the Japanese, who wanted to provide for their fishermen a more stable and predictable prospect. By 1984 the Soviets had also come to see the need for a new agreement, in view of the conclusion of UNCLOS III in December

fisheries problems and international cooperation) (Tokyo: Sekai no Ugokisha, 1980), p. 48.

1982 and the establishment of a 200-mile exclusive economic zone at home on 1 March 1984.[23] Talks began in Tokyo in November and by mid-December a new treaty was in operation.

The new treaty—the Agreement between the Government of Japan and the Government of the Union of Soviet Socialist Republics concerning the Fisheries off the Coasts of Japan and off the Coasts of the Union of Soviet Socialist Republics—incorporated the main components of the 1977 provisional accords. It provided that each party "shall permit, on the basis of the principle of mutual benefit and in accordance with its relevant laws and regulations, nationals and fishing vessels of the other country to engage in fishing in a zone of 200 nautical miles adjacent to its coasts of the Northwest Pacific Ocean."[24] The treaty also directed each party to determine annually "quotas of catch, species composition, and areas of fishing for fishing vessels of the other country" in its 200-mile zone with due consideration given to the "condition of resources," the coastal state's "own harvesting capacity, the traditional catch and methods of fishing by the other country, and other relevant factors."[25] This provision was in line with one of the UNCLOS-approved principles that a fish quota and other conditions pertaining to foreign fishing within a coastal state's 200-mile exclusive fishery or economic zone be determined through agreement between the coastal state and the fishing state.[26]

The bilateral agreement further provided that each party "shall take necessary measures to ensure that its nationals and fishing vessels engaging in fishing" in the other state's 200-mile zone "comply with the conservation measures of the living resources and the other terms and conditions established in laws and regulations of the other country,"[27] and that, conversely, each party "may take, in accordance with international law, necessary measures" in its 200-mile zone "to ensure that nationals and fishing vessels of the other country comply with the conservation measures of the living resources and the other terms and conditions established in its laws and regulations."[28] Moreover, the agreement called on the two governments to "cooperate in

23. The Soviet 200-mile economic zone was established by a decree of the Presidium of the Supreme Soviet on 28 February 1984. For a full English text, see *The Law of the Sea: Current Developments in State Practice* (New York: Office of the Special Representative of the Secretary-General for the Law of the Sea, United Nations, 1987), pp. 103–10.

24. Agreement between the Government of Japan and the Government of the Union of Soviet Socialist Republics concerning the Fisheries off the Coasts of Japan and off the Coasts of the Union of Soviet Socialist Republics, 1984, Art. 1. An unofficial English translation of the agreement is found in *The Japanese Annual of International Law*, no. 28 (1985): 297–99.

25. Ibid., Art. 2.

26. The relevant provision of the bilateral agreement is found in ibid., Art. 2.

27. Ibid., Art. 4, Par. 1.

28. Ibid., Art. 4, Par. 2.

the conservation and optimal utilization of the living resources" within their 200-mile zones.[29]

The treaty additionally stipulated that nothing contained in it "shall be deemed to prejudice the positions or views" of either party regarding "any question of the law of the sea or any question pertaining to mutual relations."[30] Finally, the agreement was to be effective through 31 December 1987, to be extended each year thereafter unless either side served advanced notice of termination.[31] The agreement has since been extended.

On the basis of the new agreement, a bilateral fisheries commission was established[32] and held talks on the level of fishing effort and quotas to be permitted in each country's 200-mile zone. The first round of negotiation in Tokyo was soon deadlocked. The Soviets argued that all Japanese crab, *tsubu*, and shrimp fisheries within their EEZ should be banned, the total quota for other catches set at 600,000 mt, and Japanese fishing off the Soviet Maritime Province reduced. They also demanded the same quota and reduced restrictions for their own fishermen as well as arrangements for port calls in Japan.[33]

The Japanese expectedly rejected these demands and offered a counterproposal. They proposed that Soviet port calls in Japan be discontinued because they did not accomplish the ostensible purpose of increasing the Soviet catch and also because there was strong local opposition to such visits. They maintained they could not relax fishing regulations against Soviet fishing because, in the area of intense Japanese fishing, a complex and strict array of procedures existed for coordinating Japanese fisheries and because the existing restrictions on Soviet fishing were already less stringent than those regulating Japanese fishing. Finally, the Japanese argued the proposed conditions for Japanese fishing in Soviet waters would devastate Japanese fishing in the North Pacific.[34]

When the second and third rounds of talks failed to break the deadlock, Tokyo dispatched its Agriculture, Forestry, and Fisheries Minister, Sato, to Moscow to meet with the Fisheries Minister, Kamentsev, the Soviet counterpart. The meeting was successful and a compromise was reached. The outcome was close to the compromise position that Tokyo had offered during the third round of negotiations: In the Soviet EEZ, Japan's quota was set at 600,000 mt (a reduction of 100,000 mt from the previous year), and fishing

29. Ibid., Art. 5.
30. Ibid., Art. 7.
31. Ibid., Art. 8 of the agreement. For a summary of the main provisions of the agreement, see Suisancho, *1988-nen Nisso Gyogyo Kankei Shiryo* (Documents on Japan-Soviet fisheries relations in 1988) (Tokyo: Suisancho, 1988), pp. 3–4.
32. The bilateral commission was established in accordance with Art. 6 of the Agreement of 1984 (n. 24 above).
33. Nobuhiko Nagasugi, "Nisso Gyogyo Iinkai ni tsuite" (On the Japan-Soviet Fisheries Commission), *AFF* (Agriculture, forestry, and fisheries) (August 1985): 39.
34. Ibid., pp. 39–40.

conditions remained about the same as in 1984 except that one of the fishing areas off the Maritime Province was reduced in size in return for the designation of a new area for bottom fish operations. The Soviets retracted their earlier demand concerning fishing equipment, and Japanese crab, *tsubu,* and shrimp fisheries would be discussed at a nongovernmental level. In the Japanese fishery zone, the Soviet quota was fixed at 600,000 mt (an increase of 40,000 mt from 1984), fishing conditions remained about the same as in the previous year except that a ban on saury stick-held dip-net fishing to the east of Hokkaido was lifted, and Soviet fishing boats would be allowed to call at Shiogama, Miyagi prefecture, for refueling and resting purposes. Based on the principle of reciprocity, Japanese fishermen were allowed to visit the port of Nevelsk, Sakhalin, in 1985.[35]

The second Japan-Soviet Fisheries Commission meeting concerning fishing in mutual fishery zones in 1986 took similar twists and turns. The talks began on 23 December 1985, were suspended twice, and, following a high-level negotiation between the Japanese Minister of Agriculture, Forestry, and Fisheries, Haneda, and the Soviet Fisheries Minister, Kamentsev, were concluded on 26 April 1986.[36] Disagreements centered on the Soviet desire to equalize the actual catch by Japanese and Soviet fishermen within each other's 200-mile zone. As table 1 indicates, the Japanese catch in the Soviet coastal waters has consistently exceeded Soviet catch in Japanese waters. During the suspension of the bilateral talks from the end of 1985 through March 1986, Japanese fishermen had to sit idle except for the brief period of 1–5 January, when they were temporarily allowed to operate within the Soviet EEZ.[37]

The final agreement, signed on 26 April, put the catch quota for each side at 150,000 mt because a good part of the fishing season had already passed. The number of fishing boats allowed was 1,600 for Japan and 200 for the Soviet Union (respectively 5,623 and 531 in 1985). Several fishing bans were newly instituted in the Soviet EEZ. Japan was not required to make any monetary payments for its fishing in the Soviet waters, and port calls by the two countries' fishermen were permitted on a reciprocal basis, at Hitachi, in Ibaraki prefecture and at Nevelsk, Sakhalin.[38]

In contrast to the second commission meeting, the third meeting concerning the 1987 fishing season went much more smoothly. Begun on 25 November in Tokyo, the meeting reached its conclusion on 10 December. Following the Soviet proposed idea of a distinction between fee-free and fee-based fishing, the agreement set Japan's quota in the Soviet EEZ at

35. Ibid., p. 41.
36. For a brief summary of these talks, see Ikuro Toyota, "Nisso 200-kairi Gyogyo Kosho to Nisso Sake-masu Gyogyo Kosho no Keika to Kekka ni tsuite" (On the process and results of the Japan-Soviet 200-mile fisheries negotiation and the Japan-Soviet salmon fisheries negotiation), *AFF* (July 1986): 16–20.
37. Ibid., pp. 17–18.
38. Ibid., pp. 19–20.

TABLE 1.—JAPANESE AND SOVIET FISH QUOTAS IN EACH OTHER'S
200-MILE ZONE, 1978–89

Year	Japanese Quotas in Soviet 200-Mile Zone (mt)	Soviet Quotas in Japanese 200-Mile Zone (mt)
1978	850,000	650,000
1979	750,000	650,000
1980	750,000	650,000
1981	750,000	650,000
1982	750,000	650,000
1983	750,000	650,000
1984	700,000	640,000
1985	600,000	600,000
1986	150,000	150,000
1987	300,000 (100,000)[a]	200,000
1988	310,000 (100,000)[a]	210,000
1989	310,000 (100,000)[a]	210,000

SOURCES.—Suisancho, *1988-nen Nisso Gyogyo Kankei Shiryo* (Documents on Japan-Soviet fisheries relations in 1988) (Tokyo: Suisancho, 1988), pp. 5, 10; "Nisso Ryokoku 200-kairi Nai Nyuryo Kosho: Musho, Yusho tomo Zennen Nami Wariate" (Negotiations over fishing within Japanese-Soviet 200 miles: Quotas for fee-free and fee-charged [fishing] the same as the previous year), *Suisankai*, no. 1248 (January 1989): 22, 23.
[a] Fee-based quotas in parentheses are included in the total quotas.

200,000 mt, for which Japan would not be required to pay fishing fees, and 100,000 mt, for which Japan would pay 1,290 million yen. The number of Japanese boats allowed to operate in the Soviet EEZ remained the same as in 1986, that is, 1,600. Soviet fishing in Japan's 200-mile zone was limited to a 200,000-mt quota and 305 boats. Arrangements for reciprocal port calls also were made.[39]

The fourth commission meeting, held in Moscow between 24 November and 11 December and again between 24 and 25 December 1987, went not quite as smoothly, with the Japanese arguing for a substantially larger catch quota than the Soviets were willing to accept and the Soviets in turn demanding more fishing fees than the Japanese were willing to pay.[40] In the end, Japan's fee-free fish quota was set at 210,000 mt, with an additional 100,000 mt for which the Japanese would pay 1,710 million yen in fishing fees. Japan also agreed to a Soviet proposal for setting up "checkpoints" to inspect Japanese fishing boats on their entry into and departure from designated fishing areas. The Soviets were concerned about the continuing Japanese violations of fishing regulations.[41] The checkpoints thus became a new feature of the bilateral fisheries regime.[42] In the designated Soviet waters,

39. *Suisan Nenkan* (n. 21 above), p. 29.
40. For a summary of the talks, see "Nisso Gyogyo Kosho Daketsu" (Japan-Soviet fisheries negotiation concluded), *Suisankai*, no. 1237 (February 1988): 43–47.
41. The number of Japanese boats that were caught for alleged violations of fishery regulations had increased from about 70 in 1986 to 180 in 1987 (ibid., p. 43).
42. The "checkpoints" were set up at the coordinates of 48° north latitude–154°

1,520 Japanese fishing boats were allowed. The Soviet fish quota in Japan's fishery zone was increased from 200,000 mt in 1987 to 210,000 mt in 1988. In Japanese waters 300 Soviet boats were permitted to operate with no requirement of fishing-fee payment.[43]

The negotiations for 1989 and 1990 fishing seasons were smoother and showed fewer surprises. The fifth bilateral Fisheries Commission meeting in Tokyo from 29 November to 11 December 1988 produced an agreement similar to the 1987 agreement, adopting a quota of 210,000 mt for each side.[44] In addition, Japan was permitted to catch another 100,000 mt of fish in return for fishing fees totalling 1,980 million yen. In actuality, however, the Japanese were expected to catch about 50% of their fee-free quota and 70% of their fee-based quota and the Soviets, 80,000 mt out of their total allocation of 210,000 mt. In the new agreement Japanese quotas for high-value species, that is, *madara* (*Gadus macrocephalus*), *hokke* (Atka mackerel or *Pleurogrammus azonus*), and *ainame* (rock trout), were increased. The agreement included an additional 100,000 mt of Alaska pollack. Japanese fishing effort was reduced by 50 boats to 1,470 boats, while Soviet fishing effort in Japanese waters remained the same as in the previous year, 300 boats. Japanese and Soviet fishing quotas are shown in table 2.

The sixth commission meeting began in Moscow on 28 November, and by 29 December 1989 some preliminary results had come out. Japan's fee-free fishing in the Soviet waters in 1990 was reduced to 182,000 mt, and the fee-based quota was set at 35,000 mt in exchange for 884 million yen and 1,000 mt of refrigerated saury to be supplied to the Soviets.[45] The Japanese were to reduce the total number of fishing boats operating in Soviet waters to 1,459 in 1990, and a new checkpoint was added to the existing three. The Soviets were allowed a quota of 182,000 mt and 300 fishing boats in Japanese waters. Soviet boats were allowed to call in the port of Hitachi in Ibaraki prefecture for refueling and recreation. Species-by-species quotas for the two sides in 1990 are shown in table 2 and Japanese fishing areas within the Soviet waters in figure 2.

In the mid-1980s, the salmon component of the bilateral fisheries regime also saw some important developments. In view of the 1982 UN Convention on the Law of the Sea (LOS Convention) and the establishment of the Soviet 200-mile EEZ on 1 March 1984, Moscow proposed negotiation for the conclu-

---

east longitude and 43° north latitude–146° 10′ east longitude, east of the Kurile Islands, and 45° 25′ north latitude–144° east longitude, east of Sakhalin (ibid., p. 43).

43. Ibid.

44. The account of the fifth meeting is found in "Nisso Ryokoku 200-kairi Nai Nyuryo Kosho: Musho, Yusho tomo Zennen Nami Wariate" (Negotiations over fishing within Japanese-Soviet 200 miles: Quotas for fee-free and fee-charged [fishing] the same as the previous year), *Suisankai,* no. 1248 (January 1989): 22–27.

45. "Nisso Chisen Okiai Gyogyo Kosho Daketsu" (Japan-Soviet offshore fishing negotiation concluded), *Suisankai,* no. 1261 (January 1990): 8–13.

TABLE 2.—JAPANESE AND SOVIET FISH QUOTAS BY SPECIES IN EACH
OTHER'S 200-MILE ZONES, 1989–90

JAPANESE QUOTAS IN SOVIET EEZ

Species	Total (mt) 1989	Total (mt) 1990	Fee-charged (mt)[a] 1989	Fee-charged (mt)[a] 1990
Sukesodara (Alaska pollack)	121,010	40,736	67,530	15,000
Karei (flatfish)	1,840	1,694	1,700	1,600
Menuke (ocean perch)	5,740	1,471	4,510	1,100
Madara (bridled dolphin)	10,800	10,252	6,000	6,000
Komai (wachna cod)	600	0	600	0
Sanma (saury pike)	65,210	64,290	0	0
Hokke, ainame (Atka mackerel, rock trout, etc.)	2,730	1,059	2,020	500
Ikanago (sand launce)	4,670	4,670	0	0
Katsuo, maguro (skipjack, tuna, etc.)	3,600	3,456	0	0
Same (shark, etc.)	600	560	0	0
Kichiji (*Sebastolobus macrochir*)	3,300	5,355	3,300	5,300
Other	11,660[b]	10,302	7,700	5,050
Ika (squid)	78,040	73,046	6,500	400
Tako (octopus)	200	109	140	50
Total	310,000	217,000	100,000	35,000

SOVIET QUOTAS IN JAPANESE FISHERY ZONE

	1989	1990
Maiwashi, saba (sardine, mackerel)	188,000	169,000
Itohikidara (a type of cod)	8,000	5,000
Sanma (saury pike)	12,000	7,000
Other	2,000	1,000
Total	210,000	182,000

Source.—Suisancho, *1990-nen Nisso Gyogyo Kankei Shiryo* (Documents on Japan-Soviet fisheries relations in 1990) (Tokyo: Suisancho, 1990), pp. 5, 10.
[a] The fee-charged quotas are included in the total quotas for each species.
[b] The "other" species include the fee-free quota for kichiji.

sion of a new agreement to replace the 1978 fisheries cooperation agreement. Tokyo responded affirmatively, believing that a new, long-term agreement would be in the interest of stable Japanese salmon fisheries in the Northwest Pacific.[46]

After six rounds of negotiations lasting an entire year, Tokyo and Moscow finally reached an agreement on 12 May 1985. The new agreement,

46. Moscow informed Tokyo on 26 June 1984 that it intended to terminate the existing agreement effective the end of the year. See Nobuhiko Nagasugi, "Nisso Sake-masu Gyogyo Kosho ni tsuite" (On the Japan-Soviet salmon-fishing negotiation), *AFF*, (July 1985): 26.

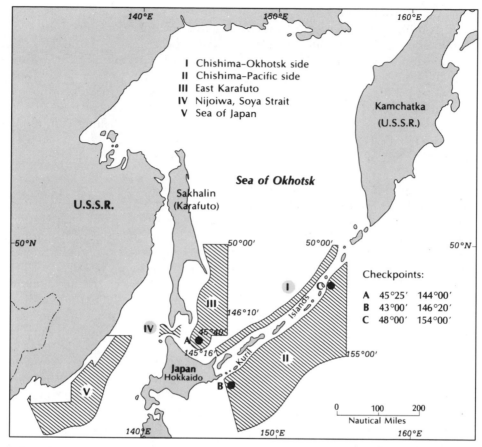

FIG. 2.—Japanese fishing areas in Soviet fishery zone, 1990. ("Nisso Chisen Okiai Gyogyo Kosho Daketsu" [Japan-Soviet offshore fishery negotiation concluded], *Suisankai*, no. 1261 [January 1990]: 8.)

effective 13 May, was to stay in force until 31 December 1987 and to be extended each year thereafter unless either party notified the other of its intent to terminate the agreement.[47]

The most difficult and divisive issue concerned the Soviet insistence on its soveriegn right, outside its 200-mile EEZ, to the anadromous species of fish of Soviet origin. Tokyo rejected the Soviet claims. In the end, however, the agreement recognized the Soviet Union's "primary interest and responsi-

47. Art. 9 of the Agreement between the Government of Japan and the Government of the Union of Soviet Socialist Republics concerning Cooperation in Fisheries (my translation of the Japanese). For a summary of the agreement, see Suisancho (n. 31 above), pp. 12–14.

bility" concerning anadromous species originating in its rivers[48] and its right to institute regulatory measures vis-à-vis salmon within and outside its economic zone.[49] This was in line with a similar provision of the 1982 LOS Convention.[50] As well, salmon fisheries were to be permitted only within the 200-mile zone. This also corresponded to the LOS Convention.[51] The agreement did recognize Japan's financial contribution to the reproduction of salmon stocks in the Soviet Union and stated that Moscow would give a special consideration to Japan and determine the latter's fishing conditions through bilateral consultation.[52] Furthermore, the control of salmon fisheries beyond the Soviet 200-mile zone was to be executed on the basis of agreement between the two countries.[53]

The agreement also provided for bilateral cooperation on scientific research of fisheries, for consultation of scientists and experts concerning the coordination and execution of cooperative research,[54] and for cooperation on the improvement of technology and methods for catching, propagating, and culturing fisheries resources.[55] It also called upon the two governments to cooperate on the conservation and management of living resources of the Northwest Pacific outside the 200-mile zones.[56] Finally, the agreement charged the bilateral fisheries commission with the responsibility for assessing the status of the fisheries and the fish stocks within the scope of the new agreement, including the determination of Japan's quotas within and outside the Soviet 200-mile zone.[57]

Thus, the new 1984 salmon fisheries agreement represented a mutually acceptable, if not totally satisfactory, arrangement. A conflict of interest clearly existed between the Soviet Union, whose position as the "state of origin" had been substantially strengthened by the LOS Convention, and Japan, which nonetheless managed to maintain sufficient salmon catches to support its fishing industry.

The Joint Japan-Soviet Fisheries Committee held a series of discussions in Moscow from May to June to determine the scope of Japanese salmon fishing in the Northwest Pacific in 1985. The two sides presented conflicting assessments of Japanese catch of salmon stocks of Soviet origin in 1984. Another controversy arose over the notice that the Japan Fisheries Association had sent to the Soviet Union in February 1985 that Japan would pay a re-

48. Art. 2, Par. 1, Agreement of 1985 (n. 47 above).
49. Ibid., Art. 2, Par. 2.
50. See Art. 66, Par. 1, 1982 UN Convention on the Law of the Sea (LOS Convention).
51. See, ibid., Art. 66, Par. 2.
52. Art. 2, Par. 3, Agreement of 1985 (n. 47 above).
53. Ibid., Art. 2, Par. 4.
54. Ibid., Art. 3, Par. 1.
55. Ibid., Art. 3, Par. 2.
56. Ibid., Art. 4.
57. Ibid., Art. 7.

TABLE 3.—JAPANESE SALMON FISHERIES IN THE NORTHWEST PACIFIC[a]

Year	Quotas (mt)	Actual Catch[b] (mt)	Fishery Cooperation Fee (in 10 million yen)
1978	42,500	41,500	176
1979	42,500	42,400	325
1980	42,500	42,500	375
1981	42,500	42,500	400
1982	42,500	42,400	400
1983	42,500	42,100	425
1984	40,000	35,400	425
1985	37,600	34,300	425
1986	24,500	20,300	350
1987	24,500	20,300	370
1988	17,700	—	335
1989	15,000	—	335

SOURCE.—*Suisan Nenkan* (Fisheries yearbook) (Tokyo: Suisansha, 1988), p. 29; Suisancho, *1988-nen Nisso Gyogyo Kankei Shiryo* (Documents on Japan-Soviet fisheries relations in 1988) (Tokyo: Suisancho, 1988).
[a]Japan-Soviet agreement.
[b]Rounded to the nearest hundred.

duced fisheries cooperation fee because it had not exhausted its fish quota for 1984. The Soviets at one time threatened to suspend the talks over this issue but were eventually persuaded to proceed with the negotiation.[58]

Japan and the Soviet Union also differed over Japan's quota, fishing areas, and cooperation fee for the 1985 fishing season.[59] The Soviets proposed introduction of a quota for each type of fish in the Japanese and U.S. 200-mile zones on top of the existing numeric species-by-species restrictions on the High Seas. Included in this proposal was a Soviet demand that Japan reduce its catch of Soviet-originated salmon in the Sea of Japan by 50%. This proposal resulted from the Soviet concern that the salmon stocks originating in Soviet rivers were showing an unmistakable deterioration, due in large measure to Japanese fishing outside the Soviet 200-mile zone.

In the end, Japan agreed to a reduction of its total catch from 40,000 mt and 32,600,000 fish in 1984 to 37,600 mt and 29,775,000 fish in 1985 (including a catch reduction outside the Soviet 200-mile zone from 22,100 mt and 16,950,000 fish to 18,750 mt and 13,965,000 fish; see table 3). For the first time in the postwar history of bilateral fisheries relations, the total quota was further broken down into subquotas for each of the five types of fish in all areas of the sea subject to the 1984 salmon fisheries agreement: 22,240 mt of pink salmon (*karafuto masu*), 9,700 mt of dog salmon (*shirozake*), 2,950 mt of red salmon (*benizake*), 1,910 mt of silver salmon (*ginzake*) and

58. For a summary of the discussions and their results, see Nagasugi, (n. 46 above), pp. 26–29.
59. Ibid.

chinook salmon (*masunosuke*).[60] Fishing areas were virtually identical to those permitted in 1984. Japan's fisheries cooperation fee remained the same as in 1984, that is, 4,250 million yen. Arrangements were also made for a Soviet scientific observer and an interpreter to be assigned to each of the four Japanese mother boats, to verify Japanese compliance with the agreement.[61]

The second (May 1986), third (February 1987), fourth (February and April 1988), and fifth (April 1989) meetings of the Joint Japan-Soviet Fisheries Commission exhibited similar patterns. Throughout the negotiations, the Soviets stressed the deteriorating resource condition within its EEZ and blamed it mostly on Japanese salmon fishing outside the 200-mile zone. Japan's assessment of the status of salmon resources was consistently more optimistic. Moscow proposed a total ban by 1992 on all Japanese catch of salmon of Soviet origin outside the Soviet EEZ, including that within Japan's 200-mile fishery zone. Tokyo countered that Japanese salmon fisheries outside the Soviet EEZ were strictly within the framework of the 1984 bilateral agreement and that a ban on certain salmon fisheries outside the Soviet EEZ would conflict with both the 1984 bilateral agreement recognizing salmon fishing on the High Seas and with the LOS Convention providing for the continued fishing of anadromous species in cases where a ban on such fishing would result in economic difficulties in the fishing country.[62]

The final agreement in 1986 included (1) a Japanese total quota of 24,500 mt and 19,340,000 fish; (2) shortened fishing periods; (3) virtually the same fishing areas as in 1985; (4) Japanese fisheries cooperation fee of 3,500 million yen; and (5) new arrangements for Soviet observers to go on board the Japanese patrol boats, and continued assignment of Soviet scientific inspectors on four Japanese mother boats, etc.[63] The final agreement in 1987 saw the Japanese quota set at 24,500 mt. Japanese fishing areas were virtually identical to those permitted in 1985, except for a ban, in accordance with the Japan-U.S.-Canada agreement on North Pacific fisheries, on mother-boat fishing in the Bering Sea east of 178° west longitude. The Japanese fisheries cooperation fee was set at 3,700 million yen. Tokyo and Moscow also agreed on the assignment of Soviet officials on four Japanese patrol boats.[64] In 1988 Japan accepted a further reduction in its salmon quota, by 3,700 mt. Japan also agreed to pay 3,700 million yen in fisheries cooperation fees. Fishing control measures were also expanded.[65]

The 1989 negotiation resulted in a further reduction of the Japanese salmon quota outside of the Soviet EEZ by 2,668 mt, to 15,000 mt, and an increase in the Japanese quota within the Soviet EEZ from 2,000 mt to 5,000

60. Ibid., p. 28.
61. Ibid., p. 29.
62. Ibid., p. 20.
63. Ibid., pp. 21–22.
64. *Suisan Nenkan* (n. 21 above), 29–31.
65. Ibid., pp. 67, 69, 72.

TABLE 4.—JAPANESE SALMON QUOTAS IN THE NORTHWEST PACIFIC BY
SPECIES AND AREA, 1988 AND 1989

Species	1988 (mt)	1989 (mt)
Karafuto masu (pink salmon)	11,295	9,935
Shirozake (dog salmon)	4,086	3,410
Benizake (red salmon)	949	744
Ginzake (silver salmon)	954	588
Masunosuke (chinook salmon)	384	323
Total	17,668	15,000

Area	1988 (mt)	1989 (mt)
Area 1	7,984	6,485
Area 2	2,655	2,981
Area 3	—	—
Area 4	1,851	1,234
Area 7	2,878	2,356
Area 8	2,300	1,944
Total	17,668	15,000

Source.—"Nisso Sakemasu Kosho, Yoyaku Daketsu" (Japan-Soviet salmon negotiations concluded at last), *Suisankai*, no. 1252 (May 1989): 36.

mt.[66] The latter quota increase was agreed to as an integral part of a Japanese-Soviet joint venture in salmon reproduction, an arrangement that was established in 1988. These changes were seen in Japan as a reflection of the Soviet effort to prohibit by 1992 all fishing for salmon of Soviet origin outside of the Soviet EEZ. Japan also agreed to pay 3,350 million yen in fisheries cooperation fees. Japanese salmon quotas in the Northwest Pacific in 1989 are shown in table 4.

JOINT VENTURES AND OTHER NEW FEATURES OF COOPERATION

The establishment of a Japan-Soviet salmon joint venture in 1988 resulted from a Soviet proposal in 1987 for Japanese capital and technological participation in the development of salmon hatcheries on Sakhalin Island. Following preliminary government-level discussions and private-level consultations, on 9 June 1988 representatives of three Soviet and six Japanese salmon fisheries organizations and associations signed a memorandum of understanding on the establishment of the proposed joint venture company, Pilenga-Godo.[67]

66. For an account of the 1989 negotiation, see "Nisso Sakemasu Kosho, Yoyaku Daketsu" (Japan-Soviet salmon negotiations concluded at last), *Suisankai*, no. 1252 (May 1989): 36.

67. Suisancho (n. 31 above), p. 29; "Nisso Sake-masu Fukajo Kensetsu, Kotoshiju

Accordingly, a Japanese corporation—Hokuyo Godo Suisan Kabushikikaisha (North Pacific Joint Fisheries Company, Inc.)[68]was founded on 1 July.

The Pilenga-Godo was formally established on 13 July with a total capital of 21.6 million rubles (about 4,700 million yen), of which 49% would be contributed by the Japanese side and 51% by the Soviets. The first of the hatcheries would be built on the Pilenga River in northeast Sakhalin. To generate funds for the joint venture, Japan was allowed to catch up to 2,000 mt of salmon in the Soviet EEZ east of the Kuriles in July 1988. In return, Japan was required to pay 1,750,000 rubles (about 380 million yen) in compensation.[69] This was the first time in 11 years that the Japanese would be allowed to catch salmon within the Soviet 200-mile EEZ.

The first hatchery, constructed in 1989, was projected to release 3,000 *shirozake* (dog salmon) in 1990.[70] Eventually the joint venture plans to build salmon hatcheries on five other Sakhalin rivers. In addition to the hatching of salmon, the new binational enterprise is scheduled to conduct scientific studies with a view to developing scallop culture farms in Sakhalin.[71]

Pilenga-Godo[72] is the second Japanese-Soviet joint venture in the fisheries area. The first joint venture firm, Sonico, was established in Khabarovsk on 26 May 1988. A joint venture between a Japanese trading company (Nisso Boeki), with 49% capital participation, and a Soviet association of trading cooperatives, a Khabarovsk association of fishery consumer cooperatives, and a Khabarovsk association of fishery *kolkhozes* (collective farms), the company is capitalized at 122.5 million yen and is engaged in the processing and marketing of fishery products.[73]

The Soviet Union is interested in developing more joint venture arrangements in the production, processing, and marketing of highly valued fisheries resources including invertebrates (e.g., shrimp) and kelp, as well as in coastal

---

ni Gobengaisha Setsuritsu e," (The construction of Japan-Soviet salmon hatcheries, towards the establishment of a joint venture company within the year), *Suisankai*, no. 1238 (March 1988): 55–56.

68. My translation of the Japanese name of the firm.

69. Suisancho (n. 31 above), p. 30.

70. The agreement initially called for totally Japanese-funded construction of the first hatchery. However, when the construction was found to cost an estimated 5 billion yen or more, the agreement was renegotiated and the Japanese contribution was reduced to a maximum of 1.7 billion yen, with the rest of the cost being borne by the Soviet side ("Saharin no Sake-masu Fukajo" [The salmon hatchery in Sakhalin], *Suisankai*, no. 1251 [April 1989]: 55).

71. "7-gatsu Jojun, Nisso Gobengaisha ga Hossoku" (The Japan-Soviet joint venture company to be launched early July), *Suisankai*, no. 1242 (July 1988): 53.

72. For a summary description of the joint venture, see "Soren 200-kairi Nai de 11-nen buri ni Sake-masu Sogyo" (Salmon fishing operations within Soviet 200 miles for the first time in 11 years), *Suisankai*, no. 1243 (August 1988): 59–62.

73. This information was supplied to me by the Japanese embassy in Moscow, 14 March 1990.

fishery cultures.[74] Since the establishment of Pilenga-Godo, five other Japanese-Soviet joint ventures have been established in the fisheries area.[75]

There has been phenomenal growth in the number of Soviet-foreign joint ventures in the Soviet Union since 1987. As of 10 October 1989, there were 947 Soviet-foreign joint ventures in the Soviet Union, representing 2,520 million rubles in total authorized capital. Of these, 21 involved Japanese firms. Fishery joint ventures were a very small part of the whole picture, with only 62 million rubles in capital authorized for 16 USSR-foreign fishery joint ventures registered in the Soviet Union as of 10 November 1989.[76]

Complementarity of interests between the Soviet Union and Japan is apparently pushing both sides to explore these joint ventures. The Soviet Union needs to increase its food supply for domestic consumption. Fishery exports will also generate badly needed foreign exchange for the Soviets. Japan also wants to increase its fish supply to meet the growing domestic demand and has capital to invest and technology to transfer.

Other elements of the bilateral fisheries regime are (1) consultation and joint research programs among scientists and fisheries experts;[77] (2) Japanese crab fisheries off Sakhalin, in the Seas of Japan and Okhotsk, in exchange for fisheries cooperation fees;[78] (3) Japanese kelp and sea-urchin production around the Soviet-controlled Kaigara Island, east of Hokkaido;[79] (4) Japanese purchase at sea of Alaska pollack and herring;[80] and (5) Japanese *madara* dragnet fishing in the Soviet EEZ.[81]

74. For a Soviet expression of interest in future expansion of Japanese-Soviet fishery cooperation in these areas, see N. A. Kotov, "Sonichi Gyogyo Kyoryoku no Isso no Hatten o Mezashite" (For further development of the Soviet-Japan fisheries cooperation), *Jiyu*, no. 325 (March 1987): 80–82.

75. The joint ventures are Diana in Yuzhno-Sakhalinsk, Sakhalin (capitalized at 7 million rubles); Okhotsk Suisan in Khabarovsk (3 million rubles); Sakhalin Tairiku in Kholmsk, Sakhalin (1 million rubles); Amur Trade in Khabarovsk (500,000 rubles); Aniwa in Yuzhno-Sakhalinsk (1 million rubles); and Magadan Gyogyo Godo in Magadan (20 million yen). The information was provided by the Japanese embassy in Moscow, 14 March 1990.

76. Dimitri Kuzmin, "Joint Ventures: The Upward Trend," *Vestnik* (USSR Ministry of Foreign Affairs) (March 1990): 51–53.

77. Suisancho (n. 31 above), pp. 26–28.

78. In 1988, Japan was allotted a total quota of 9,900 mt and 20–21 fishing boats in these areas (ibid., p. 33).

79. *Suisan Nenkan* (n. 21 above), p. 32. In 1987, Japan collected 1,031 mt of sea kelp in exchange for 110,500,000 yen (Suisancho [n. 31 above], p. 36). Japan harvested 260 tons of sea urchin in 1987, in exchange for 57 million yen (ibid., p. 37).

80. Japan bought a total of 38,950 mt of Alaska pollack in 1987 through this arrangement. Japan started buying herring from the Soviets at sea in 1988, purchasing 361 mt that year (ibid., p. 38).

81. Japan's 1988 quota was 25,500 mt, and the number of Japanese boats permitted to operate within the Soviet waters was 17–18 (ibid., p. 35). In 1989, a maximum of 8 Japanese boats were allowed at any one time. The total allowable catch of *madara*

A striking feature of the postwar Japan-Soviet fisheries regime is that, despite the rather acrimonious overall relations between the two countries in the earlier years, the bilateral fisheries agreements and their associated institutional arrangements eventually provided the degree of stability and even predictability that one could not have anticipated in the 1950s. This is particularly noteworthy in view of the fact that a peace treaty between Tokyo and Moscow has been prevented by the dispute over the northern islands and their surrounding sea area, which is very rich in fish and other marine resources.

In the initial phases (1957–61) of Japanese-Soviet negotiations over the Japanese salmon catch in the Northwest Pacific, Japanese proposals and Soviet counterproposals were wide apart, as were the agreed quotas from either side's initial proposals. However, as the years went by and the two sides gathered experience in dealing with each other, the gaps began to decline, particularly that between Japanese proposals and agreed quotas. Although the Soviets persistently proposed smaller quotas than the Japanese, after 1961 initial Soviet proposals for annual negotiations also moved closer to eventual agreements.

In conclusion, one can characterize the postwar Japan-Soviet fisheries relations as a regime based on mutual policy adjustments, or cooperation.

FUTURE COOPERATION BETWEEN THE SOVIET UNION AND JAPAN

For cooperative international ventures to develop between two parties who are equally interested in exploiting fishery resources to meet their national needs, several factors must be present. First, the parties must clearly recognize the need to cooperate in the management of exploitation and conservation of those resources. Second, the parties must individually or collectively possess sufficient technical expertise to meet the substantive requirements of fisheries resource management. Third, there must be sufficient domestic institutional support to initiate and then sustain cooperative efforts. Fourth, there must be political will on both sides to overcome obstacles that may stand in the way of cooperative arrangements. Fifth, the parties must be persuaded that cooperation will produce tangible benefits. Sixth, the parties must be willing to share the benefits and costs, both procedurally and substantively, in an equal and fair manner. In other words, reciprocity is an important ingredient in stable cooperation. Finally, each party must have sufficient confidence in the willingness and capability of the other to carry out its obligations arising out of the eventual cooperative programs.

---

was set at 25,000 mt ("Kotoshi no Nisso Haenawa Kyodo Jigyo" [This year's joint Japan-Soviet dragnet enterprise], *Suisankai,* no. 1244 [September 1988]: 60).

The analysis of the postwar Japanese-Soviet fisheries regime has amply demonstrated the need to develop effective international means to manage fisheries resources. The regime was based on the shared recognition that uncontrolled exploitation of fisheries resources would bring about irreversible consequences.

First and foremost, fisheries resource management—the rational use of fisheries resources through regulated and efficient exploitation and effective conservation of those resources for the attainment of a long-term stable fish supply—requires reliable information on the status of the resources in demand. The analysis of the postwar bilateral fisheries regime has demonstrated that the Soviet Union and Japan often came up with different resource assessments. This fact alone argues strongly for coordinated efforts in resource studies and improved information exchange between the two countries. This observation is particularly applicable to those anadromous species that spend part of their migratory life cycle in the Seas of Japan and Okhotsk, especially *karafuto masu* (pink salmon or *Oncorhynchus gorbuscha*) and *sakuramasu* (cherry salmon or *Oncorhynchus masou*).

Since the establishment of the Soviet and Japanese 200-mile zones in 1977, Japanese pink salmon fisheries have been restricted to the southern limits of the species' distribution in the Sea of Japan. This has caused a great deal of difficulty on the part of Japanese drift-net and long-line fisheries in the area.[82] If more reliable assessments of the stock indicate any significant improvement in the resources, Japan would certainly like to see its share increased. Japan has recently succeeded in the artificial propagation of cherry salmon larvae and is now exploring ways to develop them to sustainable levels on a stable and efficient basis.[83] Certainly, the technology developed by Japan will be of interest to the Soviet Union as well.

International cooperation will also enhance each coastal state's ability to assess the status of other valued stocks of fish found in the Seas of Japan and Okhotsk. These include *masaba* (mackerel or *Scomber japonicus*), *gomasaba* (spotted mackerel or *Scomber tapeinocephalus*), *sanma* (saury or *Cololabis saira*), *maiwashi* (sardine or *Sardinops melanosticta*), *katakuchi-iwashi* (anchovy or *Engraulis japonica*), *urume-iwashi* (round herring or *Etrumeus micropus*), *maaji* (common horse mackerel or *Trachurus japonicus*), *buri* (yellowtail), *surumeika* (Japanese common squid or *Ommatostrephes sloani pacificus*), *kuromaguro* (bluefin tuna), *sukesodara* (Alaska pollack or *Theragra challcogramma*), and *nishin* (herring or *Clupea pallasi*).[84] Japanese fisheries scientists call for careful moni-

82. Nihonkaiku Suisan Kenkyujo, *Saikin no Nihonkai ni okeru Gyogyo Shigen Doko* (Recent trends in the fishery resources in the Sea of Japan) (Niigata, Japan: Nihonkaiku Suisan Kenkyujo, 1985), p. 23.

83. Ibid., p. 25.

84. A convenient set of maps showing the distribution and fishing grounds of these and other fish stocks around Japan can be found in Suisancho Kenkyubu, *Nihon Kinkai ni okeru Shuyo Gyorui no Bunpu oyobi Gyobazu* (Maps showing the distribution

toring of these stocks in view of their erratic fluctuations and, in some cases, deterioration.[85]

Japan's visible commitment to effective management of fisheries resources is a relatively recent phenomenon. In the reconstruction of the war-devasted Japanese fishing industry, quantitative expansion of fishery production was given top priority. This soon resulted in overfishing of some coastal stocks. The government took two measures. The first was reduction of fishing effort. Outstanding examples include the 30% reduction in fishing effort in *isei sokobiki* (bull trawl) fisheries in 1950 and a similar measure taken in small-scale coastal dragnet fisheries in 1952–54. A more recent—and rare—example of self-imposed fishing reduction is the 20% cut in the number of fishing boats used in bull trawl in 1971–72.[86] Second, as the wartime ban on Japanese offshore fishing was lifted in 1952, the government encouraged Japanese fishermen to move into offshore fishing grounds. When this proved inadequate in meeting the nation's growing food and employment needs, Tokyo promoted the development of a major distant-water fishery. The dramatic growth of Japanese fisheries along these lines needs no repeating.[87] What is important is that in the fervent desire to expand their fishing grounds and fishery production, the Japanese long neglected rational and effective management of fisheries resources.[88]

With the exceptions just noted, virtually all reductions of Japanese fishing effort in the postwar period came as a result of pressure from Japan's partners in its bilateral or multilateral fisheries regimes. Former Director General of the Fisheries Agency Yoshihide Uchimura has acknowledged that the Japanese fundamentally lack a conservationist mind.[89]

The one field in which Japan has recently made some notable progress is the development of fish propagation technology. Here Japan may be able to make an immediate and significant contribution to international collaboration in the Seas of Japan and Okhotsk and elsewhere. Efforts are currently underway to develop technology for seedling production, resource management, and environmental control as part of marine farming.[90]

---

and fishing grounds of major types of fish in the areas of the sea around Japan) (Tokyo: Suisancho Kenkyubu, 1982).

85. See, for example, Nihonkaiku Suisan Kenkyujo (n. 82 above).

86. Yutaka Hirasawa, *Nihyaku-kairi Jidai to Nihon Gyogyo: Sono Henkaku to Saisei no Michi* (The 200-mile age and Japanese fisheries: Its road to reform and renovation) (Tokyo: Hokuto Shobo, 1978), p. 188.

87. Akaha (n. 14 above), pp. 19–20.

88. For this verdict, see, for example, Hirasawa (n. 86 above), pp. 187–97; and Akira Hasegawa, *Gyogyo Kanri* (Fisheries management) (Tokyo: Koseisha Koseikaku, 1985), pp. 213–14.

89. Yoshihide Uchimura, "21-seiki no Wagakuni Suisangyo no Tenbo" (The prospect for Japanese fishery industry in the 21st century), *AFF* (April 1986): 24.

90. Toru Amano, "Toru Gyogyo kara Shigen Baiyogata Gyogyo e: Kaiyo Bokujo Gijutsu no Kaihatsu" (From exploitative fisheries to resource incubation-type fisheries: The development of marine ranching technology), *AFF* (March 1978): 38–42.

The most successful so far has been the Fisheries Agency's program, Comprehensive Research concerning the Development of Technology for the Propagation of Anadromous Salmons, which is aimed at the development of salmon hatching technology. The government project, undertaken between 1977 and 1981, succeeded in improving the homing rate of dog salmon (*shirozake*) from the natural rates of 0.5% (in the Sea of Japan), 1% (on the Pacific side of Honshu), and 2% (in Hokkaido) to 3–4%. With the aid of a newly developed technology, Japan successfully harvested 40 million dog salmon in 1985–86.[91]

The Fisheries Agency's Marine Ranching Project, started in 1979, is an even more ambitious program. The 10-year project is exploring the possibility of artificially creating a diversity of living resources for optimal output.[92] The project involves ecological studies of tuna, common horse mackerel, flounder, cherry salmon, bay scallop, ark shell, sea trumpet (*Ecklonia cava*), and *arame* (*Eisenia bicyclis*, a seaweed).

Is there sufficient institutional support for cooperative efforts in the Seas of Japan and Okhotsk? In answering this question, it should be noted that both the Soviet Union and Japan have accepted the LOS Convention by signing both the Final Act of UNCLOS III and the LOS Convention.[93] Neither the Soviet Union nor Japan had ratified it. However, many of the provisions of the LOS Convention have already become customary international law, and the Soviet Union and Japan observe them in their national and international fisheries practice.

One may question the competence of the two countries to comply fully and precisely with the provisions of the convention. Even Japan, with its scientific, technical, financial, and bureaucratic competence, finds many obstacles to full implementation of the 1982 convention at the national level.[94]

Difficult questions remain as to the adequacy of some of the provisions of the convention in stipulating a sufficiently precise guide for the formulation and execution of domestic laws and regulations in the coastal states that are parties to the new Law of the Sea.[95] Coastal states may find it difficult to

91. Hisashi Kanno, "Shigen Baiyo: Kaiyo Bokujo Gijutsu Kaihatsu Kenkyu no Keika" (Resource propagation: The status of research for the development of marine ranching technology), *AFF* (April 1986): 6–7. See also Tetsuo Kobayashi, "Jiniteki Kontororu ni yoru Sakuramasu Shigen no Zodai" (The expansion of dog salmon resources by artificial control), *AFF* (April 1986): 11–18.

92. Kanno (n. 91 above), pp. 7–9.

93. The Soviet Union signed the LOS Convention on 10 December 1982, and Japan on 2 July 1983.

Editor's note.—See table 1G, "Status of the UN Convention on the Law of the Sea," in this volume.

94. See, for example, Tsuneo Akaha, "Internalizing International Law: Japan and the Regime of Navigation under the UN Convention on the Law of the Sea," *Ocean Development and International Law* 20 (1989): 113–39.

95. See, for example, Office of the Special Representative of the UN Secretary-General for the Law of the Sea, *The Law of the Sea: National Legislation on the Exclusive*

carry out their obligation to determine the allowable catch of the living resources in their EEZs.[96] Even Japan did not begin a systematic effort to gather data on the living resources within its own 200-mile fishery zone until after the fishery zone had become law. Since then, many new species have been discovered in the area concerned, and further research continues.

To promote "the optimal utilization of fisheries resources," under the LOS Convention, the coastal state is obligated to "determine its capacity to harvest the living resources of the exclusive economic zone" and "give other states access to the surplus of the allowable catch" when it "does not have the capacity to harvest the entire allowable catch."[97] However, the precise meaning of the "capacity to harvest" living resources is unclear. Does "capacity" mean actual or potential ability, or even intended ability?[98]

The convention obligates the coastal state to take into account a number of factors in determining foreign access to the surplus of the allowable catch. The subjective nature of such determination has been pointed out by many international legal scholars.[99] Moreover, serious questions have been raised about the adequacy of the provisions of the LOS Convention concerning the settlement of disputes arising out of the interpretation and implementation of the convention. For example, many types of disputes are left out of the dispute settlement provisions of the LOS Convention and are left to the discretion of the coastal state.[100]

To prevent potential complications resulting from these problems, the Soviet Union and Japan must engage in frequent and frank consultations and negotiations. The extensive experience between the two countries, documented in this paper, augurs well for such a development. There are further requirements, however.

In Tokyo, greater political will must be cultivated if Japan is to fulfill its potential as a reliable and fair partner. Fortunately, at the highest level of the Japanese fishery policy system there is today the recognition that Japan must develop mutually beneficial fisheries relations with neighboring countries.[101]

Economic Zone, the Economic Zone, and the Exclusive Fishery Zone (New York: United Nations, 1986).

96. The obligation is provided for in the LOS Convention, Art. 61, Par. 1 (n. 50 above).

97. Ibid., Art. 62, Par. 2.

98. See Shigeru Oda, "Fisheries under the United Nations Convention on the Law of the Sea," *American Journal of International Law* 77, no. 4 (October 1983): 743–44.

99. See, for example, ibid., pp. 744–45.

100. See ibid., pp. 746–54. See also Hideo Takabayashi, "Haitateki Keizai Suiiki ni okeru Gyogyo Funso no Shori" (The settlement of fisheries disputes in the exclusive economic zone), *Hosei Kenkyu* 51, no. 2 (February 1985): 255–64; 51, nos. 3–4 (March 1985): 571–614; and 52, no. 1 (September 1985): 89–110.

101. A good example of complementary interests is found in the Soviet reliance on herring, which is not so highly valued in Japan, and the Japanese appreciation of Alaska pollack, which the Soviets tend to undervalue.

There are some potentially disruptive factors in Japanese-Soviet relations. The most troublesome is the territorial dispute noted earlier. However, the past record tends to indicate that pragmatic considerations on both sides are likely to prevent the issue from chilling their fisheries relations beyond control.

The next question is, What tangible benefits can the Soviet Union and Japan expect from cooperation? Experience with this question is still quite limited, and the exact nature of benefits to each side is difficult to estimate. However, there are indications that both sides see mutual benefits in increased cooperation. Recent developments in the Japan-Soviet fisheries regime, with new elements of bilateral cooperation in fisheries resources production and conservation, indicate increasing Soviet interest in cooperative ventures in fisheries resource management. It is generally recognized that the Gorbachev government is keen on firmly establishing Soviet sovereign jurisdiction over the living and nonliving maritime resources within its EEZ and on the rational management of the Soviet fishing industry. The Soviet effort to use fishery resources as a part of its design to develop a more efficient national economy is apparent in the Soviet-proposed principle of equivalence in fish catch, in the rigorous effort to control and phase out foreign fishing within the Soviet 200-mile zone, in the purported intent eventually to ban all catching of salmon of Soviet origin beyond its 200-mile limit, in the introduction of fishing and fisheries cooperation fees, and in the promotion of joint ventures in fisheries.

Finally, Japan now realizes that whatever fisheries relations may develop between Japan and its neighboring countries in the future must be beneficial not only to its own fishermen but also to the economies of neighboring countries.

Today, the USSR-Japan fisheries relations more closely approximate an equitable and fair regime than at any time in the past. Japan must fully commit itself to the idea of fisheries resource management and thus cultivate sufficient confidence on the part of its neighbor to the north.

SUMMARY AND CONCLUSIONS

In summary, the review of the postwar Japanese-Soviet fisheries regime reveals several interesting features. First, the construction of a bilateral fisheries regime took place within the context of the normalization of bilateral diplomatic relations. The fisheries talks affected and were affected by the state of overall diplomatic relations. In contrast to the Republic of Korea–Japan fisheries negotiations, which took 13 years before they came to a successful conclusion in 1965, the Japanese-Soviet negotiations took less than 1 year (from 29 April to 15 May 1956). This was largely due to Japan's pragmatism, as represented by its decision early on to separate the bilateral negotiations on fisheries from those for the normalization of diplomatic relations.

Second, the bilateral fisheries regime was established initially to deal with the resumption of Japanese fishing off Soviet coasts. It was the Japanese fishing that had to be brought under some regulatory control. Simply in terms of fish catch Japan's experience in the earlier years was largely positive, with Japanese fishermen harvesting substantially more in their neighbors' waters than Soviet fishermen in Japan's coastal areas. Subsequently, however, the expansion of Soviet fishing in Japanese coastal waters made the balance more equitable.

Third, the Japan-Soviet fisheries regime has enjoyed a modicum of stability and predictability. The annual fisheries committee meetings under the bilateral regime have allowed Soviet and Japanese fisheries experts to exchange their respective assessments of fisheries resources and to recommend acceptable, if not optimal, levels of fishing effort and regulatory measures. When scientific assessments differed substantially, as was often the case, high-level officials were called upon to break the deadlock. The whole process thus forced the two governments to coordinate their respective policies. In other words, "cooperation" as defined at the outset of this paper was achieved.

Fourth, notwithstanding the sensational—largely alarmist—media coverage of the postwar fisheries regime in Japan, the Japan-Soviet fisheries regime has shown a remarkable degree of adaptability in the face of changing needs of the two countries. Probably the single most important development that has affected the bilateral fisheries relations has been the expansion of the Soviet fishing industry since the 1970s. The fisheries regime has shown enough flexibility to accommodate these trends, with painful consequences to the Japanese.

Finally, as for future prospects, the two sides clearly understand the need to cooperate. Technical expertise for fisheries resource management is growing in Japan, although so far the most promising developments have been limited fisheries production rather than conservation and rational use. Institutional arrangements for bilateral cooperation should not be too difficult, as the two countries have a long history of dealing with each other. Although future tangible benefits of the bilateral cooperation are hard to gauge, cooperation will certainly prove more beneficial than the lack of it. Equity and fairness will be key to any successful cooperative arrangements.

If Japan and the Soviet Union are to maintain their status as two of the most important fishing countries of the world, it is imperative that they continue to cooperate. The alternative—conflicting claims and unbridled competition—will surely bring about irreversible consequences, unwanted deterioration of the marine ecosystem, and in the end harm to those whose livelihood depends critically on the stable supply of marine living resources.

Sustainable Development of Fisheries

John E. Bardach*
Environment and Policy Institute, East-West Center

THE FACTS

The question of fisheries sustainability is germane even at a time when world fisheries yields have increased overall, but catches of several stocks, especially coastal ones, have declined (fig. 1). Nor are the yield increases uniform; many developing countries have begun to exploit their exclusive economic zones (EEZs) more vigorously while long-established fisheries have not produced higher volumes (fig. 2).[1] The overall increase may well reflect that estimates of traditionally fishable resources could be revised upwards, as the result of recent research activities based on primary production in the various oceanic provinces.[2]

Fishable renewable aquatic resources can be partitioned with a view to their distribution in the sea and to certain aspects of their management (figs. 3, 4). Inside the 200-mile EEZ where the vast majority of fisheries occur, one may usefully separate, according to the volume of catches, the near- and offshore demersal (bottom and near bottom) species among which the gadoid fishes (cods, hakes, and pollack) predominate. They make up about 32% of the renewable resources of the sea, including mollusks and crustaceans. Higher in the water column and nearer the surface are the pelagic fishes (and squids), comprising about 64% of the harvestable ocean biomass. They are most prevalent within the 200-mile limit, often far closer to shore, and they furnish most of the world's store of fishmeal and fish oil derived, in the

---

*EDITORS' NOTE.—This is the third article on fisheries that John Bardach has written for *Ocean Yearbook*. His earlier articles are "Pacific Tuna: Biology, Economics, and Politics," with Penelope J. Ridings, *Ocean Yearbook 5*, ed. Elisabeth Mann Borgese and Norton Ginsburg (Chicago: University of Chicago Press, 1985), pp. 29–57; and "Fish Far Away: Comments on Antarctic Fisheries," *Ocean Yearbook 6*, ed. Elisabeth Mann Borgese and Norton Ginsburg (Chicago: University of Chicago Press, 1986), pp. 38–54.

1. Food and Agriculture Organization (FAO), *Production Yearbook 1985* and *1987* (Rome: FAO, 1987, 1989).

2. J. H. Martin, G. A. Knauer, D. M. Karl, and W. W. Broenkow, "Vertex: Carbon Cycling in the Northeast Pacific," *Deep Sea Research*, 34 (1987): 267–85.

FIG. 1.—Estimated annual global catch of all aquatic species, 1950–87. (Food and Agriculture Organization [FAO], *Production Yearbook 1985* and *1987* [Rome: FAO, 1987, 1989].)

main, from the so-called clupeid fishes like herring, sardines, and anchovies. Salmon, although a special case because of their freshwater spawning habit, also belong to the EEZ. The upper waters of the oceans at large harbor only about 4% of species that contribute to the total harvest.[3] These are largely the tunas, which bring high returns per unit of volume and pose certain

3. John E. Bardach, "Technology and the Exclusive Economic Zone," *Proceedings of EEZ Resources: Technology Assessment Conference*, 22–26 January 1989, 2:1–14, Honolulu, Hawaii.

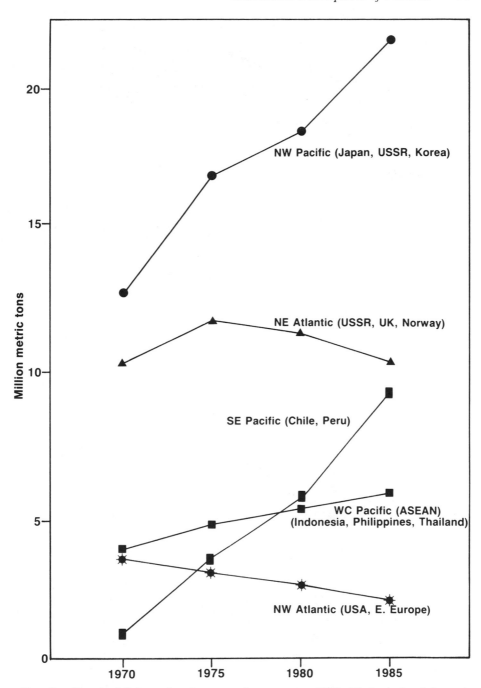

NW Pacific (Japan, USSR, Korea)

NE Atlantic (USSR, UK, Norway)

SE Pacific (Chile, Peru)

WC Pacific (ASEAN)
(Indonesia, Philippines, Thailand)

NW Atlantic (USA, E. Europe)

Million metric tons

FIG. 2.—Nominal fish catches by countries or areas, 1970–85. Includes fish, mollusks, and crustaceans. (Food and Agriculture Organization [FAO], *Production Yearbook 1985* and *1987* [Rome: FAO, 1987, 1989].)

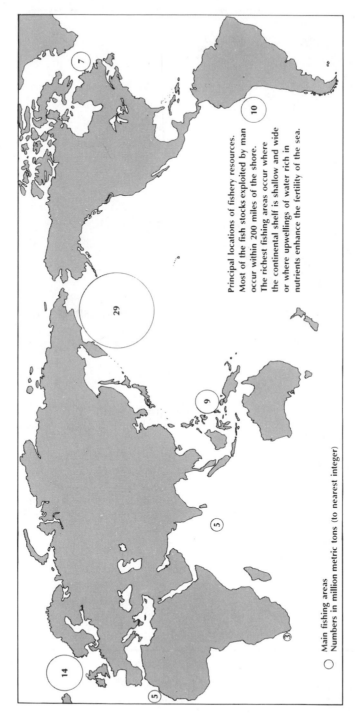

Fig. 3.—Principal locations of fishery resources. (Food and Agriculture Organization [FAO], *Production Yearbook 1985 and 1987* [Rome: FAO, 1987, 1989].)

Principal locations of fishery resources. Most of the fish stocks exploited by man occur within 200 miles of the shore. The richest fishing areas occur where the continental shelf is shallow and wide or where upwellings of water rich in nutrients enhance the fertility of the sea.

○ Main fishing areas
Numbers in million metric tons (to nearest integer)

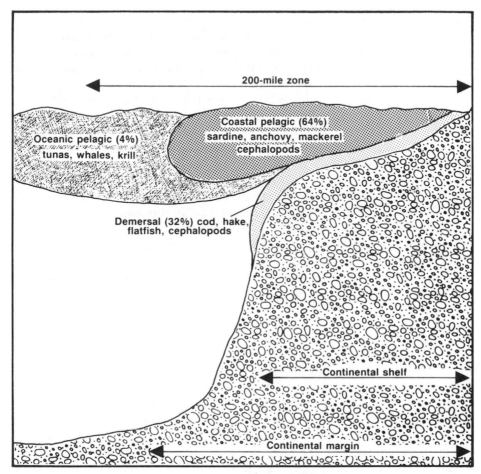

Fig. 4.—Basic distribution of marine catches, by regions in the sea. (Food and Agriculture Organization [FAO], *Production Yearbook 1985* and *1987* [Rome: FAO, 1987, 1989]; John E. Bardach, "Technology and the Exclusive Economic Zone," *Proceedings of EEZ Resources: Technology Assessment Conference,* 22–26 January 1989, 2:1–14 [Honolulu, Hawaii].)

management problems because they swim freely through the artificial 200-mile boundaries.

Biologic and economic management of tuna fisheries thus requires the cooperation of several nations. Injunctions embodied in the articles of UNCLOS III to deal with highly migratory species demand international cooperation over their entire ranges. The Inter-American Tropical Tuna Commission (IATTC) was intended to regulate tuna management in the eastern Pacific. It was an attempt at cooperation, although it was eventually doomed by various influences.[4] It antedated the 1982 Law of the Sea Conven-

4. The IATTC has operated under a convention originally concluded by Costa

tion by decades. Since then, the South Pacific Forum Fisheries Agency and the South Pacific Commission have operated in the main fishing grounds for various tuna species. The agency is trying to bring about economic equity and the commission to set the stage for biological management, when and as the status of the stocks involved require it.[5]

When examining certain fisheries with regard to their sustainability, a prime consideration will have to be the coincidence of human harvesting activities with the influences of climate. Sometimes a balance between the two can be achieved, resulting in years or decades of well-sustained fisheries, but all too frequently this balance is disturbed and fisheries suffer from severe economic dislocations. The natural regenerative capacities of aquatic food chains on which fisheries rely are affected by several forces that may act singly or jointly, as will be examined, albeit cursorily, in this article.

The first such force is climate; it is clearly beyond the influence of man, acting on the marine biosphere mainly through variations in winds, temperature, and atmospheric pressure. One of the best examples of purely climate-influenced fluctuations in stock strength for fishes is that of the Japanese sardine (fig. 5). The fishery has existed for several centuries and has experienced several ups and downs. It is now believed that shifts in the Tsushima and the Kuroshio currents away from the Japanese islands lead to a temporary decline of the species.[6] That such variations can indeed be climate-dependent is also well illustrated by tree growth in California and a prevalence index of albacore tuna (fig. 6).

Well-organized research in fishery oceanography coupled with a reliance on past experience has important, but rarely sufficient, predictive powers to soften the economic impact of a failing, strongly climate-dependent fishery and to divert a fleet to alternative catches. When there is a rise in one species, the biomass of another often falls, as is the case in a quasi balance between a decline of sardines and a rise of anchovies, jack mackerel, and other fish to which Japanese fishermen could turn (fig. 7).[7] In general, coastal pelagic species, especially clupeid, tend to show pronounced, apparently climate-dependent fluctuations (for instance, anchovies and sardines on the coasts of Chile and Peru).

Rica and the United States and open to adherence by other governments whose nationals fished for tropical tunas in the eastern Pacific Ocean. The convention was adhered to by seven countries (Panama [1953], Ecuador [1961], Mexico [1964], Canada [1968], Japan [1970], France [1973], and Nicaragua [1973]). However, four of these have since withdrawn from IATTC (Ecuador [1968], Mexico [1978], Costa Rica [1979], and Canada [1984]).

5. Biliana Cicin-Sain and Robert Knecht, "The Emergence of a Regional Ocean Regime in the South Pacific," Working Paper 14 (Honolulu, Hawaii: Environment and Policy Institute, East-West Center, 1989).

6. S. Chikuni, "The Fish Resources of the Northwest Pacific," *FAO Fishing Technology Paper* 266 (Rome: FAO, 1985).

7. Ibid.

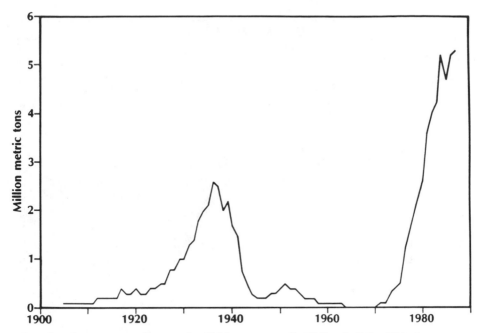

FIG. 5.—Japanese sardine catch, 1900–present. (S. Chikuni, "The Fish Resources of the Northwest Pacific," FAO Fishing Technology Paper 266 [Rome: FAO, 1985].)

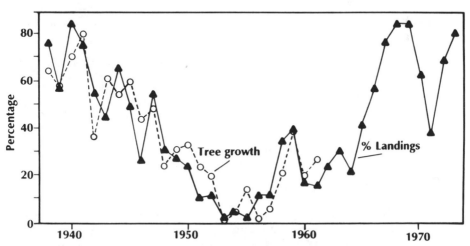

FIG. 6.—The influence of climate on tree growth and fish abundance. Correlations of albacore catches off the coast of southern California and the annual width of conifer tree rings in western North America. (N. E. Clark, T. J. Blasing, and H. C. Fritts, "Influence of Interannual Fluctuations on Biological Systems," *Nature* 256 [1975]: 302–5.)

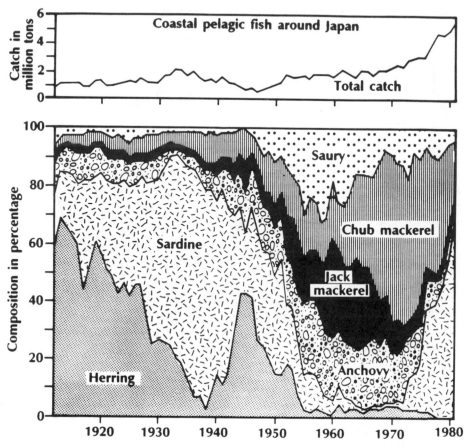

FIG. 7.—Changes in the total catch and species composition in the major coastal Japanese pelagic fisheries, 1920–80. (S. Chikuni, "The Fish Resources of the North-west Pacific," FAO Fishing Technology Paper 266 [Rome: FAO, 1985].)

A second set of forces acting on the prevalence of fish in many areas of the ocean is entirely due to the activities of man. Even so, these forces often escape our management grasp or are managed too late.

An example is the introduction of new technologies into the fishery. More powerful boats and new and more effective gear, such as purse seines instead of gill nets or pole and line, and drift nets which are indeed destructive, are cases in point. These advances are so effective that they usually go hand in hand with the apparently inevitable tendency to overcapitalize in boats and gear that seems to prevail in man's approach to the use of common properties. Fisheries may collapse in the wake of these trends; the North Sea herring fishery is a prime example, as it declined precipitously soon after the introduction of purse seines (fig. 8).[8] The British Columbia herring fishery suffered a similar fate aggravated by the tendency of the declining herring schools to occupy smaller and smaller regions of their former distribution. Thus, it is difficult to assess these reductions as long as catches per unit of effort remain adequate. In the North Sea and British Columbia waters, rigorous management of quotas, times, and regions of fishing were instituted, and the fishery began to recover.

Other technology-dependent influences on fish stocks occur when new processing techniques coupled with effective harvesting methods lead to great increases in catches, if not new fisheries altogether. The post–World War II development of the fishmeal industry, essentially for livestock feed, is a classic example. Another is the more recent perfection of the *surimi* or gel extrusion technique with its great upswing in the fishery for Alaska pollack. It is true that demersal gadoid fish populations are more or less abundant because of fluctuations in climate, like any other wild fish stock, yet it seems that their fecundity per female, greater than that found in herring and their relatives, helps make climate-influenced swings in population size less pronounced. They also show less pronounced movements for spawning and feeding; both behavior traits help detect influences of fishing pressure on their abundance. The grounds where gadoid fish are found can be delineated more or less, and industry-wide regulation of their fishery, in essence a single-species one, is easier. With further improvements of fish location devices and methods, it may even be possible to rely on such an industry to partially control itself through a kind of privatization, a method of vesting management measures of the fishery fully or partially in the industry itself.

A very important set of forces that influence, often to depress, coastal fisheries in particular has without a doubt its roots in terrestrial activities. Industrial and domestic effluents and even oil flows notwithstanding, building dams on large and small rivers can have the most pervasive effect on fisheries nearshore and sometimes further offshore. (Luckily salmon culture is sufficiently developed so that stocks of blocked spawning streams can be and, in many places, have been saved.) Where the marine biota relies on the fertilizing effect of a river that is reduced or cut off by a dam, however, fisheries can be wiped out, albeit after a delay of several years while the shortfall in

---

8. A. Saville and R. S. Bailey, "The Assessment and Management of Herring Stocks in the North Sea and to the West of Scotland," *Rapports et procès—verbaux des réunions (ICES)* 177 (1979): 112–42.

Editors' note.—For more information about the North Sea herring fishery, see James R. Coull, "The North Sea Herring Fishery in the Twentieth Century," *Ocean Yearbook 7*, ed. Elisabeth Mann Borgese, Norton Ginsburg, and Joseph R. Morgan (Chicago: University of Chicago Press, 1988), pp. 115–31.

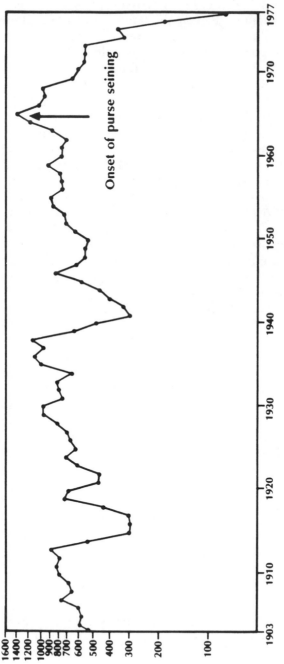

FIG. 8.—Total landings of herring from the North Sea, including the Skagerrak and English Channel, 1903–77. (A. Saville and R. S. Bailey, "The Assessment and Management of Herring Stocks in the North Sea and to the West of Scotland," *Rapports et procès—verbaux des réunions [ICES]* 177 [1979]: 112–42.)

land-derived nitrates and phosphates reduces plankton and bottom organisms on which the fish stocks depend. The sardine fisheries of the southeastern Mediterranean relied on the Nile; after the Aswān High Dam was built they declined severely, to revive somewhat when irrigated agriculture, made possible by the dam, again funneled some nutrients into the sea.[9]

A more complex example comes from the Indus River; upstream dams now greatly reduce the flow into the eastern Arabian Sea, and therefore certain stretches of coastal waters experienced substantial increases in salinity. Mangrove swamp associations were affected and shrimp populations were decimated because the coastal waters were too saline for their spawning requirements.[10] It may well be that national economic gains due to irrigated agriculture upstream and/or the generation of electricity, made possible by the dams, more than made up for the dislocation in the fishery. In any case, that is little consolation to the many thousands of small-scale fishermen who, most likely, cannot find alternative employment—even as meager as the one they may have derived from harvesting the seas around them. It indicates also that years may elapse before effects of land-based activities on marine life are noticed. If economic planning of major river regulations is done with foresight and considers marine effects (as rarely happens), the delay might even make planning easier.

Concurrent with influences on the nearshore marine environment, and therefore on fisheries due to land-based, usually development-oriented activities, there is competition among fishermen. In many countries they operate with little capital, and therefore they have a limited reach. These small-scale fishermen are growing in numbers much too fast for the resources available to them (fig. 9).[11] Often the competition is not only among themselves but also with wealthier and more mobile components of national fishing industries; strict partitioning between size of boats and kinds of gear is the answer, in theory, but it is only effective with good surveillance and control by authorities, or by internal or self-policing. Needless to say, that is not often practicable.

THE PROBLEMS

The lives of aquatic organisms and thus the size of fish stocks are, as mentioned above, subject to a number of influences that are sometimes negative.

9. Editors' note.—This point is elaborated on in Adalberto Vallega's article, "A Human Geographical Approach to Semienclosed Seas: The Mediterranean Case," *Ocean Yearbook 7*, ed. Elisabeth Mann Borgese, Norton Ginsburg, and Joseph R. Morgan (Chicago: University of Chicago Press, 1988), pp. 372–93.

10. E. Goldsmith and N. Hildgard (eds.), *The Social and Environmental Effects of Large Dams*, vol. 2, *Case Studies* (Wadebridge, United Kingdom: Wadebridge Ecological Centre, 1986).

11. International Center of Living Aquatic Resources Management (ICLARM), *Annual Report 1987* (Makati, Metro Manila, Philippines: ICLARM, 1987).

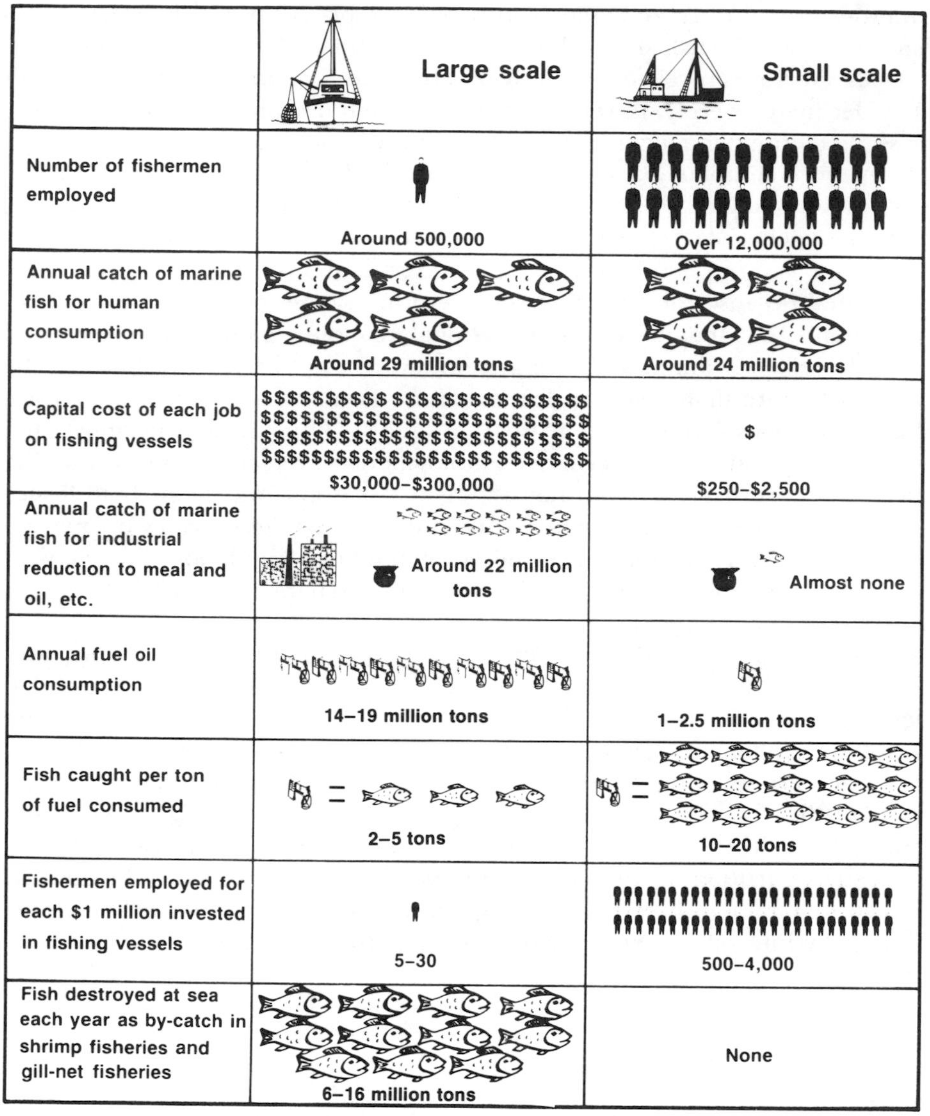

	Large scale	Small scale
Number of fishermen employed	Around 500,000	Over 12,000,000
Annual catch of marine fish for human consumption	Around 29 million tons	Around 24 million tons
Capital cost of each job on fishing vessels	$$$$$$$$$$$$ $$$$$$$$$$$$$$$$$$$ $$$$$$$$$$$$$$$$$$$$$$$$$$$$$$$ $$$$$$$$$$$$$$$$$$$$$$$$$$$ $$$$ $$$$$$$$$$$$$$$$$$$$$ $$$$$$$$$ $30,000–$300,000	$ $250–$2,500
Annual catch of marine fish for industrial reduction to meal and oil, etc.	Around 22 million tons	Almost none
Annual fuel oil consumption	14–19 million tons	1–2.5 million tons
Fish caught per ton of fuel consumed	2–5 tons	10–20 tons
Fishermen employed for each $1 million invested in fishing vessels	5–30	500–4,000
Fish destroyed at sea each year as by-catch in shrimp fisheries and gill-net fisheries	6–16 million tons	None

FIG. 9.—Large-scale and small-scale fisheries compared. (International Center of Living Aquatic Resources Management [ICLARM], *Annual Report 1987* [Makati, Metro Manila, Philippines: ICLARM, 1987].)

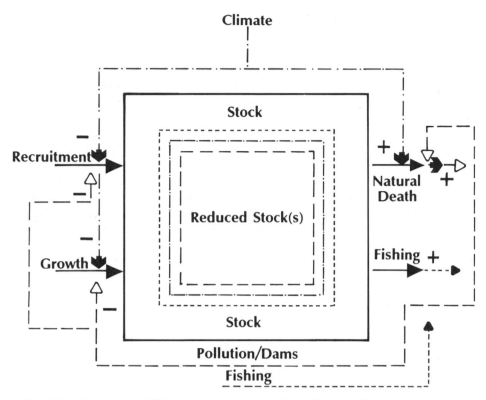

FIG. 10.—Dynamics of fisheries environment. (John E. Bardach, "Technology and the Exclusive Economic Zone," *Proceedings of EEZ Resources: Technology Assessment Conference,* 22–26 January 1989, 2:1–14 [Honolulu, Hawaii].)

Figure 10 shows the interaction of climate and human influences on fish stocks and by implication the ecosystems in which they occur. Climate may directly favor recruitment and growth, and accelerate or retard natural death, while fishing essentially reduces numbers unevenly and indirectly affects growth and perhaps recruitment. Pollution acts on growth and recruitment as well as death and thus strongly influences fishing.

An unexploited fishery has a more or less lasting balance between recruitment and growth on one side and natural mortality on the other. Adding fishing pressure to effect optimum, or in some cases maximum, sustainable yields (the latter with greater risks of unbalance) can result in a healthy, maintainable fishery, everything else being equal. But, as has been pointed out, everything else is usually not equal, and the managing of fish stocks calls for application of imaginative biological and technical regulatory measures and for the use of economic and social remedial techniques.

Climatic events, like an El Niño, can severely depress fisheries, as the collapse of the Peruvian anchovy fishery in 1972 illustrated, even though an overcapitalization of fleet and fishmeal plants also apparently played a role in the collapse of what was the world's largest recorded fishery. If and when improvements in climate prediction are attained, they must be coupled with very good coordination among climatologists, fisheries services, and the industry itself. This would permit the necessary reduction in fleet strength by retiring older vessels in time to avoid serious overcapacity, or the adjustment in use of grounds or shifts to other species as the presaged changes in climate begin to occur. But climate forecasting is an uncertain business at best, as is well exemplified by the present range in projected rate and levels of a future global warming.[12] While some warming is likely, its anticipated time frame is such that it is of little use to know that an increase in ocean temperatures is likely to bring northward and southward extension of commercial fish species ranges and, in some cases, most certainly a reduction in stocks. To set in motion, and achieve in time, the complex set of economic and social adjustments in a fishery about to be influenced by a climate change would optimally require fairly exact regional climate forecasts within a 5- to 10-year range. Though a great deal of progress has been made in the last decade in assessing interaction between the oceans and the atmosphere, such forecasting ability still eludes us. Even so, the fishing industry or profession will obviously benefit from having internal organization—unions, cooperatives, village associations, and the like—that enables dissemination of information and makes it possible for industry members to adjust to or counteract the various outside forces or pressures on the environment. Some such measures will be possible or more effective with certain fisheries than with others. In general, self-policing in the taking of a vulnerable resource that belongs to the gatherer only after it is captured is more easily achieved where the quarry is not highly vagrant but stationary, if not attached, such as algae in the wild, lobsters, or abalone.

Where the nature of the resource facilitates entitlement to portions of its distribution sites, such internalization over the control of the take (i.e., privatization of one or another functional component of the fishery) may create a bridge to enhancement of the stocks, if not to aquaculture proper.

Aquaculture offers a wide range of possibilities for sustainable production of marine organisms. Suffice it to state here that aquaculture has grown faster than capture fisheries in the last two decades, and lately it has produced about 13% in weight of directly consumed aquatic products.[13] In terms of value, aquaculture makes an even higher contribution because the cultured

12. Editors' note.—The author is an internationally recognized science and policy consultant on global warming and sea-level rise. He has coauthored an article for *Ocean Yearbook* on this subject. See Lynne Rodgers-Miller and John E. Bardach, "In Face of a Rising Sea," *Ocean Yearbook 7,* ed. Elisabeth Mann Borgese, Norton Ginsburg, and Joseph R. Morgan (Chicago: University of Chicago Press, 1988), pp. 177–90.

13. John E. Bardach, "Aquaculture: From Craft to Industry," *Environment* 30, no. 2 (1988): 6–11, 36–40.

species (e.g., shrimp and salmon) tend to be among the more expensive sea-food items. Given ultimate production limits of the impending aquatic environment and world demographic trends, an even greater role can be anticipated for aquaculture in the future. Like coastal fisheries, though, it has been depressed and at times destroyed by nearby or off-site activities on land. How one attempts to deal with balances between renewable resource use and other economic activities touches the very heart of sustainable development issues.

The goal of sustainability is obviously desirable; it has become part of the development rhetoric since the report of the World Commission on Environment and Development was issued in 1987.[14] That it may more easily be applied to renewable rather than stock resources such as minerals is self-evident. Management for sustained yields of single renewable resources such as fisheries or forestry—as opposed to a more general notion of sustainable development—found its way into the practice of renewable resource use much earlier.[15] If sustainability is extended from a single resource to the ecosystem(s) on which that resource relies, great difficulties are introduced. Ecosystem functions and components may change at different rates and in different directions while they are also influenced by forces outside the system. Simultaneous control or mastery over all of them is usually not attainable for physical, social, or economic reasons. Examples pertaining to coastal fisheries have been given earlier, that is, the Nile and the eastern Mediterranean. While fishery intensities and locations might have been controllable, the system of the Nile would not have been amenable to concurrent regulation with fisheries, either in time or extent.

In practical terms the regulation of off-site effects or forces (externalities, as they are called by economists) demands that quantification of costs and benefits of all activities in the system be assessed and weighed against one another. Often that requires longer time frames to be employed in the calculations than are involved in many economic enterprises. Furthermore, such assessment will demand not only that the system boundaries be enlarged but also that social as well as economic components be quantified, as they change in systems interactions.[16] Far broader environmental impact statements are required than are generally done, and the expression of all inputs and outputs, including the off-site ones, should be made in monetary terms. This would include, for instance, not only loss of income for artisanal fishermen but also the accounting for changes that could occur in their social welfare.

14. World Commission on Environment and Development, *Our Common Future* (Oxford: Oxford University Press, 1987).

15. R. S. Beverton and J. J. Holt, *On the Dynamics of Exploited Fish Populations,* Publication of the Great Britain Ministry of Agriculture, Fish, and Food, Fishery Investigations, ser. 2, vol. 19 (London: Her Majesty's Stationery Office, 1957).

16. John A. Dixon, "Project Appraisal Involving Applications of Environmental Economics," Working Paper 19 (Honolulu, Hawaii: Environment and Policy Institute, East-West Center, 1990).

We have only recently seen the needs for such a holistic approach, and as yet there are only searches for satisfactory techniques to deal with these exigencies. What is more, the larger the system and the longer the time frame over which sustainability is desired the more difficult things become. This is well illustrated by the introduction of the concept of intergenerational equity in the discussion of international adjustment, if not abatement, of global warming.

A few facts about systems sustainability have emerged, however, and one cannot do better in describing them than to quote two economists who have given the matter considerable thought.[17]

> It would be easy if sustainability were a "motherhood" issue, easily defined and clearly desirable. Unfortunately, although there is near unanimity on its desirability, sustainability and sustainable development have proven especially difficult to define. If these terms are to escape becoming the empty "buzzwords" of the late 1980s, careful thought must be given to clarifying their exact meaning or alternate meanings; only in that way can they be useful as a touchstone for sound policy making. Clearly, we favor a socioeconomic definition of sustainability—one that revolves around social and economic well-being for the present generation and retention of future options for our children. Futher debate may not settle the question of what sustainability is, but it certainly will help all who are involved to understand what is at issue.

As far as fisheries are concerned, the socially desirable goal is to increase yields as demand for renewable aquatic resources increases. It is self-evident that the use of wild stocks has natural limits, and the use of cultured ones, mostly economic and social limits. At present, fisheries science has fashioned tools and techniques to manage single stocks, but adequate ecosystem management still eludes us. To attain true systems sustainability—in our case that of the living resources in the oceans and its coastal waters, probably in a mosaic fashion, but eventually to encompass the biosphere—may seem like a lofty goal. Given the now-recognized human role in altering the environment, efforts to attain true systems sustainability must not slacken.

17. John A. Dixon and Louise A. Fallon, "The Concept of Sustainability: Origins, Extensions, and Usefulness for Policy," *Society and Natural Resources* 2 (1989): 83.

Leatherback Sea Turtles: A Declining Species of the Global Commons*

Karen L. Eckert
U.S. National Marine Fisheries Service

INTRODUCTION

In virtually every corner of the globe, wild sea turtle populations are declining as a result of habitat loss or degradation, overexploitation, incidental catch, and pollution. Japan's demand for hawksbill sea turtle (*Eretmochelys imbricata*) shell, known as "tortoiseshell" or "bekko," removed nearly 700,000 hawksbills from the world's oceans between 1970 and 1986.[1] Indiscriminate fishing technologies, particularly shrimp trawls, capture tens of thousands of turtles each year and are held largely responsible for the persistent decline in loggerhead sea turtles (*Caretta caretta*) nesting on the southeastern coast of the United States.[2] The ingestion of plastic and other persistent marine debris presents a significant hazard to sea turtles, as do the thousands of miles of drift nets, hanging ghostlike in the water column, which ensnare uncounted numbers of marine mammals, turtles, and birds annually.[3] Oil and other toxic wastes

---

*The opinions expressed herein are those of the author and are not necessarily shared by the U.S. National Marine Fisheries Service.

1. T. Milliken and H. Tokunaga, *The Japanese Sea Turtle Trade 1970–1986*, report prepared by Traffic/Japan for the Center for Environmental Education, Washington, D.C. (1987).

2. T. A. Henwood and W. E. Stuntz, "Analysis of Sea Turtle Captures and Mortalities during Commercial Shrimp Trawling," *Fishery Bulletin* 85, no. 4 (1987): 813–17.

3. G. H. Balazs, "Impact of Ocean Debris on Marine Turtles: Entanglement and Digestion," in *Proceedings of the Workshop on Fate and Impact of Marine Debris*, ed. R. S. Shomura and H. O. Yoshida, U.S. Department of Commerce, National Oceanic and Atmospheric Administration, Tech. Memo. NMFS, NOAA-TM-NMFS-SWFC-54 (1985), pp. 387–429; K. O'Hara, N. Atkins, and S. Iudicello, *Marine Wildlife Entanglement in North America* (Washington, D.C.: Center for Environmental Education, 1986); Center for Environmental Education, *Plastics in the Ocean: More Than a Litter Problem*, U.S. Environmental Protection Agency, Contract 68-02-4228 (1987).

Editors' note.—See article by Paul G. Sneed, "Controlling the 'Curtains of Death': Present and Potential Ocean Management Methods for Regulating the Pacific Drift Net Fisheries," in this volume.

present as yet unquantified hazards, particularly to eggs and young turtles.[4] Furthermore, important nesting beaches are sometimes densely developed for tourism or industry. Artificial beachfront lighting can dissuade female turtles from nesting and disorient hatchlings away from the sea.[5]

Marine turtles are global citizens. Most are highly migratory and thus require international protection to ensure their survival. Too often, despite protection in the waters of one country, they are hunted and killed when they wander into the waters of another. The leatherback sea turtle (*Dermochelys coriacea*) is, arguably, one of the luckier ones. In contrast to the green sea turtle (*Chelonia mydas*), its flesh is not widely savored. Lacking a bony shell and carapace scutes, it is of no use to the Japanese tortoiseshell industry. Its smooth, scaleless skin is not tanned into leather, a fate that awaits some 75,000 olive ridley sea turtles (*Lepidochelys olivacea*) every year in Mexico alone.[6] Nonetheless, the leatherback is hunted for meat and oil, and the species has not escaped the general downward trend of modern sea turtle populations.

BIOLOGY OF THE LEATHERBACK SEA TURTLE

The leatherback is pelagic by nature.[7] Consequently, information on behavior away from the nesting beach is scant, and virtually nothing is known about males or juveniles. This elusive species migrates further,[8] dives deeper,[9] and ventures into colder water[10] than does any other reptile.

4. T. H. Fritts and M. A. McGehee, *Effects of Petroleum on the Development and Survival of Marine Turtle Embryos*, U.S. Fish and Wildlife Service, FWS/OBS-82/37 (1982).

5. P. W. Raymond, *Sea Turtle Hatchling Disorientation and Artificial Beachfront Lighting: A Review of the Problem and Potential Solutions* (Washington, D.C.: Center for Environmental Education, 1984); K. L. Eckert, "Wildlife Resource Management Plan: Sea Turtles," in *The Southeast Peninsula Project in St. Kitts*, vol. 1, *Resource Management Plans*, prepared by Tropical Research and Development, Inc., for U.S. Agency for International Development, Contract DHR 5438-C-00-6054-00 (1989).

6. H. Aridjis, "Continuing Sea Turtle Problems in Mexico," *Marine Turtle Newsletter* 50 (July 1990): 1–3.

7. J. Hendrickson, "The Ecological Strategies of Sea Turtles," *American Zoologist* 20 (1980): 597–608.

8. P. C. H. Pritchard, "Post-nesting Movements of Marine Turtles (Cheloniidae and Dermochelyidae) Tagged in the Guianas," *Copeia* (1976): 749–54.

9. S. A. Eckert, K. L. Eckert, P. Ponganis, and G. L. Kooyman, "Diving and Foraging Behavior of Leatherback Sea Turtles (*Dermochelys coriacea*)," *Canadian Journal of Zoology* 67 (1989): 2834–40.

10. J. D. Lazell, Jr., "New England Waters: Critical Habitat for Marine Turtles," *Copeia* (1980): 290–95; G. P. Goff and J. Lien, "Atlantic Leatherback Turtle, *Dermochelys coriacea*, in Cold Water off Newfoundland and Labrador," *Canadian Field Naturalist* 102, no. 1 (1988): 1–5.

Physical Characteristics

Leatherbacks are black or dark gray, mottled with pale spots. The front flippers are proportionately larger and more powerful than those of other sea turtles, and the flexible carapace is raised into seven streamlining ridges or keels. The core body temperature, at least for adults in cold water, has been shown to be several degrees centigrade above the ambient.[11] This may be due to several features, including the thermal inertia of a large body size (adult females typically weigh 250–500 kg; a record male weighed 916 kg[12]), an insulating layer of subepidermal fat, and countercurrent heat exchangers in the flippers.[13] Recent studies have suggested that the turtle may also have heat-generating brown adipose tissue[14] and a relatively low freezing point for body lipids.[15] These characteristics, in concert with certain chondro-osseous developmental features,[16] combine to render the leatherback a most unusual reptile.

Migrations

The evidence currently available from tag returns and strandings in the western Atlantic suggests that adult females engage in routine migrations between boreal, temperate, and tropical waters, perhaps to optimize foraging and nesting opportunities. The composition and growth[17] of barnacle communi-

11. W. Frair, R. G. Ackman, and N. Mrosovsky, "Body Temperature of *Dermochelys coriacea:* Warm Turtle from Cold Water," *Science* 177 (1972): 791–93.

12. P. J. Morgan, "Occurrence of Leatherback Turtles (*Dermochelys coriacea*) in the British Islands in 1988 with Reference to a Record Specimen," in *Proceedings of the Ninth Annual Conference on Sea Turtle Conservation and Biology*, compiled by S. A. Eckert, K. L. Eckert, and T. H. Richardson, U.S. Department of Commerce, National Oceanic and Atmospheric Administration, Tech. Memo. NMFS-SEFC-232 (1989), pp. 119–20.

13. N. Mrosovsky and P. C. H. Pritchard, "Body Temperatures of *Dermochelys coriacea* and Other Sea Turtles", *Copeia* (1971): 624–31; A. E. Greer, J. D. Lazell, and R. M. Wright, "Anatomical Evidence for a Counter-current Heat Exchanger in the Leatherback Turtle (*Dermochelys coriacea*)," *Nature* (London) 244 (1973): 181; W. H. Neill and E. D. Stevens, "Thermal Inertia versus Thermoregulation in 'Warm' Turtles and Tunas," *Science* 184 (1974): 1008–10.

14. G. P. Goff and G. B. Stenson, "Brown Adipose Tissue in Leatherback Sea Turtles: A Thermogenic Organ in an Endothermic Reptile?" *Copeia* (1988): 1071–74.

15. J. Davenport, D. L. Holland, and J. East, "Thermal and Biochemical Characteristics of the Lipids of the Leatherback, *Dermochelys coriacea:* Evidence of Endothermy," *Journal of the Marine Biological Association U.K.* 70 (1990): 33–41.

16. A. G. J. Rhodin, J. A. Ogden, and G. J. Conlogue, "Chondro-osseous Morphology of *Dermochelys coriacea*, a Marine Reptile with Mammalian Skeletal Features," *Nature* 290 (1981): 244–46.

17. K. L. Eckert and S. A. Eckert, "Growth Rate and Reproductive Condition of the Barnacle *Conchoderma virgatum* on Gravid Leatherback Sea Turtles in Caribbean Waters," *Journal of Crustacean Biology* 7, no. 4 (1987): 682–90.

ties colonizing leatherback turtles at Caribbean nesting grounds suggests that gravid females embark from and subsequently return to temperate latitudes.[18] Direct evidence of long-distance movement is scarce, but is available from leatherbacks tagged while nesting and then seen again at a later date. For example, nesters tagged in French Guiana have been recaptured in Mexico and along the Atlantic coast of the United States.[19] Individuals tagged on Caribbean islands have been subsequently reported in Mexico,[20] as well as further north in New York and New Jersey.[21] Typically these animals wash ashore dead after having been entangled in fishing gear or harvested by local fishermen. The longest known movement is that of an adult female leatherback who traveled some 5,900 km to Ghana after nesting in Suriname in 1970.[22]

Feeding and Diving

Beyond fragmented, often opportunistic distributional and movement data, little information exists on the leatherback's habits at sea. "Schooling" behavior is sometimes reported and may be related to densely distributed prey items.[23] Leatherbacks are believed to rely primarily on medusae and, perhaps to a lesser extent, other soft-bodied invertebrates such as squid and planktonic tunicates, siphonophores, and salps.[24] This specialized medusae diet places them atop a distinctive marine food chain based on nannoplankton, and largely independent of the more commonly recognized trophic systems

18. K. L. Eckert and S. A. Eckert, "Pre-reproductive Movements of Leatherback Sea Turtles (*Dermochelys coriacea*) Nesting in the Caribbean," *Copeia* (1988): 400–406.
19. Pritchard (n. 8 above).
20. R. H. Boulon, "Virgin Islands Turtle Recoveries Outside of the U.S. Virgin Islands," in *Proceedings of the Ninth Annual Conference on Sea Turtle Conservation and Biology*, compiled by S. A. Eckert, K. L. Eckert, and T. H. Richardson, U.S. Department of Commerce, National Oceanic and Atmospheric Administration, Tech. Memo. NMFS-SEFC-232 (1989), pp. 207–9.
21. I. Lambie, "Two Tagging Records from Trinidad," *Marine Turtle Newsletter* 24 (March 1983): 17; R. H. Boulon, K. L. Eckert, and S. A. Eckert, "*Dermochelys coriacea* (Leatherback Sea Turtle) Migration," *Herpetological Review* 19, no. 4 (1988): 88.
22. P. C. H. Pritchard, "International Migrations of South American Sea Turtles (Cheloniidae and Dermochelyidae)," *Animal Behavior* 21 (1973): 18–27.
23. T. Leary, "A Schooling of Leatherback Turtles, *Dermochelys coriacea coriacea*, on the Texas Coast," *Copeia* (1957): 32.
24. L. D. Brongersma, "Miscellaneous Notes on Turtles: IIA and IIB," in *Koninklijke Nederlandse Akademie van Wetenschappen-Amsterdam*, Proceedings, Series C, 72, no. 1 (1969), pp. 76–102; J. C. den Hartag and M. M. Van Nierop, "A Study on the Gut Contents of Six Leathery Turtles, *Dermochelys coriacea* (Linnaeus) (Reptilia: Testudines: Dermochelyideae) from British Waters and from the Netherlands," *Zoologische Verhandelingen* (Leiden) 209 (1984): 1–29.

supporting whales or tuna, for example.[25] Reports of the turtles feeding on jellyfish (*Cyanea, Aurelia, Rhizostoma, Stomolophus, Physalia*) are available from the Atlantic, Pacific, and Indian oceans. In addition, individuals have been hooked on longlines baited with squid or octopus in Caribbean[26] and Australian[27] waters. Leatherbacks lack teeth, but two cusps separate a median from two lateral notches on the upper jaw and facilitate the grasping and tearing of soft-bodied prey. Posteriorly directed papillae scattered over the lining of the mouth and esophagus presumably aid in the retention of gelatinous items.

It is possible that not all feeding occurs in surface waters. Hartog speculated that foraging may occur at depth, after finding nematocysts from deepwater siphonophores in leatherback stomach samples.[28] In northeastern Caribbean waters, dense plankton strata (the "deep scattering layer," or DSL) migrate toward the surface at dusk and are composed largely of siphonophores, salps, and medusae.[29] Recent studies of gravid females in the northeastern Caribbean Sea have revealed diel periodicities in dive depth and duration which suggest internesting nocturnal foraging in the DSL.[30] In the latter study, time-depth recorders (TDR) were secured to turtles during nesting on Sandy Point beach, St. Croix. As each turtle returned some 10 days later to deposit another clutch of eggs, her TDR was removed. The data revealed that the turtles were diving both day and night, averaging 100 dives per 24-hour period. The typical surfacing interval was less than 2 minutes, and often just long enough to recharge oxygen stores for another dive. Dives were shallower and more frequent at night, averaging about 60 m; very deep dives (e.g., 1100 m and 1300 m) were sometimes recorded during the day and may have represented predator avoidance.[31] Brief studies of diving behavior in subadult leatherbacks in New England waters have suggested iterative diving in northern latitudes as well.[32]

25. Hendrickson (n. 7 above).

26. K. L. Eckert and B. B. Lettsome, *WIDECAST Sea Turtle Recovery Action Plan for the British Virgin Islands,* prepared by the Wider Caribbean Sea Turtle Recovery Team, with support from UNEP Caribbean Environment Programme, Contract CR/5102-86 (1988).

27. C. J. Limpus, "A Benthic Feeding Record from Neritic Waters for the Leathery Turtle (*Dermochelys coriacea*)," *Copeia* (1984): 552–53.

28. J. C. den Hartog, "Notes on the Food of Sea Turtles: *Eretmochelys imbricata* (Linnaeus) and *Dermochelys coriacea* (Linnaeus)," *Netherlands Journal of Zoology* 30, no. 4 (1980): 595–610.

29. H. B. Michel and M. Foyo, "Caribbean Zooplankton, Part 1," (Washington, D.C.: Department of Defense, Navy Department, Office of Naval Research, Government Printing Office, 1976).

30. Eckert et al. (n. 9 above).

31. Ibid.

32. E. A. Standora, J. R. Spotila, J. A. Keinath, and C. R. Shoop, "Body Temperatures, Diving Cycles, and Movements of a Subadult Leatherback Turtle, *Dermochelys coriacea,*" *Herpetologica* 40 (1984): 169–76; J. A. Keinath, "A Telemetric Study of the

Leatherbacks appear to sustain this demanding exercise regime by partitioning their oxygen reserves a little differently than do other marine turtles. Leatherback lung volumes are less than half what we might expect from examining other sea turtle species, but blood and muscle stores are more than double.[33] The advantages of this include reduced buoyancy, a reduced threat that lung compression will limit dive time, and a reduced risk of decompression sickness. Interestingly, the apparent reliance on tissue oxygen stores (versus lung oxygen stores) is strikingly similar to some marine mammals. In deep-diving marine mammals most oxygen is stored in body tissues, with the largest proportion found in the blood, somewhat less in the muscles, and the least in the lungs.[34] Thus, since there is some evidence that other marine turtle genera rely primarily on lung oxygen stores for diving,[35] leatherback diving behavior and physiology resemble the adaptations of deep-diving marine mammals more closely than they do the adaptations of other turtles.

Nesting

Despite recent advances in our knowledge of behavior in free-swimming leatherbacks, we know considerably more about the behavior of gravid females on nesting beaches. It is generally believed that mating occurs away from the nesting grounds, presumably prior to (or during) the nesting migration, though isolated observations of courtship and/or mating are reported from tropical latitudes.[36] Fertilization is internal, and the female maintains viable spermatozoa in her reproductive tract, permitting the fertilization of each clutch of eggs as needed without subsequent mating during the long nesting season. Gravid females emerge from the sea nocturnally in nearly all cases. Since individuals lack a bony shell and are vulnerable to abrasion and injury, they prefer high-energy beaches (often steeply sloping to the sea)

Surface and Submersion Activities of *Dermochelys coriacea* and *Caretta caretta*," master's thesis, Department of Zoology, University of Rhode Island, 1986.

33. S. A. Eckert, "Diving and Foraging Behavior of the Leatherback Sea Turtle, *Dermochelys coriacea*," Ph.D. diss., Department of Zoology, University of Georgia, Athens, 1989; Scott Alan Eckert, Scripps Institution of Oceanography, University of California at San Diego, telephone conversation, 1 May 1990.

34. C. Lenfant, K. Johansen, and J. D. Torrance, "Gas Transport and Oxygen Storage Capacity in Some Pinnipeds and the Sea Otter," *Respiratory Physiology* 9 (1970): 277–86; R. Elsner and B. Gooden, *Diving and Asphyxia* (New York: Cambridge University Press, 1983).

35. P. L. Lutz and T. B. Bentley, "Respiratory Physiology of Diving in the Sea Turtle," *Copeia* (1985): 671–79.

36. T. Carr and N. Carr, "*Dermochelys coriacea* (Leatherback Sea Turtle) Copulation," *Herpetological Review* 17, no. 1 (1986): 24–25.

Fig. 1.—A gravid leatherback sea turtle selects a nesting site, Sandy Point National Wildlife Refuge, St. Croix, U.S. Virgin Islands. (Photo by Scott Alan Eckert.)

with deep, unobstructed submarine access.[37] Unaware of the complexities of temperature-dependent sex differentiation and the theoretical aspects of risk assessment (see below in section on incubation), leatherbacks regularly haul ashore under the cloak of darkness to excavate a nest cavity in the womb of a tropical beach. After choosing an appropriate site, the fore flippers sweep aside dry sand and debris. The rear flippers then methodically scoop sand from a deepening flask-shaped hole under the cloaca, tossing it over the carapace in an effort to keep it from sifting back into the hole. When the flippers can reach no further, the chamber is complete. There is a pause before the snow-white spherical eggs, bathed in lubricating mucous, begin dropping from the cloaca. The parchment-shelled eggs bounce gently against one another and come to rest at the bottom of the chamber. Viable (yolked) eggs average 50–55 mm in diameter, but a variable number of smaller and sometimes misshapen eggs are also laid. These latter eggs are essentially "yolkless," although flecks of yolk are sometimes observed. From these eggs there will be no hatchling, and their function (if any) remains a mystery.

37. H. G. Hirth, "Some Aspects of the Nesting Behavior and Reproductive Biology of Sea Turtles," *American Zoologist* 20 (1980): 507–23; N. Mrosovsky, "Ecology and Nest-Site Selection of Leatherback Turtles, *Dermochelys coriacea,*" *Biological Conservation* 26 (1983): 47–56.

During nesting, copious tears flow from postorbital glands. These tears, once thought to cleanse sand from the females' eyes (or to represent pain or sorrow), actually serve an osmoregulatory function.[38] The tears flow continuously, on land as well as at sea, to discharge excess salt from the turtle's body. Nevertheless, the obvious exertion of nest preparation appears particularly hard on a physiological system so finely tuned to retain heat under nearly all circumstances (the leatherback being predominately a temperate latitude species), and a witness to the event cannot help but feel that the tears are an appropriate response to such a workout. On average the female is ashore for 60–90 minutes, but several hours may pass in cases where very dry sand or other untoward conditions retard progress. Vasodilation is employed to radiate muscle-generated heat, and while on land the throat and undersides of the turtle turn bright pink. After the eggs are laid, the cavity is meticulously backfilled with sand by the rear flippers. Finally, the area is camouflaged by violent backward sweeps of the fore flippers. The female then returns to the sea, leaving the nest behind her, unattended.

Incubation

Frequently the nesting beaches are unpredictable with respect to erosion, and it is not uncommon for a large number of eggs to be lost prior to hatching. This is the case on Sandy Point National Wildlife Refuge on St. Croix, U.S. Virgin Islands, where 45% to 60% of the eggs laid per annum are washed to sea by shifting currents.[39] A similar pattern is reported from the Guianas[40] and Trinidad,[41] as well as southern Africa and Madagascar where "thousands of tons" of sand are seasonally lost to storms.[42]

In an attempt to mitigate egg loss at Sandy Point National Wildlife Refuge on St. Croix, the local Division of Fish and Wildlife (with support from Earthwatch, Inc.) maintains a conservation program of nocturnal monitoring where biologists collect eggs (at deposition) laid in erosion risk zones and rebury them in stable beach areas.[43] In some parts of the world, "doomed" eggs justify a limited local protein harvest. In Suriname, eggs deposited in

38. D. M. Hudson and P. L. Lutz, "Salt Gland Function in the Leatherback Sea Turtle, *Dermochelys coriacea*," *Copeia* (1986): 247–49.

39. K. L. Eckert, "Environmental Unpredictability and Leatherback Sea Turtle (*Dermochelys coriacea*) Nest Loss," *Herpetologica* 43 (1987): 315–23.

40. P. C. H. Pritchard, "The Leatherback or Leathery Turtle, *Dermochelys coriacea*," International Union for the Conservation of Nature Monograph 1 (1971), pp. 1–39; Mrosovsky (n. 37 above).

41. P. R. Bacon, "Studies on the Leatherback Turtle, *Dermochelys coriacea* (L.), in Trinidad, West Indies," *Biological Conservation* 2 (1970): 213–17.

42. G. R. Hughes, "The Sea Turtles of South-East Africa 2," Investigational Report no. 36 (Durban, South Africa: Oceanographic Research Institute, 1974).

43. K. L. Eckert (n. 39 above).

FIG. 2.—A female leatherback sea turtle carefully camouflages her nest before returning to the sea, Sandy Point National Wildlife Refuge, St. Croix, U.S. Virgin Islands. (Photo by Scott Alan Eckert.)

erosion risk zones, usually below the high-tide line, are collected by government biologists and sold in local markets to generate revenue for conservation projects. The collection and reburial of eggs (such as to a protected hatchery) always entails a risk. Soon after an egg is laid, the embryo migrates atop the yolk and attaches to the shell membrane to facilitate respiration (in contrast to avian eggs, there is no internal air pocket). Moving the egg can dislodge the tiny embryo, an event that proves fatal. The risk of movement-induced mortality increases as time passes, necessitating that eggs be collected for conservation purposes at (or very soon after) oviposition.[44]

The temperature at which marine turtle eggs incubate largely determines the sex of developing hatchlings. Thus, while it is important to move eggs immediately, perhaps the most significant challenge is to duplicate the natural circumstances (cavity depth and dimension, proximal vegetation, etc.) as closely as possible in order to avoid skewed thermal regimes. As ambient temperature increases, the duration of incubation is shortened, and increas-

44. C. J. Limpus, V. Baker, and J. D. Miller, "Movement Induced Mortality of Loggerhead Eggs," *Herpetologica* 35 (1979): 335–38; C. E. Blanck and R. H. Sawyer, "Hatchery Practices in Relation to Early Embryology of the Loggerhead Sea Turtle, *Caretta caretta*," *Journal of Experimental Marine Biology and Ecology* 49 (1981): 163–77.

ing numbers of female hatchlings are produced. Under relatively cooler regimes, as when a nest is shaded by vegetation, eggs take longer to hatch, and
more male hatchlings emerge. The "pivotal temperature" (that temperature
at which the sex ratio of hatchlings emerging from a nest approximates 1:1)
may differ with species and locale. For example, in Suriname leatherbacks
require slightly higher (ca. 0.5°C) temperatures for female differentiation
than do green turtles on the same beach, and thus it is not surprising that
leatherbacks nest in relatively greater numbers during the warmer parts of
the season.[45] The pivotal temperature for leatherback eggs has been estimated at approximately 29°C (28.75°C in Suriname;[46] 29.50°C in French
Guiana[47]).

In order to provide a variety of incubating temperatures and a degree
of security in terms of unpredictable erosion, gravid leatherbacks disperse
their clutches widely along the length and width of a nesting beach. Favorable
nest placement is crucial, since the immediate nest environment (including
proximity to tidal zones and supralittoral vegetation) affects the survivability
of developing embryos. Eggs fail to develop in nest cavities saturated with
seawater, effected by placement too near the sea.[48] Similarly, nest cavities
excavated in tangles of creeping beach vines (e.g., *Ipomea*) can invite the en-
tombment and suffocation of eggs by nutrient-seeking roots.[49] Nest success
would obviously be enhanced if the appropriate environmental information
were available at the time of nesting. However, in uncertain or inconsistent
environments, information relative to the future state of the environment
may not be available at the required time. A recent study of leatherbacks
nesting on St. Croix concluded that nest dispersal, or the spreading of risk,
reduced the likelihood that a high proportion of reproductive effort would
be lost with the destruction of any particular zone of habitat.[50] By spreading
the risk of reproductive failure in space (i.e., nest dispersal) and time (i.e.,
iteroparity), the leatherback has prospered in the face of uncertainty.

45. N. Mrosovsky, P. H. Dutton, and C. P. Whitmore, "Sex Ratios of Two Species
of Sea Turtles Nesting in Suriname," *Canadian Journal of Zoology* 62 (1984): 2227–39.
46. Ibid.
47. F. Rimblot-Baly, J. Lescure, J. Fretey, and C. Pieau, "Sensibilité à la température de la différenciation sexuelle chez la tortue luth, *Dermochelys coriacea* (Vandelli,
1761): Application des données de l'incubation artificielle à l'étude de la sex-ratio
dans la nature," *Annales des Sciences Naturelles, Zoologie* (Paris), series 13, 8 (1986–87):
277–90.
48. T. Adams, "Tagging and Nesting Research of Leatherback Sea Turtles (*Dermochelys coriacea*) at Manchineel Bay Beach, St. Croix, U.S. Virgin Islands, 1984,"
annual report to the U.S. Virgin Islands Division of Fish and Wildlife, St. Thomas
(1984); C. P. Whitmore and P. H. Dutton, "Infertility, Embryonic Mortality, and
Nest Site Selection in Leatherback and Green Sea Turtles in Suriname," *Biological
Conservation* 34 (1985): 251–72.
49. Personal observation.
50. K. L. Eckert (n. 39 above).

Sea turtle eggs incubate for roughly 60 days, longer during the rainy season, shorter during the heat of summer. Thermal gradients defined by tidal wash, the depth of the water table, and the distribution of vegetation also influence incubation duration.

The Hatchling

When development is complete, each hatchling pips the egg using an egg tooth. The tiny turtle wiggles free, stretches for the first time, and pulls inside the yolk sac protruding from its belly. The yolk secure, the opening in the plastron seals, and an essential source of energy is available for the climb to the surface and the first several days at sea. The climb to the beach surface is a cooperative effort, each hatchling doing its part to collapse the overburden and pull sand beneath the ever ascending cluster of baby turtles. Those left behind may be stranded, unable to make it out alone. As the hatchlings near the surface, the tropical heat slows their metabolism.[51] Literally unable to move, they are suspended several centimeters below the surface. Here they are vulnerable to beach umbrellas punched in the sand and the crushing weight of vehicles bearing "joy riders" and fishermen.

When the sun sets and the beach cools, the little turtles become active again, and a final push exposes them to the night air. Blinking and eager, they gauge the direction of the sea by the bright, open horizon which is in stark contrast to the dark curtain of vegetation behind them.[52] Dodging nocturnal birds and crabs, they scurry to an ocean they have never seen and are engulfed by the waves.

The Juvenile Leatherback

The travels of young leatherbacks are uncharted. Studies that document the occurrence of young sea turtles of other genera in oceanic convergence zones or other pelagic habitats[53] do not mention *Dermochelys*. It is not known where the young turtles go, what they eat (or what eats them), how fast they grow, or whether they interact significantly with one another. The question of growth has been addressed in some other species of sea turtle either by cap-

51. N. Mrosovsky, "Nocturnal Emergence of Hatchling Sea Turtles: Control by Thermal Inhibition of Activity," *Nature* 220 (1968): 1338–39.
52. N. Mrosovsky, "Individual Differences in the Sea-Finding Mechanism of Hatchling Leatherback Turtles," *Brain, Behaviour, and Evolution* 14 (1977): 261–73.
53. Archie Carr, "New Perspectives on the Pelagic Stage of Sea Turtle Development," *Conservation Biology* 1, no. 2 (1987): 103–21.

FIG. 3.—A newly hatched leatherback sea turtle scurries to the sea, Sandy Point National Wildlife Refuge, St. Croix, U.S. Virgin Islands. (Photo by Scott Alan Eckert.)

tive study[54] or by capturing young animals at sea, tagging them, and then recapturing them at a later date.[55] These opportunities are largely denied the student of leatherbacks. The species has a notoriously poor survival record in captivity, and if there are feeding grounds that predictably attract young animals, they have not been discovered. Growth rates reported from a handful of captive studies are disparate.[56] Records of fast-growing individuals, coupled with evidence of chondro-osseous development conducive to rapid growth, have led to speculations that leatherbacks may reach sexual maturity in 2 or 3 years.[57] Others challenge this hypothesis with descriptions of healthy but slow-growing individuals.[58] For the time being, no one knows how long it takes for a leatherback turtle to reach sexual maturity, but, as is the case for other sea turtle species, it may be measured in decades.

Reproduction and Life Span

Sexual maturation in sea turtles arises from a complicated set of circumstances frequently dependent on ambient temperature and the quality of food available. Maturation cannot be equated with a particular age or size, though average values (such as the average length of females nesting on a particular beach) are useful. The average number of clutches laid per female

54. S. F. Hildebrand and C. Hatsell, "On the Growth, Care, and Behavior of Loggerhead Turtles in Captivity," *Proceedings: National Academy of Sciences* 13 (1927): 374–77; W. N. Witzell, "Growth of Captive Hawksbill Turtles, *Eretmochelys imbricata*, in Western Samoa," *Bulletin of Marine Science* 30 (1980): 909–12; I. N. S. Nuitja and I. Uchida, "Preliminary Studies on the Growth and Food Consumption of the Juvenile Loggerhead Turtle (*Caretta caretta* L.) in Captivity," *Aquaculture* 27 (1982): 157–60; N. B. Frazer and F. J. Schwartz, "Growth Curves for Captive Loggerhead Turtles, *Caretta caretta*, in North Carolina, USA," *Bulletin of Marine Science* 34 (1984): 485–89.

55. C. J. Limpus and D. G. Walter, "The Growth of Immature Green Turtles (*Chelonia mydas*) under Natural Conditions," *Herpetologica* 36 (1980): 162–65; M. T. Mendonca, "Comparative Growth Rates of Wild Immature *Chelonia mydas* and *Caretta caretta* in Florida," *Journal of Herpetology* 15 (1981): 447–51; G. H. Balazs, "Growth Rates of Immature Green Turtles in the Hawaiian Archipelago," in *Biology and Conservation of Sea Turtles*, ed. K. A. Bjorndal (Washington, D.C.: Smithsonian Institution Press, 1982), pp. 117–25; N. B. Frazer and L. M. Ehrhart, "Preliminary Growth Models for Green, *Chelonia mydas*, and Loggerhead, *Caretta caretta*, Turtles in the Wild," *Copeia* (1985): 73–79.

56. E. Birkenmeier, "Rearing a Leathery Turtle, *Dermochelys coriacea*, in Captivity," *International Zoological Yearbook* 14 (1970): 204–7; E. J. Phillips, "Raising Hatchlings of the Leatherback Turtle, *Dermochelys coriacea*," *British Journal of Herpetology* 5 (1977): 677–78; V. Bels, F. Rimblot-Baly, and J. Lescure, "Croissance et maintien en captivité de la tortue luth, *Dermochelys coriacea* (Vandelli 1761)," *Revue Française d'Aquariologie* 15, no. 2 (1988): 59–64.

57. A. G. J. Rhodin, "Comparative Chondro-osseous Development and Growth in Marine Turtles," *Copeia* (1985): 752–71.

58. Bels et al. (n. 56 above).

per season lies between five and seven, depending on locale, and the maximum number of documented nestings in a single season is 10.[59] An average of 60–85 yolked eggs are laid per clutch, though some clutches may contain more than 100. An adult leatherback may nest (typically on the same beach) for 2 decades or more, commonly at 2- to 5-year intervals, and deposit thousands of eggs. So why isn't the world awash in leatherback turtles? In addition to mortality factors on the beach (the loss of eggs to erosion, depredation by crabs and small mammals, etc.), the consumption of hatchlings and small juveniles by predator fishes at sea is high. It would not be unreasonable to predict that less than 1% of the hatchlings produced survive to adulthood.

Historically, high mortality in young age classes has been offset by high fecundity and favorable prospects of survival upon reaching adulthood (estimated at 0.81 in loggerhead sea turtles nesting on the Georgia coast[60]). Large juveniles and adults, able to evade all but a handful of ocean predators (large sharks and orcas), live long, fruitful lives. But the harvest of breeding adults by humans is significantly altering the balance. Survival in the later age classes is arguably the most important component in the equation describing the survival of sea turtle species.[61] Leatherbacks cannot easily discard or adjust a reproductive "strategy" that has evolved over millennia. Persistent harvesting of breeding age turtles on nesting beaches and killing large juveniles at sea (debris ingestion, entanglement, harvest) hastens the extinction of this ancient species.

CONSERVATION OF SEA TURTLES

All seven sea turtle species are listed in Appendix I of the Convention on International Trade in Endangered Species of Wild Fauna and Flora (CITES), which prohibits signatory nations from engaging in the international trade of marine turtles or their products. In the United States, all marine turtles (with the exception of the flatback, endemic to Australia) are protected by the Endangered Species Act of 1973 (Public Law 93-205), which

59. K. L. Eckert, S. A. Eckert, and D. W. Nellis, "Tagging and Nesting Research of Leatherback Sea Turtles (*Dermochelys coriacea*) on Sandy Point, St. Croix, U.S. Virgin Islands, 1984, with Management Recommendations for the Population," annual report to the U.S. Fish and Wildlife Service (1984); A. D. Tucker, "A Summary of Leatherback Turtle (*Dermochelys coriacea*) Nesting at Culebra, Puerto Rico, from 1984–1987 with Management Recommendations," report to the U.S. Fish and Wildlife Service (1988).

60. N. B. Frazer, "Survivorship of Adult Female Loggerhead Sea Turtles, *Caretta caretta*, Nesting on Little Cumberland Island, Georgia, USA," *Herpetologica* 39 (1983): 436–47.

61. D. T. Crouse, L. B. Crowder, and H. Caswell, "A Stage-Based Population Model for Loggerhead Sea Turtles and Implications for Conservation," *Ecology* 68, no. 5 (1987): 1412–23.

provides for the conservation of wild fauna and flora actually or potentially in danger of becoming extinct. Protective legislation is in place in many other countries as well, but law enforcement is often inadequate.

In some areas, leatherbacks have been nearly extirpated as a result of local harvest for oil and meat (e.g., British Virgin Islands) or eggs (e.g., Terengganu, Malaysia). The flesh is either consumed in local villages or used for bait. Yellow oil rendered from the dismembered animal is administered medicinally, burned in oil lamps, or employed as boat caulking. Even where populations remain large (e.g., Pacific Mexico), concern is voiced for the increasing number of animals harvested in waters adjacent to nesting beaches, the continuing collection of eggs, and commercial development on important nesting beaches.[62]

Some progress has been made toward minimizing the killing of adults, especially at the breeding grounds. In Guyana, a country where leatherbacks are routinely harvested on the nesting beach for meat, native Amerindians have suggested that a portion of the gravid females be tagged as "brood stock." In this way, all turtlers would recognize the protected animals and agree to leave them alone when they came ashore to nest.[63] This innovative program has since been implemented and represents a considerable improvement over the previous situation.[64] In adjacent Suriname and French Guiana, ongoing research and conservation projects serve to minimize the slaughter of nesting animals because scientists and volunteers are on the beach throughout the night. Much has been learned about the biology of leatherbacks from these persistent efforts.[65] Similarly, on beaches from Michoacan (Pacific coast of Mexico) to Sri Lanka to Venezuela, nightly patrols of nesting beaches reduce the mortality of nesting adults and the theft of eggs, while enhancing our knowledge of this mysterious reptile.

Turtle excluder devices (TEDs) are now required by law on U.S. shrimp

62. P. C. H. Pritchard, "Nesting of the Leatherback Turtle, *Dermochelys coriacea*, in Pacific Mexico, with a New Estimate of the World Population Status," *Copeia* (1982): 741–47.

63. P. C. H. Pritchard, "WIDECAST Sea Turtle Recovery Action Plan for Guyana," prepared by the Wider Caribbean Sea Turtle Recovery Team, with support from UNEP Caribbean Environment Programme, Contract CR/5102-86 (1988).

64. P. C. H. Pritchard, "Turtles in Guyana," paper presented at the Tenth Annual Workshop on Sea Turtle Biology and Conservation, Hilton Head Island, South Carolina (20–24 February 1990).

65. J. Fretey, "Causes de mortalité des tortues luths adultes (*Dermochelys coriacea*) sur le littoral Guyanais," *Le Courrier de la Nature* 52 (1977): 257–66; J. Lescure, F. Rimblot-Baly, J. Fretey, S. Renous, and C. Pieau, "Influence de la température d'incubation des oeufs sur la sex-ratio des nouveaux-nés de la tortue luth, *Dermochelys coriacea*," *Bulletin Société Zooligiquede France* 110, no. 3 (1985): 355–59; J. Fretey and M. Girondot, "L'activité de ponte de la tortue luth, *Dermochelys coriacea* (Vandelli 1761), pendant la saison 1988 en Guyane Française," *Revue d'Ecologie (la Terre et la Vie)* 44 (1989): 261–74.

trawlers. These devices provide a safe exit route for turtles caught in the trawls. The National Marine Fisheries Service has estimated that a total of 11,179 turtles are killed annually in shrimp trawls operating in U.S. waters alone.

In some countries, existing or proposed sanctuaries include important nesting or offshore habitats. In Terengganu, Malaysia, the number of leatherbacks coming ashore to nest has declined some 98% in little more than 3 decades.[66] The impetus behind this dramatic decline has not been the harvest of adults, but rather unrelenting egg collection.[67] Nonetheless, efforts to restore the population depend not only on protecting eggs (eggs are now collected for burial in guarded beach hatcheries) but also on eliminating the incidental catch of gravid females as they approach the nesting beach. After radio tracking a sample of nesters and determining their internesting range, an offshore sanctuary was proposed wherein certain types of fishing gear would be banned during the nesting season in order to improve the survival prospects of breeding adults.[68] On the Pacific coast of Costa Rica, a sanctuary (Las Baulas de Guanacaste National Park) has been proposed in the Playa Grande/Playa Langosta area specifically for nesting leatherback turtles.[69] The species' most important nesting beach under U.S. jurisdiction, located at Sandy Point, St. Croix, became the first nesting beach of any marine turtle to be proposed as critical habitat,[70] and in September 1978 the U.S. Fish and Wildlife Service so designated it. Nearly a decade before, in March 1979, the National Marine Fisheries Service had designated the waters surrounding Sandy Point as critical habitat.[71] In September 1984, Sandy Point became a national wildlife refuge.[72]

66. S. Aikanathan and M. Kavanaugh, "The Effects of Fishing on Leatherback Turtles," report prepared by World Wildlife Fund-Malaysia, Kuala Lumpur (1988).

67. B. B. Salleh, E. H. Chan, and A. R. bin Kassim, "An Update on the Population Status and Conservation of the Leatherback Turtles of Terengganu," in *Proceedings of the Eleventh Annual Seminar of the Malaysian Society of Marine Sciences* (1987), pp. 69–77.

68. The sanctuary was proposed on the basis of data obtained by radio tracking gravid leatherbacks during internesting intervals off Terengganu. The project was staged during June and July 1989 and was a collaborative effort between Scott Eckert and Karen Eckert (then of the University of Georgia) and Chan Eng Heng and Liew Hock Chark (Universiti Pertanian Malaysia, Kuala Terengganu).

69. Maria Teresa Koberg, Director, Sea Turtle Program, Neotropica (Costa Rica), letter of 15 March 1990.

70. C. K. Dodd, Jr., "Terrestrial Critical Habitat and Marine Turtles," paper presented at the symposium "Recovery and Management of Marine Turtles in the Southeast Region," 26–27 June 1978, Tampa, Florida.

71. G. S. Baker, "Mini Recovery Plan for the Leatherback Sea Turtle at Sandy Point, St. Croix," report to the U.S. Fish and Wildlife Service, Titusville, Florida (1981).

72. U.S. Fish and Wildlife Service, "Leatherback Turtle Nesting Beach Becomes Wildlife Refuge," *Endangered Species Technical Bulletin* 9 (November 1984): 3, 11.

In addition to progress made at the national level, essential multilateral cooperation toward the conservation of sea turtles is emerging (though adequate law enforcement is often lacking). For example, CITES has been ratified by 107 nations (including the United States in 1974) and forbids international commerce in marine turtles, including their parts or products.[73] The UNEP Convention on the Protection and Development of the Marine Environment of the Wider Caribbean ("Cartagena Convention") has been ratified by 16 nations (as of 16 January 1990),[74] including the United States in 1984. Signatory nations agree to "take all appropriate measures to protect and preserve the habitat of . . . endangered species." The MARPOL prohibition of the disposal of plastics at sea, which went into effect in December 1988, has the potential to dramatically reduce the availability (and thus the ingestion) of persistent debris. In January 1990, at a conference of plenipotentiaries convened in Kingston, Jamaica, 13 nations (including the United States) adopted a Cartagena Convention Protocol on Specially Protected Areas and Wildlife. The protocol contains very specific language on the subject of national and international measures to be taken to protect endangered Caribbean species.

CONCLUSION

Leatherback sea turtles, like all other marine animals, are facing great dangers in this modern world. In addition to the direct harvest of individuals, important nesting beaches continue to be densely developed without taking into account the often simple modifications, architectural and otherwise, that would encourage turtles to continue to nest there. Artificial lighting shines on beaches that were once dark, dissuading females from coming ashore and disorienting hatchlings so that many never find the sea. Lounge chairs and pleasure craft left on beaches block access to suitable nesting sites, confuse females, and present insurmountable obstacles to the tiny, newly hatched turtles. Some nesting beaches have simply disappeared, having been mined for aggregate or dredged for piers and marinas.

There are more subtle problems as well. Even if turtle fishermen were given satisfying alternative livelihoods and important nesting beaches were protected from adverse development, leatherbacks would remain at risk. High incidental mortality occurs at sea due to entanglement in active or dis-

---

73. Editors' note.—See Appendix G, "List of Parties to CITES," and "Globally Protected Marine Species (CITES and CMS)."

74. UN Environment Programme (UNEP), "Report of the 5th Intergovernmental Meeting on the Action Plan for the Caribbean Environment Programme and 2nd Meeting of the Contracting Parties to the Convention for the Protection and Development of the Marine Environment of the Wider Caribbean Region, January 17–18, 1990, Kingston, Jamaica," UNEP(OCA)/CAR IG.6/6 (1990).

carded fishing nets and the ingestion of persistent marine debris. Thus, if the world's oceans are not freed from plastics and other marine debris, toxic wastes, and indiscriminate fishing gear, the leatherback may yet slide inexorably toward extinction.

Extinction of sea turtles is a red flag: a warning that life-support systems are failing. Whatever the cause, we too are ultimately vulnerable. Our dependence on living marine resources is high, and as the world population increases, so will our dependence on these resources. Many marine species, once relied upon for food and other essential items, have been overexploited and are now severely depleted. As we seek now to conserve many of these species, we discover that we cannot be successful without protecting the ecosystem of which they are a part. The oceans themselves must be safeguarded; ecosystem conservation must be our goal. Only in this way can we ensure that future generations have their opportunity to appreciate not only the marine turtles, but all the splendid creatures of the earth.

Indonesia's National Policy on Offshore Mineral Resources: Some Legal Issues[1]

Mochtar Kusuma-Atmadja
Center for Law of the Sea, Archipelago, and Development Studies, Indonesia

INTRODUCTION

The Indonesian national policy on the exploration and exploitation of mineral resources (including hydrocarbons) in the archipelagic waters constitutes a system of mineral and hydrocarbon production development as part of a national energy and mineral resources development policy. This approach is made possible by Indonesian law, which vests the right to minerals (including oil and gas) in the state, unlike U.S. law, for example, where the right or title to the land includes the right and title to the minerals (including hydrocarbons) found underneath the surface of the land (right of capture).

The Indonesian concept of a right to mineral resources separate from the rights to the land is a continuation of a legal tradition introduced by the Dutch in 1899.[2] Indonesia replaced the colonial law on mining with two national laws, one dealing with hard minerals (Undang-Undang Pertambangan, 1960, amended 1967) and one with oil and gas (Undang-Undang Minyak dan Gas Bumi, 1960). Both laws retained the basic principle that the right to the minerals remains vested in the state rather than in the holder or owner of the land. This is the first factor in Indonesia's national policy on mineral resources (hard minerals) and energy (oil and gas) development.

As Indonesia views its marine space in the same way as its land space, both being integral parts of the nation's territory, the laws on mining both

---

1. Developed from a paper presented at the Penataran Hukum Internasional (International Law Upgrading Courses), 8–20 January 1990, conducted jointly by the Netherlands Institute for the Law of the Sea (NILOS), the University of Utrecht, the Netherlands, and by the Indonesian Center on the Law of the Sea (ICLOS), a project of the Konsorsium Ilmu Hukum (Indonesian Law Consortium) entrusted to the Padjadjaran University Law School, Bandung, Indonesia.
2. Indische Mijnwet (Mining Act of the Indies), 23 May 1899, *State Gazette* no. 214 (1899), last amended on August 1938, *State Gazette* nos. 618 and 652 (1938). See Arts. 1, 4, 7, 13, and 35.

hard minerals and hydrocarbons (oil and gas) are equally applicable to operations on land and offshore.

The second factor in the system concerns the method of extraction. The Dutch licensing system (*concessie*) was replaced by the 1960 Mining and Oil and Gas Law. This uses a completely different system, based on native Indonesian legal thinking and taking into account the realities the nation faced in the early 60s, when it aspired to explore and exploit its mineral resources and hydrocarbons without possessing the financial resources and technology for that purpose.

The *kontrak karya* (for hard minerals)[3] and the production-sharing contract (for oil and gas; PSC) are essentially based on the concept of the owner of the resources (the state) engaging a third party (a mining company in the case of hard minerals and an oil company in the case of hydrocarbons) as contractors. The proceeds of the contractors' work or activity (i.e., the production) are shared between the state and the contractor on the basis of a previously agreed formula, after the subtraction of costs. In assessing the nature of the *kontrak karya* and PSC, production sharing should not be confused with profit sharing. Profit sharing is often not advantageous to owners of the resource, as they have no control at all over cost. To simplify the discussion in the following pages, I will concentrate on describing the PSC used in the development of hydrocarbon resources.

The regulatory agency at the Ministry of Mines in charge of supervising oil-and-gas production is the Directorate General Migas (Minyak dan Gas Bumi). This agency ensures that hydrocarbon production and development are in line with the government's program and policy. The operational agency in charge of hydrocarbon production and development is the state oil company, Pertamina. It is Pertamina, on behalf of the state, that holds the right to mine. Consequently, by law it is also Pertamina that holds the oil-and-gas work area (*wilayah kerja pertambangan migas*). The work area is offered to prospective contractors through a bidding system, the successful bidder obtaining a contract area (*wilayah kontrak kerja*). Usually a signature fee is paid by the contractor at the signing of a PSC.

Up to this point there appears to be no big difference between a production-sharing system of oil-and-gas production and a concession system. The difference appears in the structure and content of a PSC.

BASIC FEATURES OF THE PRODUCTION-SHARING CONTRACT[4]

As already stated, a PSC is a contract between the owner of the resources, that is, the Indonesian state represented by the state oil company Pertamina,

---

3. See infra, "Basic Features of the Production-Sharing Contract."
4. This section is largely based on my *Mining Law* (Bandung, Indonesia: Padjadjaran University Law School, 1974). For a detailed comparison of terms and conditions

and the contractor, that is, the oil company assigned a certain contract area. The production-sharing concept has been developed since 1966, initially comprising the following principles:

1. The state oil company has control over management.
2. The contract is to be based on the sharing of production rather than profit.
3. The foreign company—as contractor to the state oil company—is to bear preproduction risks, cost recovery to be limited to 40% of the oil produced annually if oil is discovered and produced.[5]
4. The remaining 60% of production (or more when cost amortization is below the 40% maximum) is to be split, with 65% going to the state oil company and 35% to the contractor.
5. Title to all project-related equipment bought by the contractors is to pass to the state oil company upon entry into Indonesia, the cost to be recovered from the 40% of the oil produced and set aside for recovery of cost.

These five principles, reflecting some basic ideas on petroleum resources development, are not embodied in the *kontrak karya* on oil signed in 1963 between the Government of Indonesia and Caltex and Stanvac Oil companies.[6] Management of the operation should be controlled by the owner of the resource, that is, the state or the state enterprise, rather than by the contractor. As it is the contractor who bears the risks, this means in practice that the relationship between Pertamina and the contractor is one of joint management. Second, to ensure a role for the state enterprise in marketing and to eliminate disputes over prices, the product is to be shared, rather than the profits. The limitation of cost recovery is a logical corollary of these two basic principles.

It should be noted here that, although Pertamina is given the power of management, in the early years it used this power only to keep tight control over costs. There are several reasons for this: (1) Pertamina felt that cost is the most important aspect of management from their point of view, because the incentives (especially the 40% of output allowed for cost recovery) are

of PSCs of the first (1965–75), second (1976–88), and third (since 1988) generations, see A. S. Moch. Anwar, F. X. Sujanto, and D. Zahar, *Pertamina: Indonesia Production Sharing Contract: Its Development and Current Status* (Jakarta, Indonesia: Pertamina, 1989).

    5. This ceiling was lifted when oil prices were low in early 1989 and replaced by a 20% government share taken from the first *tranche* (phase) of production. The split between Pertamina and the contractor now is usually 85:15, and in some cases 80:20. A higher percentage in the production split and other incentives are provided in marginal cases and in remote areas, considered "frontier areas."

    6. These were the first and last *kontrak karya* on oil and gas. This type of agreement was later used exclusively for hard minerals.

such that the contractor's interest in keeping costs down may not be as great as Pertamina's; (2) Pertamina had to economize on the use of its trained personnel, for it did not have an unlimited number; (3) Pertamina management felt that foreign investors, who are risking their capital, must in fairness have an important say in management matters, especially those in which the interests of the foreign investor and of Pertamina are similar or do not conflict.

At least 3 months prior to each contract year the contractor must submit to Pertamina a work program and budget. From the point of view of Pertamina, the work program is a basic management tool, containing the essential elements for an exploration program. It defines the proposed types of exploration activities, the duration of the program, disposal of data and information, estimated expenses and cost, plans for the relinquishment of the area, and so forth.

Pertamina obtains all the data gathered by the contractor during the exploration stage. If the contractor finds a promising location or structure, the matter is discussed with Pertamina after test drilling has ascertained the viability of production.

It is at this stage that the difference between the production-sharing system and the concession system becomes apparent. Under the concession system, the decision is entirely up to the oil company. They usually base their decision on the company's production policy and program, which in the case of big oil companies are integrated into their worldwide production program. They may proceed with actual production or cap the well. As an oil company's decision may not always be a good one from the point of view of Indonesia's energy development program, the preproduction consultations between Pertamina and the contractor are very useful. If the size of the deposit and the nature and composition of the hydrocarbons are attractive, a decision may be made to enter the next stage of development, which is the actual production of oil and gas.

Recently, incentives have been given to oil companies concluding exploration contracts with Pertamina in areas that are marginal both in the sense of productivity potential and/or accessibility and convenience (remoteness).

The original PSC stipulated that *all equipment* purchased by the contractor and brought into Indonesia for purposes of exploration and production becomes the property of Pertamina. As a cost element it is written off according to a formula agreed upon between Pertamina and the contractor. Experience has shown, however, that even at the exploration stage the contractor very often subcontracts the drilling to other companies, so that this provision is no longer considered advantageous to Pertamina. One other consideration making the equipment provision of dubious advantage to Pertamina is the cost of removal after a work area or contract area has been abandoned.

Other provisions in the contract contain stipulations on the transfer of technology, the Indonesianization of staff, and other ways to ensure that

Indonesia will be able in a reasonably short time to run its own oil industry. The contracts also contain provisions on marketing and on fulfilling the needs of the domestic oil market.

To a large extent the objectives set out by the designers of the PSC have been achieved. Not only is Pertamina doing its job as a manager of Indonesia's oil-and-gas development program, but many service companies supporting or supplying Pertamina and the oil contractors are now run or owned by Indonesians, including drilling contractors. Work or services requiring high technology or large capital input, however, are still provided by foreign firms or operate as joint ventures. At present there are no private Indonesian oil companies working as independent contractors.

With the introduction of the new tax law some years ago, the tax provisions in the PSC have been largely rewritten. In the original PSC no tax was paid; tax payments were included in the cost component deducted from gross revenue to arrive at net production (revenue) to be shared between Pertamina and the contractor. This was gradually changed over the years, and at present the picture is much clearer. The oil industry is one of the big tax contributors to the Indonesian Treasury, in all stages of operations up to and including the gas pump. At the production stage no consolidation of taxes is allowed as a matter of general principle, as indeed no consolidation of costs is allowed even between different contract areas worked by one company operating within Indonesia.

One PSC provision that merits special attention is the clause on mutuality, which states that "Pertamina and Contractor undertake to carry out the terms and provisions of this contract in accordance with the principles of mutual good will and good faith and to respect the spirit as well as the letter of said terms and provisions." One can read this mutuality clause in a number of different ways. It is certainly unusual, especially when thought of in the context of a transaction involving millions of U.S. dollars. Nevertheless, the clause reflects the atmosphere surrounding the PSCs in the 1960s, which should be viewed as laying down the rules for a partnership in search of oil and gas, with one partner, Pertamina, as the custodian of the resources providing the opportunity and facilities and taking the responsibility for all burdens involving administrative, fiscal, and government-related matters, and with the foreign partner, i.e., the contractor, concentrating on the financial, technical, and commercial aspects of oil exploration and production.

One area in which this mutuality clause could become very important is the decision whether to exploit a marginally economic oil deposit. Here the interests of the partners may be different: the contractor's interest is in recovering costs plus a good return, whereas Pertamina's interest is in getting its fair share. This conflict has not yet given rise to disputes, however, because so far all of the deposits found have been fairly substantial.

There is certainly room for improvement in the PSC, especially for a more precise formulation of some of the rights and duties of both parties.

However, improvements should not disturb the mutuality of trust and good-will that so far have characterized these contracts in actual performance.

Having given a brief sketch of the PSC, I next present some legal issues pertaining to mineral (oil and gas) resources development in the Indonesian marine space.

SOME LEGAL ISSUES IN MINERAL RESOURCES DEVELOPMENT IN INDONESIA

The Form of Contracts Employed

Kontrak Karya Compared to Concession Contract

From a practical point of view the *kontrak karya* gives the contractor rights and obligations similar to the concession holder. As far as the implementation of the safety regulations contained in Mijn Politie Reglement is concerned, for example, all the provisions are considered applicable to the contract simply by substituting mining-rights holder, and on the holder's behalf the contractors, for concession holder. In other aspects of the *kontrak karya,* for example, the payment of land rent, royalties, and other taxes and levies, and the relation-ship between contractor and the holder of surface land rights, the similarity is indeed striking. The work contract further gives the contractor actual control over the minerals found and produced.

These facts have led people to state that a *kontrak karya* is nothing but a concession in disguise, and that the rights of the contractor, although deriva-tive in character, are in fact similar to those of a mining-rights holder (conces-sion or license). As one supporter of this view has written,

> although the agreements entered into in the field of mining are identified as work contracts, *kontrak karya,* it appears to the writer that there is little difference between the substance of these agreements and the more traditional mineral concession agreements; the only difference being that in the traditional concession agreements the concessionaire obtains title to the minerals in the concession area at the time the concession is granted, while in the Indonesian work contract agreements the "contrac-tor" apparently obtains title only upon exploitation. The importance of this distinction appears to be minimal. What is of far greater importance is that under the Indonesian agreements the "contractor" maintains the authority over virtually all management, production, and marketing de-cisions, at least within the very liberal time limits imposed by the agreements.[7]

7. Timothy Manring, "Comments on Agreements in the Fields of Mining and Timber," cited in Mochtar (n. 4 above), p. 58.

I cannot agree with such a view, however persuasive it may seem, for two reasons. The conditions for tenure in a *kontrak karya* are more restrictive than those usually found in concession agreements. And legally more important, the juridical nature of the contractor's right under a *kontrak karya* is radically different from rights under a concession.

At least under the Netherlands Indies legislation, that is, the Indische Mijnwet, the concession right was a "zakelijk recht" (right in rem), which could be mortgaged. Related to this was the obligation of the concession holder to have the concession right made public by registering it in the same manner as land rights. The rights of a contractor under a *kontrak karya,* on the other hand, are contractual rights that cannot be subject to mortgage under Indonesian law, nor are they freely assignable like concession rights under the Indische Mijnwet. This legal difference between rights of the contractor under a *kontrak karya* and a holder of concession rights is not inconsiderable, especially with respect to possibilities of acquiring third-party financing for the mining undertaking in question.

PSC Compared to Concession Contract

What has been said so far applies to *kontrak karya* related to minerals other than petroleum, that is, mining as regulated by the Basic Mining Law of 1967 and related government regulations. With regard to petroleum, the development of the *kontrak karya* concept has—as has already been shown—followed a different road, ultimately leading to the PSC.

Like rights under *kontrak karya* (for hard minerals), rights under a PSC are personal rights (rights in personam) and as such are distinguished from rights under a concession. Moreover, there are other important legal differences between a PSC and the former concession contracts. Whereas under a concession contract the concessionaire (i.e., the license holder or licensee) had the exclusive right to manage the enterprise, under a PSC the right of management belongs to Pertamina. The PSC is no longer based on profit-sharing (like the concession contract) but on a production-sharing scheme, thus minimizing the occurrence of conflicts over price-setting and accounting procedures. Unlike the concession contract, which granted the concessionaire title to oil at the wellhead, the PSC grants title to the contractor only at the point of export. Other points of difference, intended to increase the benefits to Indonesia, obligate the contractor to contribute some of the oil produced to the domestic market, to submit its geological and other data to Pertamina, to offer participation in the enterprise to an Indonesian national, to relinquish a certain portion of the contract area at specified intervals, to give title to its imported equipment to Pertamina, to consider local processing, and to employ and train Indonesian personnel.

Work Contract (Kontrak Karya) Compared to PSC

As has been shown, rights under work contracts as well as rights under PSCs are personal rights. Yet it is important to bear in mind the major differences between work contracts and PSCs.

1. Under the PSC the government, through Pertamina, retains the management of the operations and control over the resources; under the work contract, management of the operations and control over the resources is in the hands of the contractor, with the government (i.e., the Department of Mines) only supervising.

2. Under the PSC the work program and budget are annually submitted for approval by Pertamina, and Pertamina may propose revisions; most work contracts concerning minerals do not contain such provisions. Where such provisions do exist, their function is not to give the government management rights, as it is in the PSCs, but merely to indicate whether the contractor is adhering to the mining laws, regulations, and guidelines of the department.

3. Responsibility for taxes and other government levies under the PSC is Pertamina's, since Pertamina is the holder of the authority to mine; under the work contract, however, it is the contractor who must pay land rent, royalties, taxes, and other levies.

4. The PSC explicitly states that "title to the [contractor's share of] crude oil passes at the point of export"; no such provision exists in the work contracts in which title to oil presumably passes at the wellhead.

Paradoxical as it may seem, it could be argued from the above differences that the PSC is more truly a "contract of work" (*kontrak karya*), with the foreign company acting as a contractor, than is the *kontrak karya,* where the contractor, although indirectly, does have actual mining rights.

One legal issue that could be important for purposes of liability (general), risk (insurance), and computation of damages to third parties is the question, At what point does title to the oil lifted pass to the contractor (oil company)? In the concession system, title passes at the wellhead, as opposed to the PSC, where title passes at the point of export.

Removal of Equipment

Ownership of oil-and-gas production equipment, especially drilling platforms, has recently given rise to problems for Pertamina.[8] After the expiration of the contract, the contractor is free to leave, abandoning the equipment, including that to which Pertamina has title. This may cause problems, as removal of drilling platforms is quite costly. The alternative would be to leave the platforms intact for other purposes or for use if resumption of production at the same well becomes commercially feasible.

In the meantime, however, Pertamina is responsible for the maintenance

8. At present there are approximately 200 offshore wells temporarily abandoned, and some 130 wells permanently abandoned in the Indonesian offshore areas.

of these platforms and for the damage or mishaps they can cause to passing vessels. At least Pertamina has to bear the cost not only for maintaining the drilling platforms but also for providing them with warning lights in accordance with safety and navigation requirements. The cost to Pertamina is considerable, since in the Java Sea alone there are about 100 drilling platforms, many no longer active. In negotiations an attempt is made to build these costs into the PSC by including maintenance and lighting costs for abandoned platforms in the cost component of the contract. A more radical solution would be to revise the PSC provision relating to passage of title to *all* equipment purchased and brought into Indonesia by the contractor. In negotiating new PSCs, this particular provision could be rewritten to solve the problem.

Another alternative would be to sell the platforms to third parties, which may or may not be feasible as drilling technology keeps advancing. Unusable platforms can always be sold for scrap. On the other hand, experience has proven that the presence of drilling platforms in shallow seas like the Java Sea is conducive to the creation of a favorable marine environment for the sea's living resources. In particular, the oil/gas rigs may be used wholly or partially as artificial reefs for fish aggregation.

Protection of the Marine Environment

The production of hydrocarbons offshore naturally entails the risk of marine pollution. The oil industry's record on this matter in Indonesia, however, is good, thanks to the offshore production safety measures prescribed by the Directorate General Migas, which has issued offshore regulations dealing with the matter, including provisions to protect the marine environment.[9] Oil companies operating in Indonesian offshore areas have an excellent record of adherence to regulations related to protection of the marine environment.

Boundary Delimitation

Boundary delimitation gives rise to disputes between oil companies and between countries.

9. The two main regulations are Peraturan Menteri Pertambangan (Minister of Mines Regulation) no. 04/P/M/Pertamb./1973, 22 March 1973; Surat Keputusan Bersama Dirjen Perhubungan Laut dan Dirjen Migas (Joint Decision [Regulation] Director-General of Sea Communication, and Director-General of Oil and Gas) no. DKP/49/1/1-no. 01/KPTS/DM/MIGAS, 1981 *Tetang Prosedur Tetap (Protap) Pencegahan dan Penanggulanagan Pencemaran Laut oleh Minyak Bumi di Selat Malaka dan Selat Singapura* (Permanent Procedure concerning Protection of the Straits of Malacca and Singapore against Pollution).

Boundary disputes that arise between two oil companies assigned adjoining contract areas may have their source in the delimitation or demarcation of the contract area originally negotiated between the contractors and Pertamina. The natural way to resolve the problem would be to appeal to Pertamina for a settlement. This is, of course, easier said than done, especially when the disputed area holds great promise of oil or gas deposits. At least one such case occurred in the early years of the PSC,[10] but the situation has improved over the years, and hardly any disputes are occurring at present over faulty delimitation of boundaries of contract areas.

Another problem may arise when a promising geological structure straddles a boundary between contract areas held by two oil companies. Such a situation is usually solved by joint development of the structure or prospective hydrocarbon deposit, applying the principle of unitization, which involves contractual cooperation between licensees/concessionaires of a field straddling the boundary.

Delimiting boundaries between two countries creates another set of problems. It is because of this potential for conflict that Indonesia has paid great attention to negotiating its Continental Shelf boundary with its neighbors, starting with Malaysia in 1969, when the first Continental-Shelf Agreement was signed. At present a total of 15 boundary delimitation agreements between Indonesia and other countries have been signed.[11] Most of them relate to Continental Shelf or seabed boundaries, with a few Territorial Sea boundaries included, one with Malaysia in 1970 and one with Singapore in 1972. The Indonesia-Australia Boundary Treaty of 1974 also involves a Territorial Sea delimitation.

The principle followed in relation to areas held by oil companies under contract to Pertamina is that, in cases of a boundary agreement concluded between Indonesia and a foreign country, the contract area boundary is in consequence corrected as a matter of law. Usually there is a provision in the PSC related to this matter, to prevent a legal issue or dispute arising between Pertamina and the contractor.

Not all boundary questions that Indonesia has with its neighbors have been solved, however. Two exceptions are the Continental Shelf boundary with Vietnam, which is still under negotiation, and the Continental Shelf or seabed boundary between Indonesia and Australia in the area that was previously Portuguese East Timor, popularly called the "Timor Gap."

With regard to the latter case, the solution has drawn much attention because of the new ideas and concepts it has created. It is a very interesting

10. A dispute between Union Oil Co. (Indonesia) and Shell (Indonesia) (1971–73) over the Sangkulirang (Bay), situated on the east coast of Kalimantan.

11. For a detailed examination, see regional overviews by Choon-ho Park (on central Pacific/east Asia) and J. R. V. Prescott (on Indian Ocean), in Jonathan I. Charney and Lewis M. Alexander, eds., *International Maritime Boundaries* (Dordrecht, Netherlands: Martinus Nijhoff Publishers, forthcoming).

example of seeking a temporary solution for an intractable boundary question, through the establishment of a joint development zone. Because of its importance as a novel legal concept, it will be dealt with in some detail.

THE INDONESIA-AUSTRALIA JOINT COOPERATION ZONE

The joint development zone has frequently been employed in cases where countries were unable to solve boundary disputes, in order that the development of resources could be undertaken pending the final settlement of the boundary question. The most well known are (1) the Japan-Korea Joint Development Zone, (2) the Saudi–United Arab Emirates Joint Development Zone in the Persian Gulf, (3) the U.K.-Norwegian Joint Development Zone in the North Sea, and (4) the Thai-Malaysia Joint Development Zone in the Gulf of Thailand.

The Indonesia-Australia case is unique, however, because it involves the creation of new institutions and new law, thereby setting a precedent of sorts. The basic principles for the treaty were contained in a joint statement signed by the Foreign Minister of Indonesia and the Secretary of External Affairs and Trade of Australia on 25 October 1988 in Jakarta. The operative paragraphs state,

3. The Zone of Cooperation will be delineated in the northern side by a simplified bathymetric axis line, in the southern side by the 200 nautical mile line measured from the Indonesian archipelagic baselines, and in the eastern side and western side by equidistance lines. The establishment of the Zone and its delineation will not prejudice the respective positions of the two Governments on a permanent continental shelf delimitation in the area and will not in any way be construed as affecting the respective sovereign rights claimed by each side in the Zone of Cooperation.

4. The Zone of Cooperation will comprise three component areas, namely Areas A, B and C as in the attached sketch map. A joint development regime will apply in Area A, and there will be established a Ministerial Council and a Joint Authority. In Area B the relevant Australian legal regime will apply, and in Area C the relevant Indonesian legal regime will apply, subject to a regime of sharing in tax returns applicable in each of the two areas and a process of notification and consultation between the two Governments through the Joint Authority on petroleum exploration and development activities.[12]

12. Joint Statement of Jakarta, 25 October 1988, Pars. 2 and 3; see also Mochtar Kusuma-Atmadja, "Indonesian-Australian Joint Cooperation Zone South of East Timor (the 'Timor Gap')," paper presented at the IRR Conference, Singapore, 30–31 May 1989.

On 11 December 1989, Indonesia and Australia signed the Treaty on the Zone of Cooperation in an Area between the Indonesian Province of East Timor and Northern Australia (the so-called Timor Gap), embodying the principles contained in the joint statement referred to above. The treaty, which was signed by the foreign ministers of Indonesia and Australia in an aircraft circling the area, has three annexes: Annex I delimiting the zone of cooperation, consisting of Area A (joint development zone), Area B, and Area C; Annex II, a mining code especially designed for the joint development zone (Area A); and Annex III, a model PSC, applicable exclusively to the joint development zone (Area A). In other words, oil-and-gas companies operating as contractors in Indonesia or companies holding rights under Australian law in Australia cannot use or claim rights based on the provisions contained in this special PSC.

A thorough discussion of the provisions of the treaty, the mining code, and the model PSC is clearly beyond the scope of this article.[13] Suffice it to say that the system employed in the exploration and production of oil and gas in the Timor Gap area of the seabed between Indonesia and Australia is closely following the production-sharing system pioneered by Indonesia and now widely used throughout the world.

THE PSC AS A SYSTEM OF OIL-AND-GAS RESOURCES DEVELOPMENT

As described in this article, the PSC system is truly a system of oil-and-gas *resources development* as opposed to a system of *mineral resources extraction* followed under the concession or licensing system, the difference being that management of and title to the oil and gas in the PSC system are both held by the state company (Pertamina) on behalf of the state as the custodian of the nation's resources (Article 33 of the Indonesian Constitution). This system, which is a manifestation of the principle of sovereignty over national resources if properly handled and implemented, provides a balance between the public interests represented by Pertamina and the private-enterprise interests represented by the oil company (contractor).

The system has worked well in Indonesia, because Pertamina has since its inception handled its mandate with good sense, adhering to generally accepted good oil practices. The same can be said of the foreign oil companies, which have learned to behave like partners in development rather than

---

13. Mochtar Kusuma-Atmadja, "Perjanjian Indonesia-Australia di Celah Timor" (The Indonesia-Australia Treaty on the Timor Gap), paper read at the University of Gajah Mada, Yogyakarta, 9 February 1990, p. 32; Prescott (n. 11 above), report no. 6-2(5); J. R. V. Prescott, "Maritime Boundary Agreements: Australia-Indonesia and Australia-Solomon Islands," *Marine Policy Reports* 1 (1989): 37–45.

like profiteers. The mutuality clause expresses the spirit of cooperation and partnership for mutual benefit.

The PSC system of oil-and-gas resources development, now nearly 30 years old and widely adopted around the world, benefits developing countries because it provides the basis for sound mineral resources development as opposed to mineral resources extraction. On the other hand the PSC system for oil and gas and the *kontrak karya* system for hard minerals have provided a stable environment for investment and production for foreign companies. These companies are treated as partners in developing Indonesia's resources, not only by the government but also by the communities in which they operate. In remote areas this can be very important for the successful operation of a mining venture.

The various provisions on personnel development and the transfer of skills, know-how, and the experience gained in management and environmental protection are useful from a developing country's point of view. The opportunity the PSC system gives to Indonesia to husband its scarce mineral resources is perhaps the most important advantage from a resource-development point of view.

The difficulties experienced by the mining companies at the Ok Tedi and Bougainville copper mines in Papua New Guinea are a telling example of troubles arising from a mineral resources extractions system, which can perhaps be explained to a large extent by the different system employed and different philosophy underlying it.

It may be said with some justification that the PSC system for oil and gas and the *kontrak karya* system for hard minerals are important contributions made by Indonesia to the concept of mineral resources development.[14] It is a system in keeping with finite mineral resources and with concerns for the environment in a rapidly shrinking world.

14. Developing states that follow Indonesia's PSC system include Malaysia, India, Egypt, Mexico, Algeria, and recently, China and Vietnam. The PSC, although more beneficial than a concession system, may be applied only by those countries that have adequate legal means to effect involvement of the state or state enterprise, as discussed in this article.

A Regional View of Shipping: Southeast Asia

Hal F. Olson
East-West Center, Honolulu

Oh, East is East, and West is West, and never the twain shall meet.

Kipling

Southeast Asia today, contrary to Kipling's observation, does provide a meeting place between East and West, for commerce if not for culture. Trade between the industrialized countries of Europe and the economic powers of East Asia moves through the water passages of Southeast Asia. Crude oil, liquified gases, and other petroleum products from the Middle East, bulk cargos from Africa and Australia, and manufactured goods and machinery from Europe move east and north, while other manufactured goods, vehicles, and machinery move in the opposite directions.

Singapore, the highly developed city-state, is the focal point of much of this activity, supporting through traffic and serving as a transshipment point for regional cargos. Other countries of the region are seeking a share of these activities, and their approaches are as varied as the countries themselves: socialist and capitalist, outward and inward looking, developed and developing, rich and poor, centralized and fragmented, ethnically homogeneous and diverse, "conventional" and archipelagic. All are affected by the "container revolution" and the changing patterns of shipping that are developing from this.

Southeast Asia has historically occupied an advantageous position, there being no practical alternate water route between East and West. The region is characterized by the intermingling of land and water to such an extent that all countries of the region must of necessity look to the sea as a primary means of transportation. This applies not only to international trade, but to much of the domestic traffic as well. Even those countries occupying portions of the Asian mainland lack the road and rail networks that elsewhere may provide alternative transportation choices.

International marine traffic converges in three major straits in the region—Malacca, Sunda, and Lombok (fig. 1). Malacca Strait provides the shortest route between East and West and until recent years was the only one normally used. During the period when the Suez Canal was closed, vessels

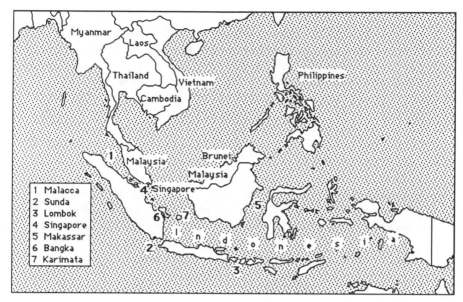

Fɪɢ. 1.—Southeast Asia straits.

sailing directly from Europe around the Cape of Good Hope, not calling at intermediate Asian ports, took a route through Sunda Strait. At about the same time, the very large crude carrier (VLCC) and ultralarge crude carrier (ULCC) came into common employment for the transportation of petroleum. Their extreme drafts, of as much as 90 feet, barred them from using Malacca Strait, so between the Persian Gulf and Japan, these vessels are routed through Lombok and Makassar Straits.

Malacca Strait lies between the Malay Peninsula and the Indonesia island of Sumatra, connecting the Andaman Sea with Singapore Strait and the South China Sea. Approximately 500 miles in length, Malacca Strait narrows to a width of about 6 miles at its eastern end, where it merges with Singapore Strait. Governing depths in the strait are about 72 feet, but because of the sandy nature of the bottom, frequent changes occur, and that depth may not always be present. In addition, a requirement has been established by the International Maritime Organization (IMO) for an underkeel clearance of 3.5 m, thus further limiting, for safety, the maximum draft of ships plying the strait.[1]

Sunda Strait, between the islands of Sumatra and Java, serves as an alternate to Malacca Strait for vessels sailing to and from southern Africa or

1. Defense Mapping Agency Hydrographic and Cartographic Center, *Sailing Directions (Planning Guide) for the Indian Ocean,* 3d ed. (Washington, D.C.: Defense Mapping Agency Hydrographic and Cartographic Center, 1988), p. 429.

around the Cape of Good Hope. Physically it is less desirable as a shipping route, for even though it is deep enough and wide enough for large vessels, it contains strong currents, and the Indonesian waters to the north are relatively shallow and dangerous. Also, the active volcano Krakatau occupies the center of the strait and is considered to present a significant danger. Vessels proceeding on through the Java Sea to Makassar Strait find that this route, 150 miles shorter than the one through Lombok Strait, is difficult and provides no significant advantage.[2]

Lombok, the third of the major straits, lies between Lombok and Bali. It is wide, more than 11 miles at its southern end; deep, greater than 450 feet in most places; and free of hazards.[3] As such it provides the safest route for even the largest of the ULCCs, which may exceed 500,000 dwt.

Four other straits are also significant for international traffic. Singapore Strait connects the eastern end of Malacca Strait with the South China Sea, and the two together form the complete passage. Singapore Strait itself is approximately 60 miles in length, narrows to less than 2.5 miles, and like Malacca Strait has a naturally occurring depth of about 72 feet.[4] Makassar Strait, between Borneo and Celebes, forms a 400-mile-long continuation of the ULCC route north from Lombok Strait. Though deep and wide, Makassar Strait contains a number of detached shoals and coral reefs that are hazardous to the unwary. The preferred east channel has a least width of 22 miles between the 100-fath curves.[5] Karimata Strait (96 feet) and Bangka Strait (36 feet) are both relatively shallow. The route usually taken by ships between the eastern part of the Java Sea and the South China Sea passes through the wide Karimata Strait,[6] while the best route between Singapore and Sunda Strait includes Bangka Strait.[7] Both are used by Southeast Asia–Australia traffic and by traffic to and from Singapore and Indonesian ports.

The archipelagic countries of Indonesia and the Philippines have been recognized as having specific concerns that differ from island states. These countries extend over large areas of the earth's surface and include a great number of islands among which are passages that can be and are used by

2. Defense Mapping Agency Hydrographic and Cartographic Center, *Sailing Directions (Planning Guide) for Southeast Asia,* 2d ed. (Washington, D.C.: Defense Mapping Agency Hydrographic and Cartographic Center, 1987), p. 297.

3. Ibid.

4. *Sailing Directions for the Indian Ocean* (n. 1 above), p. 431.

5. Defense Mapping Agency Hydrographic and Cartographic Center, *Sailing Directions for Java; Lesser Sundas; South, Southeast, and East Coasts of Borneo; and Celebes,* 5th ed. (Washington, D.C.: Defense Mapping Agency Hydrographic and Cartographic Center, 1962, rev. 1976), p. 351.

6. Defense Mapping Agency Hydrographic and Cartographic Center, *Sailing Directions for Soenda Strait and the Western and Northeast Coasts of Borneo and Off-Lying Islands,* 5th ed. (Washington, D.C.: Defense Mapping Agency Hydrographic and Cartographic Center, 1951, rev. 1975), p. 134.

7. Ibid., p. 80.

ships engaged not only in domestic, but in international, trade. The special nature of these two countries in particular has been recognized in Part IV of the United Nations Convention on the Law of the Sea (LOS Convention) with respect to the movement of foreign-flag ships through their waters. Island states, on the other hand, have the same rights to Territorial Seas and exclusive economic zones as do any other countries. The issue of navigation "through" their water areas remains the same as elsewhere for passage through Territorial Seas. Hence the distinction here between "archipelagic" and "island" states.

Singapore and Brunei Darussalam are effectively island states, even though technically they are in physical contact with other land areas. Singapore is connected by causeway to peninsular Malaysia, while Brunei occupies a portion of the west coast of the island of Borneo. Land connections between the latter and adjacent territory are absent as far as trade is concerned.

Malaysia has both peninsular and insular characteristics, occupying parts of both the Malay Peninsula and the island of Borneo. The southern part of Thailand is peninsular, giving it a coastline on the Andaman Sea as well as on the Gulf of Thailand, but otherwise its territory is firmly embedded in the Southeast Asian landmass. The remaining mainland states of the region, Myanmar (Burma), Vietnam, and Cambodia, have little active current involvement in maritime trade.

NATIONAL EMPHASIS

Whether or not a country can be considered a "maritime" state can be judged in several ways. These include the size of the national merchant fleet; the number of modern, specialized vessels; ports open for international and domestic trade; adequacy of port facilities; vessel-replacement programs; and compliance with international conventions that facilitate marine transportation or contribute to its safety. Objective standards are difficult, if not impossible, to quantify, but the attitude of a government becomes quite clear when a number of these criteria are examined.

Possession of a merchant fleet has costs as well as advantages. The mere desire to move cargo in a country's own ships, or to enhance its international prestige, should not govern a country's decision to develop and support a fleet. Capital costs involved in the purchase or construction of ships are high, especially for developing countries where alternative investments in infrastructure can more directly promote economic growth.

Benefits can be derived from a national merchant fleet, though not all of them can be quantified. A country has an accepted right to move cargos in ships flying its own flag, a right recognized in the United Nations Conference on Trade and Development (UNCTAD) 40-40-20 formula. (Each of a pair of trading partners would carry 40% of that trade, with 20% carried in

TABLE 1.—MERCHANT FLEETS, 1988

	Number of Vessels	Gross Tons	Average Size
Brunei	34	354,313	10,411[a]
Cambodia	3	3,558	1,186
Indonesia	1,736	2,126,016	1,225
Malaysia	499	1,608,155	3,222
Myanmar	120	272,665	2,275
Philippines	1,483	9,311,555	5,279
Singapore	715	7,208,974	10,082
Thailand	258	515,314	1,996
Vietnam	164	337,875	2,061

SOURCE.—*Lloyd's Register of Shipping Statistical Tables 1988*. (London: Lloyd's Register of Shipping, 1988), pp. 7–9.
[a] Includes seven vessels averaging 49,782 GT.

third-country vessels.) This should, ideally, produce a profit, especially in foreign exchange. Should a government subsidy be necessary to operate the fleet, there will be an added drain on financial resources above and beyond acquisition costs. Additional benefits include the employment opportunities for ship crews; the promotion of export trade, in both traditional and newly developed cargos; and the assurance that ships will be available for use in case of national emergency.[8] Employment in marine-related fields including ship repairing, cargo handling, bunkering, and the variety of ship support operations associated with international trade is also in national interests. These may not be as clearly recognized by governments, yet require less financial investment.

NATIONAL MERCHANT FLEETS

The Philippines and Singapore lead all other Southeast Asian countries in the size of their merchant fleets in terms of gross tonnage. As of the end of 1988 these totaled 9.3 and 7.2 million GT, respectively, and ranked 11th and 17th in the world. Both Indonesia (2.1) and Malaysia (1.6) also exceeded 1 million GT in their fleets. Fleet size decreased sharply for all other states in the region, as shown in table 1. The distribution of ship types in the five largest fleets is presented in table 2.

Singapore's encouragement of a shipping industry to compete in international trade has produced a balanced distribution of vessel types in its fleet. Several of its crude and chemical tankers, while not among the world's largest vessels, are of more than 150,000 GT and average more than 14,000 GT;

8. Thomas R. Leinbach and Chia Lin Sien, *South-East Asian Transport* (Singapore: Oxford University Press, 1989), p. 99.

TABLE 2.—MERCHANT FLEETS, SHIP TYPE DISTRIBUTION (number/1000 gross tons)

	Tankers	LPG/LNG[a] Carriers	Bulk Carriers	Container Ships	General Cargo	Other	Total
Indonesia	209/679	6/8	11/129	4/60	717/867	789/383	1736/2126
Malaysia	77/240	6/342	17/378	16/190	177/364	206/94	499/1608
Philippines	95/483	18/18	263/6906	10/49	529/1385	568/471	1483/9312
Singapore	177/2497	23/291	82/2287	43/720	171/1187	219/227	715/7209
Thailand	68/77	21/17	0/0	6/36	108/357	55/28	258/515

SOURCE.—*Lloyd's Register of Shipping Statistical Tables 1988* (London: Lloyd's Register of Shipping, 1988), pp. 10–17.
[a]Liquified petroleum gas/liquified natural gas.

bulk carriers average nearly 28,000 GT, and containerships 16,700 GT. Liquified gas carriers and general cargo vessels are also the sizes generally employed on international routes.

Though larger than that of Singapore, the Philippine fleet has almost three-fourths of its tonnage in bulk carriers that average over 26,000 GT. All other classes average only 2,000 GT per vessel, indicating primary employment in domestic trade, though some larger ships are undoubtedly present among them.

Six liquified gas carriers make up 21% of all Malaysian registered tonnage, averaging 57,000 GT each. Seventeen bulk carriers compose another 23% of the fleet, though smaller at an average of 22,000 GT each. The remainder of the fleet averages less than 1,900 GT, again an indication of their employment generally in domestic or intraregional trade.

All classes of vessels flying the Indonesian or Thai flag are relatively small, averaging 1,200 and 2,000 GT per vessel, respectively. Both fleets appear best suited to service on domestic or intraregional routes.

Among the other countries of the region, only Brunei has ships intended for use on international routes. Seven liquified gas (liquified petroleum gas/liquified natural gas, LPG/LNG) carriers in the 50,000-GT range have been added to the fleet since 1986 and make up 98% of its total of 354,000 GT! For all practical purposes, they are the Brunei fleet. (The other 27 vessels in the fleet total only 5,839 dwt, averaging only 216 dwt.)

Vietnam and Myanmar had 338,000 and 273,000 gross tons of shipping under their flags in 1988, both with vessel size averaging between 2,000 and 2,300 GT. Cambodian registry includes only three vessels totalling 3,558 GT.

FLEET DEVELOPMENT

The growth of national fleets reflects differing and changing government policies (figs. 2 and 3). Whereas the Indonesian and Malaysian fleets grew steadily in numbers and tonnage, average vessel size remained about the same. Singapore experienced an increase of more than 700,000 GT per year for 6 consecutive years (1973 to 1978), much of which can be attributed to its "open registry" policy. Initiated by passage of the Merchant Shipping Act of 1968, tax incentives were offered to owners of Singapore-flag vessels. These included exemption from Singapore income tax on profits derived from the operation of those ships, and on crew members' pay when their employment was substantially outside of Singapore's waters. A rebate of 50% of the annual tonnage tax was also available to shipowners employing crews that were more than 25% Singapore seamen.[9]

9. Marine Department, Singapore Ministry of Communications, *The Singapore Registry of Ships* (Singapore: Marine Department, Singapore Ministry of Communications, 1976), pp. 7, 12.

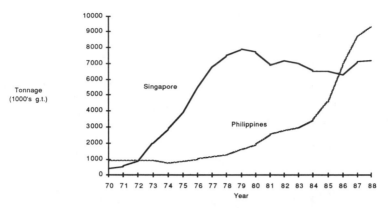

Fɪɢ 2.—Merchant fleets, Singapore and the Philippines, 1970–88 (*Lloyd's Register of Shipping Statistical Tables 1988* [London: Lloyd's Register of Shipping, 1988], pp. 33–34).

Amendments tightened up the law in 1979, and Singapore-registered tonnage decreased in 6 of the next 7 years. Foreign-owned vessels then had to be larger than 1,600 GT and less than 15 years old. Singapore-owned vessels were not to be more than 22 years old, decreasing to no more than 20 years by 1982. It was intended to further reduce the age of all vessels to less than 15 years by 1984.[10]

In contrast to the restrictive shipping policies of other ASEAN countries, Singapore has encouraged the growth of its fleet and shipping in general by a nearly "hands-off" policy, allowing the free play of the market to determine the size and form of its fleet. If anything, the establishment of an open register has been the single most significant government action in the shipping field.

The Philippine fleet has grown principally through private investment rather than by government regulation of and subsidies for shipping. Although an open register policy was considered, it was not adopted. Instead, greater foreign investment was encouraged by increasing the permitted percentage of foreign ownership of shipping enterprises from 25% to 40%.[11] Also, the UNCTAD 40-40-20 cargo reservation scheme was adopted as the minimum apportionment of cargos to Philippine-flag vessels. An additional requirement is that all government cargo, the definition of which is notably broad, must move in ships of Philippine registry.[12]

Whether from this or a combination of factors, the average size of Philippine-flag vessels increased from 3,129 GT in 1969 to 5,279 GT in 1988.

10. Mary R. Brooks, *Fleet Development and the Control of Shipping in Southeast Asia* (Singapore: Institute of Southeast Asian Studies, 1985), p. 26.

11. Leinbach and Chia (n. 8 above), pp. 112–13.

12. Brooks (n. 10 above), pp. 18–20.

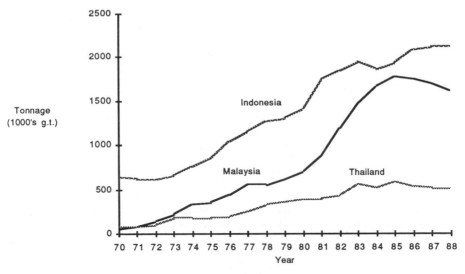

Fig. 3.—Merchant fleets, Indonesia, Malaysia, and Thailand, 1970–88 (*Lloyd's Register of Shipping Statistical Tables* [London: Lloyd's Register of Shipping, 1970–86], pp. 4–5, [London: Lloyd's Register of Shipping, 1987–88], pp. 6–7).

This indicates a greater emphasis on vessels normally employed in international trade.

Indonesian policies to encourage the growth of a national shipping fleet have been confused and inconsistent. Restrictive policies were established that, among other things, required all nonpetroleum exports to be shipped from one of four ports: Jakarta, Belawan (Medan), Surabaya, or Ujung Pandang. The entry of foreign ships into those ports was then controlled by the issuance of short-term permits. Forty-eight ports were open to foreign vessels for the discharge of cargo only, with loading prohibited. Indonesian-flag vessels were to be employed not only for the movement of cargo within Indonesia but also for Indonesian cargo destined for ports in Thailand and Cambodia. Domestic interisland traffic was divided into five zones, with freighters assigned to liner service in one zone prohibited from working into another zone.[13] The 1984 Five-Year Development Plan banned the employment of vessels over 30 years old from any export/import operations after 1 May 1984, and of 25-year-old vessels from those operations from 1 January 1985. Foreign vessels over that age would be barred from Indonesian ports.[14]

The intent of these policies was to encourage the growth of the Indonesian fleet and promote the domestic shipbuilding industry. The actual result was to restrict both foreign and domestic trade and increase the cost of what remained.

13. Ibid., p. 15.
14. Ibid., p. 62.

Recognition of an extreme imbalance in the way in which Thai–South Korean cargos were being carried (45% in Korean and 1.3% in Thai vessels) led Thailand to identify five areas in which improvements could be made to encourage enlargement of its merchant fleet. These included

1. elimination of rigid and wasteful bureaucracy,
2. training of seagoing and shore-based personnel,
3. government assistance to reduce interest rates on capital for ship financing,
4. tax exemptions, and
5. official cargo reservations: Thai government cargo in Thai ships.

Some of these measures had previously been in effect but had not significantly encouraged prospective investors.

While government cargos are required to be carried in Thai-flag ships, waivers have been so easily obtained that this has provided negligible encouragement for the expansion of the fleet. Tax credits and other cargo-sharing schemes have also been less than successful. The goal is to handle 40% of all foreign trade, reduce reliance on foreign lines, and reduce the added expense of transshipment of cargo in Singapore. This was especially aimed at bulk trade in tapioca to Europe (6 Mt/year) and the import of oil (10 Mt/year). However, there are presently few advantages and many disadvantages to putting a ship under Thai registry.[15]

Despite some ill-advised and unevenly applied policies in some ASEAN countries, all have experienced substantial growth in their merchant fleets.

PORTS AND PORT DEVELOPMENT

Singapore

Singapore owes its domination of maritime activity in Southeast Asia to three principal factors: favorable location on main trade routes to which there are limited alternatives, excellent port facilities and services essential to support the operation of merchant fleets, and government policies that encourage efficient and economical operation. These combine to make Singapore a logical place for the transshipment of cargos to and from smaller regional ports.

Requirements that Singapore, or any other port, must meet to be competitive in the transshipment of cargos include[16]

15. Michael Westlake, "Flag of Inconvenience," *Far Eastern Economic Review* (25 May 1989): 67.

16. Keri Phillips, "Jurong Port Invests S$530 Millions to Double Its Capacity," *International Bulk Journal* (September 1989): 9.

1. flexibility on the part of terminal operators, enabling them to respond quickly to changing conditions;
2. use of the port by a large number of liners, ensuring greater utilization of port facilities;
3. ability to handle multipurpose cargos, permitting redistribution of bulk cargos in smaller lots in bulk, bags, or containers;
4. availability of financial, communication, and transportation services;
5. availability of land for the conduct of port-related activities;
6. appointment of agents, regionally and worldwide, to obtain cargos for transshipment;
7. keeping up with the latest packaging technology;
8. keeping current knowledge of competing ports; and
9. maintaining a flexible tariff structure and transshipment rates cheaper than domestic cargo-handling rates.

More than 33,000 ships totalling over 343 million GT called in Singapore in 1987. The port handled more than 2.6 million 20-foot equivalent units (TEUs) of container traffic that year, and expansion of existing facilities is intended to maintain its regional dominance in that specialty. Plans call for the construction of two new container terminals in Jurong to double its capacity, and the addition of an extensive bulk cargo–handling terminal. The latter will include the construction of the first deep-water berths in Southeast Asia capable of handling bulk carriers of 200,000 dwt, and will provide water-depths of 16 m in the approach channels and alongside the berths. A coal transshipment facility is the first stage of the expansion and is scheduled for completion by late 1991 or early 1992.[17] The coal would be transshipped into barges and "handy-sized" bulkers for further distribution to other Southeast Asian ports. There has been no coal handled until now, since there is no significant local market for it.

Private company operators will be responsible for design, construction, and operation of the expanded facilities and will also arrange for the necessary financing. It is intended that gantry cranes will be designed and built that can serve the dual purposes of bulk and container handling. The new container terminal will also be configured to serve semicontainer ships that may find use of dedicated container terminals less suitable for their mix of container and conventional cargo.[18]

Indonesia

Jakarta, Surabaya, and Belawan are the major ports handling general cargo traffic in Indonesia. Development plans for these ports include a new con-

17. Ibid., p. 4.
18. Ibid., p. 8.

tainer terminal for Jakarta, which will provide additional berthing space and a container yard area of 60,000 m². Five additional berths are planned for Belawan, the port for Medan, and Surabaya intends to add two more container berths, install gantry cranes for handling containers, and construct 220,000 m² of warehousing facilities.[19] Traffic in these three ports is summarized as follows:

Jakarta	3,654 vessels	8 Mt of foreign and 3.1 Mt of domestic cargo. 265,437 TEUs handled in 1986.
Belawan	988 vessels	10.5 million dwt of shipping in 1987. Some container facilities are provided
Surabaya	10,062 vessels	9.7 Mt of cargo handled in 1987. 4 container berths are available.

A number of other ports handle substantial volumes of traffic but should more properly be regarded as terminals for the export of petroleum and LNG (table 3).

Philippines

Manila dominates the foreign trade of the Philippines, particularly in containers and general cargo. Major improvement projects include the expansion and updating of the International Container Terminal by adding 190,000 m² of container yard, 220 m of quay, two container freight terminals, repair and maintenance shops, and improved road access. Improvements are also under consideration for Batangas to enable it to serve as a backup for Manila and to ease congestion there.[20] A summary of traffic in the largest Philippine ports (1985 figures), includes[21]

Batangas	390 foreign ships	5.8 Mt of cargo handled, principally oil and LPG
Cagayan de Oro	389 foreign ships	6.4 Mt of cargo
Davao	77 foreign ships	1.9 Mt of cargo (3 container berths, no gantries)
Limay	109 ships	7.3 million dwt (oil and LNG)
Manila	2,293 ships	8.2 Mt of cargo (gantries installed for container handling)

19. *Lloyd's Ports of the World* (Colchester, Essex, United Kingdom: Lloyd's of London Press, 1989), pp. 290, 287, 297.
20. Ibid., pp. 381, 376.
21. Ibid., pp. 375–84.

TABLE 3.—INDONESIAN PORTS SERVING AS OIL AND LNG TERMINALS, 1987

Port	Cargo[a]	Volume (millions dwt)	Port	Cargo	Volume (millions dwt)
Arjuna	Oil	6.08	Dumai	Oil	21.25
Blang Lancang	Oil, LNG	19.3	Pulau Sambu	Oil	6.89
Balikpapan	Oil	6.56	Santan	Oil	7.22
Bontang	LNG	10.79	Semangka Bay	Oil	6.66
Cilacap	LNG	19.55	Senipah	Oil	9.78

SOURCE.—*Lloyd's Ports of the World* (Colchester, Essex, United Kingdom: Lloyd's of London Press, 1989), pp. 285–99.
[a]LNG = liquified natural gas.

Malaysia

Port Klang (Kelang) is the largest of Malaysia's ports, handling twice as much cargo (15.9 Mt) as Johore (7.3 Mt) and Penang (8.4 Mt) combined in 1988. Traffic through Klang Container Terminal (KCT) showed a continued expansion of 325,661 TEUs in 1988, an 18% increase over the previous year. A bulk coal facility was added in 1988, principally to supply fuel to a new power plant. It handled over 0.5 Mt in its first year of operation, but that was surpassed by bulk fertilizer imports of 867,000 tons. By 1990, more than 100,000 tons per year of grain, another bulk cargo, was required just for the manufacture of feed for Kentucky Fried Chicken (Malaysia). Bintulu, the second-ranking Malaysian port in terms of cargo tonnage handled (9.5 Mt), is essentially a terminal operation. The export of LNG accounts for 9.3 Mt of that total.[22]

Johore port facilities are being expanded to reduce the amount of Malaysian cargo that is now transshipped through Singapore. Commercial operations were started only in 1977, and the port is now embarking on a M$10 million expansion program to be completed in 1991. This was to include a new container wharf, a break-bulk cargo wharf, and an additional dangerous cargo jetty. In 1988, 31,502 TEUs of containers were handled, and plans were to increase that capacity to 110,000 annually.[23]

KCT was the first port terminal in Malaysia to be privatized, in 1986, and while it now handles primarily Malaysian imports and exports, there are plans to attract calls by container ships on otherwise through routes, to compete with Singapore as a transshipment port. The Malaysian government is examining the feasibility of additional privatization in Kelang, Johore, and Penang to encourage the continuing growth of traffic.[24]

22. Keri Phillips, "Malaysian Port Expansion Gains Freedom from Singapore," *International Bulk Journal* (September 1989): 15, 16.
23. Ibid., p. 10.
24. Ibid., p. 15.

Thailand

Congestion in the port of Bangkok has long been a problem: 1987 traffic figures included 3,681 ships of over 36 million dwt. Installation of three new container gantry cranes is planned for Bangkok's Klong Toey Terminal, and new container facilities are to be constructed at Bangsue at an estimated cost of $10 million to further ease congestion there. Bangsue plans to handle 150,000 TEUs per year initially, 250,000 TEUs within 5 years, and 400,000 within 10 years. Also, State Railways of Thailand and American President Lines have worked out an arrangement to transport containers between Bangkok and the port of Sattahip, 115 miles to the south.[25] New deep-sea and industrial ports at Laem Chabang and Mab Tapud, southeast of Bangkok, are under construction.[26] The Laem Chabang project is expected to take 15 to 20 years before completion, at which time cargo-handling capacity is expected to be more than 10 Mt per year.

INTERNATIONAL MARITIME CONVENTIONS

A number of international conventions have been developed to facilitate shipping and to enhance its safety. The most important of these apply to ships of any size, description, or employment, while others are of more limited application. Some require action on the part of masters of vessels; others are of concern primarily to agencies of various governments.

Southeast Asian countries have a mixed record in being parties to the principal conventions, as is apparent from an examination of table 4. Conventions dealing directly with safety and pollution issues are by far the most important of these from the mariner's point of view, including the regulations for

1. preventing collisions at sea (COLREGS 1972),
2. the safety of life at sea (SOLAS 1974/78), and
3. the prevention of pollution from ships (MARPOL 1973/78).

Seven of the nine coastal/archipelagic states of Southeast Asia are parties to the first two of these, exceptions being Vietnam and Cambodia. Only Indonesia, Brunei, and Myanmar were signatories to MARPOL 1973/78 by the end of 1988, however. The Load Line Convention of 1966, which prescribes the maximum draft to which a ship may be loaded, has been formally accepted by all but Thailand and Cambodia. Only Singapore, Malaysia, and the

25. Jacob Indran and Christian Mayer, "Container Plan to Ease Thai Congestion," *Lloyd's List* (16 May 1989): 3.
26. *Lloyd's Ports of the World* (n. 19 above), p. 403.

TABLE 4.—PARTIES TO INTERNATIONAL MARITIME CONVENTIONS

	Brunei	Indonesia	Malaysia	Myanmar	Philippines	Singapore	Thailand
Convention on the Intergovernmental Maritime Consultative Organization, 1948[a]	X	X	X	X	X	X	X
International Convention on Load Lines, 1966[b]	X	X	X	X	X	X	
Convention on the International Hydrographic Organization, 1967		X	X		X	X	X
International Convention on Tonnage Measurement of Ships, 1969	X		X	X	X	X	
Convention on the International Regulations for Preventing Collisions at Sea, 1972 (COLREGS)	X	X	X	X	X	X	X
Convention on the Prevention of Marine Pollution by Dumping of Wastes and Other Matter, 1972					X		
International Convention for the Prevention of Pollution from Ships, as amended by the Protocol of 1978 (MARPOL 1973/78)	X[c]	X[c]		X[c]			
International Convention for the Safety of Life at Sea, as amended by the Protocol of 1978 (SOLAS 1974/78)	X	X	X	X	X[d]	X	X[d]
Convention on the International Maritime Satellite Organization, 1976 (INMARSAT)		X	X		X	X	
International Convention on Standards of Training, Certification and Watchkeeping for Seafarers, 1978					X		

Source.—International Maritime Organization, *Status of Multilateral Organizations and Instruments as of December 31, 1988.* (London: International Maritime Organization, 1989).
[a] Cambodia and Vietnam are also parties.
[b] Vietnam is also a party.
[c] Not a party to optional Annexes III, IV, and V.
[d] Not a party to the Protocol of 1978.

Philippines, among the five largest fleet registries, have signed the Tonnage Measurement Convention of 1969.

Although all states of the region have not signed all of the conventions, there is de facto recognition that some, such as COLREGS 1972, must be observed by vessels of whatever flag, especially those engaged in international trade. Observance of others is essential if vessels are to be permitted to enter the ports of many trading partners. One convention lacking signatures but probably observed is the International Convention for Safe Containers (1972). This is intended to enhance the safety of human life in the handling of containers and to facilitate their international movement and intermodal use.

Lack of signatures is notable on conventions for pollution prevention and control. Only the Philippines is a party to the Convention on the Prevention of Marine Pollution by the Dumping of Wastes and Other Matter (1972). Of the countries with the five largest merchant fleets, only Indonesia has signed the basic MARPOL 1973, but it has not signed the Protocol of 1978. Unsigned by any ASEAN country are the International Convention Relating to Intervention on the High Seas in Cases of Oil Pollution Casualities (1969) and the International Convention on Maritime Search and Rescue (1979).

In view of their locations bordering the heavily traveled and congested Malacca and Singapore Straits, it is strange that Singapore, Malaysia, and Indonesia have not signed the 1969 Intervention Convention. The convention affirms the right of a coastal nation to take all necessary measures on the High Seas to prevent or mitigate danger to its coastline, or related interests, from actual or threatened pollution by oil following a maritime causalty. The convention predates the LOS Convention, and its applicability to straits used for international navigation, and to the Territorial Seas in those straits, may be in question. The LOS Convention, however, permits coastal states to adopt laws and regulations for the prevention, reduction, and control of marine pollution, which may serve the same purpose.

As to the training of officers and crews for their ships, only the Philippines has become a party to the International Convention on Standards of Training, Certification, and Watchkeeping for Seafarers, 1978 (STCW). Singapore and Malaysia have the facilities and training courses that meet the requirements of STCW and could be in compliance with the intent of the convention upon signing. Singapore has been training some 400 seamen annually not only for employment aboard Singapore-flag ships, but also as a labor pool from which foreign ships can fill their crew needs.[27]

Indonesia and the Philippines, notwithstanding the latter being a signatory, have some deficiencies in their programs but could attain full compliance with a dedicated effort. Thailand, on the other hand, has been deficient

27. *Singapore Registry of Ships* (n. 9 above), p. 28.

in almost all respects as far as STCW standards are concerned, and is unlikely to have made significant improvements in recent years.[28]

SUMMARY

Shipping serves as a vital link in the economies of all nations in Southeast Asia, with respect to both their domestic and international trade. Changes in the mechanics of marine transportation; the development of new, specialized ships; and the economics of their operation have resulted in changes to the patterns followed by international traffic. Ports that in the past were on the main routes are now being relegated to roles as secondary ports, whose international cargo is being more and more often routed to Singapore and Malaysian ports for transshipment.

This routing for transshipment is especially significant with respect to the "container revolution." Ships are expensive, and any reduction in turnaround time, as has occurred with the shift from break-bulk to container loading, will help to minimize transportation costs. Unfortunately, rapid handling of containers, particularly in large numbers, requires the investment of millions of dollars in gantry cranes, straddle carriers, large forklifts, and other equipment, as well as extensive marshalling areas for the containers themselves. This will result in an ever-increasing concentration of such facilities in only a few major ports.

It does not appear from a purely objective viewpoint that all countries need a merchant fleet for international trade. This will undoubtedly be considered a heretical statement, contrary to the popular view that all countries should participate in the carriage of their own foreign trade, for political if for no other reasons. Instead, emphasis should be on the provision of vessels for domestic trade, with possible exceptions for some specialized bulk trades.

Merchant fleets will become more important than ever as economic development continues in Southeast Asia. How appropriate they are in numbers, sizes, and types of ships and in their method of operation will determine how effectively they will serve national purposes and how closely they will come to meeting national goals.

28. *ESCAP Report of the Joint ESCAP/IMO Regional Meeting of Experts in Maritime Training and Certification, 26 April–6 May 1980* (Bangkok: Economic and Social Commission for Asia and the Pacific, 1980), p. 104.

Transportation and Communication

Seaborne Oil Trade in the 1990s: Rising Volumes, Rising Concerns

Nancy D. Yamaguchi
East-West Center

INTRODUCTION: WHAT IF THE *EXXON VALDEZ* HAD BEEN CARRYING WHEAT?

The recent spate of oil tanker accidents and spills has made the topic of seaborne oil transport quite fashionable in both the popular press and in the trade press. On 24 March 1989, the grounding of the *Exxon Valdez* in Alaska's Prince William Sound caused the largest spill ever in North American waters, 240,000 barrels of Alaska North Slope (ANS) crude oil.[1] In the public's eye (and, perhaps more interesting, in the eyes of many other oil companies), Exxon's response to the spill lacked good judgment, essential promptness, and the expected corporate penitence. Cleanup activities have continued for over a year now, but, despite the fact that Exxon has expended approximately $2 billion to clean Alaskan coastal areas, many remain dissatisfied with the job, and Exxon's public image has suffered perhaps irreversible damage. The *Exxon Valdez* itself has been repaired, but it is being renamed and transferred to service in another area.

More recently, in the Gulf of Mexico on 8 June 1990, a series of explosions aboard the Norwegian supertanker *Mega Borg* caused the ship to catch fire, and it was feared that a major spill might result that would threaten the coast of Texas. In the end, very little of the ship's 1-million-barrel cargo of Angolan crude oil was spilled, and most of this either burned or evaporated.

---

1. Essentially all major newspapers and trade journals covered the *Exxon Valdez* spill. For a thorough reporting of the spill, including a chronology of events, the following series of articles from the *Oil and Gas Journal* are helpful: "Huge Cargo of North Slope Oil Spilled," *Oil and Gas Journal* (April 3, 1989), pp. 26–27; "Political, Economic Fallout Spreads from Exxon Valdez Crude Oil Spill," *Oil and Gas Journal* (April 10, 1989), pp. 13–16; and "Alaskan Cleanup Campaign Pressed," *Oil and Gas Journal* (April 17, 1989), pp. 20–22 (includes Coast Guard log chronicle of the spill and response). Also see "Smothering the Waters," *Newsweek* (April 10, 1989), pp. 54–57 (includes map); and "Alaska after Exxon," *Newsweek* (September 18, 1989), pp. 50–62 (includes map and details of the cleanup efforts during the 6-month period following the spill).

By 14 June, the Coast Guard estimated that less then 300 barrels of oil remained on the water. However, unlike the *Exxon Valdez* spill, the *Mega Borg* explosion and fire killed two crewmen, left two others missing and presumed dead, and injured 17 others.[2]

Growing concern over environmental quality and public safety, both on and off shore, have focused attention on oil tanker operations and have galvanized public opinion against tanker traffic. Yet the oil transport industry remains poorly understood for a variety of reasons. The oil shipping business is an old and complex one. For many, shipping is not merely a job but a way of life. Shipping incorporates elements of engineering, economics, business, climatology, geography, oceanography, electronics, chemistry, physics—even disciplines as arcane as international and marine law. Small wonder that the casual observer can barely understand the confusing patter of acronyms and trade jargon heard when shippers speak with one another. There are a number of excellent trade journals on shipping, such as *Lloyd's Shipping Economist, Fairplay International Shipping Weekly,* and *Drewry's Shipping Statistics and Economics,* yet these journals typically assume a level of familiarity with the business of shipping that most people do not possess.

This article is written from the perspective of a geographer and energy analyst who was forced to learn about oil transport because of its key role in linking a raw resource, crude oil, with processing centers and centers of demand. No analysis of world trade is complete without some discussion of oil. Despite the fact that oil trade volumes declined by 25% between 1979 and 1989 (the reasons for which are discussed in a later section), the tonnage of oil trade remains larger than the tonnage of iron ore, coal, and grain trade combined.[3]

Of course, petroleum poses a different set of hazards than other bulk commodities. What if the *Exxon Valdez* had been carrying wheat instead of oil? Higher prices for bread might have been annoying, but it scarcely would have catapulted the ship and the ship's captain into the ranks of the infamous. According to *Golob's Oil Pollution Bulletin,*[4] there were five major tanker incidents in 1989 involving spills of 10,000 gallons or more, slightly below the average of 6.5 spills per year witnessed for the 1978–1989 period. However, since the *Exxon Valdez* lost such a huge amount of oil, 1989 ended up being the second-worst year in terms of total volume of oil spilled. The worst year had been 1979, when there were 16 major spills, including the *Burmah Agate* spill of over 10 million gallons. The *Exxon Valdez* spill has sensitized the public

2. "Tanker Spills, Safety Again Chief Industry Environmental Worries," *Oil and Gas Journal* (June 18, 1990), pp. 13–16; "More Oil on the Waters," *Newsweek* (June 25, 1990), pp. 60–61.

3. Shell Briefing Service, *International Oil Movements* (London: Shell International Petroleum, 1989), p. 1.

4. Cited in "Study Shows Fewer U.S. Tanker Spills in 1989," *Oil and Gas Journal* (April 23, 1990), pp. 36–37.

once again to the potential hazards of oil transport; ever since, even minor spills receive intense public scrutiny. In essence, a passing familiarity with the oil transport industry has become a prerequisite for anyone concerned with marine pollution and coastal zone preservation. To provide a comprehensive overview, this article discusses both the supply side and the demand side of the oil transport industry.[5] The first section provides a historical overview of supply by examining the structure and development of the world oil fleet; the second section assesses the elements of demand for seaborne oil transport and develops a forecast of demand for tanker ton-mileage through the year 2000 based on country-by-country forecasts of oil production capability, refinery capacity, and oil demand.

THE EVOLUTION OF THE WORLD OIL FLEET

World oil tanker and combined carrier fleet tonnage, 1972–90, is displayed in figure 1.[6] Combined carriers are ships capable of handling more than one type of cargo, chiefly ore and oil (O/Os), bulk cargoes and oil (B/Os), or ore, bulk cargoes, and oil (O/B/Os). A typical trade pattern might involve bringing iron ore from South America to Japan, then returning via the Persian Gulf to load a cargo of oil for delivery back to Brazil. This reduces the amount of time spent without a cargo, or "in ballast," and therefore helps cut shipping costs. As shown in figure 1, fleet tonnage expanded rapidly during the 1970s, rising from 171.1 million deadweight tons (DWT)[7] in 1972 to a peak of 331.9 million DWT in 1978. Thereafter, fleet tonnage contracted steadily for a period of 10 years, before leveling off in 1988 and beginning to rise again in 1989 and 1990.

To understand the forces behind this cycle, it is necessary to think back to the late 1960s and early 1970s, when the conventional wisdom held that the demand for oil would continue to grow unabated, and that the Persian Gulf would be the principal source of supply. A huge fleet was considered essential for the transport of oil on long-haul routes from the Persian Gulf to the United States, western Europe, and Japan. The number of ships in

---

5. Readers interested in detailed background on oil shipping may wish to investigate Lane C. Kendall, *The Business of Shipping* (Centreville, Md.: Cornell Maritime, 1973); and Alex Marks, *Elements of Oil-Tanker Transportation* (Tulsa: Pennwell, 1982).

6. The charts in this section rely on historical data from Fearnley's *World Bulk Fleet* and *Review* (Oslo: Fearnley, 1982–88 editions for 1972–88 data). Where possible, the data are updated to 1989–90 from Drewry's *Shipping Statistics and Economics* (London: Drewry Shipping Consultants, 1989, various issues).

7. A ship's deadweight tonnage is the combined weight of ship, cargo, and stores such as fuel and water. In an oil tanker, the cargo itself is by far the largest component of the deadweight tonnage. A good rule of thumb is to assume that 95% of a supertanker's deadweight tonnage is available for cargo.

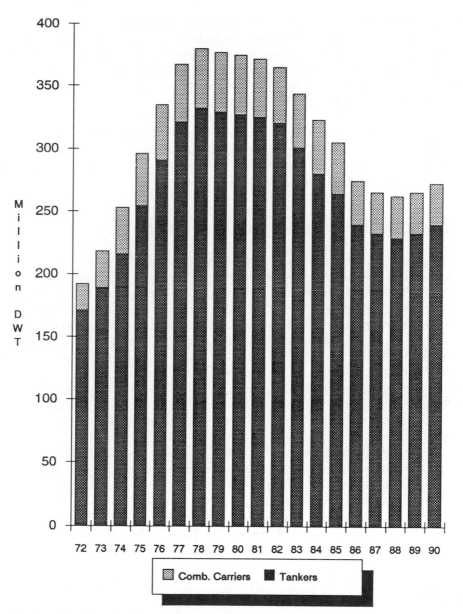

Fɪɢ. 1.—World tanker fleet, 1972–90. Sources.—Fearnley's *World Bulk Fleet* (1985–89) and Drewry's *Shipping Statistics and Economics* (March 1990). Figures are as of beginning of year.

the fleet grew, and average ship size also increased, in part because per-barrel transport costs were significantly lower on larger ships. For example, a cargo of fuel oil shipped from the Persian Gulf to Singapore aboard a 30,000-DWT tanker might have a transport cost of $2.34 per barrel, whereas the transport cost aboard a 120,000-DWT ship might be only $0.91.[8] These economies of scale prompted a flurry of orders for larger tankers, or "supertankers," most of which were fuel-inefficient turbine tankers designed for long voyages at fairly high speeds. Another key factor in the decision to build larger ships was political instability in the Middle East. The Suez Canal closures of 1957 and 1967 forced Europe-bound cargoes of Middle Eastern crude to travel all the way around the Cape of Good Hope, and memories of these closures were still fairly fresh when the Arab oil embargo was announced in 1973. According to David T. Isaak, "The conflict in the Middle East in 1967 provided the catalyst for a full-scale move to supertankers. Given the prospect of growing instability in the region, in the long term, abandoning the Suez Canal in favor of the Cape route seemed to be the most likely scenario. When these political and logistical considerations were coupled with the demonstrated economics of large ships then afloat, the move to supertankers appeared to be the only logical response."[9]

As tanker sizes increased, supertankers of 160,000–319,999 DWT came to be referred to as very large crude carriers (VLCCs), and vessels of 320,000 DWT and above were called ultra-large crude carriers (ULCCs). The massive overordering of tankers in the early 1970s is clearly shown in figure 2, which shows both the number of tankers on order and the total tonnage on order, 1972–90. The reader may note that interest in tanker construction appears to be picking up once again; as of May 1990, over 38 million DWT were on order, including 66 supertankers greater than 225,000 DWT in size.[10]

The trend in deliveries of newly built tankers, or "newbuildings," 1972–89, is presented in figure 3. During the 1970s, tankers totaling an incredible 232 million DWT were added to the fleet—a figure nearly equal to the entire fleet as it exists today. According to Martin Stopford of Chase Manhattan Bank, "As the 'bull' market became more firmly established in the early 1970's the investment machine got out of control and the ordering of new ships escalated to a level that bore no relation to any conceivable future level of ship demand. . . . When the bubble finally burst in 1973, the total

8. Fereidun Fesharaki and Nancy Yamaguchi, *The New Saudi Export Refineries: The Role of Shipping Costs in Determination of Product Prices and Their Competitiveness* (Honolulu: East-West Center, Resource Systems Institute, 1984), p. 15.

9. David T. Isaak, "The Future of the Supertanker," in *Shipping, Energy and Environment: Southeast Asian Perspectives for the Eighties,* ed. Mark J. Valencia, Edgar Gold, Chia Lin Sien, and Norman G. Letalik (Halifax, Nova Scotia: Dalhousie Ocean Studies Programme, 1982), p. 40.

10. Drewry Shipping Consultants, *Shipping Statistics and Economics* 236 (June 1990), p. 17.

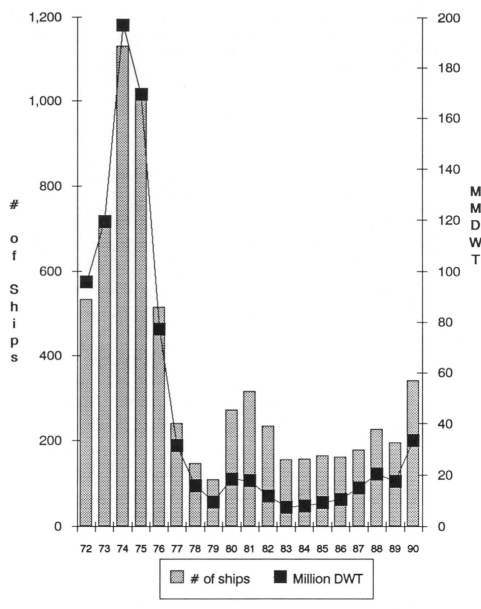

FIG. 2.—World tanker order book, 1972–90. Sources.—Fearnley's *Review* (1983–88) and Drewry's *Shipping Statistics and Economics* (June 1990).

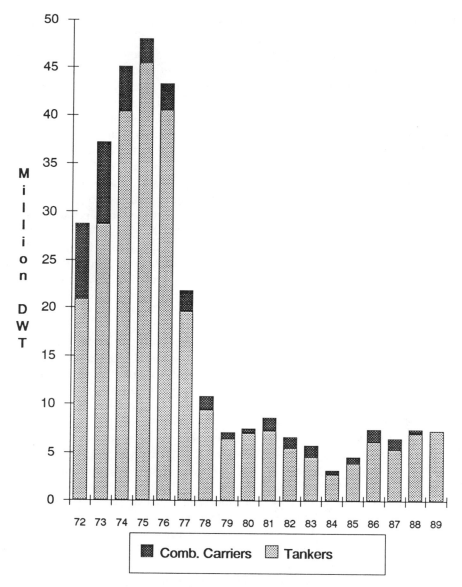

FIG. 3.—Deliveries of newbuildings, 1972–89. Sources.—Fearnley's *Review* and Drewry's *Shipping Statistics and Economics* (June 1990).

orderbook accounted for 95% of the fleet, clear evidence that the industry had completely lost touch with reality!"[11]

The bubble had burst by 1973, but the time lag between ship order and actual ship delivery resulted in a great many new ships entering the fleet during the 1974–77 period, as figure 3 illustrates. In 1975 alone, tankers totaling over 45 million DWT were delivered into the already-overbuilt fleet. In contrast, during the 1980s, deliveries averaged less than 6 million DWT per annum. Shipowners who could not find even marginal employment for their ships were often forced to place them in designated lay-up at sites such as Brunei Bay, Malagag Bay in the Philippines, the fjords of Norway, Hampton Roads in the United States, and other lay-up sites scattered around the world. Laid-up tankers are kept in a state of semireadiness so that they may reenter the fleet if freight markets improve. Figure 4 depicts tonnage in lay-up, 1972–89. As mentioned above, tankers totaling 45 million DWT were delivered in 1975. In that same year, tankers totaling over 40 million DWT were placed in lay-up. Tonnage laid up remained at high levels until freight markets improved during the 1979–80 oil price shock, which was precipitated by the Iranian Revolution. By 1981, freight markets had collapsed once again, and in 1982 tanker tonnage in lay-up reached a record-high 56.4 million DWT.

The gradual drop in laid-up tonnage seen during the 1983–86 period was not caused by any particular increase in demand for tankers. Market conditions remained weak, and soon many of the surplus ships began to be taken out of lay-up and sold to ship-breakers for the value of the salvageable scrap steel. Figure 5 portrays the trend of tonnage broken up or lost, 1971–89. Ship-scrapping activity picked up noticeably in 1975, yet the oversupply lingered for more than a decade. The decision to scrap a ship is a painful one, and many owners postponed the decision in hopes that supply and demand would soon balance. Eventually, the cost of maintaining uneconomic tonnage outweighed the potential benefits, and more ships were sold to the breakers. During the 5-year period from 1981 to 1985, approximately 105 million DWT were broken up and lost. Scrapping and lay-ups are currently at low levels, and the renewed interest in newbuilding appears to indicate that the tanker supply/demand situation is in far better balance than it has been for over 15 years.

The current fleet's size-age distribution is detailed in table 1, and the fleet's breakdown by size is charted in figure 6. The aging of the fleet is visible; 361 tankers in the fleet are over 20 years old, having been built prior to 1970. Most of these ships are in small and medium size ranges; while they

11. Martin Stopford, "Forecasts for the International Shipbuilding Market—Demand, Pricing, and Capacity," paper presented at the Shipbuilder's Council of America seminar "The Shipbuilding Marketplace in the 1990's," Washington, D.C., January 31, 1990, p. 1.

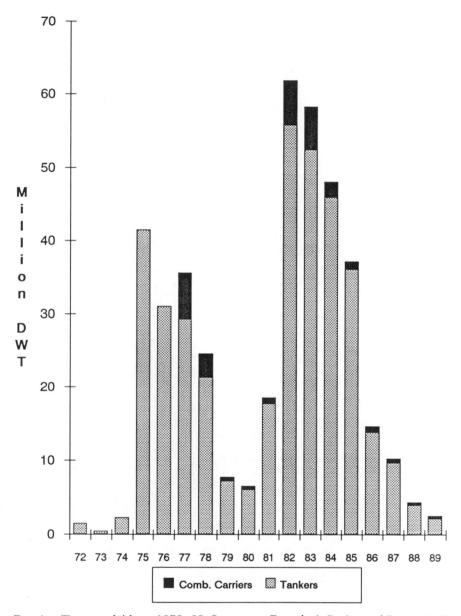

FIG. 4.—Tonnage laid up, 1972–89. Sources.—Fearnley's *Review* and Drewry's *Shipping Statistics and Economics* (April 1990). Figures are as of year end.

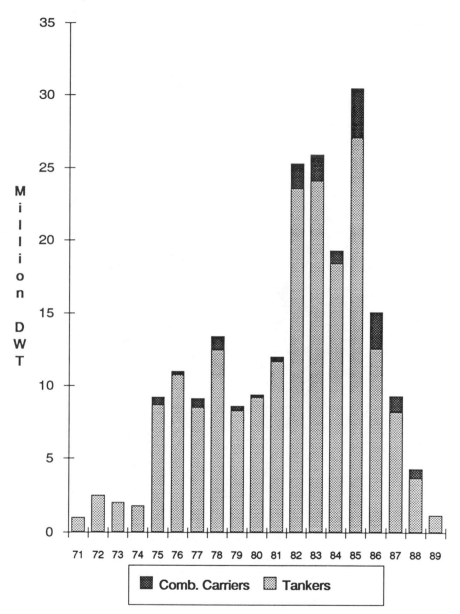

FIG. 5.—Tonnage broken up or lost, 1971–89. Sources.—Fearnley's *Review* and Drewry's *Shipping Statistics and Economics* (May 1990).

TABLE 1.—WORLD TANKER FLEET BY AGE AND SIZE, END OF DECEMBER 1989

Ship Size (MDWT)	Year of Build													
	Pre-1965		1965–69		1970–74		1975–79		1980–84		1985–89		Total	
	Number	MMDWT	Number	MMDWT	Number	MMDWT	Number	MMDWT	Number	MMDWT	Number	MMDWT	Number	MMDWT
10–25	61	1.1	93	1.8	96	1.8	87	1.5	103	1.9	49	0.8	489	8.9
25–45	55	1.9	30	1.0	145	4.5	199	6.7	169	5.8	127	4.6	725	24.5
45–65	20	1.0	22	1.2	11	0.6	40	2.2	116	6.6	42	2.3	251	13.9
65–90	3	0.2	43	3.3	45	3.6	85	7.1	101	7.9	74	6.1	351	28.2
90–125	1	0.1	27	2.9	43	4.7	68	7.1	25	2.6	30	3.1	194	20.5
125–175	—	—	3	0.3	46	6.3	113	16.1	11	1.6	17	2.3	190	26.6
175–300	—	—	3	0.6	143	36.5	147	37.3	11	2.7	35	8.6	339	85.7
300 +	—	—	—	—	6	2.0	71	26.8	5	1.8	2	0.6	84	31.2
Total	140	4.3	221	11.1	535	60.0	810	104.8	541	30.9	376	28.4	2,623	239.5
% of fleet	5.3	1.8	8.4	4.6	20.4	25.1	30.9	43.8	20.6	12.9	14.3	11.9	100	100

SOURCE.—Drewry's *Shipping Statistics and Economics*, March 1990, p. 18.
NOTE.—MDWT = Thousand DWT; MMDWT = million DWT.

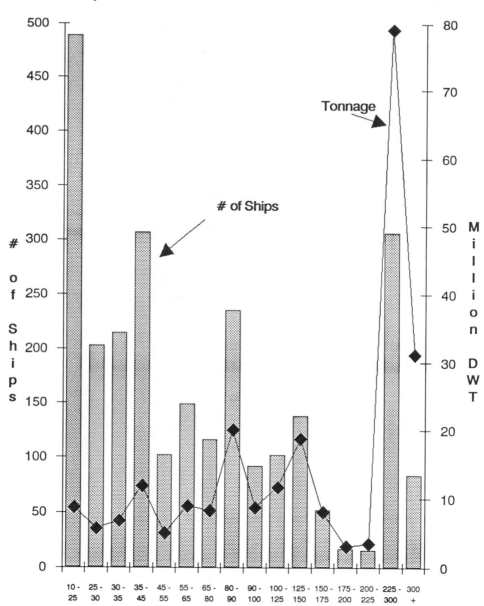

FIG. 6.—Tanker fleet by size, December 1989. Source.—Drewry's *Shipping Statistics and Economics* (January 1990), p. 16.

represent 13.7% of the total number of ships, they account for only 6.4% of the fleet tonnage. For the most part, the supertankers did not begin to enter service until the 1970s. Of the VLCC/ULCC ships 175,000 DWT in size and larger, 367 of the total 423 vessels currently in the fleet were built during the 1970s. Overall, three-quarters of the total fleet tonnage was built prior to 1980. For some of the ships, chronological age is not the best indicator of seaworthiness. For example, a tanker which has been placed in lay-up, has been used as a floating storage tank, or was otherwise inactive for a period of years may be in significantly better condition than a newer ship that has been trading actively with only infrequent maintenance and repairs. On the other hand, while such evaluations are wholly qualitative, some marine engineers insist that the ships built during the spree of the 1970s—which account for the largest share of lay-ups—were less well built than the ships of the 1960s, and they logically should be expected to have a shorter useful life. An analogous question might be, Is a 1960 Mercedes-Benz still a better car than a 1975 Fiat? A lot depends on the individual vehicle and the patterns of use and maintenance, so generalizations based on age are not always appropriate.

Examination of the size breakdown in figure 6 reveals some interesting patterns. There are several peaks in the line tracing tonnage by size range: first, the 10,000–25,000-DWT size range, which includes the general purpose (GP) class; second, the 30,000–45,000-DWT range, which includes medium-range (MR) class ships; third, the 70,000–90,000-DWT size range, which includes "Panamax" ships designed for traversing the Panama Canal; fourth, the 125,000–150,000-DWT size range, which includes the "Suezmax" tankers designed for transiting the Suez Canal (Suezmax ships are also known as the "million-barrel" ships, owing to their approximate cargo size); and fifth, the 225,000–300,000 VLCC class, which emerged as the main mode of crude oil shipping for the large markets of America, Europe, and Japan. Refined petroleum products, such as motor gasoline, naphtha, kerosene, jet fuel, diesel fuel, and fuel oil, are usually carried on smaller ships, since markets and product-receiving terminals are smaller. Larger tankers are devoted to trade in crude oil and, sometimes, in fuel oil. Crude oil and fuel oil are considered "dirty" products, whereas the other refined products listed above are classified as "clean" products. Clean-product tankers are typically more sophisticated than crude carriers in terms of internal cargo-tank configuration and pumping systems. Ships of the GP and MR size are fairly flexible and are able to call at most major ports around the world. Clean tankers of the GP size range are often referred to as "handy clean" tankers. The VLCC/ULCC class is far less flexible and far less maneuverable. The largest ULCCs are dedicated to only a handful of trade routes, since ships of this size can be accommodated at very few receiving terminals, and, of course, only the largest markets purchase crude cargoes of the size carried on ULCCs, which can be well over 2 million barrels in volume. Terminals capable of receiving

ULCCs are located in the Persian Gulf, Japan, Freeport (Bahamas), Rotterdam, and the Louisiana Offshore Oil Port (LOOP).

A detailed breakdown of tankers currently on order by ship size is provided in figure 7. Not surprisingly, interest seems to focus on the same size ranges already preferred in the existing fleet: the 25,000–45,000-DWT product carriers, the 80,000–175,000-DWT crude carriers, and the VLCCs of 225,000 DWT and above. Strong interest is seen in the 25,000–40,000-DWT size range, indicating that shippers and marketers expect to trade relatively large volumes of refined products during the 1990s.

ELEMENTS OF OIL TANKER DEMAND

Following the oil crises of 1973–74 and 1979–80, it became almost standard to distrust "big oil." Now, a blanket distrust is becoming common for the oil transport industry as well. Each accident is followed by the assignment of fault to oil companies, ship operators or captains, negligent port authorities, the Coast Guard, or any variety of government agencies charged with environmental protection. Yet it is not entirely fair to place blame on these entities, nor is it reasonable to denounce seaborne oil trade as an unmitigated evil. The oil transport industry exists because the oil industry exists, and the oil industry exists because consumers have a huge appetite for inexpensive energy and the goods and services that depend on energy as an input.

In the aftermath of the oil crises of the 1970s, higher oil prices brought new, non-OPEC oil exporters into the market, which diversified sources of supply for importers. Many oil-importing countries also reduced dependence on OPEC oil by developing domestic oil and natural gas resources.[12] Moreover, increased emphasis was given to energy conservation, efficiency, and alternative fuels. Many alternative fuels projects were begun under the premise that oil prices would soon reach $50 or even $100 per barrel, and that, at these prices, alternatives—such as solar power, wind power, shale oil, synthetic fuels, or ocean thermal energy conversion—would become cost-competitive. But oil prices hit their peak in 1980 at around $42 per barrel and subsequently began to subside. Many plans for alternatives and renewables were subsequently shelved, and many marginally economic oil and gas fields were also put on hold.

By 1986, oil prices crashed, with spot prices dipping in some cases below $10 per barrel—which, when corrected for inflation, was a price comparable in real terms to the prices paid in the mid-1970s. In all, oil was extremely cheap, particularly when its versatility as compared to other energy resources

12. In light of the fact that U.S. domestic oil production was expanded to reduce the need for "dangerous" imports, it is ironic that the *Exxon Valdez* spill involved a domestic crude.

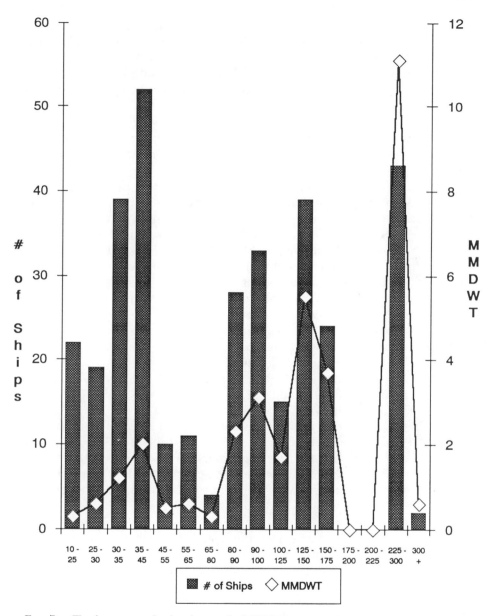

Fig. 7.—Tankers on order by size, end of 1989. Source.—Drewry's *Shipping Statistics and Economics* (January 1990), p. 17.

is considered. Demand for oil rebounded strongly, and rapid demand growth is now being witnessed in the developing and newly industrialized countries. Not only has oil not been supplanted by more environmentally benign energy sources, but also the growth in demand for it has continued. Moreover, an increasing percentage of U.S. oil demand is being met by imports. The main importing regions are western Europe, the United States, and Japan. Japan has virtually no domestic oil resource, and although western Europe and the United States are significant producers of oil, levels of consumption consistently outstrip production, and prospects for new oil discoveries in the developed world are very limited.

Figure 8 displays percentage shares of the world's proven oil reserves compared with percentage shares of refining capacity and oil consumption.[13] Immediately apparent is the fact that the Middle East is the site of over two-thirds of the world's proven oil reserves. The United States, western Europe, and the Asia-Pacific region have wide disparities between their percentage shares of oil reserves and their refining capacities and demand. The United States, for example, has less than 3% of the reserves, yet has 22% of the refining capacity, and 26% of the demand. Currently, the two largest consumers of oil—the United States and the USSR—are also the two largest producers. However, the reserves situation indicates that this status cannot continue forever. The United States is the most thoroughly prospected place on earth, and very few new discoveries can be expected. Many of the remaining oil-prospective basins are in environmentally sensitive areas, such as the California Outer Continental Shelf (OCS) area and the Alaskan Arctic National Wildlife Refuge (ANWR).

Declining production in the U.S., coupled with rising demand, has resulted in a steady increase in oil imports. Figure 9 displays U.S. crude imports by source, 1971–89. During the three years from 1977 to 1979, imports averaged approximately 6.5 million barrels per day (MMB/D), with the Middle East and Africa serving as the dominant sources of supply. The 1979–80 oil price shock caused imports to drop sharply, particularly from the Middle East and Africa, where many of the governments were considered unstable and possibly hostile. Higher prices and fears of dependency on OPEC oil then shaped imports in three different ways. First, absolute demand for oil declined, partly as a result of the economic slowdown caused by higher prices, and partly because of energy conservation and fuel switching. Second, renewed emphasis was given to domestic exploration and production. A large part of the impetus behind the Trans-Alaska Pipeline System (TAPS) was the

13. Data on proven reserves and crude distillation capacity are from the "Worldwide Report," *Oil and Gas Journal* (Dec. 25, 1989); 1988 consumption figures are from British Petroleum, *BP Statistical Review of World Energy* (London: British Petroleum, 1989).

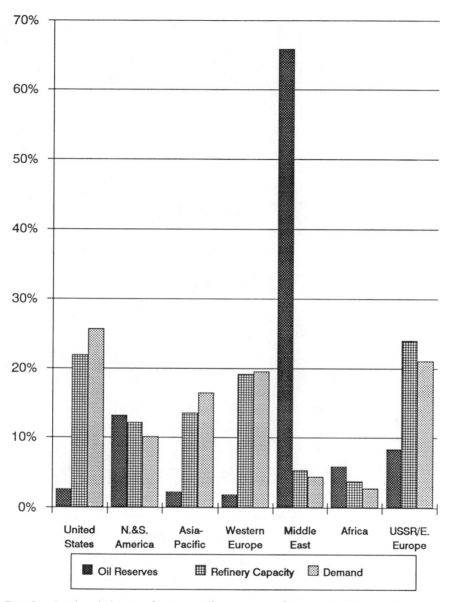

Fig. 8.—Regional shares of proven oil reserves, refinery capacity, and oil demand. Sources.—"Worldwide Report," *Oil and Gas Journal* (December 25, 1989), for reserves and crude capacity; British Petroleum, *BP Statistical Review of World Energy* (London: British Petroleum, 1989), for 1988 demand figures.

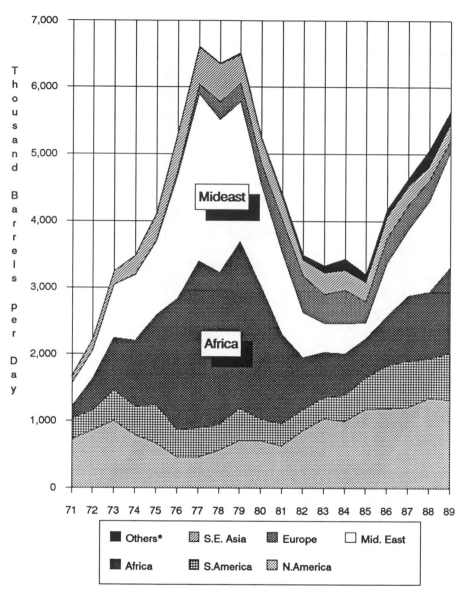

Fig. 9.—U.S. crude oil imports by source, 1971–89. *"Others" include China, Australia, and the Caribbean. Sources.—U.S. Department of Energy, *Petroleum Supply Annual* and *Petroleum Supply Monthly*.

oil price shock of 1973–74, and the TAPS line came on stream in 1977.[14] Production of ANS crude grew steadily, from 17.3 thousand barrels per day (MB/D) in 1961—the first year of production—to 463.6 MB/D in 1977, and then to a peak in 1988 of over 2 MMB/D.[15] Production grew also in other major oil-producing states such as California, Texas, and Louisiana, which reduced the need for imports. Third, a number of non-OPEC oil exporters emerged, since the higher prices had boosted exploration and development worldwide. For example, imports from Canada and Mexico grew from 456 MB/D in 1977 to 1,328 MB/D by 1989.[16]

The net effect was to reduce the amount of oil in seaborne trade. As displayed in figure 10, crude oil shipments declined steadily from 1979's peak of 1,497 Mmt to 1985's low of 871 Mmt—a drop of over 40%. In 1979, petroleum and petroleum products accounted for nearly one-half of all seaborne trade; by 1988, this had fallen to just over 37%—though absolute quantities traded have been rising since 1985. Refined products trade has in fact grown quite steadily since the mid-1970s. To a certain degree, some amount of product trade is essential to balance supply and demand; since petroleum products are jointly produced at the refinery, regions may have surpluses of, for example, a product such as fuel oil and a deficit of a product such as diesel fuel. A given country's oil-refining industry may be unable to match exactly the national pattern of demand, and it then becomes economical to trade products on the open market. In addition, many countries that traditionally have been exporters of crude oil have been building "export-oriented" refineries, so that they can capture the value added from processing their own resource. Product trade is therefore expected to continue to grow.

The changing trade pattern in the aftermath of the oil price shocks is even more visible in figure 11, which tracks the ton-mileage of crude, products, and other commodities during the 1971–86 period. A ton-mile is one mt of cargo shipped one mile, and it is therefore a better indicator of demand for oil tankers than mere volume of oil trade. The drop in crude oil shipments is even more dramatic when judged in terms of ton-mileage—less oil was being shipped overall, and, in particular, less oil was being shipped on long-haul routes from the Persian Gulf. In the early 1970s, ton-mileage of oil in seaborne trade represented nearly two-thirds of total ton-mileage of all

14. George Geistauts and Vern Hauck, *The Trans-Alaska Pipeline* (Honolulu: East-West Center, Resource Systems Institute, 1979), p. 6.

15. Historical crude production figures are from DeGolyer & MacNaughton's *Twentieth Century Petroleum Statistics*, 45th ed. (Dallas: DeGolyer & MacNaughton, 1989); current data are from the U.S. Department of Energy, Energy Information Administration, *Petroleum Supply Annual* (Washington, D.C.: U.S. Department of Energy, Energy Information Administration, 1989).

16. U.S. Department of Energy, Energy Information Administration, *Petroleum Supply Annual* and *Petroleum Supply Monthly* (Washington, D.C.: U.S. Department of Energy, Energy Information Administration, various issues).

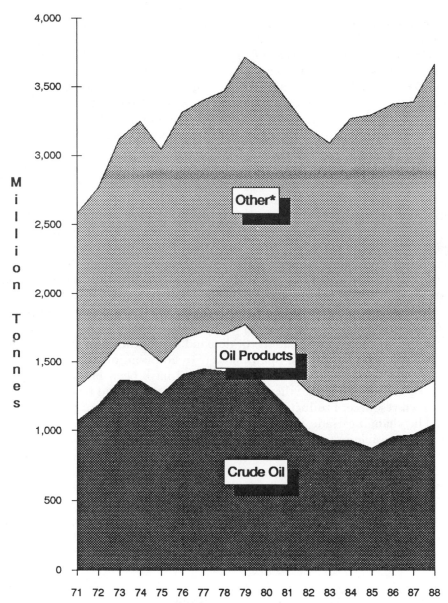

FIG. 10—World seaborne trade, 1971–88. *"Other" includes all nonoil seaborne trade. Source.—Fearnley's *Review* (1983–88 annual issues).

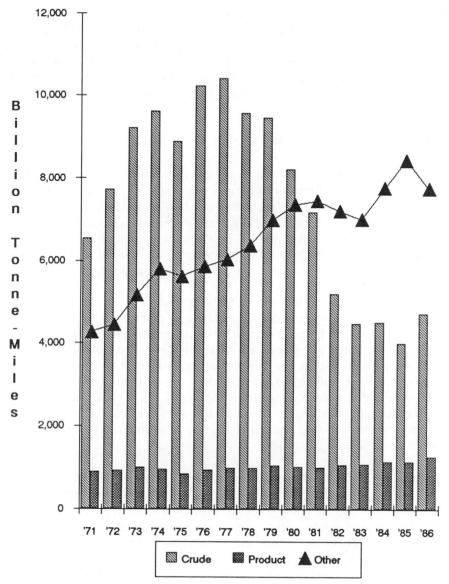

Fig. 11.—World seaborne trade, 1971–86, in ton-miles. Source.—Fearnley's *Review* (various annual issues).

commodities combined. By 1985, this figure had fallen to less than 38%. Crude oil ton-mileage peaked at 10,408 billion ton-miles in 1977, falling by 1985 to 4,007 billion ton-miles. Refined products trade fell slightly after the 1973–74 price shock, then expanded from 1975 to 1979, and then declined again somewhat after the 1979–80 price shock. In general, demand for product shipping has proven far more stable than demand for crude shipping.

FORECASTING FUTURE OIL TANKER DEMAND

Any forecasts of future oil production levels, oil demand levels, or trade patterns are open to question. There is no agreed-upon methodology for forecasting oil demand; trade patterns are often as much a result of politics as of economics; and future levels of oil production depend not only on politics and economics, but also on potential resources that are presently unknown. Thus, all of the driving inputs to an analysis of demand for oil shipping are assumptions; there are no rigorous analytical techniques for deriving such numbers.

While the foregoing may seem to be a statement of the obvious, it is important to keep in mind that the most important variables in analysis of shipping demand are the external assumptions rather than the finer points of shipping calculations. The values assumed for part-cargo voyaging, steaming speed, or canal utilization are minor factors compared to the assumptions about how much oil will be moved from one region to another. Now that the evolution of the oil fleet and the major characteristics of tanker demand have been discussed, the next logical step is to develop a forecast of future demand for crude oil ton-mileage. The forecasting procedure used in this analysis involved seven key steps:

1. Future oil demand and production levels were adopted from a recent regional forecast which presents historical data from 1970 to 1989 and develops forecasts for 1995 and 2000.[17]
2. Each major world region was divided into subregions served by primary oil ports. For areas with small total flows of oil, central ports were selected to represent surrounding minor facilities. A listing of the chosen ports is provided in table 2. The marine distances between the ports were determined from standard references[18] and compiled into a source/destination matrix.

17. Fereidun Fesharaki, David Isaak, and Nancy Yamaguchi, "World Oil Supply and Demand Outlook to 2000," *Petroleum Advisory No. 43* (Honolulu: East-West Center, Resource Systems Institute, 1989).

18. British Petroleum Tanker Company, *BP World-wide Marine Distance Tables* (London: British Petroleum Tanker, 1958); and *Reed's Marine Distance Tables*, 4th ed. (London: Thomas Reed, 1978).

TABLE 2.—REGIONS AND REPRESENTATIVE PORTS
USED IN OIL TRADE MATRIX

	Selected Port
Crude oil sources:	
Canada	Montreal
Mexico	Vera Cruz
Trinidad and Tobago	Port of Spain
Ecuador	Guayaquil
Peru	Callao
Colombia	Maracaibo
Venezuela	Maracaibo[a]
Norway	Oslo
U.K.	Dublin
Syria	Beirut
Yemen Arab Rep	Jeddah
Persian Gulf OPEC	Bahrain
Oman	Oman
Yemen, P Dem Rep	Aden
Angola	Luanda
Nigeria	Port Harcourt
Gabon	Luanda[a]
Other central western Africa	Luanda[a]
Algeria	Algiers
Egypt	Alexandria
Libya	Tripoli
Other North Africa	Tunis
Indonesia	Jakarta
Malaysia	Labuan
Other Southeast Asia	Labuan[a]
Australia/New Zealand	Melbourne
China	Dalian
USSR	Leningrad
Crude oil destinations:	
Canada/U.S. East	New York
U.S. West	Los Angeles
U.S. Gulf	Galveston
Caribbean	Havana
Western Latin America	Guayaquil
Eastern Latin America	Rio de Janeiro
Northwest Europe	Rotterdam
Southwest Europe	Genoa
Mideast-Mediterranean	Beirut
Mideast-Gulf	Bahrain
West Africa	Luanda
East Africa	Mombasa
South Africa	Capetown
North Africa	Tunis
South Asia	Bombay
Southeast Asia	Bangkok
Australia/New Zealand	Melbourne
China	Hong Kong
Japan	Yokohama
Korea, Rep	Inchon
Taiwan	Kaohsiung
USSR/Eastern Europe	Rostack

[a] A single port is used for calculation of marine distances for certain proximal nations.

3. Flows of crude oil along each route were determined for a base year from standard industry references.[19] An abbreviated form of this flow matrix is shown in table 3.

4. For future flows, the base-year pattern was used to allocate production to demand centers until either (a) supplies were exhausted, or (b) base-year flows were satisfied. The additional crude needs were then met by allocating the required additional volumes so as to minimize transit distances.

5. Quantities of oil trade in barrels were converted into cargo tonnage by a simplified set of standard industry conversion factors listed in table 4. This compensates for the different weights of the traded crudes.

6. Finally, a matrix of canal connections was assembled. This consists of nothing more than a matrix identifying which source and destination ports are linked by canals as the shortest possible routing.

7. Based on the flow assumptions and distance matrix, calculation of implied tanker demand is fairly straightforward. The equations involved are standard[20] and are provided in the following section.

Equation for Determining Oil Shipping Demand

First, the annual oil delivery capacity, A, of a given tanker on its given route is the tanker's oil-carrying capacity, O, multiplied by the number of voyages per year, V:

$$A = OV. \tag{1}$$

The number of voyages per year is the number of operational days per year, D, divided by the duration of a roundtrip voyage, L:

19. Asian Development Bank, *Energy Indicators of Developing Member Countries of ADB* (Manila: Asian Development Bank, 1989); International Energy Agency, *World Energy Statistics and Balances, 1971–1987* (Paris: Organization for Economic Co-operation and Development, International Energy Agency, 1989); Organization of Petroleum Exporting Countries, *OPEC Annual Statistical Bulletin, 1988* (Vienna: Organization of Petroleum Exporting Countries, 1988); United Nations, *Energy Statistics Yearbook, 1987* (New York: United Nations, Department of International Economic and Social Affairs, 1989); and U.S. Department of Energy, Energy Information Administration, *Petroleum Supply Annual, 1988* (Washington, D.C.: U.S. Department of Energy, Energy Information Administration, 1989).

20. Similar calculations have been made by D. Hawdon, *World Energy Transport to 1985* (London: Staniland Hall, 1980), and were perfected by D. Isaak, "World Oil Shipping Demand: An Operational Analysis" (M.A. thesis, University of Hawaii at Manoa, 1981).

TABLE 3.—SUMMARY OF CRUDE OIL IMPORT MATRIX, 1987 (figures in MB/D)

Source	Destination						
	North America	South America	Western Europe	Asia-Pacific	Africa	Mideast	USSR/ Eastern Europe	Total
North America	1,847	6	1	5	0	0	0	1,859
South America	1,361	531	603	227	0	0	0	2,722
Western Europe	573	0	1,842	0	0	0	0	2,415
Mideast	1,064	302	2,889	3,883	438	384	371	9,331
Africa	1,040	80	2,257	75	133	65	36	3,686
Asia-Pacific	414	68	18	1,397	0	0	29	1,927
USSR	1	87	808	162	35	80	—	1,173
Total	6,300	1,074	8,418	5,750	606	529	436	23,113

SOURCE.—Compiled from standard references published by the Asian Development Bank, the International Energy Agency, the Organization of Petroleum Exporting Countries, the United Nations, and the U.S. Department of Energy. For complete citations see note 19.

TABLE 4.— CONVERSION FACTORS FOR CRUDE OILS IN TRADE MATRIX

Source of Crude	Barrels per Ton
Alaska	7.10
Canada	7.40
Mexico	7.10
Trinidad and Tobago	7.00
Ecuador	7.60
Peru	7.50
Colombia	7.10
Venezuela	7.00
Norway	7.30
U.K.	7.30
Syria	7.30
Yemen Arab Rep	7.40
Persian Gulf OPEC	7.30
Oman	7.30
Yemen, P Dem Rep	7.30
Angola	7.20
Nigeria	7.40
Gabon	7.20
Other West Africa	7.40
Algeria	7.70
Egypt	6.90
Libya	7.60
Other North Africa	7.70
Indonesia	7.30
Malaysia	7.70
Other Southeast Asia	7.30
Australia/New Zealand	7.40
China	7.30
USSR	7.30

$$V = D/L. \tag{2}$$

The number of operational days per year is the number of days in a year minus the days spent on repairs and maintenance, R:

$$D = 365 - R. \tag{3}$$

R is roughly constant, since most of the scheduled maintenance consists of annual safety checks, inspection of welded seams and cargo tanks, and other routine operations.

The voyage length, L, is simply the distance of the voyage, M, in nautical miles, divided by the speed, S, in nautical miles per day, plus the number of days spent in port, P, plus the number of days spent in canal transit, C (if any):

$$L = (M/S) + P + C. \tag{4}$$

The equations can now be rewritten to yield

$$V = D/L = (365 - R)/[(M/S) + P + C]. \tag{5}$$

As defined previously, a ship's DWT is the weight of the ship plus cargo and stores. An explicit formula for O is

$$O = \text{DWT} - (F_s \times L) - (F_p \times P) - T - W, \tag{6}$$

where F_s is the tanker's oceangoing bunker fuel consumption per day at speed S, F_p is the fuel consumption per day in port, T is the reserve fuel tankage, and W is the weight of water, ship stores, and other miscellaneous cargo. Equation (6) can be simplified, since it is common practice for a ship to be fully bunkered (i.e., the fuel tanks are filled) at the onset of a voyage. Since the oil-carrying capacity pertains only to the laden leg of the voyage, the ship's cargo capacity is reduced by a constant percentage, K (the percentage of DWT not used for cargo), which may be determined for each size category of tanker. When the constant is used, equation (6) may be reduced to

$$O = \text{DWT} - (\text{DWT} \times K). \tag{7}$$

Therefore, equation (1) may be fully elaborated as

$$A = OV = \{[\text{DWT} - (\text{DWT} \times K)] \times (365 - R)\}/[(M/S) + P + C] \tag{8}$$

Since A has been defined as the total tonnage of oil delivered by the tanker over the course of a year, the deadweight tonnage required to deliver a ton of oil per year on the chosen route is merely the deadweight tonnage of the tanker divided by its annual delivery:

$$\text{Tonnage required} = \text{DWT}/A =$$

$$\{\text{DWT} \times [(M/S) + P + C]\}/\{[\text{DWT} - (\text{DWT} \times K)] \times (365 - R)\}. \tag{9}$$

The assumptions made in this forecast were as follows: $R = 20$ days/year; $S = 12.0$ knots $= 288$ nm/day; $P = 4$ days/voyage; $C = 3$ days/voyage (Suez), 5 days/voyage (Panama); $K = 3.35\%$.

The value of K declines only slightly as ship size increases; for example, noncargo uses of deadweight tonnage have been estimated at 4.54% for a medium-range ship of 50,000 DWT versus 3.23% for a ULCC of 410,000 DWT.[21] The impact of K on equation (9) is accordingly minimal; if we extract from equation (9) the entire term

$$X = \text{DWT}/\{[\text{DWT} - (\text{DWT} \times K)] \times (365 - R)\}$$

and then insert the value 20 for R, an upper and lower bound can be determined for X above. For a 50,000 DWT ship, $X = 0.00304$, whereas for a 410,000 DWT ULCC, $X = 0.00300$.

Since, across a wide range of tanker size categories, the variation in the value of K is minimal, equation (9) may now be written in a greatly simplified form:

$$\text{Tonnage required} = 0.003[(M/S) + P + C]. \tag{10}$$

In other words, total oil shipping demand can be calculated as a function of (*a*) the shipping distances, (*b*) the steaming speed at sea, (*c*) the number of days spent in port, (*d*) the number of days spent in canal transit, and (*e*) the amount of oil shipped.

The equations and assumptions listed above were applied in a series of spreadsheet models that calculated required tanker ton-mileage for 1995 and 2000. Figure 11 charted the ton-mileage of crude oil in seaborne trade, 1971–88. Figure 12 now extends the historical data with a forecast of demand for crude ton-mileage to the year 2000. While demand in the year 2000 is not expected to surpass the peak of 10,408 billion ton-miles seen in 1977, growth in demand is nonetheless strong and dramatic, particularly when compared to the 1985 low value of 4,007 billion ton-miles. In the forecast, demand reaches 6,878 billion ton-miles in 1995, growing to 8,289 billion ton-miles by the year 2000, implying average annual growth rates of 4.1% during the 1990–95 period and 3.8% for the 1995–2000 period.

At this juncture, it is also worthwhile to discuss some of the changes in projected trade patterns. Figure 13 displays interregional crude imports by the four major importing regions (North America, South America, western Europe, and the Asia-Pacific) for the base year 1987 and the projections for 1995 and 2000. Western Europe remains the largest oil importer, with imports growing from 6,576 MB/D in 1987 to 9,227 MB/D in 1995 and to 10,423 MB/D in the year 2000. Booming growth in Asia causes the Asia-Pacific region to surpass the United States/Canada region by 1995 in imports, growing from 4,353 MB/D in 1987 to 8,021 MB/D in 1995 and to 8,970 MB/D by the year 2000.

As one would expect, given the oil reserves situation detailed in figure 8, OPEC oil is forecast to grow increasingly dominant in world trade. Figure

21. Isaak (n. 20 above), p. 18.

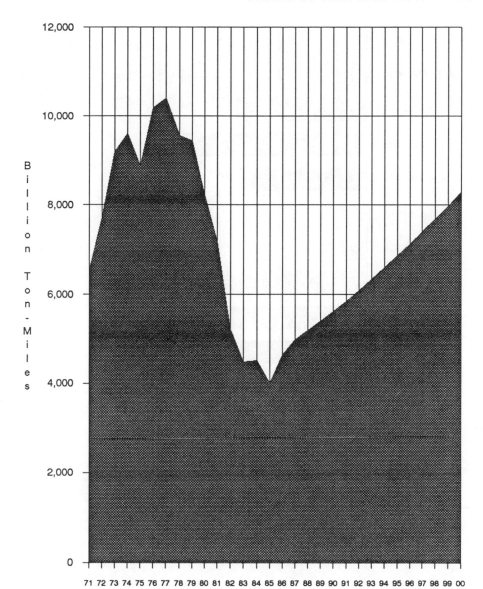

F<small>IG</small>. 12.—Historical and projected ton-mileage, crude oil. Source.—Fearnley's *Review* for historical data.

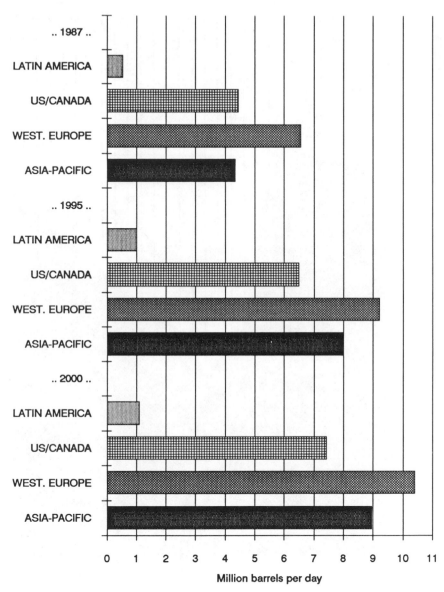

FIG. 13.—Projected interregional crude imports (MMB/D).

14 displays the overall changes projected in crude import patterns by the four major importing regions. The North American continent retains a fairly diverse supply pattern, but the increasing role of Middle Eastern oil is dramatically visible in western Europe and the Asia-Pacific region. In 1987, 24% of Asia-Pacific crude import requirements were met with other Asia-Pacific crudes; by 2000, this share is forecast to drop to only 5%, while the share of oil imports from the Mideast rises to 89% in the year 2000 from 68% in the 1987 base year. Western Europe's situation is somewhat similar, if less pronounced. In 1987, 22% of western Europe's import requirements were met from within the region, while the Middle East was the source of 34% of imports; by the year 2000, this relationship is projected to change to only 10% western European crudes and 68% Middle Eastern crudes. Figure 15 displays the growing percentage of OPEC oil in world oil trade, 1987–2000. During the 1990s, all major regions will become increasingly dependent on OPEC oil, with the Asia-Pacific region remaining the most dependent. On a worldwide level, OPEC oil is expected to account for over 80% of crude imports in the year 2000.

CONCLUSION

The key points that will define the outlook for oil shipping in the 1990s may be summarized as follows:

- First, **demand for oil will continue to grow,** resulting in larger volumes of crude, products, and blending feedstocks being moved on the high seas and in coastal areas. This will happen against a backdrop of increased concern about the environment and increased opposition to tanker traffic.
- Second, non-OPEC oil reserves are limited, and oil production is already falling in many regions, indicating that **the call on OPEC oil will rise.** This will translate into more OPEC oil moving on longer-haul voyages to consumer nations.
- Third, fleet tonnage has bottomed out and is now growing once again, since demand has picked up, many ships have been scrapped, and many ships in the remaining fleet are old and in need of replacement. However, the debate over regulations and tanker design—notably the proposed requirement for double-hulled ships in the United States—has made investors wary of building ships that might be restricted from trade in the near future. Also, fear of unlimited liability for oil spills has already prompted several oil and oil shipping companies to limit or withdraw from shipping activities in U.S. waters. **Shipbuilding and ship operating costs will rise as regulations become more stringent.**

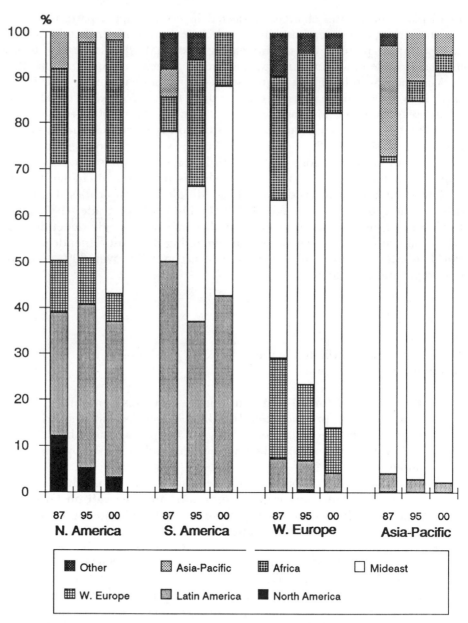

Fig. 14.—Projected changes in crude oil import patterns, 1987–2000.

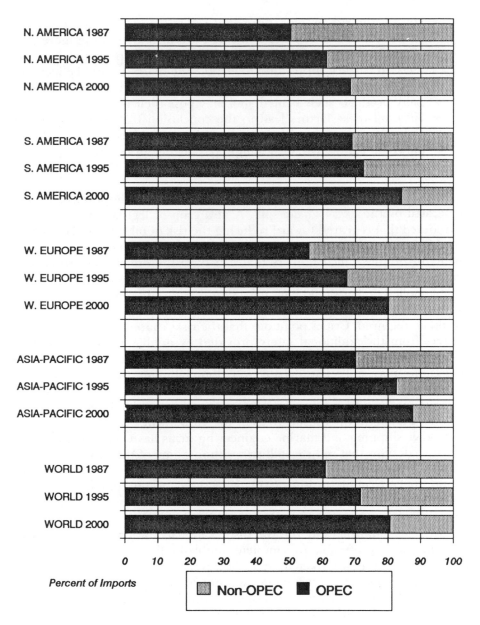

FIG. 15.—Projected OPEC shares of world oil trade by importing region, 1987–2000.

- Fourth, the rise in export-oriented refining and changing petroleum product specifications and demand patterns may result in **more trade of products,** specialty blending feedstocks, and perhaps alternative fuels, to balance supply and demand. The net result will be increased product and chemical tanker traffic, particularly in coastal areas. Specialty products such as methanol may pose their own set of hazards.
- Fifth, all these factors lead to the conclusion that **shipping activity and shipping costs will rise, and ultimately the consumer will bear the cost.** Yet even if we willingly pay higher prices to safeguard the environment, there is no guarantee that the risk of oil spills will lessen.

Few events in recent U.S. history have provoked the level of outrage generated by the *Exxon Valdez* spill. A new crop of legislation is now being considered that is geared toward reducing the risk of oil spills and enhancing capabilities for spill containment and cleanup. One possibility is a requirement for double-hulled tankers that can have the outer hull breached by an accident without necessarily losing any cargo. Obviously, a completely double-hulled tanker would have to be significantly larger than a conventional ship to carry the same amount of cargo. This implies that more voyages would be required. Critics point out that the risks caused by increased traffic detract from the additional safety provided by double-hulling. Also, if the outer hull of a double-skinned tanker is flooded, the entire ship, with crew and cargo, may sink. The debate over double-hulling therefore continues, though one company, Conoco, has already decided that double-hulling offers a greater margin of safety, and has ordered five double-hulled barges for product trading in the coastal areas of the Gulf of Mexico.[22]

Other shipping and marine engineering firms have been developing alternative designs that reduce spill risk without some of the drawbacks of full double-hulling. A recent public forum on tanker design sponsored by the American Petroleum Institute (API) showcased twenty-four different concepts in tanker design and spill limitation that their inventors—including organizations such as Wartsila Marine, Mitsubishi Heavy Industries, Shell International Marine, and the Swedish National Maritime Administration—claim are all preferable to complete double-hulling.[23] Even the strongest advocates of the new tanker designs, however, point out that even the best designs and construction techniques will not completely eliminate the risk of marine pollution in a serious collision or accident. Proper training of the bridge crew and watch officers may be a far more valuable safety tool, since

22. "Views on Double Hulls," *Oil and Gas Journal* (June 4, 1990), p. 36; and "Conoco to Construct Double Hulled Barges," *Oil and Gas Journal* (June 11, 1990), p. 20.

23. "API Conference Highlights Tanker Design, Spill Plans," *Oil and Gas Journal* (June 18, 1990), p. 16.

prevention of the accident renders moot the sophistication of ship design or spill response.[24]

To reduce the risks of oil spills, some coastal communities severely restrict tanker traffic in their areas. For example, in California's Santa Barbara area, only one tanker at a time is allowed within 3 miles of the coast. This reduces the chance of tankers colliding, but it also promotes inefficient operations and higher shipping costs. In terms of energy and the environment, we have reached a point where solutions are never simple. Quite often, the issues are so complex and intimately intertwined that it is impossible to identify, let alone evaluate, all of the trade-offs. Many goals concerning energy and economic security and environmental protection seem noble in their own right, yet they often conflict with one another. Numerous examples from the United States spring to mind:

- Shall they ban oil exploration and production in environmentally sensitive areas? If so, they may see increased imports, perhaps on ships of foreign flag that may or may not be up to the safety standards required for ships of U.S. flag.
- Shall they increase domestic oil production to reduce the need for imports? Then their long-term energy security may be jeopardized, since those oil reserves will play out long before reserves in the OPEC nations. Moreover, judging from the *Exxon Valdez* spill, domestic oil on a ship of U.S. flag seems no safer than foreign oil when it is spilled.
- Shall they restrict coastal tanker traffic and rely on pipeline transport instead? Then pipeline infrastructure will have to be built, and local communities along the proposed routes may oppose construction.
- Shall they restrict oil refining activity in, for example, the Los Angeles Basin? Pollution from the refineries may decline, and crude oil deliveries may decline, but shipping of products into the area could increase substantially.
- Shall they switch to alternative transport fuels such as methanol to reduce air pollution? If so, a massive increase in tanker traffic will be needed to deliver the new fuel, and it should be kept in mind that methanol is an extremely toxic substance. Methanol is also water soluble and colorless; a spill would be virtually impossible to contain. Some spills may even go unreported.

Despite the rhetoric, the outlook for the 1990s is for more oil shipping, not less. Taxing or regulating the shipping industry will not make oil shipping disappear. Demand for oil transport is relatively inflexible; it is a function of demand for oil products. We often fail to make the conceptual link between

24. "Double Hulls, Officer Training Key Tanker Safety Issues," *Oil and Gas Journal* (June 18, 1990), p. 15.

oil transport and our own patterns of consumption. A person may loudly denounce tanker operations, then get into his car and drive to a fast-food restaurant for a Styrofoam-packaged lunch, complaining all the while about the high prices for the gasoline and the lunch, yet never see the conflict between word and deed. In this article, both the supply side and the demand side of oil shipping have been examined. The size, structure, and composition of the future fleet will remain unclear, pending the adoption of new regulations concerning ship design, operation, and insurance requirements. On the demand side, however, the tonnage of oil trade will increase over the coming decade. The strong growth anticipated in demand for tanker tonnage is a function of both the overall increase in oil demand and the growing frequency of long-haul voyages from the Middle East to consuming areas. Therefore, oil shipping can be cut in two ways: first, by finding sources of oil closer to home, and, second, by cutting overall demand for oil. Barring an act of God that magically relocates oil resources to oil-consuming areas, the first option is limited. Barring an act of man that significantly changes the way the world economy has grown accustomed to operating, the second option is also limited—but it is probably more likely than the act of God. And in the interim, there are numerous opportunities to reduce the risk of future oil spills.

Marine Science and Technology

Marine Scientific Research and Policy Issues in East Asia

Daniel J. Dzurek[1]
East-West Center, Honolulu

INTRODUCTION

Marine scientific research, which is broadly interpreted here to include all physical and biological sciences focused on the marine environment, is undergoing a revolution in the way in which it is conducted at sea. Extensions of coastal state jurisdiction beyond the Territorial Sea and other developments in international law are modifying the way research is undertaken. Although there are studies of individual countries' policies on marine scientific research and analyses of the evolution of the global legal regime, there are few treatments of marine science and policy issues at the regional scale. The following analysis is an attempt to use a regional framework to discern trends in the evolution of marine science and related policy issues, for coastal states as well as researching countries outside the region. This approach has the advantage of focusing at the scale of the scientific phenomena of interest, the marine ecosystem, and addressing the international issues in a context suggested by the 1982 UN Convention on the Law of the Sea (hereafter the 1982 Convention), where there are frequent calls for regional solutions. With state practice in flux, the final regime governing scientific research is yet to evolve, but some tendencies are evident. In addition, certain research modes and topics appear more promising than others, at least in East Asian seas.

Marine scientific research in East Asian seas seems to have made little progress under recent developments in the law of the sea. Many of the coastal states have restricted capacities to explore the marine environment. These limitations and the nature of the scientific problems under study suggest the advisability of international cooperation to supplement current programs, which appear to be spatially constrained to areas of national jurisdiction.

---

1. Research Associate, Environment and Policy Institute, East-West Center, Honolulu. For a more extensive treatment of this topic, see Sequoia Shannon and Daniel J. Dzurek, "Marine Scientific Research," in *Atlas for Marine Policy in East Asian Seas,* ed. Joseph R. Morgan and Mark J. Valencia (Los Angeles: University of California Press, in press).

However, the membership of East Asian countries in international marine science organizations and their legislation regulating access by foreign research vessels reveal many impediments to such cooperation.

GEOGRAPHY

The Yellow and East China Seas and the Sea of Japan form the semienclosed seas of East Asia and cover an area the size of Alaska plus Texas. They are the locus of this study, which also includes that portion of the western Pacific that washes Japan and Taiwan. Early research was undertaken in the region by expeditionary efforts of Russia/Soviet Union, European countries, and the United States (figs. 1 and 2). However, these historical cruises gave little coverage to the Yellow Sea or, with the exception of Soviet work, the Sea of Japan.

In the Yellow and East China Seas physical oceanographic data have been collected for more than 50 years, but most data were obtained from hydrographic casts, which do not yield continuous temperature and salinity measurements. There are no long-term (tidal cycle or longer) continuous current-meter measurements and few synoptic or spatially extensive data sets. Geological and sedimentary data for the Yellow Sea are also deficient, in part due to the unusual nature of this epicontinental sea. Available seismic data emphasize the deep structure of the seabed. Sediment data were derived primarily from grab samples. Long core samples were few and not taken in conjunction with high-resolution seismic profiles. Therefore, the sediment structure and geology of the Yellow Sea require further study.[2]

Fisheries research in the East China and Yellow Seas is difficult to assess because of the variety of national programs, under which most relevant data are collected, and reluctance to share this information with competitors. Overfishing clearly exists and has affected species composition, but the basic data for stock assessment are lacking for the ecosystem as a whole. In terms of biological resources or pollution considerations, the East China and Yellow Seas form a single large marine ecosystem—as does the Sea of Japan—but the coastal countries lack a common scientific reference system on which management decisions should be based.[3]

2. John Milliman, "The Yellow Sea: Geological and Practical Considerations," paper presented at the International Conference on the Yellow Sea, 23–26 June 1987, East-West Center, Honolulu, as summarized by Mark J. Valencia in *International Conference on the Yellow Sea,* Occasional Paper of the East-West Environment and Policy Institute, no. 3 (Honolulu: East-West Center, 1987), pp. 132–35.

3. Joseph R. Morgan, "Large Marine Ecosystems: An Emerging Concept of Regional Management," *Environment* 29, no. 10 (December 1987): 4–30; Daniel J. Dzurek, "Prospects for Development of the Yellow Sea Regime," in *The Regime of the Yellow Sea: Issues and Policy Options for Cooperation in the Changing Environment,* ed. Dalchoong Kim, proceedings of a conference, Seoul, 19–20 June 1989, (forthcoming).

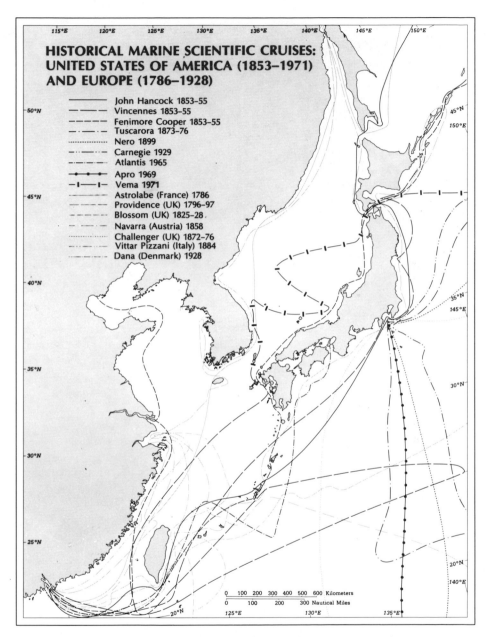

Fig. 1.—Historical marine scientific cruises: United States (1853–1971) and Europe (1786–1928).

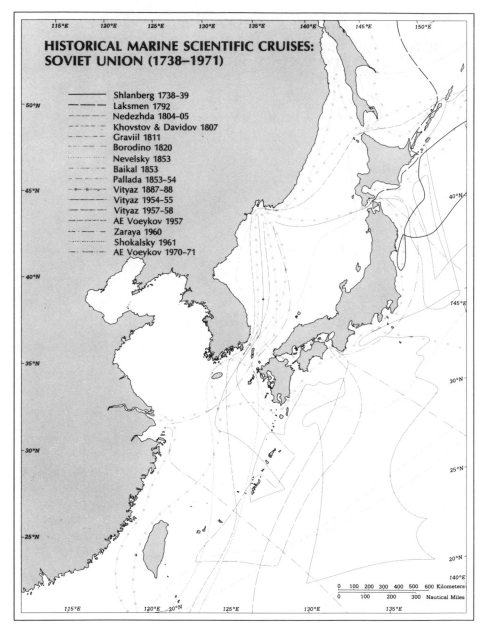

FIG. 2.—Historical marine scientific cruises: Soviet Union (1738–1971).

Because it is bordered by Japan and the USSR, which have significant marine scientific research capabilities, the Sea of Japan appears to be better studied than the East China and Yellow Seas. Areas of high interest to these two states have received considerable attention. For instance, the southeast portion of the Sea of Japan has been extensively surveyed for geologic and seismic data because of Japan's search for offshore hydrocarbon resources.[4] The extension of maritime jurisdiction has reportedly diminished the area of Japanese fisheries research, however, with a corresponding loss of important scientific data.[5]

The marine scientific research capabilities of the other countries in the area are not as well developed for use in the Sea of Japan. For example, the work of the Korea Ocean Research and Development Institute (KORDI) is restricted to the shallow Yellow and East China Seas by the limited size of its research vessel.[6] Scientific data exchange is still constrained, and the western Sea of Japan, especially in North Korea's claimed zones, is not well studied.

NATIONAL PROGRAMS

The great variance found among countries of East Asia in technological development and political culture is not reflected in fundamental concerns that condition their national marine research programs. There is surprising similarity in their emphasis on practical studies, on a strong governmental role in coordinating marine research, and on research focused within areas of national jurisdiction.

China

Marine scientific research in the People's Republic of China shares the same driving philosophy as other scientific disciplines, that of "serving economic

4. Y. Tomoda, "Comments on Scientific Research," paper presented at the International Conference on the Sea between Japan, Korea, and the Soviet Union, Niigata, Japan, 11–14 October 1988, summarized in *International Conference on the Sea of Japan*, ed. Mark J. Valencia, Occasional Paper of the East-West Environment and Policy Institute, No. 10 (Honolulu: East-West Center, 1989), p. 194.

5. Tsuyoshi Kawasaki, "International Cooperation for the Scientific Research in the Sea of Japan," paper presented at the International Conference on the Sea between Japan, Korea, and the Soviet Union, Niigata, Japan, 11–14 October 1988, as summarized in *International Conference on the Sea of Japan*, ed. Mark J. Valencia, Occasional Paper of the East-West Environment and Policy Institute, No. 10 (Honolulu: East-West Center, 1989), p. 194.

6. Byong-Kwon Park, "Comments on Scientific Research," paper presented at the International Conference on the Sea between Japan, Korea, and the Soviet Union, Niigata, Japan, 11–14 October 1988.

reconstruction."[7] Emphasis on applied research and exploitation of marine resources, which has continued since systematic marine investigation began during the 1920s in China, has promoted a focus on nearshore areas and led to developing practical knowledge of fisheries, hydrocarbon resources, pollution, and other subjects of immediate economic or social benefit. Only in the last two decades have the Chinese undertaken distant-water investigations.[8] In part, these were in support of long-range missile development and in preparation for participating in the International Sea-Bed Authority, which will be established under terms of the 1982 Convention.[9]

Nevertheless, there appears to be some backpaddling in Chinese policy. An author from the Institute of Marine Scientific and Technological Information (Tianjin) has recently reemphasized nearshore research by recommending that China undertake deep-sea research only as appropriate.[10] This would continue a pattern of activity depicted in figure 3, which demonstrates China's predominantly nearshore research activity in East Asian seas. Despite the lack of agreed maritime boundaries, Chinese research activities appear to be circumscribed by approximate median lines in disputed marine areas. There is surprisingly little overlap of research areas with neighbors, even vis-à-vis the authorities on Taiwan.

As of 1985, China had 75 research institutions involved in marine-related scientific work,[11] with principal facilities in the cities of Beijing, Tianjin, Qingdao, Nanjing, Shanghai, Hangzhou, Xiamen, and Guangzhou (see table 1). Many of the institutions are academic or provincially sponsored. However, the dominant marine scientific research organizations in China are the Chinese Academy of Sciences (CAS) and State Oceanic Administration (SOA, formerly known as the National Bureau of Oceanography).

Responsibility for conducting marine scientific research and jurisdiction over foreign research activities are centralized in the SOA.[12] Established by the People's Republic of China State Council in 1964 and expanded in 1965, the SOA took over some of the facilities formerly operated by the CAS, the

7. Niu Yinyi, "The Four Decades of Marine Science in China," *Collected Oceanic Works* 12 (August 1989): 165; see also Yann-huei Billy Song, "Marine Scientific Research and Marine Pollution in China," *Ocean Development and International Law* 20 (1989): 607.

8. C. K. Tseng, preface to *Oceanus* (issue devoted to oceanography in China) 26, no. 4 (Winter 1983/84): 8.

9. Song (n. 7 above), p. 613.

10. Niu (n. 7 above), p. 166.

11. Ibid., p. 153.

12. Baruch Boxer, "Marine Science in China: Development and Prospects," in *Ocean Yearbook 6*, ed. Elisabeth Mann Borgese and Norton Ginsburg (Chicago: University of Chicago Press, 1986), p. 237. This agency is called the State Administration of the Sea in Song (n. 7 above), p. 608.

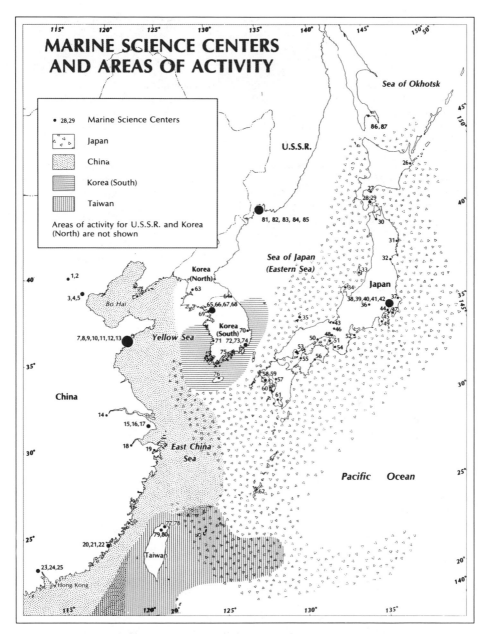

Fig. 3.—Marine science centers and areas of activity. Table 1 tells what organizations the numbers represent.

TABLE 1.—EAST ASIAN MARINE RESEARCH CENTERS

	Institute	City
	CHINA	
1	Marine Hydrological and Meteorological Forecasting Center, SOA	Beijing
2	State Oceanic Administration (SOA)	Beijing
3	Institute of Marine Scientific and Technological Information, SOA	Tianjin
4	Institute of Oceanographic Instrumentation, SOA	Tianjin
5	National Oceanographic Data Center	Tianjin
6	Institute of Marine Environmental Protection, SOA	Dalian
7	First Institute of Oceanography, SOA	Qingdao
8	Huanghai Sea Fisheries Research Institute	Qingdao
9	Institute of Oceanography, CAS	Qingdao
10	North China Sea Monitoring Station, SOA	Qingdao
11	Ocean University of Qingdao	Qingdao
12	Shandong University School of Oceanography	Qingdao
13	Yellow Sea Fishery Institute	Qingdao
14	Nanjing University	Nanjing
15	East China Normal University	Shanghai
16	East China Sea Monitoring Station, SOA	Shanghai
17	Shanghai Normal University	Shanghai
18	Second Institute of Oceanography, SOA	Hangzhou
19	East China Sea Field Laboratory, SOA	Ningbo
20	Third Institute of Oceanography, SOA	Xiamen
21	Xiamen General Ocean Station	Xiamen
22	Xiamen University	Xiamen
23	Chinese Academy of Sciences (CAS)	Guangzhou
24	Marine Environmental Protection Center, SOA	Guangzhou
25	South China Sea Institute of Oceanology, CAS	Guangzhou
	JAPAN	
26	Akkeshi Marine Biological Station, Hokkaido University	Akkeshi
27	Institute of Algological Research, Hokkaido University	Muroran
28	Hakodate Marine Observatory, JMA	Hakodate
29	Hokkaido University	Hakodate
30	Asamushi Marine Biological Station, Tohoku University	Aomori
31	Otsuchi Marine Research Center, ORI	Otsuchi
32	Tohoku University	Sendai
33	Sado Marine Biological Station, Niigata University	Sado Island
34	Noto Marine Laboratory, Kanazawa University	Noto
35	Oki Marine Biological Station, Shimane University	Oki Island
36	Suwa Hydrobiological Station	Suwa
37	Itako Hydrobiological Station, Ibaraki University	Itako
38	Hydrographic Department, Maritime Safety Agency	Tokyo
39	Marine Department, Japan Meteorological Agency (JMA)	Tokyo
40	Ocean Research Institute, University of Tokyo	Tokyo
41	Tokai University	Tokyo
42	Tokyo University of Fisheries	Tokyo
43	Maizuru Marine Observatory, JMA	Maizuru

TABLE 1.—(*CONTINUED*)

	Institute	City
44	Japan Marine Science and Technology Center	Yokosuka
45	Misaki Marine Biological Station, University of Tokyo	Miura
46	Otsu Hydrobiological Station, Kyoto University	Otsu
47	Tateyama Marine Laboratory, Ochanomizu University	Tateyama
48	Kobe Marine Observatory, JMA	Kobe
49	Shimoda Marine Research Center, University of Tsukuba	Shimoda
50	Ushimado Marine Lab	Ushimado
51	Iwaya Marine Biological Station, Kobe University	Awaji
52	Sugashima Marine Biological Laboratory, Nagoya University	Sugashima Island
53	Mukaishima Marine Biology Station, Hiroshima University	Mukaishima
54	Seto Marine Biological Laboratory, Kyoto University	Wakayama
55	Nakajima Marine Biology Station, Ehime University	Nakajima
56	Usa Marine Biology Institute, Kochi University	Kochi
57	Aitsu Marine Biological Station, Kumamoto University	Kumamoto
58	Nagasaki Marine Observatory, JMA	Nagasaki
59	Nagasaki University	Nagasaki
60	Amakusa Marine Biological Laboratory, Kyushu University	Amakusa Island
61	Kagoshima University	Kagoshima
62	Sesoko Marine Science Center, University of the Ryukyus	Okinawa

KOREA, NORTH

63	Hydrometeorological Service	Pyongyang

KOREA, SOUTH

64	Chumunjin Branch, FRDA	Chumunjin
65	Hanyang University	Seoul
66	Hydrographic Office, Ministry of Transportation	Seoul
67	Korea Ocean Research and Development Institute (KORDI)	Seoul
68	Seoul National University	Seoul
69	Inha University	Inchon
70	P'ohang Branch, FRDA	P'ohang
71	Kunsan Branch, FRDA	Kunsan
72	Fisheries Research and Development Agency (FRDA)	Pusan
73	Korea Maritime University	Pusan
74	National Fisheries University of Pusan	Pusan
75	Yosu National Fisheries College	Yosu
76	Marine Research Institute, Cheju National University	Cheju

TAIWAN

77	Taiwan Fisheries Research Institute	Keelung
78	National Taiwan College of Marine Science and Technology	Keelung
79	Institute of Oceanography, National Taiwan University	Taipei
80	Institute of Earth Sciences, Academia Sinica	Taipei

TABLE 1.—(*CONTINUED*)

	Institute	City
	USSR	
81	Pacific Oceanological Institute, USSR Academy of Sciences (Soviet AS)	Vladivostok
82	Marine Biology Institute, Soviet AS	Vladivostok
83	Pacific Ocean Geography Institute, Soviet AS	Vladivostok
84	Pacific Research Institute of Fisheries and Oceanography, Ministry of Fisheries	Vladivostok
85	Department of Oceanology, Far East State University	Vladivostok
86	Marine Geology and Geophysics Institute, Soviet AS	Yuzhno-Sakhalinsk
87	Far East Oil Geophysics Agency	Yuzhno-Sakhalinsk

SOURCES.—*Oceanus* 26 (Winter 1983/84): 13–18, and 30 (Spring 1987): 5; *Oceanographic Atlas of Korean Waters* (KORDI, 1987), passim; *Bulletin of the Marine Research Institute* (Cheju, South Korea: Cheju National University, various); *IOC Assembly Report,* 15th session (Paris: Unesco, 1989); *ROC-ROK Seminar on Oceanography* (Taipei: National Science Council, 1979); *Acta Oceanographica Taiwanica* (Taipei: National Taiwan University, 1989), pp. 1–17; *World of Learning* (London: Europa Publications, 1986); *International Research Centers Directory* (Detroit: Gale Research, 1990–91); *RNODC Newsletter for WESTPAC* (Tokyo: Japan Oceanographic Data Center) no. 3 (December 1983) and no. 6 (February 1987), passim.

Chinese navy, and the National Meteorological Bureau.[13] The SOA accounts for about 0.8% of the scientists and engineers working for all State Council ministries and the same percentage of the total funds appropriated for all ministries under the State Council in 1987.[14]

In 1950, the CAS established the first marine research institution in China, which became the current Institute of Oceanography in 1957. Its staff numbers 900. The CAS has a regional oceanographic institute in Guangzhou that studies the South China Sea and houses a staff of 500.[15]

Chinese marine research faces formidable challenges if China is to participate fully in understanding and utilizing what lays beyond its shores. For example, Chinese instrumentation is estimated to lag 10 to 15 years behind that of the West.[16] Not the least of these challenges may be continued funding for research. The author of a recent journal article on China's marine policy observed that

> in order to survive, most of the scientific and technological institutions [of China] must depend on their success in transforming the results of their work to economic benefits. Only a few, dealing with frontline science, can expect continued state support for their funding. As a result of this policy change in scientific research some people express serious

13. James Churgin, "The Structure of Oceanography in China," *Oceanus* 26, no. 4 (Winter 1983/84): 13–19; Song (n. 7 above), pp. 608–10.

14. *China Statistical Yearbook* (Beijing: China Statistical Information and Consultancy Services Centre, 1988).

15. Song (n. 7 above), pp. 608, 610.

16. Ibid., p. 610.

concerns about possible setbacks, including many ongoing marine research projects.[17]

Japan

Japanese marine scientific research began in 1871 with the establishment of the Hydrographic Department's antecedent. The visit in 1953 of the *Baird* of Scripps Institute of Oceanography rekindled interest in ocean studies. An extensive distribution of marine research facilities in Japan demonstrates the development of ocean science in that country. In addition, the geographic scope of research has also widened from nearshore to oceanwide.

Japanese organizations engaged in marine research include the Meteorological Agency with associated observatories, the Maritime Safety Agency, various universities and affiliated marine stations, and the Fisheries Agency. The Ocean Research Institute (ORI) was established at the University of Tokyo in 1962 with a mandate for basic scientific research. In 1971, the Japan Marine Science and Technology Center (JAMSTEC) was founded to promote marine science and technology.

Japan's ORI investigates physical oceanography, marine meteorology, geophysics, sedimentation, geotectonics, inorganic chemistry, biochemistry, marine organism physiology, marine ecology, planktonology, microbiology, population dynamics of marine organisms, and fisheries biology. The broadening of Japanese marine research interests is suggested by an increase in the types of measurements made, as reported to the IOC Regional Committee for the Western Pacific (WESTPAC) (fig. 4). Although more than one type of measurement is usually undertaken during any particular cruise, the number and diversity in recent years suggest a maturing marine research capability. One puzzling aspect is the relatively small number of geology and geophysics programs reported to WESTPAC, perhaps because such measurements are more frequently undertaken by private-sector vessels.[18]

Since the late 1960s, marine science–related activities in Japan have increasingly concentrated on applied science and technological innovation, rather than on basic science, as the distinction is often made in the West.[19]

17. Yu Huming, "Marine Policy in the People's Republic of China," *Marine Policy Reports* 1 (1989): 244.

18. It should also be borne in mind that the WESTPAC publications do not include all marine research activities in the area; the data available are only indicative and not exhaustive.

19. Analyses of marine scientific policy in Japan frequently commingle science and technology. See, for example, Mamoru Koga and Hiroyuki Nakahara, "Japanese Ocean Science and Technology Policy and the National Budget," in *Japan and the New Ocean Regime*, ed. Robert L. Friedheim et al. (Boulder, Colorado: Westview Press, 1984), pp. 108–29; Noriyuki Nasu, "Introduction: Japan and the Sea," *Oceanus* 30, no. 1 (Spring 1987): 3–8.

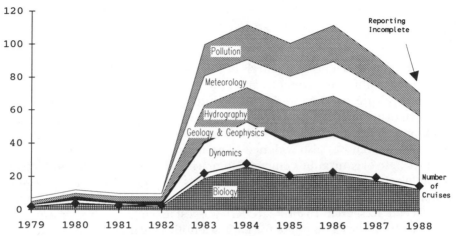

Series of
Measurements

FIG. 4.—Types of measurements made during WESTPAC Japanese cruises, 1979–88.

This shift in focus paralleled the reorganization of Japan's Ocean Science and Technology Council into a Council for Ocean Development in 1971.[20] In this emphasis on practical application, Japan mirrors most of its neighbors across the East Asian seas. It also reflects them in evolving a strong governmental role in coordinating marine research. In the Japanese case, however, this function harmonizes commercial, as well as academic and governmental, efforts.

The section of the Japanese budget for marine science *and technology* grew from 13.8 billion yen in 1975 to 64.4 billion yen for Japanese fiscal year 1986. This latter figure represents 0.03% of the total government budget or 20% of the 330.8 billion yen allocated for the promotion of scientific research.[21] The budget growth parallels an impressive increase in the estimated ship-days spent on Japanese scientific cruises. Japan also supports the Responsible National Oceanographic Data Center for the Western Pacific (RNODC-WESTPAC), at the Japan Oceanographic Data Center of the Maritime Safety Agency.

20. Hiroyuki Nakahara, "Consensus Building in the Council for Ocean Development," in *Japan and the New Ocean Regime,* ed. Robert L. Friedheim et al. (Boulder, Colorado: Westview Press, 1984), p. 79.

21. Koga and Nakahara (n. 19 above), p. 114; Nasu (n. 19 above), p. 5; Takashi Mayama, "The Japan Marine Science and Technology Center," *Oceanus* 30, no. 1 (Spring 1987): 28; Takahisa Nemoto, "Japan's Ocean Research Institute," *Oceanus* 30, no. 1 (Spring 1987): 49.

North Korea

Undoubtedly, North Korea conducts some research in its offshore areas. However, scant literature is available in the West to document its program.

South Korea

Although KORDI was established in 1973, it was not until the 1980s that Korean marine science began significant offshore work. Before that time, broad reconnaissance investigations of the Yellow and East China Seas were conducted by foreign scientists.[22] As of 1983, Korea had yet to complete a systematic and detailed characterization of its adjacent seas. The publication of a detailed oceanographic atlas of the Yellow Sea 4 years later demonstrates the effort devoted to remedying this deficiency.[23] Korean research strengths appear to lie in chemical and physical oceanography.

Korea first reported cruises to WESTPAC in 1984 (fig. 5). The pattern of Korean research focuses on the marine areas immediately adjacent to South Korea. Many of the research cruises are of short duration (a few days) because of vessel limitations.

Taiwan

Research results published in Taiwanese journals exhibit a spectrum of marine science interests, ranging from fisheries dynamics to manganese nodule occurrence. Much of the work focuses on the Kuroshio, geology and sedimentology in the vicinity of Taiwan, and general oceanography. Some Taiwanese scholars have used data obtained from their distant-water fishing fleet in analysis of fisheries off northwestern Australia and in the South Pacific.[24]

Several oceanographic cruises have been reported in the literature. In partnership with Taiwan Fisheries Research Institute (Keelung), the National Taiwan College of Marine Science and Technology (Keelung), and the Institute of Earth Sciences (Academia Sinica, Taipei), the Institute of Oceanogra-

22. K. O. Emery, "Korean Oceanography in the Yellow Sea," in *Marine Geology and Physical Processes of the Yellow Sea,* Proceedings of the Korea-U.S. Seminar and Workshop, Seoul, 19–23 June 1984 (Seoul: Korea Institute of Energy and Resources, 1984), p. 2.

23. Korea Ocean Research and Development Institute (KORDI), *A Study on the Marine Environmental Atlas of the Adjacent Seas to Korea* (KORDI, 1983), p. 7; KORDI, *Oceanographic Atlas of Korean Waters,* vol. 1, *Yellow Sea* (KORDI, 1987).

24. See, for example, *Acta Oceanographica Taiwanica* (Taipei: Institute of Oceanography, National Taiwan University, December 1986), pp. 1–17, (January 1989), pp. 36–45, and (March 1989), pp. 46–47.

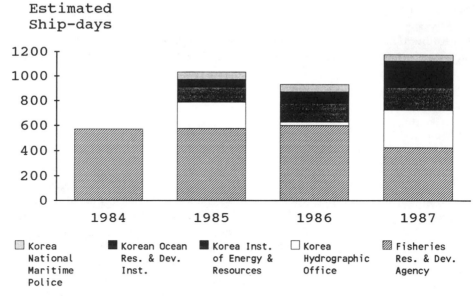

Estimated
Ship-days

FIG. 5.—Korean scientific cruises reported to WESTPAC, 1984–87.

phy of the National Taiwan University (Taipei) organized a Cooperative Hydrographic Investigation of the Philippine Sea (CHIPS) with a series of cruises beginning in 1985. Cruises have also been made in the northern South China Sea. However, most Taiwanese marine research continues to focus on nearshore areas.

USSR

Although the general capacity of Soviet marine scientific research, on the national level, is on a par with American or Japanese capabilities, there appears to be increased Soviet attention to the development of marine science in the Far East. Offshore research activities there preceded the Soviet revolution, but because of the lack of local support centers along the east coast, early Russian and Soviet investigation was expeditionary in nature (fig. 2). In marine biological studies in the Soviet Far East, for example, there has been an evolution of research infrastructure, dating from the founding of the Pacific Research Institute of Fisheries and Oceanography's forerunner in 1925, to establishment of various institutes of the USSR Academy of Sciences at the Far East Science Center in Vladivostok/Nakhodka in the 1970s.[25]

25. Robert H. Randolph and John E. Bardach, "Soviet Science in the Pacific: The Case of Marine Biology," in *Nature in Its Greatest Extent: Western Science in the Pacific,* ed. Roy MacLeod and Philip F. Rehbock (Honolulu: University of Hawaii Press, 1988), pp. 245–74.

The USSR is more dependent on fisheries than the United States is, in both absolute and relative terms. It has devoted much scientific effort to fisheries production and management—conditioning both long-term and operational plans of the Soviet fishing industry. Initial fisheries research in the Soviet Far East focused on salmon and fish in the coastal waters of the Japan, Okhotsk, and Bering Seas. After the Second World War, research shifted to large-scale and oceanic fisheries. As a result, Soviet research vessels are more massive than American counterparts because they evolved with large-scale Soviet fishing vessels. The Pacific Research Institute of Fisheries and Oceanography, headquartered in Vladivostok, maintains branches in Kamchatka, Sakhalin, Magadan, and Khabarovsk. Its investigations range from the Pacific to the Antarctic.[26] Soviet marine research has also focused on offshore hydrocarbons and deep-sea minerals.

LEGAL FRAMEWORK

The national programs show both the capabilities and deficiencies of marine science within East Asia. To complete the picture, one must examine the way these programs interact with and are augmented by extraregional researchers. This is conditioned by the international legal framework governing marine research. Eight years after the signing of the 1982 Convention, the legal regime governing the conduct of marine scientific research remains in a state of flux. Since ratifications to date are insufficient for the treaty to come into force—a situation likely to persist in the near term—state practice is the deciding, if chaotic, factor that determines procedures governing marine research.[27]

1982 UN Law of the Sea Convention

Although the 1982 Convention has yet to come into force, many of its provisions, including those dealing with marine scientific research, could be viewed as codifying customary international law. Certainly, they provide a standard against which to judge East Asian activities and national regulations for marine scientific research in the exclusive economic zone (EEZ) or extended fisheries zone and on the Continental Shelf.[28] In that vein, an overview of

26. USSR Ministry of Fisheries, Pacific Research Institute of Fisheries and Oceanography (TINRO), pamphlet prepared for the 14th Pacific Science Congress (Vladivostok: TINRO, 1979).

27. Some of the material in this section is adapted from Nien-tsu A. Hu, "Marine Scientific Research in East Asia," paper prepared at the Environment and Policy Institute, East-West Center, Honolulu, 3 October 1988.

28. The 1982 Convention recognizes the coastal state's exclusive rights to scientific research in the Territorial Sea and concentrates on provisions for the other

the 1982 Convention's relevant parts provides an analytic framework for issues germane to the present study.[29]

The first global instrument to regulate marine scientific research, the 1958 Geneva Convention on the Continental Shelf, established a coastal state consent regime, outlined in a single paragraph.[30] The 1982 Convention devotes over 32 articles to expanding the consent regime and extending it to the newly defined EEZ or fisheries zone, which can reach up to 200 nm (370 km) from the baseline from which the Territorial Sea is measured. Convention provisions also deal with the Continental Shelf where it lies beyond 200 nm from the national baseline and research in the area beyond national jurisdiction (the deep seabed), but these provisions are not further considered here because the configuration of East Asian semienclosed seas places all maritime areas within actual or potential 200-nm jurisdictional claims.

The framers of the 1982 Convention sought to balance extended coastal state jurisdiction with scientific freedom and to harmonize multiple uses of ocean space through a zonal approach to regimes for conducting scientific research.[31] Among other innovations for research in the EEZ or on the Continental Shelf, the 1982 Convention

1. requires the researching state to request permission 6 months before the starting date of a project within another state's jurisdiction;
2. provides for implied consent if, within 4 months of receipt of the request, the coastal state has not denied permission or asked for more information;

---

jurisdictional zones. See *The Law of the Sea: United Nations Convention on the Law of the Sea with Index and Final Act of the Third United Nations Conference on the Law of the Sea* (hereafter cited as 1982 Convention) (New York: United Nations, 1983).

29. The primary section dealing with these issues is ibid., Pt. XIII, Marine Scientific Research (Arts. 238–65); however, other sections of the 1982 Convention are relevant, including arts. 21, 40, 56, 87, 143, 200, 266–69, 275, 277, 297, and Annex VIII, arts. 1, 2. For an article-by-article commentary on the principal entries, see David A.Ross and Thérèse A. Landry, *Marine Scientific Research Boundaries and the Law of the Sea* (Woods Hole, MA: Woods Hole Oceanographic Institution, 1987).

30. "The consent of the coastal State shall be obtained in respect of any research concerning the continental shelf and undertaken there. Nevertheless the coastal State shall not normally withhold its consent if the request is submitted by a qualified institution with a view to purely scientific research into the physical or biological characteristics of the continental shelf, subject to the proviso that the coastal State shall have the right, if it so desires, to participate or to be represented in the research, and that in any event the results shall be published" (1958 Geneva Convention of the Continental Shelf, Art. 5, Par. 8).

31. For a discussion of the development of the marine scientific provisions of the 1982 Convention, see Alexander Yankov, "A General Review of the New Convention on the Law of the Sea: Marine Science and Its Application," in *Ocean Yearbook 4*, ed. Elisabeth Mann Borgese and Norton Ginsburg (Chicago: University of Chicago Press, 1983), pp. 150–75.

3. calls on states and competent international organizations to publicize information on *proposed major programs and objectives* as well as results from marine research;

4. requests states and international organizations to promote the flow of data and the transfer of knowledge resulting from marine research to developing states;

5. calls on states to strengthen the autonomous research capabilities of developing states by providing education and training to their technical and scientific personnel; and

6. provides for a dispute-settlement mechanism.[32]

The 1982 Convention does not define marine scientific research, however, nor does it distinguish between pure (fundamental) and applied (resource-related) research as the 1958 Geneva Convention on the Continental Shelf implicitly did. On the contrary, framers of the 1982 Convention eventually dismissed this distinction, and in the treaty text science is often compounded with technology, especially in provisions promoting the development and transfer of marine technology.[33] Moreover, a coastal state has the discretion to withhold its consent to conduct a research project if that project "is of direct significance for the exploration and exploitation of natural resources, whether living or non-living."[34] Such a broad formulation may permit a coastal state to withhold consent for virtually any scientific research project.

The 1982 Convention offers several approaches to reduce bilateral obstacles to marine scientific research. It calls on countries to cooperate through competent international organizations, through regional arrangements, and even in the absence of diplomatic relations.

Presumptive rights are given to international organizations to conduct research under certain conditions. In the event that an international organization wants to conduct a scientific project in the EEZ or on the Continental Shelf of a coastal state, and that state is a member or has an arrangement with the organization, then the coastal state is deemed to have authorized the research if it approved the detailed project when the organization decided to undertake the project or if it is willing to participate and expressed no objection within 4 months of notification by the organization. In effect, international organizations are to be treated as researching countries for purposes of requesting permission to conduct marine research.[35] Further, the convention calls on competent international organizations to establish general crite-

---

32. 1982 Convention, Arts. 245–65, 266–78 passim.
33. Ibid., Arts. 266–77; Yankov, (n. 31 above), pp. 159–60.
34. 1982 Convention, Art. 246.
35. Ibid., Art. 247.

ria and guidelines to assist coastal states in assessing scientific research re-quests.[36]

Regional arrangements are frequently invoked in the 1982 Convention, including the marine scientific research section. However, the treaty provides little on the point beyond hortatory language. Each region, however defined, is left to its own devices to evolve cooperative arrangements on a host of maritime issues, including scientific research.

One unusual aspect of the recent convention is the request that coastal states grant their consent for research projects in spite of the absence of diplomatic relations between the coastal state and the researching state.[37] However, the 1982 Convention requires that "communications concerning the marine scientific research projects shall be made through appropriate official channels, unless otherwise agreed."[38] It is unclear how to reconcile these two requirements. In East Asia, one finds governments with conflicting claims to sovereignty over the same country and countries that do not recognize each other, in addition to those that do not have diplomatic relations. If the 1982 Convention's requirements are resolved in East Asia, the solutions may be relevant for other regions.

The state of ambiguity in both the status of the 1982 Convention and the interpretation of its relevant provisions (Part XIII) is compounded in East Asia, where the most distinctive feature of statutes controlling marine scientific research is their absence (table 2). Despite historic sensitivity to foreign maritime presence, only Japan and the Soviet Union have explicit legislation governing the conduct of marine scientific research. Even the formal legal basis for regulating research is lacking for those countries that do not claim an extended fisheries zone or EEZ (China and South Korea). Despite the absence of EEZ/fisheries claims, however, both China and South Korea appear to assert effective jurisdiction on marine scientific research in the water column beyond their Territorial Seas. The recognition by other countries of this de facto jurisdiction, absent EEZ/fisheries claims, is a new development in state practice.

With regard to research on the Continental Shelf and in the Territorial Sea, customary international law requires no explicit claim, because all countries are assumed to have jurisdiction over these areas. In other words, under customary and conventional international law, a coastal state is assumed to have sovereign rights to its Continental Shelf even if it has made no formal shelf claim. The extent of the shelf may be in dispute, but the right to jurisdiction is not. A review of the legislation, regulations, and state practice of the East Asian coastal states relating to the activity of foreign scientific research

36. Ibid., Art. 251.
37. Ibid., Art. 246, Par. 4.
38. Ibid., Art. 250.

TABLE 2.—NATIONAL MARINE RESEARCH REQUIREMENTS

	China	Japan	USSR
Dates of regulations	a	1977, 1988	1968, 1969, 1982, 1984, 1985
Implied consent			After 4 months
Detailed report	Yes	Yes	Yes
Lead time		2–3 months	6 months
National participation	Yes		Yes
Permission to publish	Required		

SOURCES.—Office for Ocean Affairs and the Law of the Sea, United Nations, *The Law of the Sea: National Legislation, Regulations, and Supplementary Documents on Marine Scientific Research in Areas under National Jurisdiction* (New York: United Nations, 1989); Su Jilan, "Practice of China in Regulating Marine Scientific Research in Zones Subject to Chinese Jurisdiction," letter to Satya N. Nandan, Under-Secretary-General, United Nations, 25 August 1989; Office of Marine Science and Polar Affairs, U.S. Department of State, "Notice to Research Vessel Operators," various dates.
ᵃChina is in the process of formulating regulations.

vessels in their jurisdictions demonstrates the ambiguity of the current legal regime.

China

The People's Republic of China has no explicit legislation regulating marine scientific research within its area of jurisdiction, but it is in the process of formulating regulations and has supplied the United Nations with an outline of present practices. China claims the right to approve all marine scientific research activities; the exclusive right to the original data and samples obtained within internal waters and the Territorial Sea, with data and samples obtained in "other areas" shared according to mutual agreement; and the requirement for its consent prior to the publication of data and samples.[39] During the course of negotiations at UNCLOS, China maintained the view that

> marine research always serves military, economic and political purposes, directly or indirectly, and therefore constitutes a danger to national security and sovereignty. Scientific research, therefore, in a coastal state's territorial sea or area under its jurisdiction, is to be subject to its approval and appropriate control.[40]

39. The Chinese authorities do not define the area subject to Chinese jurisdiction except to distinguish "inland seas and territorial sea" from "the rest of the zones under her jurisdiction" (Su Jilan, "Practice of China in Regulating Marine Scientific Research in Zones Subject to Chinese Jurisdiction," letter to Satya N. Nandan, Under-Secretary-General, United Nations, 25 August 1989).
40. Jeanette Greenfield, *China and the Law of the Sea, Air, and Environment* (Alphen aan den Rijn, Netherlands: Sijthoff and Noordhoff, 1979), p. 154.

Although China has not made a de jure claim to marine jurisdiction in the water column beyond its Territorial Sea, it has maintained a latent claim, as exemplified in its 1982 Marine Environmental Protection Law, which "applies to the internal sea and territorial sea of the People's Republic of China and *all other sea areas* under the jurisdiction of the People's Republic of China" (emphasis added).[41]

Although the law provides for environmental protection and penalties for pollution, it does nothing further specifically to address or regulate marine scientific research. The 1986 Fishery Law similarly alludes to marine research.[42]

In the absence of specific legislation or regulations, marine research undertaken by other countries in Chinese waters has been managed under standing or ad hoc bilateral agreements. Such has been the case with American investigators, working under a 1979 marine and fishery sciences protocol.[43]

Japan

Japan, which claimed an extended fishing zone in 1977, has two sets of regulations or laws governing the conduct of marine scientific research.[44] "Provisional Procedures" given in a 1988 note to the United Nations provide for research in the Territorial Sea or fishing zone and on the Continental Shelf. They accord with relevant provisions of the 1982 Convention. Law 31 of 2 May 1977 outlines the relevant procedures and penalties for fisheries research. There has been a recent indication that Japan has requested additional information on research undertakings, to cope with maritime traffic and conflicts between commercial and scientific vessels.[45]

41. *The Marine Environmental Protection Law of the People's Republic of China,* adopted at the 24th session of the Standing Committee of the Fifth National People's Congress, 23 August 1982, published in Chinese and English (Beijing: China Ocean Press, 1983), p. 21.

42. 1986 Fishery Law, Art. 8, *People's Daily* (Beijing), 22 and 23 January 1986, translated by Nien-tsu A. Hu.

43. Toufiq A. Siddiqi, Jin Xiaoming, and Shi Minghao, *China-USA Governmental Cooperation in Science and Technology,* Environment and Policy Institute Occasional Paper no. 1 (Honolulu: East-West Center, 1987).

44. Office for Ocean Affairs and the Law of the Sea, United Nations, *The Law of the Sea: National Legislation, Regulations, and Supplementary Documents on Marine Scientific Research in Areas under National Jurisdiction* (New York: United Nations, 1989), pp. 143–54.

45. Office of Marine Science and Polar Affairs, U.S. Department of State, "Notice to Research Vessel Operators" no. 87, 8 May 1989.

North Korea

There are no known North Korean laws regulating marine scientific research in its Territorial Sea or EEZ. In a decree of 1 August 1977 that proclaimed a 200-mile EEZ, North Korea appeared to prohibit most foreign activities that do not have the express consent of the government. A military security zone further restricts possible marine scientific research by prohibiting virtually all navigation and overflight.[46]

There is little information regarding North Korean state practice on permitting marine scientific research by foreign vessels. China and the Soviet Union appear to be engaged in cooperative marine scientific research with North Korea, but the modality of such cooperation appears to be governed through bilateral or multilateral agreements.[47]

South Korea

The South Korean government made provision for authorizing scientific research in its 12-nm Territorial Sea under a 1978 decree, which requires application through the Ministry of Foreign Affairs.[48] It appears that the South Korean authorities also authorize research beyond the Territorial Sea, with occurrences in 1985 and 1986. This included joint U.S.–China–South Korea research cruises in the Yellow Sea.[49]

Taiwan

In 1979 Taiwan claimed an EEZ and expanded its Territorial Sea claim to 12 nm, but the declarations did not specifically refer to marine scientific research.[50] There are no laws or regulations made exclusively to regulate

46. Office of Marine Science and Polar Affairs, U.S. Department of State, "Notice to Research Vessel Operators" no. 41, 11 August 1977; Bruce D. Larkin, "East Asian Ocean Security Zones," in *Ocean Yearbook 2,* ed. Elisabeth Mann Borgese and Norton Ginsburg (Chicago: University of Chicago Press, 1980), pp. 289–92.

47. The West Pacific Fisheries, Oceanology, and Limnology Research Cooperation Agreement, among China, North Korea, Mongolia, Vietnam, and the USSR.

48. *United Nations Legislative Series: National Legislation and Treaties Relating to the Law of the Sea,* ST/LEG/SER.B/19 (New York: United Nations, 1980), pp. 95–97. The South Korean Territorial Sea is less than 12 nm in some straits.

49. Letter from Research Vessel Clearance Officer, Office of Marine Science and Polar Affairs, U.S. Department of State, 22 August 1989.

50. Hungda Chiu, ed., *Chinese Yearbook of International Law and Affairs* (Taipei: Occasional Paper/Reprints Series in Contemporary Asian Studies, 1982), pp. 151–52; Myron H. Nordquist and Choon Ho Park, *North America and Asia-Pacific and the Development of the Law of the Sea: Treaties and National Legislation* (New York: Oceana Publications, 1981), binder 1, pamphlet for the Republic of China, p. 6.

marine scientific research, other than those that normally apply to ships in these areas.[51]

USSR

The Soviet Union claims the full panoply of maritime zones and has comprehensive legislation and regulations governing the conduct of marine research in each zone.[52] The most recent regulations (1985) generally conform with the 1982 Convention and even include the "implied consent" regime (Article 252 of the 1982 Convention). Soviet marine scientists recognize the need for access to the EEZs of other countries: "Oceanologists must be assured freedom of scientific research on all water areas of the world ocean and, even more, active cooperation of littoral countries [in] carrying out such research in the zones of national interests along their shores."[53] In practice, requests from Western countries to conduct scientific research in Soviet maritime areas have been granted only recently.

INTERNATIONAL COOPERATION

Bilateral

Data on bilateral research projects in East Asian seas are fragmentary. The USSR, for example, has conducted joint oceanographic and pollution investigations with North Korea in the Sea of Japan.[54] Soviet-Japanese studies have included fisheries research. China has bilateral agreements with several countries, including Canada, France, and the United States.[55] South Korea has cooperative programs with the United States, the United Kingdom, France, West Germany, the Netherlands, and Japan.

51. Response of the Ministry of Foreign Affairs of the Republic of China, 19 August 1988.

52. Office for Ocean Affairs and the Law of the Sea, United Nations (n. 44 above), pp. 258–68.

53. Andrei Sergeevich Monin, "Perspektivy issledovaniia i ispol'zovaniia mirovogo okeana" (Prospects for investigation and use of the world ocean), *Vestnik* (Soviet Academy of Sciences) no. 5 (1980):118–26, as quoted in Randolph and Bardach (n. 25 above), p. 266.

54. A. V. Tkalin and E. N. Shapovalov, "Investigations of Chemical Pollution of the Sea of Japan: Results and Perspectives," paper presented at the International Conference on the Seas of Japan and Okhotsk, sponsored by the Environment and Policy Institute, East-West Center, and the Pacific Oceological Institute of the Far Eastern Branch of the USSR Academy of Sciences in Nakhodka, USSR, September 1989.

55. Siddiqi et al. (n. 43 above); Song (n. 7 above), p. 615.

Despite the hortatory portions of the 1982 Convention, bilateral arrangements for marine research still appear to depend on diplomatic recognition. Informal arrangements among scientists offer a weak foundation upon which to base long-term scientific cooperation. Institutional cooperation across national frontiers is also jeopardized by lack of diplomatic relations. If China or the Soviet Union were to recognize South Korea, or there were a cross-recognition arrangement vis-à-vis the United States and North Korea, then bilateral scientific arrangements might flourish in East Asian seas. Arrangements between the Koreas or across the Formosa Strait are less likely over the near term.

Multilateral

Most international cooperation in the marine sciences in East Asia appears to take place by way of multilateral arrangements, especially through international scientific organizations and programs (table 3).[56] China participates in several international programs, including the IOC-WMO Tropical Oceans and Global Atmosphere (TOGA), the IOC World Ocean Circulation Experiment (WOCE), and the Japan East China Sea Study (JECSS).[57] South Korea is active in most of the intergovernmental and nongovernmental marine research organizations and programs.[58] Taiwanese marine scientists appear to maintain informal cooperative research relationships with Japanese, South Korean, and Western colleagues and participate in nongovernmental international programs, such as the Scientific Committee on Oceanic Research (SCOR) of the International Council of Scientific Unions (ICSU) and JECSS.[59]

Japanese marine scientists have cooperated in many international research projects. The first was probably the International Indian Ocean Expeditions (IIOE), which continued from 1960 to 1962. Also in the 1960s, Japan participated in the Cooperative Study of the Kuroshio and Adjacent Regions (CSK). Other projects include the International Phase of Ocean Drilling (IPOD) from 1975, which was transformed into the Ocean Drilling Program (ODP) in 1983; WOCE; JECSS, beginning in 1980; and with the French a

56. For an elaboration of topics for general international cooperation, see Roger Revelle, "The Need for International Cooperation in Marine Science and Technology," in *Ocean Yearbook 5*, ed. Elisabeth Mann Borgese and Norton Ginsburg (Chicago: University of Chicago Press, 1985), pp. 130–49.

57. Yu (n. 17 above), pp. 241–42.

58. "Korea Ocean Research and Development Institute," pamphlet (1986).

59. Various authors in *Acta Oceanographica Taiwanica,* issues from 1986 through 1989.

TABLE 3.—MEMBERSHIP IN INTERNATIONAL ORGANIZATIONS AND REGIONAL PROGRAMS

	China	Japan	Korea, Dem P Rep	Korea, Rep	USSR	Taiwan	U.S.A.
INTERGOVERNMENTAL							
Unesco	X	X	X	X	X		X
IOC	X	X		X	X		X
WESTPAC	X	X		X	X		X
ITSU[a]	X	X	X	X	X		X
CCOP[b] (SCOR/IOC)	X	X		X	X		X
UNEP	X	X		X	X		X
IHO	X	X		X	X		X
WMO	X	X	X	X	X		X
FAO[c]	X	X	X	X			X
IPFC[d]		X		X			
ICES[e]					X		X
NONGOVERNMENTAL							
ICSU	X	X	X	X	X	X	X
SCOR	X	X			X	X	X
PSA	X	X		X	X	X	X
JECSS	X	X		X		X	X

Sources.—Intergovernmental Oceanographic Commission, *Reports of Governing and Major Subsidiary Bodies, Fifteenth Session of the Assembly*, Paris, 4–19 July 1989 (Paris: Unesco, 1989); Henry W. Degenhardt, *Maritime Affairs—A World Handbook* (Detroit: Gale Research, 1985); U.S. Central Intelligence Agency, *The World Fact Book 1988* (Washington: Government Printing Office, 1988); Scientific Committee on Oceanic Research (SCOR), *SCOR Proceedings*, vol. 24 (Halifax: SCOR, 1989); *UNEP Regional Seas Reports and Studies*, No. 1 (Geneva: UNEP, 1982); *Indo-Pacific Fisheries Commission Report* (1984); *UNEP Report of the Governing Council* (New York: United Nations, 1987), U.N. General Assembly Official Records, A/42/25; T. Ichiye, ed., *Ocean Hydrodynamics of the Japan and East China Seas* (Amsterdam: Elsevier, 1984); *Progress in Oceanography* 17, no. 3–4 (1987).

[a] International Coordination Group for the Tsunami Warning System in the Pacific.
[b] Committee for the Coordination of Joint Prospecting for Mineral Resources in Asian Offshore Areas
[c] Food and Agriculture Organization of the United Nations
[d] Indo-Pacific Fishery Commission
[e] International Council for the Exploration of the Sea

series of geological and geophysical investigations of subduction zones near Japan.[60]

"Examination of Soviet research activities in recent years shows that Soviet science has inclined toward participation in international rather than in binational programs."[61] This trend is evident in active Soviet involvement in the Pacific Science Association (PSA), including hosting a Pacific Science Congress in Novosibirsk in 1979.[62] However, Soviet reporting to WESTPAC appears to be sporadic: no cruises have been reported to the RNODC-WESTPAC, in Tokyo, since 1982. According to Soviet sources, the national coordinator of the USSR under the WESTPAC program heads a section of 295 scientists and staff, representing 77 organizations. The Soviet national scientific WESTPAC program for 1987–90 includes 21 organizations in 35 projects.[63]

Third-Party Access: U.S. Experience

Because of the difficulties of direct cooperation among some of the coastal states in East Asia, an intermediate role by third parties may facilitate cooperation in marine scientific research. In some degree, Japan, the United States, and the Soviet Union have all played this part. They have hosted conferences attended by scientists from antagonistic countries or provided research vessels manned by mixed crews.[64] Since the U.S. experience is best documented, it can serve as an example of the advantages and pitfalls of such a role.

U.S. research interests in East Asian seas vary with the scientist but appear to focus on geophysical phenomena, such as plate tectonics, seismic studies, and the Kuroshio. However, requests for U.S. research vessels to

60. T. Ichiye, ed., *Ocean Hydrodynamics of the Japan and East China Seas* (Amsterdam: Elsevier, 1984), pp. xi–xiii; Nasu (n. 19 above), pp. 6–7; Nemoto (n. 21 above), pp. 49–51.

61. Robert Randolph and John Bardach, "Soviet-American Scientific Cooperation in the Pacific," in *Soviet-American Horizons on the Pacific,* ed. John J. Stephan and V. P. Chichkanov (Honolulu: University of Hawaii, 1986), pp. 162–63.

62. Various Pacific Science Association publications.

63. M. M. Oganov, "The Coordination of Oceanological Investigations in the Soviet Far East," *Pacific Annual 1988* (Vladivostok: Far Eastern Branch, USSR Academy of Sciences), pp. 215–16.

64. For example, the International University of Japan and the East-West Center hosted an International Conference on the Sea of Japan at Niigata (October 1988), which was cosponsored by KORDI and United Nations University. The Siberian division of the USSR Academy of Sciences was among the Soviet hosts to the International Tsunami Meetings at Novosibirsk Science Center (31 July–10 August 1989). The East-West Center and the Pacific Oceanological Institute of the Far Eastern Branch of the USSR Academy of Sciences sponsored an International Conference on the Seas of Japan and Okhotsk, in Nakhodka, USSR, September 1989.

conduct marine research within the jurisdiction of the countries of East Asia (including all Soviet areas) amounted to less than 3% of total U.S. requests in 1987–88, according to Department of State information.[65] Additionally, some requests are not handled through diplomatic channels. Lack of U.S. interest can be attributed to distance and scientific topic, as well as difficulty in obtaining clearance for ship-borne research. Although a country has exclusive rights to grant or withhold permission to conduct research in its marine areas, it does not often exercise an effective monopoly on satisfying the interests of marine scientists from other countries, who demonstrate surprising mobility concerning where they conduct research. An American marine scientist might prefer to work in East Asia, but other areas can be substituted. This was evident in the period 1981–83, when no U.S. ships applied for clearance in this area (fig. 6).

Recent American research in East Asian seas also shows a bias toward repeat cruises. For instance, Japan consistently approves clearance requests. In every year from 1984 through 1988 American vessels conducted cruises there (13 in all). The USSR denied most clearance requests until 1985, when no U.S. requests were made. In 1986, the USSR granted one clearance, and there appears to be subsequent movement toward cooperation. China granted clearances to American researchers in 1985 and 1986, but the two cruises encountered so many problems, some related to the presence of South Korean researchers on the U.S. vessels, that no U.S. requests were made in the subsequent 2 years.

Lack of diplomatic relations remain an impediment to research initiatives. U.S. researchers are advised to stay out of North Korean waters and a 50-nm buffer zone owing to a lack of diplomatic relations. Further, the U.S. Navy advises against any private U.S. vessel entering North Korean waters because of security considerations and because the navy does not want to be asked to respond to a U.S. vessel in distress in that area.

POLICY ISSUES

Two clusters of policy issues relate to marine scientific research in East Asian seas. One set confronts traditional researching countries, who must adapt to the new international restrictions on freedom of research and respond to the economic and political maturation of countries in the region. The second set challenges the coastal states to participate in marine science activities and coordinate national policy with international norms. The clusters are connected by an international bargain reached during the UN Law of the Sea

65. Mr. Thomas Cocke, the Research Vessel Clearance Officer, Office of Marine Science and Polar Affairs, U.S. Department of State, kindly provided data on U.S. clearance requests and permissions.

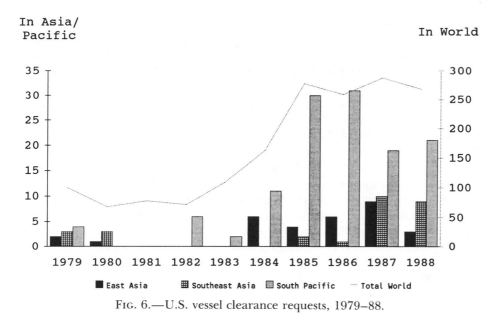

In Asia/
Pacific

In World

FIG. 6.—U.S. vessel clearance requests, 1979–88.

negotiations: the coastal state trades access to natural phenomena within its jurisdictional zone to the researching state, which provides knowledge and enhanced research capabilities. In East Asia some of the coastal states, notably the USSR and Japan, play both roles and confront both sets of issues, depending on where the research is to be conducted and who the other interested countries are.

Issues for Researching Countries

The informal, bilateral approach (i.e., scientist to scientist) to securing coastal state approval to conduct scientific research within its maritime jurisdiction does not appear to hold much promise in confined East Asian seas. The central role of government in marine scientific research in the countries of the region also accounts for a tendency toward formal procedures. Although awkward for Western researchers, the multilateral approach may be most fruitful, especially for developing continuing institutional relationships. Obviously, individual marine scientists and institutions are already moving in this direction, but intergovernmental connections should be reinforced, so that processing requests becomes routine for both researching country and coastal state.

Coastal states are wary of research requests and the potential for misusing results and data. If they do not have the scientific capacity to evaluate an application, they are more likely than not to deny a request, rather than to risk marine resources. Ignorance is a form of security, since what is unknown to all is safe from seizure. Coastal states' preoccupation with resource exploitation and sensitivity to security, no matter how ill-founded, must be accommodated by the researching countries. The task is essentially an educational one. In this regard, Western countries should foster the new role of international organizations mandated by the 1982 Convention. Competent international organizations are to be instrumental in establishing general criteria and guidelines to assist coastal countries in ascertaining the nature and implications of marine scientific research. Further, such organizations are recommended as a way to promote the transfer of marine science and technology.[66] International organizations, such as the International Oceanographic Commission (IOC) and the International Hydrographic Organization (IHO), among others, will need the assistance of scientists from developed countries to play this new role. Perhaps more critically, they will need the support of Western countries to evolve into honest brokers of marine scientific research activities.

The issue of the transfer of technology may well prove the stickiest problem in the quest for better cooperation in ocean research. As stated earlier, marine science and marine technology are often mentioned in the same breath by developing coastal countries and by the framers of the 1982 Convention. Because resource exploitation is a primary objective of the coastal states, the demand for researching countries to transfer technology will be a likely condition for coastal-states' consent to research activities within their jurisdictions. The fact that academic researchers do not own the technology that they use, such as patents on scientific equipment, must be explained to government bureaucrats in developing countries. One possible approach is to interpret "technology transfer" in its broadest sense to include training. At least training is one aspect of "technology" that Western academic researchers are capable of providing.

Although the coastal states of East Asia have taken no extreme stands on technology transfer, it seems likely that they will emphasize it in the future. This area is better prepared to assimilate marine scientific training and technology transfer than many other developing regions.

Issues for Coastal States

Like the researching countries, the coastal states must adapt to a new world order in the oceans. They must address several issues, such as how to allocate resources among competing scientific goals, how to formulate laws in accord with international norms, and how to undertake marine science beyond zones of national jurisdiction. In many ways, these issues reflect broader problems

66. 1982 Convention, Arts. 251, 266.

of science policy and international law. The coastal states must balance concerns for national development, which suggest concentration on applied science and resource exploration, with considerations of international stature and cooperation, which promote deep ocean research and cooperation with neighboring countries.

The coastal states will need to be realistic, indeed sophisticated, in demands for technology transfer. They should formulate clear regulations in accord with international standards and simplify clearance procedures. In East Asia, it would seem that virtually all the countries, except the USSR, have tilted too much toward practical, applied marine science, and only now are some going beyond parochial research and venturing outside areas of national jurisdiction.

Potential Cooperative Research Topics

Certain research topics seem well suited to serving as the object of cooperative study. These topics should be of significant interest to the coastal states to warrant investment of precious scientific resources. Further, given the sensitivities of coastal states to sharing marine resource–related information, the topics should be "nonthreatening." Topics that are not associated too closely with exploitation or that address clear, transboundary issues should be given priority. Moreover, subjects of general scientific interest or those related to the work of established international scientific organizations present greater susceptibility to international cooperation. Three such topics suggest themselves as opportune.

Pollution Management and Baseline Studies

In environmental matters, the UN Environment Programme (UNEP) is entertaining a North Pacific Initiative for a regional seas–like program to address issues in the Northwest Pacific, including East Asian seas. Because the Regional Seas Programme is generally constrained to assisting developing countries (Recommendation 8 of the 1982 Nairobi meeting) and developed countries will be primary beneficiaries of a regime for the Northwest Pacific, UNEP is addressing this region under the Oceans and Coastal Areas Programme Activity Centre (OCA/PAC).[67]

East Asian seas are prime candidates for enhanced pollution management, which can best be effected in semienclosed seas through international

67. UN Environment Programme (UNEP), *Achievements and Planned Development of UNEP's Regional Seas Programme and Comparable Programmes Sponsored by Other Bodies,* UNEP Regional Seas Reports and Studies no. 1 (1982), p. ii; UNEP Oceans and Coastal Areas Programme (Nairobi), "Oceans a Priority for UNEP," *Siren,* no. 41 (July 1989): 1–2.

cooperation. A comprehensive picture of the marine environment in East Asian seas could emerge if the respective researching countries intercalibrated instruments (where possible), coordinated marine research, and shared data, but they have demonstrated relatively little cooperation to date. Perhaps the most pressing problem is the preparation of baseline studies that comprehend entire ecosystems, such as the Yellow and East China Seas. The frequency and kinds of sampling for pollution studies depend upon the areas under study and the scientific capacities of the investigators. Multilateral or bilateral agreements, for example, should address the indicator organisms for pollution sampling. They should provide for sampling to be done jointly and for samples to be split for analysis in each country. Analytic methods should be comparable among involved laboratories. Multilateral arrangements should also provide for the exchange of scientists and data.[68] Until these measures are taken, the coastal states will have insufficient information to inform pollution abatement policy.

Global Climate Change
Current concern with global climate change may offer a second opportunity for cooperative marine scientific research in East Asia. The mechanisms coupling the ocean to the atmosphere are poorly understood for purposes of climate prediction. Observations made in East Asian seas could refine global climate models, without presenting much of a risk of strategic or economic compromise to coastal state interests. Such programs could also build working relationships under international organizations, such as World Meteorological Organization (WMO), to which most of the coastal countries belong.

Fisheries Management
Perhaps the greatest regional need is for fisheries research to inform management of overexploited stocks. Many of the fishery resources of the region are supported by large ecosystems, such as the East China and Yellow Seas, that transcend single national maritime jurisdictions. However, East Asian countries have been reluctant to share data—probably for fear that the other country will gain an advantage in negotiating fishing agreements. Nonetheless, until there is wider dissemination of fisheries data, the requisite regional management of the ecosystem will be impossible.

CONCLUSION

East Asian seas exhibit many of the currents buffeting marine scientific research elsewhere. The informal approach of researching countries and West-

68. Kwang Woo Lee, "Prospective Environmental Monitoring in the Yellow Sea," in *Marine Resource Development in the Yellow and East China Seas*, summary proceedings of a workshop sponsored by the Institute for Marine and Coastal Studies, University of Southern California, and KORDI (1984), pp. 28–29.

ern scientists will give way to institutional arrangements between governments. This is dictated by the greater role of governments in science in
developing countries and by changes in the international legal system. Technology transfer is likely to be a condition for obtaining clearance to conduct
research in coastal state jurisdiction. Western researching organizations, especially academic institutions, may be unable to transfer patented technology,
but they may be able to address the desires of the coastal states by providing
"soft technology" in the form of training to scientists from developing countries.

Because of the geographic and political complexities of East Asia, certain
multinational forums appear better suited for regional cooperation than bilateral arrangements. Most phenomena of interest cross several national jurisdictions in the region's semienclosed seas and require clearance from more
than one coastal state. East Asian marine science organizations are developing
better research capabilities, especially in resource-related investigations, and
undertaking more distant-water cruises. These tendencies suggest greater
multinational cooperation, especially through intergovernmental organizations.

For research involving North Korea, intergovernmental organizations
under the United Nations may provide the best venue for cooperation. Due
to the unique status of Taiwan, its participation appears more likely if nongovernmental forums are used. In both cases, history suggests that including
extraregional participants facilitates cooperation, perhaps by diluting direct
bilateral contact among rival countries. It seems clear from the foregoing
analysis that the future of marine science in the region will be framed by
multilateral cooperation: cooperation among neighboring coastal states and
cooperation between coastal states and researching countries. Without such
teamwork there may be rough seas ahead for marine scientific investigation
in East Asian seas.

Save the Seas: UNEP's Regional Seas Programme and the Coordination of Regional Pollution Control Efforts

Peter M. Haas
University of Massachusetts

INTRODUCTION

Marine pollution has generated widespread international concern since the early 1970s. The predominant international effort to manage this problem has been through the United Nations Environment Programme's (UNEP) Regional Seas Programme, which now administers programs in 10 regions, involving 130 countries, 16 UN bodies, and more than 40 other organizations, at a total cost of US $127 million through the end of 1989.[1] Of the 12 international arrangements explicitly developed to coordinate regional pollution control, 10 have been developed under the auspices of UNEP.[2] This article

---

1. United Nations Environment Programme (UNEP), *UNEP Oceans Programme: Compendium of Projects,* UNEP Regional Seas Reports and Studies no. 19, rev. 4 (corr) (Nairobi: UNEP, 1989–90), p. 47. This figure includes US$41.6 million from UNEP's Environmental Trust Fund, $8.8 million from participating governments, and $46.8 million from other UN agencies, largely in the form of contributions of staff time, travel, and equipment. This figure does not include the cost of national support to national marine science institutions in support of UNEP-sponsored research and monitoring projects. This money was allocated as follows: Mediterranean 20.5%, Caribbean 13.8%, West and Central Africa 9.7%, South Pacific 6.2%, Southeast Pacific 6.1%, East Asian seas 5.7%, East Africa 4.3%, Gulf of Kuwait 2.7%, South Asian seas 2.0%, Red Sea 1.1%, Southwest Atlantic 0.2%, and other global programs 27.9%. The Regional Seas Programme was originally located in Geneva. With its transfer to Nairobi in 1985, its name was changed to Oceans and Coastal Areas Programme (OCA/PAC), and its responsibilities were expanded to include coastal zone management and marine conservation.
2. For UNEP action plans and the dates they were adopted, see figure 1 and App. G, "Status of Participation in UNEP's Legal Instruments for Marine Environmental Protection," in this volume. Efforts to draft arrangements for the Southwest Atlantic (Brazil, Chile, and Uruguay) were abandoned in 1983 at Brazil's request. UNEP is currently considering developing new arrangements for the Black Sea and Northwest Pacific (China, Japan, North Korea, South Korea, and USSR). Non–UNEP-spawned agreements exist for the North Sea and the Baltic Sea. See David Edwards, "Review of the Status of Implementation and Development of Regional

analyzes and evaluates this program, describing its evolution and analyzing the determinants of successes and failures within it. In the absence of detailed information on environmental quality in the regions, and in particular the absence of time series data by which changes in quality may be determined, an evaluation of the relative successes of these arrangements must be based on the accomplishment of the region's desires as identified in the action plans, and in terms of momentum over time toward monitoring and controlling a more comprehensive set of environmental contaminants.[3]

ORGANIZATIONAL HISTORY

The general framework of each region's activities is similar, although each varies according to the specific problems encountered in its region. The model was developed for the Mediterranean and has been roughly applied in each other region.[4] Thus, UNEP pursues a common approach, replete

Arrangements on Cooperation in Combating Marine Pollution," in *International Environmental Diplomacy,* ed. John E. Carroll (Cambridge: Cambridge University Press, 1988); Lynton K. Caldwell, *International Environmental Policy* (Durham, North Carolina: Duke University Press, 1984); Peter H. Sand, *Marine Environmental Law in the United Nations Environmental Programme* (London: Tycooly Publishing, 1988). For a more extensive analysis of the development of the Mediterranean model, see Peter M. Haas, *Saving the Mediterranean: The Politics of International Environmental Cooperation* (New York: Columbia University Press, 1990). For a review of all regimes applying to the marine environment, see Boleslaw A. Boczek, "The Concept of Regime and the Protection and Preservation of the Marine Environment," in *Ocean Yearbook 6,* ed. Elisabeth Mann Borgese and Norton Ginsburg (Chicago: University of Chicago Press, 1986), pp. 271–97.

3. A fuller assessment should be based on a closer study of national practices, including investment, legislation, and policy, in order to determine whether any fuller enforcement of these public commitments has in fact occurred. This is impossible to do without extensive fieldwork, as little of such data is available for most countries. For an effort to do such an appraisal for the Mediterranean see Haas (n. 2 above), chap. 5; and in a slightly different sphere for (mixed) U.K. compliance with EEC guidelines, see Nigel Haigh, *EEC Environmental Policy and Britain,* 2d ed. (London: Environmental Data Services, 1988).

4. Stjepan Keckes, "Theory and Practice of the United Nations Environment Programme in Dealing with Regional Marine Problems," *Thalassia Jugoslavica* 13, no. 3/4 (1977): 217–38. Keckes, one of the architects of the Mediterranean strategy, notes (pp. 218–19), "The Mediterranean Action Plan, as adopted at the Intergovernmental Meeting on the Protection of the Mediterranean (Barcelona, 28 January–4 February, 1975) and developed since then is used as a model for a comprehensive programme aiming at the protection and development of regional seas. However, it should be recognized that the approach used in the Mediterranean region cannot be copied mechanically in all regions due to variations in the state of knowledge, the information and human resources available, and other regional characteristics. It must be assumed that the nature of environmental problems will vary considerably between each of the regions selected as priority areas by the Governing Council. Each regional grouping

with virtual "boilerplate clauses" of international treaties, while remaining flexible and sensitive to the different ecological conditions of each region and the needs of the region's governments. For instance, conservation is a key component for the Red Sea action plan, whereas in the Caribbean, development is stressed. In the East Asian seas, a more "holistic and interdisciplinary approach" is taken to coordinate multiple uses of the marine environment. The programs for West and East African coasts are oriented toward research and management for coastal erosion control. While the Mediterranean primarily focused on controlling multiple sources of pollution, due to the fairly high level of industrialization in the littoral zone, the South Pacific and East African regions are much more concerned about protection of species, given the relative absence of extensive pollution in those areas. Oil pollution is a common concern in all the regions.

The fundamental orientations of several of the regions also vary, as revealed by the phrasing in the preambles to the regional conventions. The Caribbean explicitly asserts the objectives of "management and development" of the environment, whereas the Red Sea and Kuwait qualify the need to minimize the impact of development on the environment "as far as possible," while the Mediterranean, South Pacific, and West African regions seem to favor the environment over development, viewing the interplay in light of a broader context of "ecological equilibrium." Monitoring in each region also tests for substances unique to the region and is adaptive to new problems as they emerge or are recognized, such as coastal erosion in Africa and seasonal jellyfish blooms and algae infestations in the Mediterranean. Table 1 indicates the variation of problems encountered among the different regions.

The 1972 United Nations Conference on the Human Environment (UNCHE) developed both the standard form for the Regional Seas Programme—concurrent assessment and management activities supported by institutional and financial arrangements—and the regional focus for controlling marine pollution.[5] At UNCHE, states adopted principles that recognized that the marine environment and all the living organisms which it supports are of vital importance to humanity and recognized that proper management is required and measures to prevent and control marine pollution must be

---

of Governments will perceive differently the common problems which they wish to resolve through co-operative programmes, and it cannot be overly emphasized that it is these Governments in whom ultimate responsibility rests to ensure wise management of the common resources."

5. The Stockholm Conference adopted an Action Plan for the Human Environment, with 109 recommendations organized within a framework consisting of environmental assessment, environmental management activities, and national, international, and educational supporting measures (United Nations, *Report of the United Nations Conference on the Human Environment* [New York: United Nations, 1973], Doc. A/Conf. 48/14/Rev. 1, chap. 2).

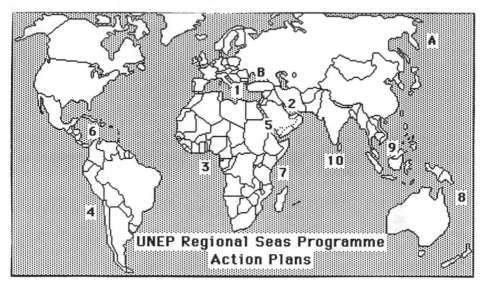

Fig. 1.—United Nations Environment Programme's Regional Seas Programme action plans (UNEP, *"Achievements,"* n.d.).

1 Mediterranean Action Plan
2 Action Plan for the Protection and Development of the Marine Environment and the Coastal Areas of Bahrain, Iran, Iraq, Kuwait, Oman, Qatar, Saudi Arabia, and the United Arab Emirates
3 Action Plan for the Protection and Development of the Marine Environment and Coastal Areas of the West and Central African Region
4 Action Plan for the Protection of the Marine Environment and Coastal Areas of the Southeast Pacific
5 Action Plan for the Protection of the Marine Environment and Development of Coastal Areas in the Red Sea and Gulf of Aden
6 Action Plan for the Caribbean Environment Programme
7 Action Plan for the Protection and Development of the Marine and Coastal Environment of the Eastern African Region
8 Action Plan for Managing the Natural Resources and Environment of the South Pacific Region
9 Action Plan for the Protection and Development of the Marine and Coastal Areas of the East Asian Seas Region
10 South Asian Seas Action Plan (1991)
   A Northwest Pacific (in preparation)
   B Black Sea (in preparation)

TABLE 1.—HOW THE SEAS ARE CONTAMINATED

Major Sources of Contamination	Baltic	North Sea	Mediterranean	Kuwait Action Plan Region	West African Areas	South African Areas	Indian Ocean Region	Southeast Asian Region	Japanese Coastal Waters	North American Areas	Caribbean Sea	Southwest Atlantic Region	Southeast Pacific Region	Australian Areas	New Zealand Coastal Waters
Sewage	X	X	X	X	X	X	X	X	X	X	X	X	X	X	X
Petroleum hydrocarbons (maritime transport)	X	X	X	X	X	X	X	X	X	X	X	X	X	X	
Petroleum hydrocarbons (exploration and exploitation)		X	X	X	X	X		X	X	X	X	X			
Petrochemical industry	X	X	X				X	X	X	X	X				
Mining	X	X	X				X	X		X			X	X	
Radioactive wastes		X								X					
Food and beverage processing	X	X	X	X	X	X	X	X	X	X	X	X	X	X	X
Metal industries	X	X	X	X	X	X	X	X	X	X	X	X	X	X	X
Chemical industries	X	X	X	X	X	X	X	X	X	X	X	X	X	X	
Pulp and paper manufacture	X				X	X		X	X	X	X	X	X	X	X
Agricultural runoff (pesticides and fertilizers)	X	X	X	X	X	X	X	X	X	X	X	X	X	X	X
Siltation from agriculture and coastal development				X	X	X	X	X			X				
Sea-salt extraction							X				X				
Thermal effluents	X		X	X		X		X	X	X	X	X	X	X	
Dumping of sewage and dredge	X	X							X	X					

SOURCE.—United Nations Environment Programme, "Cleaning Up the Seas," *UNEP Environment Brief* no. 5 (1989): 3.

regarded as an essential element in this management.[6] At its first session in 1973 and at subsequent sessions, UNEP's Governing Council identified "oceans" as one of the six priority "subject areas" in which UNEP should exercise its catalytic and coordinating role.[7] In 1974 the Regional Seas Programme was initiated, and in 1977 a Regional Seas Programme Activity Centre was established in Geneva to coordinate the efforts of those involved in the program. The UNEP Governing Council noted that "it was felt that the action taken on the Mediterranean should serve as a model for action in other marine eco-regions."[8] Scientists gradually recognized that the most urgent marine threats were in coastal areas, and that the threats to the open seas were less acute than suspected, so the program focused its energies on protecting coastal areas.[9] Since its inception, the oceans component has run between 10% and 11% of the UNEP budget.

While many of these programs are regional in scope, they seldom respond to formal collective goods problems.[10] The flow of pollutants from one country to another is seldom sufficiently great to require cooperation. Rather, regions are defined politically by the regional governments who approach UNEP and request the development of a regional program. The regional notion was not in fact even formally adopted by the Governing Council until 1978, by which time activities were already underway in six areas spanning semienclosed seas, coastal waters, and archipelagos.[11] Still, concurrent action by the regional countries assures that there is reasonably balanced momentum in the region toward environmental policy. It also assures the adoption of consistent standards within the regions and provides an opportunity for countries to compare their experiences.[12] In fact, many subsequent projects within the Regional Seas Programme—perhaps those less well conceived—have dealt solely with unique problems encountered by countries within the region, rather than with problems involving shared resources or problems facing all countries individually.

Following the preparation of the action plan for the Mediterranean in

6. Ibid., pp. 4–5.

7. UNEP/GC/Dec. 1 E. Oceans (1973); UNEP/GC/Dec./8 (II) (1974); UNEP/GC10, Annex 1 (1982).

8. UNEP/GC/55 Dec.

9. Patricia A. Bliss-Guest and Stjepan Keckes, "The Regional Seas Programme of UNEP," *Environmental Conservation* 9, no. 1 (Spring 1982): 43–49.

10. For a consideration of the ubiquitous problems typically encountered in collective environmental protection efforts at managing such common problems, see Oran R. Young, *International Cooperation Building Regimes for Natural Resources and the Environment* (Ithaca, New York: Cornell University Press, 1989).

11. Peter H. Sand, "The Rise of Regional Agreements for Marine Environment Protection," in Food and Agriculture Organization, *The Law and the Sea: Essays in Memory of Jean Carroz* (Rome: FAO, 1987), pp. 223–32.

12. Patricia A. Bliss-Guest, "Environmental Stress in the East African Region," *Ambio* 12, no. 6 (1983): 295.

1974, other regions have broadly attempted to replicate the same strategy and same set of outcomes. By 1977 UNEP was moving concurrently in the Mediterranean, Persian Gulf, Caribbean, West African region (Gulf of Guinea), Red Sea, and Straits of Malacca and Singapore (which became the "East Asian Seas"). By 1985 UNEP had added the Southeast Pacific, South Pacific, East African region, and south Asian seas.

Typically, UNEP is invited to develop a regional seas action plan. An interagency mission is sent to the region to assess local scientific capabilities, evaluate the sources of regional environmental contamination and the extent of governmental commitment, and identify primary contact points in the various governments. Regional marine science capabilities are surveyed. A meeting of regional scientists is held to agree on common regional environmental threats and major research and monitoring priorities; it then devises an action plan which it submits to the governments for adoption.[13] Figure 2 depicts this process.

Such an approach gives UNEP considerable discretion in formulating projects for the regions' governments, as well as in allocating finances during frequent budgetary shortfalls. UNEP proposes programs to governments for adoption and pressures them to comply once the programs have been approved. Many of the governments actively rely on UNEP to provide and convey information about the quality of the environment and about their own and others' activities to control pollution as well as the major sources of pollution, and to maintain the day-to-day momentum of compliance with international regimes.

Each action plan consists of a number of components:

1. *Assessment.* Regional projects assess the state of marine pollution, of the sources and trends of this pollution, and of the impact of the pollution on human health, marine ecosystems, and amenities. Such projects consist largely of baseline studies, monitoring, and research activities. UNEP has developed 51 reference methods to standardize techniques for data gathering and analysis.

2. *Management.* UNEP seeks to coordinate environmental management efforts by organizing regional training in such activities as environmental impact assessment; management of coastal lagoons, estuaries, and mangrove ecosystems; protection of endangered species; control of industrial, agricul-

---

13. UNEP, "Guidelines and Principles concerning a Comprehensive Action Plan for the Protection of Regional Seas through Environmental Sound Development," UNEP/IAMRS.1/6 Annex II (18 June 1976); UNEP, *Achievements and Planned Development of UNEP's Regional Seas Programme and Comparable Programmes Sponsored by Other Bodies,* UNEP Regional Seas Reports and Studies no. 1 (Geneva: UNEP, 1982); UNEP, *Guidelines and Principles for the Preparation and Implementation of Comprehensive Action Plans for the Protection and Development of Marine and Coastal Areas of Regional Seas,* UNEP Regional Seas Reports and Studies no. 15 (Geneva: UNEP, 1982).

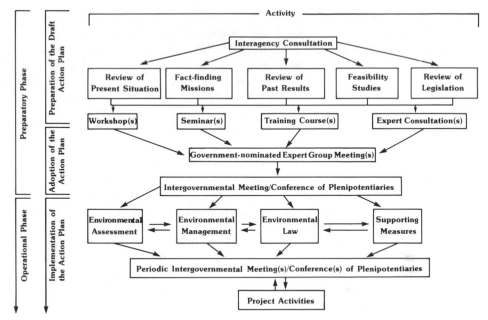

Fig. 2.—Flowchart of a simplified and generalized scheme for the development and implementation of regional action plans (United Nations Environment Programme, *Guidelines and Principles for the Preparation and Implementation of Comprehensive Action Plans for the Protection and Development of Marine and Coastal Areas of Regional Seas,* UNEP Regional Seas Reports and Studies no. 15 [Geneva: UNEP, 1982], p. 8).

tural, and domestic wastes; and formulation of contingency plans for dealing with pollution emergencies.

3. *Legal.* Each region promotes international and regional conventions, guidelines, and policies for the control of marine pollution and for the protection and management of marine and coastal resources. They typically include an umbrella convention outlining the problems to be managed, supported with more detailed protocols on specific sources of pollution. This process, although timely, allows for greater flexibility in devising protocols appropriate to the specific needs of the region, based on environmental data generated by the assessment activities. The process also creates a learning period, during which participants grow more confident at negotiating such arrangements and more sophisticated in the understanding of the scope of difficulties to be encountered. Domestic support groups are enlisted to drive the process. Legal arrangements often typically include a blacklist of banned highly toxic substances and a graylist of less toxic substances that require permits for their emission.

4. *Institutional arrangements.* An organization that acts as the permanent or interim secretariat of the action plan is established. Governments are ex-

pected to decide upon the frequency of intergovernmental meetings that will review the progress of the agreed work plan and approve new activities and the necessary budgetary support. UNEP often provides interim secretariat support for several years until a regional organization can be identified or created, and the local trust fund becomes operational.

5. *Financial arrangements.* UNEP, together with other UN agencies and organizations, provides "seed money" or catalytic financing in the early stages of regional programs. As both environmental assessment and environmental management activities are to be carried out by designated national institutions, assistance and training are provided, where necessary. As a program develops, however, UNEP expects that the governments of the region will progressively assume full financial responsibility and the program will become self-supporting, although UNEP will often continue to provide expert assistance and logistical guidance. Government financing is usually channelled through special regional trust funds to which governments make annual contributions. These funds are administered by the organization responsible for the secretariat function of the action plan. In addition, governments may contribute directly to the national institutions participating in the program or to specific project activities.

6. *Educational and support activities.* UNEP also seeks to generate public interest and awareness in member countries through environmental education, media, circulating information, educational displays, and popular brochures about the regional activities.

Due to its modest financial and staffing resources, UNEP has had to coordinate its actions with other specialized agencies. UNEP has developed amicable relations with other agencies with complementary missions, many of which have helped develop and support the Regional Seas Programme. Key among these have been the Food and Agriculture Organization (FAO)—for pollution research—at US $11.7 million; International Oceanographic Commission (IOC)—for pollution research—at $10.4 million; World Health Organization (WHO)—for public health aspects and evaluating data—$9.9 million; International Maritime Organization (IMO)—for training in oil-spill response—$5.3 million; World Meteorological Organization (WMO)—for circulation studies—$7.3 million; Unesco $8.2 million; and International Atomic Energy Agency (IAEA)—for equipment intercalibration and training in its use—$10.7 million.[14] The International Union for the Conservation of Nature (IUCN), although contributing only $1 million, has been involved with all projects relating to specially protected areas. The Group of Experts on the Scientific Aspects of Marine Pollution (GESAMP) has prepared background reports on the state of the oceans and atmospheric transport of pollution, as well as preparing methodological guidelines for monitoring marine quality and for undertaking environmental impact assessments. At times,

14. UNEP (n. 1 above), pp. 63–75.

UNEP has responded to initiatives from regional organizations such as the Arab League Educational, Cultural, and Scientific Organization (ALECSO) in the Red Sea, Association of Southeast Asian Nations (ASEAN) on East Asian seas, South Pacific Commission (SPC) and South Pacific Bureau for Economic Cooperation (SPEC) for the South Pacific, and the Comisión Permanente del Pacífico Sur (CPPS) for the Southeast Pacific.

Some organizations have different missions and institutional worldviews, and relations with UNEP have been fractious at times. For example, the World Bank, with a responsibility for promoting economic development, considers environmental quality as one factor among many to be optimized. On the other hand UNEP, largely staffed by natural scientists, adopts a more principled approach that a clean environment is to be preserved regardless of the cost. Consequently, coordinating activities and priorities between the two organizations is very difficult.

While UNEP is charged by regional countries to formulate and develop regional action plans, underlying this is a deeper set of objectives: promoting more comprehensive environmental management and planning. UNEP acknowledges that the Regional Seas Programme

> is conceived as an action-oriented programme having concern not only for the consequences but also for the causes of environmental degradation and encompassing a comprehensive approach to combating environmental problems through the management of marine and coastal areas. . . . It is designed to link assessment of the quality of the marine environment and the causes of its deterioration with activities for the management and development of the marine and coastal environment.[15]

In short, it aspires to nothing less than inculcating national policy makers with a new approach to coastal zone management and economic development, one that integrates environmental concerns at the outset of planning by applying scientific competence to technical environmental management.

If, in keeping with the action plan's emphasis on improving understanding of regional marine environmental conditions, all that UNEP required were good data on the environment, the data could have been collected more rapidly, more cheaply, and with higher quality results by contracting the task to one country in the region or to private laboratories. Rather, motivated by a broader strategy of creating support for comprehensive environmental management from the regions' scientific communities, as well as promoting indigenous technical growth, the action plans are directed at mobilizing na-

15. UNEP/IAMRS.6/INF.3, p. 6, Sec. 4.1.2; "Information on the Status and Planned Development of UNEP-Sponsored Programme for the Protection of Oceans and Coastal Areas," 31 August 1989. See also Bliss-Guest and Keckes (n. 9 above), p. 44.

tional participation and growth. The director of the Regional Seas Programme, Yugoslav oceanographer Stjepan Keckes, is an active believer in learning by doing and collective self-reliance.[16] By participating in joint assessment activities, national scientific capabilities will be improved, collective attention will be focused on a new set of problems, and governments will become more competent at managing marine environmental issues. In fact, governments have proved to be much more responsive to scientific advice offered by domestic scientists rather than by outside sources, who are suspected of giving distorted information or serving an ulterior political motive.

In support of this "masterplan," action plan components are intended to be interdependent, creating a mosaic of mutually reinforcing activities. The assessment activities generate alarm about environmental conditions and put such issues on the political agenda. Participating in assessment activities improves the technical capabilities of regional marine scientists so they can provide good data on which to base their policy proposals for their governments, which attracts them to the UNEP initiative. The management component indicates to political decision makers that the management tasks exceed their technical competence and requires policy advice from the scientific community. Directories of regional scientists compiled and published by UNEP facilitate contacts and joint research within the scientific community. The legal component creates instruments for them to enforce. Whereas networking with regional scientists is likely to have a political payoff within several years, its outreach and education efforts are aimed at creating a new environmental constituency among future generations.

The various regional seas efforts have met with differential successes. The most successful ones have satisfied three preconditions:

1. the existence of a regional community of able marine scientists interested in environmental management applications of their work;
2. the respect of political decision makers for the authority and expertise of these scientists; and
3. the existence of actual channels of contact or influence between the scientific community and national policy makers.

UNEP has actively tried to satisfy these preconditions by reinforcing scientific communities and seeking to improve their access to political decision makers.

16. Peter Hulm, "The Regional Seas Programme: What Fate for UNEP's Crown Jewels?" *Ambio* 12, no. 1 (1983): 12–13. Keckes wrote that the "key to success of any regional seas action plan is the political agreement of the governments concerned and the execution of the programme primarily by national institutions from the region, in close co-operation with the relevant specialized organizations of the United Nations system and other appropriate organizations relevant to the region" (Stjepan Keckes, "The UNEP Sponsored Regional Seas Programme," in UNEP, *Co-operation for Environmental Protection in the Pacific,* UNEP Regional Seas Reports and Studies no. 97 [Nairobi: UNEP, 1988]).

Thus, UNEP serves principally as a facilitator, putting together networks that would otherwise not know of each other or be capable of communicating with each other, and transferring information and equipment to them. In regions where tightly knit regional scientific infrastructure and networks already exist, such as the Mediterranean, UNEP merely serves as a broker. In other regions, where the indigenous scientific capability is much weaker, such as West and East Africa, UNEP attempts to create the infrastructure through the provision of equipment, training, and financial support for monitoring activities. In addition, in regional negotiations UNEP serves as a buffer between political antagonists. By negotiating arrangements through UNEP, rather than directly between governments, many north/south tensions are tempered.

ANALYSIS

The most successful programs are those in which the three preconditions above have been satisfied. The plans for the Mediterranean, South Pacific, and Southeast Pacific have actively developed legal arrangements; monitoring has been the most active for any of the regions; more management activities have been undertaken; momentum has persisted; and the overall efforts have been more comprehensive than elsewhere. Widespread scientific capability existed within each of these regions, and UNEP was able to extend its influence to national administrations. Action plans have been less successful where the scientific capability was not present or scientists lacked access to their government's decision-making processes.

Successful Action Plans

Each of the successful regions has an active network of policy-oriented regional marine scientists who had established and maintained contact with their governments.

The Mediterranean Action Plan (Med Plan) was the first action plan, adopted in 1975.[17] With more time and money than any other region, it has

17. Editors' note.—The *Ocean Yearbook* volumes have closely followed the efforts to establish environmental and conservation regimes in the Mediterranean. Some of the articles from previous volumes relevant to this article are listed here: "UN Environmental Programme: Activities for the Protection and Development of the Mediterranean Region," *Ocean Yearbook* 1, ed. Elisabeth Mann Borgese and Norton Ginsburg (Chicago: University of Chicago Press, 1978), pp. 548–97; "Recommendations for the Future Development of the Mediterranean Action Plan," in *Ocean Yearbook* 2, ed. Elisabeth Mann Borgese and Norton Ginsburg (Chicago: University of Chicago Press, 1980), pp. 547–54; Peter S. Thacher and Nikki Meith, "Approaches

progressed further than any of the other regions. Five treaties have been signed and are in force: Convention (signed 1976/in force 1978), Dumping Protocol (1976/1978), Emergency Protocol (1976/1978), Land-Based Sources Protocol (1980/1983), and Specially Protected Areas Protocol (1982/1986).[18] Joint monitoring in the region concluded a first phase conducted from 1976 to 1981, in which 86 laboratories in 16 countries coordinated measurements to generate baseline studies of contamination in the region from mercury, microbial pollution, and oil.[19] The second phase of the program, 1981–91, is designed to help implement the legal agreements in the region by generating data applicable to the development of specific emission and ambient standards to regulate land-based sources of pollution. Thirty laboratories from 8 countries now participate in the second phase.[20] In 1976, a Regional Oil Combating Centre (ROCC) was established in Malta to facilitate the flow of information about oil-spill emergencies, to promote training in control techniques, and to help draft oil-spill contingency plans. A total of 309 individuals from all countries, except Albania, have been trained in oil-spill control techniques.[21] Since 1986 a number of projects on concrete, yet generalizable, management styles, often unrelated to maritime impacts of social

---

to Regional Marine Problems: A Progress Report on UNEP's Regional Seas Programme," in *Ocean Yearbook 2,* ed. Elisabeth Mann Borgese and Norton Ginsburg (Chicago: University of Chicago Press, 1980), pp. 153–82 (the Mediterranean Action Plan is discussed on pp. 158–68); "UNEP: An Update on the Regional Seas Programme," in *Ocean Yearbook 4,* ed. Elisabeth Mann Borgese and Norton Ginsburg (Chicago: University of Chicago Press, 1982), pp. 450–61; Adalberta Vallega, "A Human Geographical Approach to Semienclosed Seas: The Mediterranean Case," in *Ocean Yearbook 7,* ed. Elisabeth Mann Borgese, Norton Ginsburg, and Joseph R. Morgan (Chicago: University of Chicago Press, 1988), pp. 372–93; Aldo Chircop, "Participation in Marine Regionalism: An Appraisal in a Mediterranean Context," in *Ocean Yearbook 8,* ed. Elisabeth Mann Borgese, Norton Ginsburg, and Joseph R. Morgan (Chicago: University of Chicago Press, 1989), pp. 402–16.

18. UNEP, Food and Agriculture Organization, and World Health Organization, *Assessment of the State of Pollution of the Mediterranean Sea by Petroleum Hydrocarbons,* MAP Technical Reports Series no. 18 (Athens: UNEP, 1987); UNEP and Intergovernmental Oceanographic Commission, *Assessment of the State of Pollution of the Mediterranean Sea by Petroleum Hydrocarbons,* MAP Technical Reports Series no. 19 (Athens: UNEP, 1988); UNEP and World Health Organization, *Epidemiological Studies Related to Environmental Quality Criteria for Bathing Waters, Shellfish-Growing Waters, and Edible Marine Organisms (Activity D): Final Report on Project on Relationship between Microbial Quality of Coastal Seawater and Health Effects (1983–86),* MAP Technical Reports Series no. 20 (Athens: UNEP, 1988); UNEP/IG. 56/5 Recs. 5 and 6; UNEP/IG. 75/5 Rec. K 3.

19. See "Status of Participation" (n. 2 above), App. G of this volume.

20. World Health Organization, "Monitoring of Land-Based Sources of Marine Pollution in the Mediterranean," report of a joint WHO/UNEP consultation, Split, EUR/ICP/CEH 044, Annex 2, 1–5 December 1987.

21. Regional Oil Combating Centre, *Directory for the Mediterranean Region of Participants to Marine Pollution Combating Training Courses* (Manoel Island, Malta: ROCC, 1988).

activities, have been undertaken.[22] A coordinating center for marine parks and endangered species proved worthless and was transferred in 1990 from Tunis to the Med Plan headquarters in Athens.[23]

There has been a great contribution to regional scientific capabilities. Equipment was provided throughout the region, and many scientists and technicians received training in monitoring and research techniques, including airborne pollution (15), monitoring (40), determination of microbiological pollution (15), and organotin compounds determination (15). Scientists also received travel grants to present findings at conferences (70) and research grants (30).[24] Table 2 indicates the equipment and assistance provided to regional laboratories participating in the first phase of regional monitoring (MEDPOL Phase 1).

Momentum for regional environmental protection has continued since the 1976 adoption of the Med Plan. Five treaties have been adopted, and in 1987 the Med Plan's direction was "reoriented" to focus more directly on integrated coastal zone management. In 1990 Mediterranean states endorsed such a move in the "Nicosia Charter." Sewage treatment plants have been built in the region (Tel Aviv, Aleppo, Athens, Naples, Genoa, Istanbul, Marseilles, Nice, Toulon, Alexandria, Tripoli, and Algiers), and oil reception facilities for tankers have been built in Greece (Piraeus, Patras, Siras, Thessaloniki), Yugoslavia (Rijeka), and Egypt (Port Said).[25] However, the Mediterranean states have yet to conclude an agreement on an interstate guarantee fund or a treaty on seabed mining.

In the Mediterranean, UNEP was able to tap a tightly knit network of marine scientists. Sixteen of the 18 countries have two or more marine laboratories staffed with a total of 12 or more scientists and technicians with at least a bachelor of science degree.[26] While the number of laboratories and technicians is clearly a proxy variable for the competence and reputation of marine science, the numbers used here do correlate with observers' impres-

22. Land-use planning in earthquake zones; rehabilitation and reconstruction of historic settlements; water resource development of small islands and isolated coastal areas, and large islands; aquaculture development; and soil protection.

23. UNEP, *The Regional Activity Centre for the Mediterranean Specially Protected Areas: Evaluation of Its Development and Achievements*, UNEP Regional Seas Reports and Studies no. 100 (Nairobi: UNEP, 1988).

24. Memo from Aldo Manos to Mr. A. T. Brough, Dr. S. Keckes, 25 September 1989, "MED Umbrella Project," p. 17.

25. Paul Evan Ress, "Mediterranean Sea Becoming Cleaner," *Environmental Conservation* 13 (Autumn 1986): 267–68.

26. Unesco, *International Directory of Marine Scientists*, 3d ed. (Paris: Unesco, 1983); UNEP, *Directory of Mediterranean Marine Research Centers* (UNEP: Geneva, 1977). The numbers for marine science capabilities presented in the following paragraphs draw from the Unesco directory and subsequent directories compiled by UNEP. Each directory is based on different surveys sent to the region's institutions, so that the information in the two sources is somewhat different.

TABLE 2.—DISTRIBUTION OF BENEFITS FROM MEDPOL PHASE 1 (1975–82)

Country	Total Assistance from MEDPOL to Participants (US$)	GC[a]	AAS[b]	Other[c]	Trainees Sent to Other Labs	Number of National Institutions in MEDPOL Projects	Number of Research Contracts Signed between UNEP and National Institutions
Yugoslavia	249,616	2	2	5	21	5	23
Egypt	201,566	2	1	1	20	2	12
Turkey	186,657	2	2	2	21	5	12
Algeria	129,905	1	1	1	3	1	4
Israel	128,275	1	0	1	7	7	12
Malta	107,883	1	1	2	5	2	8
Italy	80,700	0	0	0	8	17	19
Greece	80,621	1	0	2	15	13	24
Cyprus	77,925	1	1	1	4	1	5
Tunisia	54,505	1	0	1	11	3	5
Morocco	53,585	1	1	0	5	3	5
Spain	49,113	0	1	0	10	8	13
Lebanon	38,210	0	1	2	7	1	4
France	25,560	0	0	0	0	12	15
Monaco	6,635	0	0	0	0	1	1
Syria	0	0	0	0	0	1	0
Libya	0	0	0	0	0	0	0
Albania	0	0	0	0	0	0	0

SOURCE.—UNEP/WG 46/3, updated to February 1981 with figures from the Athens headquarters unit.
[a]GC = gas chromatograph.
[b]AAS = atomic absorption spectrophotometer.
[c]Other = other instruments worth more than $5,000 apiece.

sions of quality. As national environmental regulatory agencies were established over the 1970s, members of the scientific community were hired either as regulatory officials or as advisors to their governments, and national policies were recast in line with the broader objectives of the Med Plan.[27]

The South Pacific region adopted an action plan in 1982. Three treaties have been signed since then, although none are yet in force (Convention [1986], Dumping Protocol [1986], Emergency Oil Spills [1986], and an earlier Conservation Treaty [1976]).[28] During the first 5 years since its establishment, the action plan has given rise to a wealth of information about the region's environment and threats to it. There are 47 published reports documenting regional problems. Environmental management seminars and workshops have been conducted for governmental officials throughout the region, along with involvement in regional environmental education for over 170 regional officials. Momentum in the region has remained strong, with stable financial support for the programs and continued participation in joint monitoring projects.

In the South Pacific, a network of major universities has been widely used as consultants and policy advisors by the region's governments. While few of the countries in the region have extensive indigenous scientific capabilities, as microstates they were well aware of this weakness and accustomed to relying on foreign consultants from within the region for advice.

The Southeast Pacific action plan was adopted in 1981. Six treaties have been signed and four have entered into force (Convention [1981/1986]; Emergency Oil Spills [1981/1986], [1983/1987]; Land-Based Sources [1983/1986]; Specially Protected Areas [1989/not yet in force]; and Radioactive Dumping [1989/not yet in force]).[29] Monitoring and training have been important dimensions of the action plan. The first monitoring phase ran from 1985 to 1987, during which 50 institutions from the five countries participated in joint projects and generated baseline studies of regional contamination.[30] The second phase, approved in 1988, calls for more monitoring and research addressed toward producing policy-relevant data. During 1990 all countries received gas chromatographs and training in their use. A number of officials received training in control of oil pollution (104), techniques and methods of environmental impact assessment (104), land-based sources evaluation techniques (31), chemical contaminants (69), biological effects (62), radioactive pollution (46), and eutrophication (15).[31] An active oil-spill emer-

27. Haas (n. 2 above), chap. 5; Peter M. Haas, "Do Regimes Matter? Epistemic Communities and Mediterranean Pollution Control," *International Organization* 43, no. 3 (Summer 1989): 377–404.

28. See "Status of Participation" (n. 2 above).

29. Ibid.

30. J. Jairo Escobar Ramirez, *Programma Coordinado de Investigación y Vigilancia de la Contaminación marine en el Pacífico Sudeste: CONPACSE I* (Bogotá: Comisión Permanente del Pacífico Sur [CPPS], 1989), p. 3.

31. UNEP, "CONPACSE Evaluation and Recommendations for Its Further Development" (10 August 1989), p. 3.

gency program was developed for the region, arranging for rapid notification in emergencies, directories of available oil-spill combating equipment, and procedures for the expeditious exchange of necessary equipment between countries. All five countries have also adopted national oil-spill contingency plans. Recently, the countries have developed an interest in environmental impact assessment; they are generating a number of sample studies for broader application and are drafting a treaty on environmental impact assessment. Momentum in the region has remained strong, with the successive adoption of treaties to control new sources of pollution, the development of a regional monitoring network, and the growing concern with environmental impact assessments.

Similarly, the Southeast Pacific region is an area with a high level of scientific competence, although the environmental ministries are poorly staffed and funded. The area benefits from a common set of interests, which facilitates agreement: common problems, homogeneous economic structures, a history of maritime controls, and a political culture sympathetic to legal diplomacy.

Unsuccessful Action Plans

The Kuwait action plan was the first effort to replicate the Mediterranean model. Despite an initially ambitious budget, the Kuwait plan has been largely ineffective. In addition to adopting the action plan in 1978, two treaties were signed and ratified (Convention [1978/1979] and Emergency Oil Spills [1978/1979]).[32] While a Marine Emergency Mutual Aid Centre began operations in Bahrain in 1983, it has been underfunded and lacks the capabilities to actually engage in oil-spill controls. Monitoring has been scant and generally of poor quality. Few managment efforts have been undertaken. National financial support for the action plan dried up in 1981. Momentum has resumed slightly since the cessation of hostilities in November 1988, with the reorientation of regional monitoring to survey war-related wrecks and disposal of mines, as well as developing detailed maps showing the main ecological habitats and sensitive areas to be protected during clearance and mine-disposal operations. In an evaluation of the Kuwait action plan conducted in 1988, the region's relative failure was attributed to the absence of an indigenous network of well-trained regional scientists; "the most severe drawback in this respect seems to be lack of development of suitable institutions and legal instruments capable of coping with the environmental problems in the . . . area."[33]

32. See, "Status of Participation" (n. 2 above).
33. Regional Organization for the Protection of the Marine Environment, *In-depth Evaluation and Reorientation of the Kuwait Action Plan,* ROPME/WG-37/2 (Kuwait: ROPME, 1988), p. 17.

Cooperation in the Red Sea region has also been hampered by the absence of regional scientific support and institutional responsibilities for environmental management. The Red Sea and Gulf of Aden action plan was adopted in 1976, and two treaties were subsequently adopted and ratified: a Convention (1982/1985) and an Emergency Oil Spills (1982/1985).[34] Monitoring has been limited to a 1985 study of oil pollution in Port Sudan. Management activities have included support to Somalia and to the Yemen Arab Republic for designing environmental policies and an oil-spill equipment center in Djibouti. The Red Sea region was hamstrung by organizational and political factors. ALECSO evicted Egypt in 1979 and moved the headquarters to Tunis. Subsequently, the program has suffered from loss of organizational direction and from the fact that two of the key environmental actors in the region—Egypt and Israel—are not members of the organization.[35]

Circumstantial conditions conspired to further impair the effectiveness of the Kuwait and Red Sea regions. It is difficult to maintain environmental cooperation in the midst of a war zone, and recurrent civil wars in the Sudan, Somalia, and Ethiopia further distracted those countries from participating in the Red Sea plan. Since the August 1990 Iraqi invasion of Kuwait, regional decision makers have been further distracted from environmental issues.

The East Asian seas region adopted an action plan in 1981, although the delegates have not yet adopted any legal arrangements. About 50 national institutions and agencies participate in monitoring programs in the region, although these projects usually focus on purely national concerns rather than studying or monitoring for pollution with regional applicability.[36] Some equipment has been provided to regional institutions, and 454 scientists received training in pollution-monitoring techniques: Brunei (3), Indonesia (101), Malaysia (42), Philippines (107), Singapore (27), Thailand (88), and outside the region (86).

The East Asian seas, while well endowed with a strong regional marine science capability,[37] lacked community access to governments. Korea, Indonesia, Singapore, and Thailand, however, do have sufficiently well-staffed and -funded environmental bodies capable of enforcing policies.[38] Countries are

34. See "Status of Participation" (n. 2 above).
35. Editors' note.—Egypt acceded to both the Convention and the Emergency Protocol on 21 May 1990, and they were in force on 20 August 1990. See "Status of Participation" (n. 2 above).
36. UNEP, *The East Asian Seas Action Plan: Evaluation of Its Development and Achievements,* UNEP Regional Seas Reports and Studies no. 86 (Nairobi: UNEP, 1987), p. 3.
37. Editors' note.—See the article in this volume on marine research in the North Pacific: Daniel J. Dzurek, "Marine Scientific Research and Policy Issues in East Asia."
38. Mark Baker, Libby Bassett, and Athleen Ellington, *The World Environment Handbook* (New York: World Environment Center, 1985); Ichiro Kato, Nobuo Kumamoto, and William H. Matthews, eds., *Environmental Law and Policy in the Pacific Basin*

universally represented at intergovernmental meetings by midlevel environmental officials, thus foreclosing any opportunity to put the operating scientists in contact with higher level ministerial decision makers to better integrate scientific expertise with environmental management. The ASEAN Expert Group on the Environment of the Committee on Science and Technology, the international organization with ultimate responsibility for the region, was upgraded in 1990 to the ASEAN Senior Officials for the Environment, reporting directly to ASEAN's Policy and Decision Making Standing Committee, which may elevate the saliency of environment management in the region and reinvigorate the action plan.

The Caribbean has received the second largest amount of UNEP's attention (second to the Mediterranean). UNEP spent US$17.9 million in the development and enforcement of the Caribbean action plan. The action plan was adopted in 1981. Three treaties have been adopted: the Convention (1983/1986), an Emergency Oil Spills Protocol (1983/1986), and a Specially Protected Areas (1990/not yet in force).[39] Monitoring efforts have been poorly coordinated and have yielded little information about regional environmental conditions. Very little equipment has been transferred to governments in the region, as UNEP concentrated its support for institutional development in the Caribbean Environmental Health Institute on St. Lucia. The consequence of this choice, counter to UNEP practices in other regions, is to make much of the regional assessment dependent upon the progress of one laboratory, which has been much slower than initially desired. Overall, 576 individuals received training at regional workshops. While a wide number of management projects have been undertaken, these have largely been oriented toward relatively sui generis problems encountered by specific countries, and the benefits of participation in the program have accrued to a few countries. For instance, US$6 million went for a multiyear environmental assessment effort to manage pollution of Caribbean bays, with a sample case of Havana Harbor. Two meetings were held at which the findings were reported: 82% of the participants at these two meetings (67 of 82) were Cubans. Nor were any follow-up measures taken to circulate the findings to other Caribbean countries. More broadly, Cubans received more training than any other nationality participating in all assessment components.

Momentum in the region has been mixed. While national payments to the trust fund have only amounted to 53% of pledges, diplomatic momentum picked up after 1987. In 1989, a Specially Protected Areas treaty was adopted—although countries were unable to agree on specific areas to be included in the treaty—and a redesigned set of monitoring projects intended

Area (Tokyo: University of Tokyo Press, 1981); Colin MacAndrews and Chia Lin Sien, eds., *Developing Economies and the Environment: The Southeast Asian Experience* (Singapore: McGraw-Hill International, 1979).

39. See "Status of Participation" (n. 2 above).

to more closely integrate monitoring and management activities was approved.

Contacts between the region's scientists and decision makers has been spotty at best. The Caribbean is distinguished by the wide disparity of national capabilities and experiences: from small island states to large industrialized and technically sophisticated countries. The regionwide scientific community is very limited, and the smaller island countries have much weaker institutions for environmental management. Consequently, UNEP has had great difficulties in mobilizing or coordinating regionwide scientific involvement in the action plan.

The South Asian seas is UNEP's most recent effort. Since 1982 UNEP has been advising countries of the region through the South Asian Cooperative Environment Programme on the drafting of an action plan and framework convention for the region. Countries have remained reluctant to adopt them, although UNEP is optimistic that they will be adopted soon. Such delays are due to the absence of a significant body of regional marine scientists. India has the region's strongest marine science capability, but the science and policy communities within the country are strongly divided.

In addition to the difficulties of scientific coordination, several of the relative failures suffered from inadequate project design. Most of the projects in the Caribbean and East Asia deal only with national problems experienced by the governments involved in the projects rather than in broader regional problems.[40] Participation in broader projects leads scientists to understand their common problems, rather than merely focusing on issues sui generis to their own country and failing to develop a wider concern with problems or political identification with colleagues in the region. In both these regions, UNEP has recently recognized these shortcomings and has made efforts to redesign the projects for the regions in order to make them more effective.

Indeterminate Efforts

Two other action plans fall into an intermediate category, where an assessment of their effectiveness and success cannot be fully established. Lack of funds has been a major obstacle to implementation in these areas, and, while scientific networks are presently being created or reinforced, the outputs of the regions remain tentative. With poor developing countries heavily affected by the debt crisis of the 1980s, actual national commitments have only been about 20% of pledges in each region.

40. A 1989 evaluation of the Caribbean concluded that "funds . . . support[ed] projects that can hardly be classified as regional projects or as national projects having a regional significance" (UNEP, *The Action Plan for the Caribbean Environmental Programme: Evaluation of Its Development and Achievements,* UNEP Regional Seas Reports and Studies no. 109 [Nairobi: UNEP, 1989], p. 6).

The East African region adopted an action plan in 1985 and signed a Convention (1985/not yet in force), Emergency Oil Spill Protocol (1985/not yet in force), and Specially Protected Areas Protocol (1985/not yet in force).[41] A regional pollution-monitoring project is being developed, and in 1990, US$180,000 worth of equipment was purchased and provided to regional institutions. Momentum has been positive since, at the 1989 intergovernmental meeting, representatives called for rapid payment of national arrears to the trust fund and the immediate commencement of regional monitoring. With a relatively strong indigenous marine science capability in the region— Kenya, Madagascar, Mauritius, Mozambique, Seychelles, and Tanzania each have three or more marine science laboratories staffed by over 12 people— and moderate institutional capabilities within national administrations, future cooperation is certainly possible.

The West and Central African region signed an action plan in 1981, along with a Convention (1981/1984) and an Emergency Oil Spill Protocol (1981/1984).[42] Twenty-three institutes and laboratories from 10 countries are organized in a monitoring network to assess the quality of the marine environment. Equipment and expendable supplies worth US$540,000 were provided to institutions participating in the network.[43] Information has been gathered on oil pollution, industrial pollution from major land-based sources, and the onshore impact of offshore oil and natural gas development, along with studies of the causes and extent of coastal erosion. Training was offered to 213 scientists in the region. Ten states received assistance in developing national contingency plans for marine pollution emergencies, and Congo and the Côte d'Ivoire have adopted national plans.[44]

Scientific networks are in the process of being established and strengthened for the West and East African regions. Formal governmental environmental bodies responsible for policy making remain weak in these countries, and there is limited existing indigenous scientific capability. It is perhaps too early to offer an authoritative judgment about the utility of the regional seas strategy in these regions. Missions to the region have found very little awareness of the program among ministers, indicating that there remains a major

41. See "Status of Participation" (n. 2 above).
42. Ibid.
43. UNEP (OCA)/WACAF IG.3/4 Annex III; UNEP, *The West and Central African Action Plan: Evaluation of Its Development and Achievements*, UNEP Regional Seas Reports and Studies no. 101 (UNEP: Nairobi, 1989). This support went to Cameroon (US$54,000), Congo ($6,000), Côte d'Ivoire ($102,000), Gambia ($38,000), Ghana ($59,000), Nigeria ($69,000), Senegal ($55,000), and Sierra Leone ($47,000). Atomic absorption spectrophotometers went to Ghana, Côte d'Ivoire, Senegal, and Cameroon. Gas chromatographs went to Gambia, Nigeria, Sierra Leone, and Côte d'Ivoire.
44. UNEP/IAMRS.6/4 Annex IV, p. 20.

split between the incipient scientific community and the actual policy makers.[45]

LIFE CYCLES

Each of the action plans has evolved through a similar life cycle. Following the withdrawal of UNEP administrative and financial support to the region, a natural malaise ensues. The new regional coordinating units often take a while to get accustomed to their new responsibilities and to manage the ongoing projects, and national administrations take several years to regularize national allocations to the trust funds. During this period of adjustment, support ebbs in terms of attendance at meetings, financial support, and general levels of activity. Moreover, focus naturally shifts from institution creating under UNEP's auspices, to a consolidation of attention to enforcing and modifying the agreements that were devised earlier. While UNEP helps countries to draft regional agreements, under their own power attention is focused on developing guidelines to actually apply such commitments.

The creation of the program can be a consequence of several international forces, including the leadership of an international organization, shared national concerns, or the leadership of one country. However, a region's momentum, past the postseparation anxiety to self-sufficiency, persists only in regions in which a strong indigenous scientific capability and network has been established.

To date, regional coordinating units or regional organizations have been established for the Mediterranean (1982), Caribbean (1986), West and Central Africa (1981, although provisional headquarters remain in Nairobi), Kuwait Gulf (1982), Red Sea (1981), South Pacific (1981), and Southeast Pacific (1981). Nairobi remains responsible for East Asian seas, East Africa, and South Asian seas. In the Mediterranean it was 3 years after the headquarters was shifted from UNEP offices in Geneva to Athens before momentum was restored, in 1985. In that year the governments made up some of the arrears from the intervening years, when collections fell to 80% of pledges. After 1985, the countries renewed their commitment to regional environmental protection and to more comprehensive patterns of coastal zone management. The Genoa Declaration (1985) and a generalized reorienting of the action plan have focused directly on coastal zone management, and monitoring information has been applied to yield data relevant to littoral land-use policy. In the Southeast Pacific, funding did not become regularized until 1987. Coordination in the Caribbean was made more difficult because the transfer

45. UNEP (OCA)/WACAF IG. 3/3 Annex "Report on the UNEP/Gulf of Guinea Action Plan Contact Mission to Six Countries of West and Central Africa."

of headquarters to the Caribbean overlapped with the transfer of the Regional Seas Programme headquarters from Geneva to Nairobi in 1985. In the Caribbean, regular national commitments to the trust fund didn't stabilize until 1988, following 4 years of progressive decline. The Kuwait and Red Sea programs became moribund after they became self-sufficient.

LEARNING

Within the broad parameters of the Mediterranean "strategy," a number of specific tactical readjustments have occurred, as the UNEP staff realized that specific regions had insufficient momentum. In each case UNEP remains wedded to the broad objective of applying scientific competence to technical management; new techniques were devised to promote such ends. In response to perceived lags in momentum at intergovernmental meetings and from periodic assessments, UNEP has developed a new set of political strategies and a new set of organizational designs to compensate for perceived mistakes. Regional assessments, implemented on a regular basis since the early 1980s within UNEP and from requests, inform the governments to assess their own programs after about a decade. This has provided feedback to UNEP about the effectiveness of its master plan.

Recent monitoring projects have been redesigned in order to coordinate more closely with ongoing regional management activities. UNEP had initially presumed that governments would automatically use monitoring and research findings. A reappraisal of the Mediterranean revealed that the assessment and management activities generated discrete networks of scientists who interacted independently of one another and neither exchanged information nor coordinated their actions. Consequently, environmental assessments were conducted in a policy vacuum. Subsequently, UNEP's draft monitoring proposals for the Mediterranean, Caribbean, West Africa, and East Africa specified that monitoring be directed to substances and parameters of immediate policy relevance and should be generated in order to yield concrete management proposals within 2 years.[46]

If UNEP deemed that governments were unresponsive to its overtures, it sought to create new networks with which to influence them. Following disillusionment with the progress of intergovernmental cooperation, UNEP sponsored the creation of associations of academic institutions with an objective of applying marine science to environmental management and supporting the regions' action plans. Such associations were established in the South Pacific (Association of South Pacific Environmental Institutions, 1986), East

46. See, for example, *Marine Pollution Assessment and Control Programme for the Wider Caribbean Region: CEPPOL,* IOC/UNEP-RRW-I/8, IOC Workshop Report no. 59 Supp. (1989).

Asian (Association of Southeast Asian Marine Scientists [ASEAMS], 1986), and East African regions (approved at the First Intergovernmental Meeting of the Action Plan, November 1989).[47]

Recently UNEP has attempted to cross-subsidize some of the regions. With temporary staffing difficulties and with overlapping interests between the regions, interregional training workshops have been held open to people from different regions, particularly for training in environmental impact assessment.[48]

Overall, UNEP has become more pragmatic. Management programs in the other regions are more modest than in the Mediterranean, and monitoring programs have become more concrete and included fewer substances and parameters for monitoring, in order to integrate monitoring projects more deliberately into the broader objectives of environmental management.

Countries participating in the Regional Seas Programme may be said to have learned by virtue of their involvement. More recent conventions, building from the precedents of earlier ones, have included a wider variety of pollutants for collective control. Over time, as the individual components of the regional seas arrangements have accreted, a subtle shift of focus has occurred. By gradually and incrementally dealing with a variety of competing uses of the seas, a more coherent, comprehensive set of arrangements has emerged, handling a host of potential uses and users of the seas within a more balanced framework. In addition to the legal control of specific sources of pollution, the more successful of the regional seas programs have moved to deal with species conservation (specially protected areas), collective management schemes (the use of environmental impact assessment as a widespread technique and the long-term focus on integrated management in the Mediterranean), as well as more refined monitoring and research in the regions. An informed observer-participant notes that "the importance of this change in legal perspective, from a use-oriented to a 'resource-oriented' approach, cannot be over-emphasized; it is the essence of the new law of the sea."[49] Participating governments exposed to regional seas projects have developed a growing sophistication of thought about marine environmental management. Moreover, this corpus of law reflects an incipient shift in collective temporal conceptions on the parts of the drafting parties. Seven of the eight regional seas conventions explicitly confirm, in their preambles, a commitment to "future generations."[50]

---

47. UNEP (OCA)/EAF IG.2/4.
48. UNEP/IAMRS. 6/4, pp. 47–48.
49. Sand (n. 2 above), p. xv.
50. Ibid. See also Edith Brown Weiss, *In Fairness to Future Generations: International Law, Common Patrimony, and Inter-generational Equity* (Ardsley-on-Hudson, New York: Transnational Publishers, 1989).

FINAL ASSESSMENT

The regional seas programs have taken more than a decade to develop, and the more successful ones have cost tens of millions of dollars. Other regions, where a baseline of regional scientific competency has to be instilled, may take longer. Yet the regional seas programs have succeeded in impressing on regional elites the need for more comprehensive regional environmental policy (or coastal zone management); they have created or extended domestic constituencies for environmental protection and pollution control; they have contributed to a gowing body of international environmental law; they have contributed to the development of a baseline of knowledge about the quality of the world's regional seas, and they have transferred extensive amounts of technology, both in terms of hardware and knowledge, to many developing countries.

At present, about 250 national institutions and more than 1,000 experts actively participate in the monitoring programs of the 10 action plans. In addition, the Regional Seas Programme has contributed to the development of regional plans to coordinate responses to oil spills, including training in various techniques, devising response plans, and providing information about available equipment.[51]

The real utility of the action plans has been to transfer marine science technology to developing countries and to build up marine science capabilities in many developing countries. In keeping with the broader vision of the Regional Seas Programme, such scientific advances will generate the technical basis for closer intergovernmental coordination to manage regional environmental pollution.

---

51. International Maritime Organization and UNEP, *Catalogue of Oil Spill Response Equipment and Products,* UNEP Regional Seas Directories and Bibliographies (Rome: FAO, 1988).

Environment

Controlling the "Curtains of Death": Present and Potential Ocean Management Methods for Regulating the Pacific Driftnet Fisheries

Paul G. Sneed

Department of Geography, University of Hawaii

INTRODUCTION: THE TRAGEDY OF THE COMMONS

The traditional forms of national sovereignty are increasingly challenged by the realities of ecological and economic interdependence. Nowhere is this more true than in shared ecosystems and in "the global commons"—those parts of the planet that fall outside national jurisdictions. Here, sustainable development can be secured only through international co-operation and agreed regimes for surveillance, development, and management in the common interest. But at stake is not just the sustainable development of shared ecosystems and the common, but of all nations whose development depends to a greater or lesser extent on their rational management.[1]

As the planet Earth has become more crowded and demand on its resources has increased, there has been a corresponding increase in international concern over the management of the global, or international, commons. The international commons are those features or realms of the earth that lie outside the jurisdiction of any single nation. Examples of these global commons "include a number of major river systems, some lakes and inland seas, most of the ocean, Antarctica, the atmosphere beyond the airspace immediately above the land, all of the outer space, and the earth's weather and climate."[2]

The High Seas portion of the ocean probably represents one of humankind's oldest recognized global commons. Traditionally, the High Seas have been defined as "all parts of the sea not included in the territorial sea

---

1. World Commission on Environment and Development, *Our Common Future* (New York: Oxford University Press, 1987), p. 261.

2. S. Brown, N. W. Cornell, L. L. Fabian, and E. B. Weiss, *Regimes for the Ocean, Outer Space, and Weather* (Washington, D.C.: Brookings Institution, 1977), p. 1.

or in the internal waters of a state."[3] This definition had to be modified with the introduction of the archipelagic waters and EEZ concepts in the 1982 UN Convention on the Law of the Sea (1982 LOS Convention). According to Article 86 of the LOS Convention, High Seas provisions apply to "all parts of the sea that are not included in the exclusive economic zone, in the territorial sea or in the internal waters of a State, or in the archipelagic waters of an archipelagic State."[4] Thus, the 1982 LOS Convention expanded national jurisdiction over a large part, about 35%, of the open ocean. Nevertheless, there is still an extensive area of High Seas ocean commons.

Customary law and even the most recent international agreements guarantee freedom of the High Seas. The LOS Convention states that "the high seas are open to all States, whether coastal or land-locked";[5] that this freedom must be exercised under conditions laid down by the LOS Convention and other international law; that the freedom includes, among other things, the freedom of fishing on the High Seas; and that "these freedoms shall be exercised by all States with due regard for the interests of other States in their exercise of the freedom of the high seas."[6] "There has not always been freedom of the high seas: In the fifteenth century there were many claims to sovereignty over extensive areas of the oceans: by Sweden and Denmark in the Baltic and Norwegian Sea; by Venice in the Adriatic and Genoa and Pisa in the Ligurian Sea; and by Britain in the ill-defined 'British seas' around its coasts."[7] However, when the great expansion of European exploration and commerce began in the early 17th century, opposition to High Seas enclosure mounted. According to Churchill and Lowe, the decisive arguments for the freedom of the seas were ably presented in 1609 by Grotius in his *Mare Liberum*. The battle was won by advocates for open oceans "as the importance of free navigation in the service of overseas and colonial trade came to overshadow national interests in coastal fisheries, and as the development of real naval power displaced national claims to sovereignty over the seas."[8] By the 19th century, the notion that the High Seas were jurisdictionally distinct from national waters and not subject to annexation by any nation-state had become firmly entrenched in customary law.

---

3. Geneva Convention on the High Seas, 1958, entered into force on 20 September 1962; *United Nations Treaty Series* 450 (1958): 82; *American Journal of International Law* (1958): 842; also reprinted in *The Law of the Sea*, R. R. Churchill and A. V. Lowe (Manchester, U.K.: Manchester University Press, 1958), p. 164.

4. United Nations, *The Law of the Sea* (New York: St. Martin's Press, 1983), p. 30.

5. *The Law of the Sea: Official Text of the United Nations 1982 Convention on the Law of the Sea with Annexes and Index* (New York: United Nations, 1983), sales no. E.83.V.5, Art. 87(1).

6. Ibid., Art. 87(2). See United Nations (n. 4 above), pp. 30–31.

7. Churchill and Lowe (n. 3 above), p. 165.

8. Ibid.

For almost three centuries the High Seas have been subject to what has become known in international law as the concept of *res nullius*—that is, they have been treated as though they belong to no one. This has meant that the High Seas were treated as common property with open access and free use by anyone. Apart from coastal navigation and fishing areas, where users came into conflict and developed local institutions to manage common pool resources,[9] "the high seas seemed abundant enough for everyone's use and, therefore, not in need of any kind of regime to allocate their use."[10] As long as it could be assumed that the High Seas and marine resources were vast and abundant, open access and free use could be universally regarded as just and rational.

Unfortunately, with population growth and rapid technological change, the High Seas can no longer be considered either vast or infinitely rich in resources. For the first few centuries "freedom of the seas essentially meant freedom of fishing and navigation, [but] the concept was broadened in the twentieth century to include *the freedom to exploit and to pollute, and the freedom to do so irresponsibly*" (emphasis added).[11] This situation, where the High Seas have been treated as *res nullius,* has led to what Garrett Hardin called the "tragedy of the commons";[12] that is, where common property resources, such as those found in the High Seas, are subject to open access and free use, the likely result is diminishment of the resources and impoverishment of the users.

Overfishing of Territorial and High Seas, leading to a "tragedy of the commons," has been a recurrent phenomenon in the 20th century.[13] There has been a steady increase in catch—due to increasingly sophisticated technology—from an average of about 19 Mmt in 1948–52 to 67 Mmt in 1981–83, but "the rate of increase in the world catch has slowed down in recent years, mainly because most commercially exploitable fish stocks are now fully exploited."[14] The Pacific drift gill-net fishery, carried out since the early 1980s by Japan, Korea, and Taiwan, is probably one of the best examples of how virtually unregulated use of a very efficient fishing technology, driven by population pressure and economic imperatives, can lead to overfishing and the "tragedy of the commons."

This article describes the Pacific drift gill-net fishery[15] and some of its

9. B. M. McCay and J. M. Acheson, eds., *The Question of the Commons* (Tucson: University of Arizona Press, 1987).

10. Brown et al. (n. 2 above), p. 5.

11. L. Cuyvers, *Ocean Uses and Their Regulation* (New York: John Wiley and Sons, 1984), p. 47.

12. G. Hardin, "The Tragedy of the Commons," *Science* 162 (1968): 1243–48.

13. R. D. Eckert, *The Enclosure of Ocean Resources* (Stanford, California: Hoover Institution Press, 1979), pp. 116–20.

14. Churchill and Lowe (n. 3 above), p. 223.

15. The Pacific drift gill-net fishery is hereinafter referred to as the driftnet fishery.

impacts on the Pacific environment and resources. In addition, it reviews the recent bilateral and multilateral attempts to regulate this fishery, comments on some of the possible mechanisms for management of the High Seas commons, and proposes a scheme for international ocean management. The goal of this article is to explore how we might transform the High Seas from a realm that is *res nullius* to one that is *res communes*—a commons that belongs to everyone and is managed accordingly.

THE CASE OF THE PACIFIC DRIFTNET FISHERY

Driftnets

The so-called driftnets used in the Pacific Ocean are a type of gill net. A gill net is "a flat net whose meshes are capable of capturing fish by permitting head and gill covers to pass through the net in one direction but not be withdrawn."[16] Gill nets were formerly made of linen, which made them expensive because they required constant repair, and inefficient "because they were highly visible to their quarry."[17] Today, gill nets are made of monofilament plastic or nylon that is almost invisible and relatively inexpensive to purchase and maintain.

A typical synthetic gill net used in U.S. territorial waters is 100 yards long by 6 yards deep. Small floats keep the net perpendicular to the surface, and lead weights keep the bottom edge hanging straight down into the water. In the Atlantic, this typical net will often be joined together with up to 30 other nets to form a gill net more than 1.5 miles long. However, this is a small net compared to the giant driftnets that are used to fish squid in the North Pacific and albacore in the South Pacific.

These supernets of the Pacific "are so large they defy comprehension."[18] The squid fishery uses single-strand monofilament or nylon, which is woven into a gill-net section with mesh sizes of 4.5 inches. Heavier line, sometimes braided or twisted together, is used in making nets with mesh sizes of about 7 inches for the albacore fishery. Both squid and albacore driftnets "are in sections called *tan*, with each tan 35 to 50 meters (115 to 160 feet) long and 8 to 10 meters (26 to 33 feet) deep."[19] Anywhere from 500 to 1000 of these tan are linked together to form nets 11 to 30 miles long. Although estimates vary tremendously, there could be anywhere from 120 to 1500 vessels in the Pacific driftnet fleet depending on the time of year and kind of fishery. Drift-

16. C. Gammon, "A Sea of Calamities," *Sports Illustrated* 68 (1989): 46–53.
17. Ibid.
18. J. TenBruggencate, "The Impact of the 'Walls of Death,'" *Sunday Star-Bulletin and Advertiser* (24 September 1989): D4.
19. Ibid.

nets regularly stretch for thousands of miles across the ocean, and, at its maximum deployment in the North Pacific, the squid fishing fleet might be employing driftnets that combined end-to-end could encircle the globe.[20]

Typically, crews of the driftnet vessels set up their synthetic giant gill nets so that they are vertically suspended in the water for 10 to 14 hours at a time.[21] While the crews sleep and the net drifts, fish swim into the nearly invisible net and are trapped. The following morning, crew members relocate the nets with radar and haul the catch aboard their vessel. Unfortunately, part or all of one of these driftnets may be lost (or sometimes discarded) at any time, creating drifting "ghost nets" that go on fishing indefinitely.

Bycatch or Incidental Capture of Nontarget Species

The tons of trapped fish caught in the pelagic driftnets act as a lure for birds and marine animals searching for food in the deep Pacific (fig. 1). Since the driftnets are barely visible underwater, animals approaching the fish often become entangled in the net and drown. Those that do escape sometimes suffer severe cuts or other bodily damage and die later. Seabirds, attracted by the thrashing fish at the surface, become entangled in the net mesh and drown or suffer fatal injuries. Environmentalists have called the driftnet fishery "environmental terrorism" and the nets themselves "walls of death" (or "curtains of death") because of this apparently large-scale destruction of noncommercial marine life or bycatch.

Since the driftnet fishery is very new and largely unregulated, it is extremely difficult to get hard data on what the real impact of the fishery is on incidental species. However, there have been numerous reports and some photographs of dead dolphins, sea turtles, seals, whales, sharks, and other marine creatures, not to mention nontarget species of fish, such as marlin and salmon, in the bycatch of the driftnet fishery. Greenpeace has argued that Japanese driftnets alone "are responsible for the deaths of at least 5,000 Dall porpoises and 250,000 to 750,000 sea birds every year."[22] Japanese officials, on the other hand, say the Greenpeace figures are exaggerated:

> "When I talk to environmentalists about figures, they are always misinformed," says Ichiro Nomura, First Secretary of the Japanese Embassy in Washington, D.C. "The annual incidental ceiling [allowed by the U.S. permit] is 5,500 Dall porpoises," he says. "On the average, 2,000 to 2,500 are caught a year—that's well within the allowed amount." Nomura says

---

20. T. Glavin, "35 Asian Driftnet Boats Spotted Fishing Illegally," *Vancouver Sun* (4 July 1989): C8.
21. "No Net Gain for Ocean Life," *Sierra* 71 (1986): 14–15.
22. Ibid., p. 15.

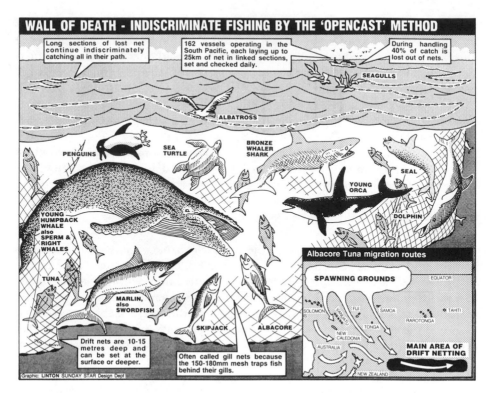

Fig. 1.—"Wall of Death" (Linton, Design Department, *Sunday Star,* Auckland, New Zealand, as published in article by Phil Twyford, "Pacific Pirates," *Sunday Star* [28 May 1989]: A12. Republished with permission).

the number of birds that die in Japanese driftnets is closer to 132,000 to 170,000 annually, and notes that the impact of these deaths on bird populations is not known.[23]

Limited research by the U.S. National Marine Fisheries Service seems to throw support behind the magnitude of the Greenpeace figures when it "estimates that about 850,000 seabirds become entangled in the nets and drown every season."[24]

Perhaps these disagreements over numbers will be resolved with new research. Recent agreements between the United States and Asian driftnet fishing countries allow for observers to be placed aboard vessels fishing the North Pacific to "collect biological samples and record information on the

23. Ibid.
24. M. Nichols, "An Alarming Catch: Critics Condemn the Use of Drift Nets," *Maclean's* 102, no. 24 (1989): 49.

catch of squid and other 'non-target' species."[25] In the South Pacific, a boat chartered by Earthtrust, a Hawaii-based international environmental organization, set out in November 1989 "to document driftnet fishing—and to confront driftnet crews if necessary—in the Tasman Sea between New Zealand and Australia."[26]

Meanwhile, there are many anecdotes that attest to the impact of the "curtains of death."

> Fishermen west of Vancouver Island reported last month that an eerie calm had settled over an area where they were accustomed to seeing seabirds, including albatrosses and puffins. Bruce Petrie of Sooke, B.C., who has fished for salmon for the past 26 years, expressed concern that seabirds, walruses and whales that he used to watch while he fished have begun to disappear. . . . He also said that drift nets are catching B.C.-bred salmon before they can return to their native waters.[27]

Thus, there is a growing concern among Canadian and U.S. fishermen and officials that much of the North Pacific driftnet bycatch consists of young salmon bred in North American waters as well as marine mammals and birds.

The North Pacific Driftnet Fishery

From about April to December anywhere from 700 to 1,500 Asian driftnet vessels move into North Pacific waters to fish for squid: "As the sea temperature rises each summer, the fleet moves north to Alaska. Every night at dusk, as the squid rise to the surface to feed, each boat casts about 20 kilometres [or more] of nets, made up of sections about 90 metres long and up to 15 metres deep. The fleet is after 'aka-ika' or flying red squid."[28] In order to protect the annual migration of Pacific salmon, which by international treaty are considered the property of American and Canadian fishermen, the Japanese have restricted the areas and times in which their vessels can operate. For example, "before July 1, the internationally recognized squid-fishing grounds lay south of 40 degrees N. or about halfway between the Hawaiian Islands and the Aleutian Islands chain. By July 1, the northern boundary has moved north 120 nautical miles to 42 degrees N."[29] However, despite these

25. "Japan Vows to Clamp Down on Rogue Squid Fishers," *New Scientist* (3 June 1989): 27.

26. D. Waite, "Expedition to Oppose Driftnets," *Honolulu Advertiser* (26 November 1989): A3.

27. Nichols (n. 24 above), p. 49.

28. " 'Wall of Death' Confronts Wildlife in the Pacific," *New Scientist* (2 June 1988): 26.

29. Glavin (n. 20 above), p. C8.

precautions to avoid "salmon interceptions," there is apparently an enormous bycatch of salmon.

According to West Coast fisheries officials, at least 20,000 tons of salmon are caught annually as bycatch to the squid fishery. More insidious, however, is the increasing "accidental" incursion of driftnet vessels, often Taiwanese, into the waters of the North Pacific. During the summer of 1989, numerous Asian "pirate vessels" were sighted fishing illegally north of the boundaries established to protect salmon. Some were even caught red-handed with the goods: "a U.S. 'sting' operation stretching from Seattle to the Pacific high seas uncovered $1.3 million worth of salmon illegally caught in international waters with drift gillnets."[30] This and other incidents, along with "the large amounts of salmon being offered at low prices on foreign markets,"[31] provides strong circumstantial evidence for the growing suspicion that a number of the Asian driftnet vessels are ostensibly fishing for squid while really concentrating on catching salmon.

Driftnet fishing in the North Pacific, whether legal or illegal, is ecologically destructive and severely impacts the commercial salmon fishery. For example, in Alaska alone "fishermen had predicted a catch of 40 million pink salmon last year, but only 12 million were taken."[32] Clearly, other factors, such as oil pollution, may be causing some of this decline in returning populations. Nevertheless, if driftnetting is allowed to go on unregulated in the North Pacific, fewer young salmon will return to their North American spawning grounds, and the fishery could collapse, posing a serious threat to the livelihood of Canadian and American fishermen.

Driftnet Fishing Moves South of the Equator

Since 1952, the South Pacific albacore, the white-meat tuna used for canning, has been harvested by Japanese, Korean, and Taiwanese longline fleets. In addition, long-liners based in the South Pacific island countries also fish albacore. With a decline in the longline fishery, a U.S.-based troll fishery was initiated, targeting surface-swimming stocks of albacore. This conflicted very little with the long-liners, and the albacore stocks seemed to be remaining healthy. However, in the 1988/89 season all of this changed when about 160–180 Asian drift gill netters were discovered fishing for albacore in the South Pacific.

This South Pacific driftnet fleet is apparently moving from the North

30. M. Quimby, "Pacific Fisheries Face 'Wall of Death,' " *Honolulu Advertiser* (22 September 1989): A15.

31. Nichols (n. 24 above), p. 49.

32. M. Eastly, "Driftnet Fishing Controversy Heats Up," *Pacific* 14, no. 6 (1989): 42.

Pacific when the fishing season ends there. It has grown from about 7 Taiwanese and 10 Japanese vessels in the 1987/88 season to at least 130 Taiwanese and 50 Japanese vessels in the 1989/90 season.[33] It is estimated that in 1988/89 the driftnet fleet took between 30,000 and 60,000 tons of albacore, worth about US$50 million to US$100 million, in a 4-month season from "the Tasman Sea between Australia and New Zealand, and east of New Zealand at a latitude of about 40 degrees south"[34] (see fig. 1). At this rate of growth, the South Pacific Forum Fisheries Agency (SPFFA) predicts a collapse of the albacore fishery within 5 years.[35]

Because the Asian driftnet fishery concentrates on catching the surface-swimming juvenile albacore, some observers argue that the species could disappear within as little as 2 years. Furthermore, driftnetters come into direct conflict with albacore fishermen using longline and trolling methods. These two methods catch fewer fish and mostly adults. Therefore they have a much less severe impact on the albacore stocks. Collapse of the albacore fishery through overfishing by Asian driftnetters "would be disastrous for the long-line industries of Fiji, Tonga, French Polynesia, Solomon Islands, the Cook Islands, and Vanuatu."[36] Ironically, long-liner fleets from Japan, Korea, and Taiwan, which also participate in the albacore fishery, will be equally hurt by the "invasion" of the South Pacific driftnetters.

Restricting the Driftnets

Drift gill-net fishing, both in the North and South Pacific, results in serious ecological and economic impacts. The case against these supernets may be summarized as follows:

1) They are indiscriminate in what they catch. Selectivity is determined only by where they are set and the size of the mesh.
2) They preempt the use of the fishing grounds by others.
3) The quality of the catch is poor because gill nets are frequently left untended for long periods.
4) Fish, mammals, and birds [that are not killed and] that fight their way out of the nets are often so severely cut in their struggles by the wire-like monofilament that they subsequently die from loss of blood or infection.
5) Because the nylon monofilament mesh is all but indestructible, nets that are lost or even . . . discarded go on "fishing" for years.[37]

33. "Trying to Stop the Slaughter," *Pacific Islands Monthly* 59, no. 18 (1989): 8–9.
34. I. Anderson, "Drift Net Peril Moves to South Pacific," *New Scientist* (17 June 1989): 35.
35. Quimby (n. 30 above).
36. "Trying to Stop the Slaughter" (n. 33 above), p. 9.
37. Gammon (n. 16 above), p. 48.

Because of the destructiveness and wastefulness of these methods, many environmentalists, fishermen, and politicians are calling for a global ban on all driftnet fishing.[38]

While Japan "banned driftnetting within 600–700 miles of its own coastline nearly a decade ago in response to concerns from coastal fishermen,"[39] it has been slow to agree to restrictions on its driftnet activities in the Pacific High Seas. The Japanese, like other Asian countries, have argued that driftnets are essential to them because they are an economical means of fishing, requiring only small crews compared to the labor-intensive long-liners. In addition, Japanese and Taiwanese officials have refused to accept SPFFA claims that the albacore stocks are being depleted, and according to SPFFA, they have been slow to cooperate in reporting full catch statistics.

The U.S. State Department was able to use the several poaching incidents that occurred during 1989 in the North Pacific to forge agreements with Japan, Korea, and Taiwan to regulate driftnet fishing. The agreement with Japan was the weakest of the three because it "generally accepted Japanese promises to better self-regulate and minimize 'by-catches' through increased patrolling." In addition, Japan agreed to accept American and Canadian enforcement and scientific observers on board its fishing vessels. The agreement with Taiwan was tougher in that it "requires Taiwanese driftnet ships to carry, beginning next year, location-fixing transmitters beaming signals to overhead satellites to allow U.S. National Marine Fisheries Service to track the ships," and it "gives the U.S. Coast Guard for the first time full rights to board and inspect Taiwanese driftnet vessels." Finally, "the U.S.–South Korean agreement, signed Sept. 26, 1989, similarly gives the Coast Guard unfettered boarding rights and requires satellite transmitters next season."[40] Thus, bilateral agreements designed to foster international cooperation and control some aspects of the North Pacific driftnet fishery were offered by the State Department as an alternative to an outright ban.

The Asian driftnet fishing nations were slower to respond to concerns about the albacore fishery expressed by the 16-nation South Pacific Forum. After several meetings during the summer of 1989, only South Korea agreed to withdraw its few boats from the region in the coming season. Taiwan and Japan refused to compromise and reduce their effort, so negotiations came to a standstill. However, during September 1989, in a surprising turnaround, Japan agreed to cut its driftnet fleet from 60 to 20 boats for the 1989/90 season.[41] Unfortunately, Taiwan did not comment on the Japanese decision

38. W. Friedenberg, "Driftnet Restrictions May Not Stretch Far Enough," *Honolulu Star-Bulletin* (10 October 1989): A10.

39. Waite (n. 26 above), p. A3.

40. Friedenberg (n. 38 above), p. A10.

41. "Japan Will Cut Drift-Net Fleet," *Honolulu Advertiser* (20 September 1989): A3–A7.

or announce any changes in plans to send its 60 or more boats back to the albacore fishery in the South Pacific.[42]

In the meantime, pressure from U.S. fishermen concerned about the loss of salmon and tuna stocks and from environmentalists upset about the dramatic destruction of marine wildlife, has caused U.S. legislators to move closer to a total ban on driftnet fishing in the Pacific. Many people feel that the bilateral agreements struck last summer to help regulate the North Pacific fishery and the voluntary reductions in effort proposed by Japan for the South Pacific are too little and too late. On 17 November 1989, the U.S. House of Representatives voted to approve a resolution calling for an end to driftnet fishing.[43]

Even if this proposed ban is implemented, only a battle, and not the war, in the international struggle to manage the global commons will have been won. First of all, the ban will have to be complete if it is to be effective. If, for example, the ban on driftnet fishing is restricted to the Pacific, "the boats will go to other areas such as the Indian Ocean where the southern blue-fin tuna, a highly valuable resource, would be threatened."[44] Secondly, even if drift gill nets are banned from all oceans, undoubtedly some new technology will be invented and deployed to exploit the living resources of the High Seas. Banning a destructive technology like driftnets may be a necessary short-term solution to the "tragedy of the commons" but a more long-term resolution to this problem will require development of more effective international ocean management.

EVOLVING INTERNATIONAL OCEAN MANAGEMENT REGIMES

The Pacific drift gill-net fishery illustrates almost all the problems associated with managing the global oceanic commons. The introduction of advanced technology in the form of gigantic synthetic gill nets and the economic incentive to use them coming from rapidly growing populations in East Asia, coupled with the tradition of the freedom of the High Seas, has led to a classic case of overfishing and environmental destruction.

The stocks of salmon and albacore are probably being seriously depleted in one of two ways or both.

Fishing is at such intensity that spawning populations are being depleted below the numbers required to produce a steady level of juvenile fish; or the average size of the fish in the stock is reduced to the point at

42. Eastly (n. 32 above).
43. "House Endorses Calls for Ending Driftnet Use," *Honolulu Advertiser* (18 November 1989): A6.
44. Anderson (n. 34 above), p. 35.

which the same number of harvested fish begins to result in lower ton-nages. Despite evidence of continuing depletion surpassing the level of maximum sustainable yield, the existing incentives for fishermen encourage them to harvest as much as they can today, regardless of the effect on future stocks.[45]

Furthermore, ecological damage in the form of wildlife destruction is apparently considered by East Asian fishing officials as part of the cost of doing business: "Fishermen come to the sea to fish, not to care about the other animals. And the marine mammal issue is out of our control. There's not much we can do. We have to operate in those grounds [North Pacific]. If we're forced out, there are no other grounds."[46] Although we have no firm data on the extent of the damage caused by driftnetting to other marine species and birds, there is reason to believe that the devastation is massive.

ALTERNATIVE MANAGEMENT REGIMES

The driftnet controversy shows that the High Seas are no longer a vast, underexploited, and self-equilibrating system, unaffected by human activities. Recent and potential developments are greatly modifying the inherited belief in the freedom of the High Seas. These new developments include

1. the revolution in marine technology,
2. the rise of ecological consciousness,
3. the new economics of ocean resource scarcity,
4. new demands for international income redistribution,
5. the new politics of the ocean, and
6. the Law of the Sea conference.[47]

In less than one generation, the entire ocean has become an arena for human activities almost as diverse as those occurring on land.

Many of the same problems associated with land use are also arising in the ocean realm. These include conflicts over use of space and resources, as well as an inadequate understanding of the complex, multispecies ecosystems of the oceans. Thus, "the shape of an ocean regime able to accommodate peacefully its multiple users, preserve its health, make optimal use of its wealth, and yet be regarded as legitimate by most of the world's peoples is currently the subject of international debate."[48] This debate has led to propos-

45. Brown et al. (n. 2 above), p. 51.
46. "No Net Gain for Ocean Life" (n. 21 above), p. 15.
47. Brown et al. (n. 2 above), pp. 20–29.
48. Ibid., p. 19.

als for three alternatives for managing the ocean commons: continuation of open access and free use, national ownership and management, and international management of the global commons.

Open Access and Free Use

The long-established principle of the freedom of the High Seas, championed by Grotius, was premised on the assumption that the open ocean was ubiquitous, that its resources were inexhaustible, and that its waters were indivisible. There are still some supporters of this alternative regime for ocean "management":

> Despite the difficulties of maintaining open access to the ocean and free use of its resources as ocean uses proliferate, this principle continues to be propounded as the most valid basis for a general regime by some interests—notably military users, shippers, long-distance fishermen, and oceanographers. These interests have been satisfied with the tradition of relatively unimpeded access to the ocean, and are generally opposed to attempts to restrict their movements or put constraints on the type of vessels or equipment they deploy.[49]

However, as shown by the driftnet case, the assumptions underlying this ocean regime are no longer valid.

Resources are becoming scarce, numerous conflicts are occurring between users, and the global ocean ecosystem is being seriously degraded. In addition, increasing economic interdependence and changing international political norms preclude continuing independent economic and political action and favor international cooperation. In short, it seems unlikely that the regime of open access and free use, based on the doctrine of *res nullius*, will survive the end of the 20th century as a viable alternative for managing the oceanic commons.

National Management

Although various maritime special interests have continued to advocate freedom of the seas, "the actual regime that has evolved since the Second World War is more accurately characterized as a regime of creeping national jurisdiction."[50] Even in Grotius's time, when European expansionism was rampant and the freedom of the seas was the ideology of the day, every country

49. Ibid., p. 31.
50. Ibid.

exercised sovereignty over some portion of its coastal waters. The question then, as now, was to decide to what extent the ocean and its resources should be owned and controlled by national jurisdictions.

Many policy advisors argue that "enclosure" of the High Seas by nation-states will lead to economic efficiency and corresponding resource conservation. According to one commentator,

> economic analysis suggests that the property arrangements associated with enclosure by coastal nations would reduce the likelihood that ocean resources would be exploited wastefully. Therefore, enclosures, according to economic criteria, will usually be superior either to the preexisting regime based entirely on high seas freedoms or to the international controls for certain resources which may evolve from the treaty negotiations that have been under way since 1974 at UNCLOS.[51]

In actual fact, the Third Law of the Sea Conference (UNCLOS III) resulted in a big step toward the national enclosure advocated above.

Many coastal states made the case at UNCLOS III that they should be given exclusive ownership and jurisdiction over resources. They argued that they could ensure economic efficiency and equity in the distribution of benefits from the resources as well as "more effectively prevent overfishing, pollution, and general ecological abuse if they were given unambiguous authority over all activities in extended territorial waters."[52] As is well known by now, these arguments carried the day. The 1982 LOS Convention extended ownership of Territorial Seas out to 12 miles and gave extensive use of resources to coastal states in exclusive economic zones (EEZs) out to 200 nm.[53]

Despite this escalating enclosure movement, there still remains a large area (about 65%) of the ocean that lies outside the limits of the national EEZs and is essentially international. Furthermore, the movement toward enclosure and exclusive national management ignores the inherent interrelationships of ocean ecosystems and resources. Finally, national management lacks the "means to assure that interdependent users [e.g., Pacific Island long-liners and East Asian drifnetters], some of which would be under different national jurisdictions, would be adequately coordinated with one another."[54] National management seems to be a rational response to environmental degradation and the growing scarcity of resources, but it does not adequately deal with global ecological and economic interdependence or with the persisting indivisibility of the ocean commons.

51. Eckert (n. 13 above), p. 5.
52. Brown et al. (n. 2 above), p. 32.
53. United Nations (n. 4 above); J. R. V. Prescott, *The Maritime Political Boundaries of the World* (London: Methuen, 1985).
54. Brown et al. (n. 2 above), p. 33.

International Management

The continuing case for increasing international management of the ocean commons derives from the premise that it is essentially indivisible and that its resources are finite. International management is required:

> To reflect adequately the far-flung, often global, interdependencies of ocean users, broadly based negotiating and decisionmaking forums would be required to implement the international management concept. Moreover, the periodic readjustment of jurisdictional boundaries, the renegotiation of exploitation and fishing quotas—all of which is inevitable as expanding technologies affect the use of the oceans—point to the importance of permanent multinational institutions.[55]

There are few examples of international management of the global commons. However, there have been rudimentary attempts at international management of fisheries resources.

The first commission was formed in 1924, and since then over 22 international commissions have been established to manage the living resources of the sea. These international commissions can be classified into two groups: "the species commissions which are concerned with one species or a group of related species, for example, the Inter-American Tropical Tuna Commission, and the regional commissions which deal with all or the most important fished species within a defined region, for example, the Indian Ocean Fisheries Commission."[56] The general stated aim of these commissions has been to conserve the various stocks of fish being exploited, but they have achieved varying degrees of success.

Some species commissions, such as the International Pacific Halibut Commission, have been relatively successful at conserving stocks and allocating sufficient catches to the fishermen of nations party to the commission.[57] On the other hand, the International Whaling Commission (IWC) is generally seen as a failure in both stock preservation and long-term economic development.[58] One observer has said, "The weakness of the IWC as a protective agency was that its voting members represented in most instances the industry that it was intended to police."[59] The limited success of the species commissions has prompted some observers to advocate different jurisdictional arrangements.

55. Ibid.

56. P. A. Driver, "International Fisheries," in *The Maritime Dimension,* ed. R. P. Barston and P. Birnie (London: George Allen and Unwin, 1980), p. 40.

57. Churchill and Lowe (n. 3 above).

58. J. E. Bardach, *Harvest of the Sea* (New York: Harper and Row, 1968); and Driver (n. 56 above).

59. L. K. Caldwell, *International Environmental Policy* (Durham, North Carolina: Duke University Press, 1984), p. 33.

Regional commissions, such as the recently established South Pacific Forum,[60] are not really new. The International Council for the Exploration of the Sea (ICES) was formed in 1902 by nations engaged in research on the fisheries of the North Atlantic. The regional approach is viewed by some as the best type of international management.

> A regional approach permits planning on a level that is neither too large (global) nor too confined (national or local), and therefore appears to be very useful for ocean management purposes. Of course, not all of the ocean can conveniently be divided up into regions, but some areas, many of which are intensively used, present themselves as good candidates: the North Sea, the Baltic Sea, the Mediterranean, the Caribbean, the Red Sea, or even much larger areas such as the Southwest Pacific. Regional efforts to develop these regions, already under way in several instances, are a step in the right direction.[61]

Unfortunately, many of these regional commissions, like the SPFFA, in the case of the Asian albacore driftnet fishery, have been unable to prevent the overfishing and depletion of migratory resources that move between their EEZs and the High Seas.

In summary, the fisheries commissions have not worked very well to date. This is so for at least three reasons.

> First, many of the commissions have inadequate constitutions which lack provisions for the control of fishing effort (particularly control of entry to the fishery), lack any worthwhile enforcement scheme, and have loopholes written into them allowing individual members to dissent. Secondly, for national and international political reasons member nations have found it difficult to agree to proposed regulations and even more difficult to agree to reductions in catch based on scientific advice. Thirdly, even with the establishment of an international Commission, the high seas nature of the fishery has remained, allowing unlimited entry to the fishery by non-member states.[62]

All of the shortcomings of management by international commissions mentioned above can be seen in the Pacific driftnet case. However, they are a necessary, but not sufficient, step in the evolution of international management for the regulation of environmental impact, overfishing, economic waste, and international conflict in the global marine commons.

60. J. P. Craven, *The Management of Pacific Marine Resources: Present Problems and Future Trends* (Boulder, Colorado: Westview Press).

61. Cuyvers (n. 11 above), p. 165.

62. Driver (n. 56 above), p. 43.

CONCLUSIONS: FROM *RES NULLIUS* TO *RES COMMUNES*

As argued above, neither the alternative of High Seas freedoms (open access and free use) nor national ownership and control is adequate for the management of global commons such as the oceans. Furthermore, although they have been a step in the right direction, regional and species international commissions have not been very successful at conserving and regulating the use of living resources in the ocean. What, then, is the alternative?

Regardless of the difficulties in establishing and maintaining effective international management of the High Seas, it is imperative that this alternative continue to be strengthened. UNCLOS III made a valiant attempt to establish the incentive for international management of living resources in the High Seas when it asserted in Articles 117, 118, and 119 that all nations have a duty to cooperate in conserving and managing them. Unfortunately, the national interests in exclusive management of ocean resources predominated at UNCLOS III, yet it provides clear evidence that a world order for ocean management is emerging.

Some of the basic principles, institutional imperatives, and policy guidelines for such a global management system might include the following:

1) Because the living resources of the ocean are shared, any group [such as a nation or regional group] with management authority or exploitation rights over a portion of these resources should be considered a trustee of the only real owners, the international community. Delegation of management authority (which subsumes authority to allocate exploitation areas and harvesting quotas) to particular communities should be based on the communities' technical and political competence to implement conservation measures. . . .

2) The controlling guidelines for implementing the above principles should be established by a Global Fisheries Commission, an institution with universal membership. One of the most important responsibilities of this commission would be to appoint a nonpolitical, scientific board to determine levels of maximum sustainable yield for specific stocks or species. . . . The Global Fisheries Commission would have direct management and allocation authority for such highly migratory species as tuna and whales, as well as for fish in internationalized areas such as the Antarctic.

3) All states should commit themselves to abide by certain specific processes for the allocation of exploitation rights and for the settlement of disputes. A treaty should also provide for countries to abide by limits on maximum sustainable yield, as determined by a nonpolitical, scientific institution. In addition to establishing the institutional pattern and committing nations to certain procedures, the treaty would establish which

countries belong in which regions, if regional ownership were decided upon.[63]

This proposal recommends the strengthening of existing regional commissions, plus the creation of new regional management units—perhaps, based on the notion of the large marine ecosystem[64]—in combination with formalized global cooperation in allocating and enforcing the use of the living resources of the High Seas.

Although by no means perfect, adoption of a model similar to the one proposed above would alleviate many of the problems exhibited in the Pacific driftnet case. More important, it would move humankind from treating the oceanic commons as *res nullius*—belonging to no one and, therefore, not the responsibility of anyone—to acting as though the High Seas are *res communes*—common property of everyone who must work together to ensure its sustainable development.

63. Brown et al. (n. 2 above), pp. 60–61.

64. J. R. Morgan, "Large Marine Ecosystems: An Emerging Concept of Regional Management," *Environment* 29, no. 10 (1987): 4–34. Editors' note.—See article by Kenneth Sherman, "Biomass Yields of Large Marine Ecosystems," in *Ocean Yearbook 8*, ed. Elisabeth Mann Borgese, Norton Ginsburg, and Joseph R. Morgan (Chicago: University of Chicago Press, 1989), pp. 117–37.

Managing Urban Coastal Zones: The Singapore Experience[1]

Chia Lin Sien
National University of Singapore

INTRODUCTION

There has been increasing research interest in the Southeast Asian coastal zone over the past two decades. Compared with North American and European countries, there is a lag of perhaps a decade in terms of the perception and action required to deal with the problems that have become so evident in developing countries.[2] Within the Association of Southeast Asian Nations (ASEAN) area, official interest has been awakened only in the last decade, and a better understanding of the concept of managing the coastal zone has come about.[3] Research and moves to manage the coastal zone as a distinct ecosystem have resulted from a number of concerns relating to developments within the region. The issues of concern include the depletion of fisheries, particularly in shallow waters; the destruction of mangrove stands and tidal

1. The data used in this paper are drawn from the research findings of the Singapore component of the ASEAN-USAID Coastal Resource Management (CRM) Project, which is coordinated by the Science Council of Singapore with the overall management of the project under the International Center for Living Aquatic Resources Management (ICLARM), Manila. I wish also to acknowledge the support given and personal interest in this research by Dr. Chua Thia-Eng, the project manager.

2. See Stella Maris A. Vallejo, "Development and Management of Coastal and Marine Areas: An International Perspective," in *Ocean Yearbook 7,* ed. Elisabeth Mann Borgese, Norton Ginsburg, and Joseph R. Morgan (Chicago: University of Chicago Press, 1988), pp. 205–22. Perhaps the first attempt at research on the coastal zone within the region was that undertaken under the auspices of the Agricultural Development Council, which resulted in the publication of *Man, Land, and Sea—Coastal Resource Use and Management in Asia and the Pacific,* ed. Chandra Soysa, Chia Lin Sien, and William L. Collier (Bangkok: Agricultural Development Council, 1982).

3. This is supported by the success of the Policy Workshop on Coastal Area Management, 25–27 October 1988, in Johor Bahru, Malaysia, involving policymakers, administrators, and scientists, and a follow-up Policy Conference on Managing ASEAN's Coastal Resources for Sustainable Development, 4–7 March 1990, in Manila and Baguio, Philippines, organized by ICLARM under the ASEAN-USAID CRM Project. High-level government officials from the environmental ministries of the ASEAN states participated in these meetings.

wetlands for conversion into fish farms; increasing landfill activities in foreshore areas for industrial and other developmental purposes; the growth of coastal tourism and recreational facilities; coastal erosion; salt intrusion; the increasing incidence of coastal pollution arising from mining activities; and waste effluents from domestic, industrial, agricultural, and commercial sources. More recently there has been concern about the possible rise in sea level, which will have serious consequences for many island communities and heavily populated coastal areas as well as coastal urban centers.[4]

An aspect of the coastal zone that has received only marginal attention has been the segments of the coast that are urbanized. Within Southeast Asia, there are a number of major coastal cities, including Singapore, Manila, Davao, Cebu, Jakarta, Surabaya, Semarang, and Songkhla. In addition, there are numerous large and smaller urban centers on the coast or on the floodplains of rivers. These conurbations represent concentrations of resources and provide the foci for development for their respective hinterlands. Many of them also function as seaports and industrial centers. Their physical presence has drastically altered the biophysical environment and resulted in permanent irreversible changes to the coastal ecosystem. They exert powerful centripetal forces on settlement patterns, land use, transport networks, economic and social functions, and other activities over long distances along the coast. Many of the concerns expressed above are enhanced through the intensified pace of development and competition for limited and valuable land and other resources close to the urban center.

Among coastal and riverine urban centers throughout North America, Europe, and other advanced countries, there are major urban waterfront redevelopment projects including those in San Francisco, Baltimore, and Honolulu (United States); Halifax (Canada); Liverpool and London (United Kingdom); Tokyo (Japan); and Sydney (Australia). In spite of the popularity and recognized value of this aspect of the urban coastal zone, there have been no similar waterfront developments for the coastal cities of Southeast Asia.

The waterfront development projects of advanced countries have in most cases come about as a consequence of changing port functions, which have resulted in the abandonment of large tracts of unused, derelict docks and shipyards, often located in prime urban locations. Redevelopment creates new high-valued complexes for a variety of uses, including shopping, recreational, educational, and cultural activities, which incorporate residential, commercial, and tourist facilities. To ensure their success, they require improved access through new transport facilities and public utilities. By and large, the problems encountered have been in the administrative, legal, and

4. See, for instance, United Nations Environment Programme, "Report of the Third Meeting of Experts on the East Asian Seas Action Plan," 7–10 February 1989, Quezon City, Philippines, UNEP(OCA)/EAS WG.3/8 (Bangkok: United Nations Environment Programme, 1989).

financial areas, with ecological and cultural issues playing only a marginal role.[5]

While most urban waterfront redevelopment schemes are generally limited in areal coverage and scope, a more comprehensive, multisectoral, integrated approach toward managing the coastal resources and activities is needed for coastal zones within which an urban center is located. This approach is essential for the intensively developed urban coastal areas, which have brought about rapid changes in coastal landscapes and uses of both waterfront land and the adjacent sea space. Within a relatively limited area, severe conflicts exist among the many interest groups represented by the users of coastal resources.

This approach has been adopted in the case of the coastal zones of Singapore and other ASEAN participants in the ASEAN-USAID Coastal Resource Management (CRM) Project.[6] Management planning takes into account the full range of biophysical, socioeconomic, and cultural factors bearing on the zone of coastal waters and waterfront land and directly or indirectly influencing each other. Consideration is given to activities that are in some ways dependent on or closely associated with the sea.[7]

Some of the preliminary results of the project research have been published, and others will follow.[8] This paper discusses the issues and options

5. See, for example, A. Church, "Waterfront Regeneration and Transport Problems in the London Docklands," pp. 5–38; Morgan Sant, "Waterfront Revitalisation and the Active Port: The Case of Sydney, Australia," pp. 69–94; Michael Goldrick and H. Roy Merrens, "Waterfront Changes and Institutional Stasis: The Role of the Toronto Harbour Commission, 1911–89," pp. 119–53; all in *Port Cities in Context: The Impact of Waterfront Regeneration,* ed. B.S. Hoyle, Transport Geography Study Group, Institute of British Geographers (Southampton: Dept. of Geography, University of Southampton, 1990).

6. The approach has also been adopted elsewhere, as in the case of Oman. See Rodney V. Salm and James A. Dobbin, "Coastal Zone Management Planning and Implementation in the Sultanate of Oman," *Coastal Zone '89,* Proceedings of the Sixth Symposium on Coastal and Ocean Management, Charleston, South Carolina, 11–14 July 1989, ed. Orville T. Magoon, Hugh Converse, Dallas Miner, L. Thomas Tobin, and Delores Clark (New York: American Society of Civil Engineers, 1989), pp. 72–78. The ASEAN-USAID CRM Project is a 4-year project that began in late 1986. It is coordinated by ICLARM in Manila and involves the participation of Brunei Darussalam, Indonesia, Malaysia, Philippines, Singapore, and Thailand. Each of the participating countries has a research team, which in the case of Singapore is administered under the Science Council of Singapore.

7. See Chua Thia-Eng and A. A. Agulto, "ASEAN/USAID Cooperative Program on Marine Sciences: Coastal Resources Management Project: Research, Training, and Information Programs and Activities," ASEAN/US CRMP Working Paper 87/13 (Manila; ASEAN, USAID, and ICLARM, 1987). In practice, the whole of the territorial waters of Singapore (a mere 600 km^2, approximately), the strip of coastal properties on the main island of Singapore, and all of the offshore 50-odd small islands form part of the coastal zone of Singapore.

8. See Chia Lin Sien, Habibullah Khan, and Chou Loke Ming, *The Coastal Environmental Profile of Singapore* (Manila: ICLARM, 1988).

for the management of the Singapore coastal zone, in particular the Southern Islands, located off the south and southeast coast of the main island, and the intervening waters surrounding them. The Southern Islands have been selected as the pilot area of the Singapore CRM Project.[9]

DEVELOPMENTAL VERSUS ENVIRONMENTAL USES OF COASTAL RESOURCES

The shelf-locked, small island city-state of Singapore, with an area of 620 km[2], is located 80 km north of the equator.[10] It is bounded to the north by peninsular Malaysia. The Johor Straits separate the two states, but they are joined by a causeway 1.5 km long (fig. 1). To the south of Singapore lie the islands of the Riau Archipelago. The coastal zone of Singapore experiences a low-energy environment as a consequence of the weak winds throughout the year and the protection afforded by surrounding land masses.[11] The tidal range of the surrounding waters is only 3.5 m. There is a strong, steady sweep of westerly tidal currents across the waters to the south of the main island of Singapore.[12]

The country's gross national product has increased tenfold since 1970, to reach S$56.3 billion, resulting in a per capita income of S$17,910 in 1989.[13]

9. See Chia Lin Sien, "A Coastal Area Management Plan for Singapore's Southern Islands," paper presented at ASEAN/US Technical Workshop on Integrated Tropical Coastal Area Management, 28–31 October 1988, Singapore (organized by ICLARM).

10. Singapore attained self-rule in 1959 from the British colonial administration and became a sovereign state on 9 August 1965. Although the government of Singapore intends to claim a 12-mile Territorial Seas limit, its current maritime limits are based on the Straits Settlements and Johore Territorial Waters (Agreement) Act 1929 (3 August 1928) in the Johor Straits and, to the south, by the Singapore-Indonesia agreement of 1973. See Daniel J. Dzurek, "Boundary and Resource Disputes in the South China Sea," *Ocean Yearbook 5*, ed. Elisabeth Mann Borgese and Norton Ginsburg (Chicago: University of Chicago Press, 1985), p. 283. At the time of writing, negotiations are ongoing with the Malaysian government to determine the boundary between the two countries.

11. See, for instance, S. B. St. C. Swan, "Coastal Geomorphology in a Humid Tropical Low Energy Environment: The Islands of Singapore," *Journal of Tropical Geography* 33 (1971): 43–61; P. P. Wong, "Singapore," in *The World's Coastline*, ed. E. C. Bird and M. L. Schwartz (New York: Van Nostrand Reinhold, 1985), pp. 797–801.

12. N. F. W. Chua and W. K. Lim, "Tides and Currents in Singapore," in *Proceedings of the Conference on the Biophysical Environment of Singapore and Its Neighbouring Countries*, 3–5 May 1985, ed. Chia Lin Sien, H. C. Lee, A. Rahman, P. L. Tong, and W. K. Woo (Singapore: Singapore Geography Teachers' Association, 1986), pp. 77–92.

13. Singapore, Department of Statistics, *Yearbook of Statistics* (Singapore: Department of Statistics, 1989).

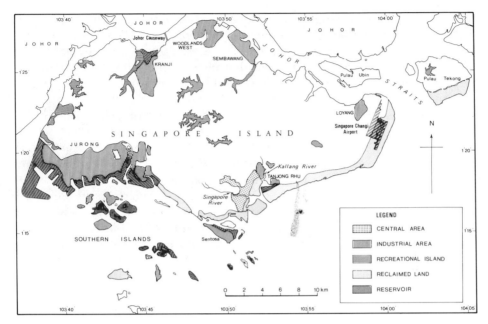

FIG. 1.—Singapore's coastal zone.

This has been achieved through developmental efforts including sound economic, social, and physical planning, a massive buildup of infrastructure, and large-scale foreign investments brought in by liberal investment schemes and nurtured by an open-door policy.

The economic development program was laid down in the first national development plan (1961–64),[14] and the main vehicle for achieving rapid growth was through massive investments in manufacturing industries both by the government directly and by foreign investors. In order to encourage investors to establish manufacturing plants quickly, planned industrial estates were developed. The Jurong Industrial Estate (now called Jurong Town), the largest such estate, was begun in the early 1960s. It is located on partially reclaimed land on the southeast coast. Other major estates on the northeast coast were developed subsequently (fig. 1). Many of the offshore islands, including those off the Jurong coast, have also been used for industrial purposes.

The program of industrial development was aimed at promoting export-oriented industries. The seaport was the other major element in the buildup of infrastructure affecting the coast. Massive investments were poured into the construction of berths, cargo-handling facilities including container quay

14. Government of Singapore, *First National Development Plan, 1961–1964* (Singapore: n.d.).

cranes and warehouses. The port comprises four gateways and is still being expanded. Sophisticated electronic equipment provides ship-shore telecommunications and the tracking of vessels within the port waters.[15] Unlike many of the major ports around the world that have undergone containerization, the port of Singapore has developed so rapidly in terms of cargo throughput that the usual dereliction of docklands did not occur; port expansion has absorbed all available land. The only exception is the waterfront along the Singapore River, where the traditional lighters (*tonkang*) that work the ships anchored in deeper waters were removed in 1983. However, the removal was made not because of disuse of the warehouses along the river but primarily because the lighters were a major source of pollution of the waterway. The operation of these lighters is now at the Pasir Panjang wharfs in the sheltered waters of Terembu Retan Laut, a long, slim island created on a shoal.

The coastal zone of Singapore has been subjected to intense developmental pressures over the last three decades, leading to rapid alteration of the biophysical coastal environment as well as changes in the pattern of use of the coastal resources. The status of the various coastal uses, the changes undergone, and the conflicts emerging therefrom have been described elsewhere.[16] Essentially, the conflicts have been due to use of the limited coastal space and resources to achieve developmental goals rather than to meet environmental concerns. Within the developmental objectives for a newly independent nation, the urgent need was to provide adequate space mainly for essential industrial, seaport, and shipping developments, as described earlier.

COASTAL CHANGES AND THEIR IMPLICATIONS

The changes in the use of coastal resources have in part been made possible through massive foreshore landfills to provide firm land for constructing piers, warehouses, and buildings for commercial, industrial, and other uses. This process had already begun soon after the founding of the colonial settlement in 1819, with the construction of wharf facilities and dockyards and expansion of the original commercial area south of the Singapore River.[17]

Low-lying and swampy land was reclaimed along the waterfront of the

15. Nearly all of the territorial waters have been designated as port waters under the control of the Port of Singapore Authority, which was established in 1964 and took over the functions of the Singapore Harbour Board.

16. Chia Lin Sien, "Utilization and Management of Singapore's Coastal Zone," in *Proceedings, MAB/COMAR Regional Seminar on Man's Impact on Coastal and Estuarine Ecosystems,* 13–16 November 1984 (Tokyo: Man and Biosphere and Committee on Marine Resources, 1984), pp. 51–54.

17. For a description of coastal changes in Singapore see P. P. Wong, "The Changing Landscapes of Singapore," in *Modern Singapore,* ed. Ooi Jin Bee and Chiang Hai Ding (Singapore: University of Singapore Press, 1969), pp. 20–51.

early settlement and from the mid-1880s further to the west toward Keppel Harbour. The major reclamation schemes covered almost the entire southern shore of the main island. With the completion of the Tuas (adjacent to Jurong) on the western tip of the main island, current reclamation activities are mainly along the northeast coast (fig. 1). The land was made available for expansion of the seaports as well as for planned industrial estates and other uses such as the airport in Changi.

For a brief period in the 1960s, some reclaimed land along the east coast (Marina Parade) was used for government low-cost housing. It was soon decided that the land was far too valuable to be put to this use. Marine Parade estate became the only truly coastal government-built housing estate until the decision to build Pasir Ris New Town on the northeast coast. These and other housing estates now accommodate some 85% of Singapore's total population and form a major element of Singapore's cultural landscape. Unused waterfront land has been allocated for coastal parks, the largest of which is the East Coast Park, with smaller ones on the west coast and along the Pasir Ris area.

Foreshore reclamation has also taken place on the offshore small islands, enlarging and joining adjacent islands to provide additional land for oil refineries, petrochemical plants, oil storage facilities, other industrial uses, and an electric power generation plant. Even Sentosa Island, which has been devoted to providing recreational facilities, has been enlarged.

Associated with foreshore reclamation is the construction of coastal reservoirs, impounding water across the mouth of the estuaries of a number of rivers, mainly in the northern portion of the coast. For flood control many of the streams have tidal gates to control the inflow of water during high tide. Additionally, most of the rivers and streams have been canalized, with sides of concrete walls.

The physical changes wrought on the coast through the activities described above have been drastic and, in many places, effectively irreversible. Foreshore reclamation has proceeded down to the 5-m mark and in the future will advance into deeper water. This has filled in the fringing reefs and changed the beach and seabed profile. Siltation of the seabed in the surrounding areas has been a major problem. Where seawalls were erected, they have prevented any direct exchanges of materials and plant nutrients across the land-water interface, thus changing the character of the substrate of the shallow waters and reducing the abundance of the marine life therein. Enlargement of offshore islands with fill material brought in from elsewhere has similarly drastically changed the coastal ecological conditions, leading to the destruction of the surrounding coral reefs.

The major hydrological changes due to the increased built-up area of the country have resulted in fast surface runoff. The incidence of floods has been reduced, however, by several major flood alleviation schemes and the use of tidal sluice gates.

Large-scale foreshore reclamation and the construction of seawalls and coastal structures cannot be avoided in a populous and intensely developed small island nation. However, it is precisely this situation that requires that coastal modifications be carefully planned and implemented to minimize the destruction and damage to the local and surrounding natural environment. Efforts should be made to create an aesthetically appealing environment, avoiding the stark harshness of concrete and steel as well as the total removal of vegetation. There has in general been a lack of awareness of the marine ecosystem and life under the sea surface.

Beginning in the late 1960s, attention was given to management of the environment and pollution control, and indeed much has been achieved in bringing about a clean and green environment and in creating a "garden city" image. The main objective has been to protect human health and safety and has not had an ecological basis.[18] In the 1980s, however, the greater affluence of the country and the attainment of the status of a developed nation have promoted the desire for a higher quality of life through creation of an aesthetically appealing built environment and the conservation and rehabilitation of the natural environment.

The 10-year project, completed in 1988, to clean up the Singapore and Kallang river basin is a major step in this direction.[19] One aim of the ASEAN-USAID CRM Project is to rehabilitate these two rivers through the release of young fish and shrimp and use of artificial sea grass. Yet another aim of the project is to build artificial reefs on selected sites within the waters of the Southern Islands. Preliminary results of these experiments have been encouraging. It is important to note that, with careful planning of developmental projects, the need and the cost of rehabilitating the natural environment and marine life can be substantially reduced. This is demonstrated in the still vigorous growth of the coral reefs around the more distant offshore islands, indicating considerable ability of the natural biota to regenerate itself, provided there are no overwhelming physical changes.[20]

18. There is a good deal of literature on the environmental management of Singapore. For a recent survey, see Chia Lin Sien and Chionh Yan Huay, "Singapore," in *Environmental Management in Southeast Asia,* ed. Chia Lin Sien (Singapore: Faculty of Science/University of Singapore Press, 1987).

19. The Singapore River Project involved a number of government agencies and was coordinated by the Ministry of the Environment. The land use of the entire watershed of the Singapore and Kallang rivers was scrutinized, and the polluting elements were either removed or controlled. This has considerably reduced the discharge of untreated waste effluents into the river system. Polluted material from the riverbeds and banks was physically removed and replaced by clean material such as sand.

20. There has been a series of studies undertaken in the Department of Zoology, National University of Singapore. Examples of these reports include Chou Loke Ming, "The Coral Reef Environment in Singapore," in *Proceedings of the Conference on the Biophysical Environment of Singapore and its Neighbouring Countries,* ed. Chia Lin Sien,

MANAGING SINGAPORE'S COASTAL ZONE

Following the general procedure provided by the CRM Project,[21] an essential part of the development of a coastal zone management plan is to formulate goals and to identify and obtain consensus on the issues relating to the Singapore situation. This is to be followed by the development of policy and specific management recommendations before finalization and implementation of the plans.

In the case of Singapore, the initial step taken toward the formulation of a coastal zone management plan was the collection of secondary and primary data. This resulted in the production of the Coastal Environmental Profile of Singapore.[22] In addition, the information gathered was put on a series of maps, and a description of the information formed part of a preliminary coastal management guideline plan for Singapore. This document was presented as background to the Second National Workshop as part of the project's activities.[23]

The workshops are an important part of the formulation of a coastal zone management plan. The first workshop was convened in November 1986 and the second enlarged workshop in November 1989. They brought together representatives from government agencies and researchers from the National University of Singapore. They provided an opportunity for the project researchers—drawn from both universities and government agencies—to present their findings, and for the representatives of many of the government and private organizations to present papers dealing with their responsibilities and activities related to the coastal area of Singapore.

The workshops established a forum for organizations to discuss matters of mutual interest and had the effect of increasing awareness of the interrelationships among the diverse activities of the coastal zone. A set of recommendations was put together after the second workshop and was circulated to all

H. C. Lee, A. Rahman, P. L. Tong, and W. K. Woo (Singapore: Singapore Geography Teachers' Association, 1986), pp. 93–102; Chou Loke Ming and Chia Lin Sien, "The Marine Environment," in *The Biophysical Environment of Singapore*, ed. Chia Lin Sien, Ausafur Rahman, and Dorothy B. H. Tay (Singapore: University of Singapore Press, 1991).

21. The ASEAN-USAID CRM Project provides guidelines that evolved during the course of the project. An important part of the process of formulating a management plan was to identify issues through consultation with private and governmental agencies concerned with the management and use of coastal resources and with activities in the coastal zone. See Chua and Agulto (n. 7 above), also Chua Thia-Eng, "Developing Coastal Area Management Plans in the Southeast Asian Region," *Coastal Zone '89* (n. 6 above), pp. 2192–2201; Alan T. White, "Comparison of Coastal Resources Planning and Management in the ASEAN Countries," ibid., pp. 2123–33.

22. Chia et al. (n. 8 above).

23. Second National Workshop on the Management of Singapore's Urban Coastal Areas, 9–10 November 1989, Singapore, organized by the Science Council of Singapore and ICLARM.

participants for their views. Amendments to the recommendations were then made on the basis of the responses obtained.[24]

The recommendations set up objectives and identified the issues, as well as suggesting a possible management framework for a coastal zone management plan for the country. The discussion highlights the main issues in Singapore. The broad objective of the plan covering the waterfront of the main island, offshore islands, and all of the territorial waters of Singapore is that they should be managed on an integrated and multisectoral basis for maximum benefit of the nation as a whole, in the short as well as in the long term.

Developing an Appropriate Framework for Coastal Management

Discussion of the management of the coastal zone can be divided into three aspects: the sea space, the waterfront land and offshore islands, and the entire coastal zone.

Of the many government agencies involved in the utilization of coastal sea space, the most important is the Port of Singapore Authority (PSA), which is charged with the control of nearly all of the territorial waters of Singapore. The importance attached by the nation to the activities of the seaport and navigation is such that the use of sea space for all port, navigation, and related activities must take first priority before other uses of sea space and resources are considered.

The ocean's waters are continually moving, carrying with each parcel of water various properties and constituent materials. In addition, the water depth varies considerably from place to place within the Territorial Sea space of the country. Thus, the same piece of sea space has the potential for being used for one or more purposes simultaneously. It is here that the unisectoral character of the agencies emerges. Administrators who are charged with the operation of a unisectoral agency will typically want to use the sea space exclusively for the uses or activity mandated by the agency's enabling legislation. The present practice requires the agreement of the PSA for all other uses of sea space or activities carried out therein. There is a good deal of consultation on an ad hoc and informal basis. Nevertheless, the scope of the deliberations must necessarily be limited, and a great deal more can be achieved by way of coordinated and multiple uses of sea space.

On the landward side, two major agencies are involved, the first being the Urban Redevelopment Authority (URA), which has recently taken over

---

24. The Preliminary Coastal Resources Management Plan for Singapore and the papers presented at the second workshop together with the set of recommendations will be published as *Proceedings of the Second National Workshop on Managing Singapore's Urban Coastal Resources,* 9–10 November 1989, ed. Chia Lin Sien and Chou Loke Ming (Manila: ICLARM, in press).

the functions of the Planning Department under the Ministry of National Development. URA is charged with executing the Statutory Master Landuse Plan, 1958. Under the master plan, all land is zoned for certain permissible uses with stipulations of the permitted intensity of development. Land uses within the Central Planning Area and New Towns are planned in great detail. An equally important planning instrument is the Long-Range Comprehensive Concept Plan, which was completed in 1971 and reviewed in 1989.[25] While this plan does not carry the force of law, it is used to guide future physical development, and its essential framework has been followed very closely. While details of the review have not been released, it is understood that the use of sea space as well as the environmental-ecological aspects of the nation have received considerably more attention than previously had been the case.

The second agency is the Land Office under the Ministry of National Development. All government-owned land, including reclaimed land, and water space comes under the ownership and responsibility of the Land Office. Land is often leased for a limited time, up to 99 years, to other government agencies such as the Jurong Town Corporation (JTC) for industrial uses, the Housing Development Board (HDB) for public housing, and the Sentosa Development Corporation (SDC) of several islands for recreational and tourism development.

Waterfront land is considered prime property and potentially of critical importance. There has been a conscious effort to reserve coastal land for such purposes as industrial, airport development, and telecommunications uses. It is implicitly understood that the large coastal parks on waterfront land along the east coast (east of the city center) and parts of the west coast in Pasir Panjang may at some date be given over to a more essential and intense form of land use. All of the islands have been earmarked for industrial, recreational/tourism, or military uses, and only Pulau Ubin in the north has retained its residents.

The Land Office and that function of the URA responsible for the master plan undertake only the allocative functions of land and basic activities that may be permitted once the land is allocated to either public or private use. It should be noted here that over 90% of the coastal land is owned by the government directly or by its public agencies. These agencies then undertake to plan and manage their respective physical resources and activities. Thus, in this regard, while there is an umbrella planning agency, the administrative system in Singapore is essentially unisectoral in nature. All development plans of the various agencies require the approval of the URA.

All submitted development plans are sent to all other concerned agencies, who then point out conflicts of land use and potential problems and register

25. For discussions of the plan, see K. Olszewski and R. Skeates, "Singapore's Long-Range Planning," *Royal Australian Planning Institute Journal* (April 1971): 57–70.

their disagreements or submit suggestions for amendments to the plans. Through this mechanism, there is considerable coordination and integration of physical planning within the country as far as the land—as opposed to the sea—is concerned.[26] The single-level system of government has worked well, although the newly established town councils may in the near future add another layer to the government structure. It should also be noted that, technically, the URA is also responsible for all sea space, although in practice this has not been effectively carried out. Setting priorities for use of coastal land and sea space as well as the coastal and marine environment is the key to more effective coastal zone planning and management in Singapore.

The management of Singapore's sea space is effectively the responsibility of the PSA. While there is informal coordination and consultation between the authority and other users, it may be desirable to formalize a mechanism for consultation, recognizing the legitimate rights of the many alternative uses, especially for recreation and tourism. Two agencies are involved in the provision of recreation and tourism facilities, the SDC and the Singapore Tourist Promotion Board (STPB), which are concerned with the development of Sentosa (already well developed) and several nearby islands to the south of it.

The Singapore Sports Council is involved in providing facilities and training in sea sports and has been active in promoting and organizing sea-sport events at various levels (local, national, and international), including wind surfing, sailing, and the annual Dragon Boat and International Powerboat races. In this context, private organizations for various sea sports and for the operation of leisure craft have been considerably disadvantaged due to the loss of waterfront sites for their clubhouses and boathouses. However, the PSA and the planning agencies are trying to accommodate their needs. This favorable development is a result of reclaiming sufficient land to satisfy developmental needs. Also, commercial opportunities and benefits have become more available for developing facilities such as marinas and tourist resorts.

The conspicuous lack of pleasure craft on the Singapore waterfront scene has often been noted. Indeed, the attractiveness and liveliness of the Singapore waterfront can be further enhanced through recreational developments. The demand for leisure outlets for a more affluent Singaporean population will certainly increase, as will pressure for the provision of suitable sites and facilities for marine sports and recreation.

26. For a recent discussion on Singapore's physical planning, see K. F. Olszewski and Chia Lin Sien, "National Development, Physical Planning, and the Environment in Singapore," in *The Biophysical Environment of Singapore*, ed. Chia Lin Sien, Ausafur Rahman, and Dorothy B. H. Tay, pp. 185–205 (Singapore: University of Singapore Press, 1991).

Recognizing the Land-Sea Linkage

There is clear recognition of the linkage between land and sea for the operation of seaports, fishing ports, and certain manufacturing and other industries. The seaports are regarded as the gateways to Singapore and are referred to as such. Many of the waterfront industrial installations have private
and/or exclusive use of cargo-handling facilities to effect the transfer of goods
and materials across the coast. Thus the oil refineries, petrochemical plants,
electric power generation plants, and oil and chemical storage facilities have
installed private berths/terminals and equipment.

Many of the industries, typically material processing plants such as flour
mills, cement factories, and iron and steel mills, are located either within the
port areas or on the waterfront, with appropriate equipment for handling
the flow of materials. Ship construction and repair and rig-building yards are
traditional occupiers of the waterfront and have been given priority for access
to suitable sites. A related industry occupying waterfront space is marine
supplies for offshore petroleum-exploration activities within the region.
Other waterfront activities are the security, customs, and immigration facilities and telecommunications and military installations.

Apart from the above uses, the land-sea linkage has not been so readily
recognized. Perhaps the most clearly demonstrable examples are coastal
parks. The major coastal parks are almost entirely artificial, since they are
reclaimed foreshore areas with planted vegetation and beaches created with
sand brought in from elsewhere. There has been little regard for preserving
or rehabilitating the indigenous coastal vegetation, although recently a small
area of mangroves has been preserved in the Seletar Park on the north coast.

The scale of foreshore reclamation has produced considerable steepen-
ing of the seabed slope off the reclaimed land. The natural flow of nutrients
and materials has been drastically altered, resulting in degradation or destruction of coral reefs, sea grass beds, and the substrate with its benthic organisms
and associated marine life. The approach to managing a coastal park or
coastal forest reserve should be integrated and include all of the adjacent
water, seabed, and marine life. Evidence that more has been done in this
regard is the preservation of a small swampy area at the western end of the
West Coast Park, in order to provide a bird sanctuary.

Environmental Concerns

The large number of tanker and other cargo ship movements daily through
the port waters creates an ever present threat of oil spills.[27] The Singapore

27. The Port of Singapore is the largest shipping port in the world. In 1989,
79,000 vessels with a total of 862 million gross registered tons (grt) entered the port
waters (Singapore, Department of Statistics [n. 13 above], p. 200).

Strait is narrow at various points, and there have been oil spills involving tankers, the best known being the *Showa Maru* in 1973.[28] A major task in coastal zone management is the control of marine pollution. The PSA coordinates an Oil Spill Contingency Plan to combat large oil spills.[29] The plan involves the participation of the major oil companies operating refineries on the offshore islands and the waterfront of the main island. Vessels larger than 500 gross registered tons must be piloted into the harbor areas.

The total amount of oil from small spills and from land-based sources entering the marine environment is far greater than the occasional larger spills. The control of effluents from industrial, commercial, agricultural, and domestic sources on the main island, including waterfront industries and facilities, is adequate, but a great deal more needs to be done in tightening controls on ships and pleasure craft. Recently, increased control was imposed on the transportation of hazardous materials. The cleanup of the catchment of the Singapore and Kallang rivers has been a major achievement, although other rivers, such as the Jurong River, need to be similarly treated.

A major environmental problem is the disposal of solid waste. There are now two incinerators operating, but there is a shortage of dumping sites with waterfront locations. There are plans to develop two offshore islands, Pulau Semakau and Saking, into dumping sites. Feasibility studies have been conducted, and attention has been given to ensuring acceptable levels of water quality and preserving the rich coral and marine life in the surrounding waters.

Within the waters of the Johor Straits and off the eastern end of the island, fish farming thrives, using floating net cages. There have been recorded incidences of oil spills resulting in fouling of the net cages and fish kills. A significant number of vessels use the strait to reach the Sembawang port, the Johor (Malaysia) port of Pasir Gudang, the Sembawang shipyards, and the naval facility. Foreshore reclamation on both sides of the strait is in progress, and industrial, commercial, residential, and recreational activities are increasing rapidly. There is clearly a need to manage the coastal zone here and to foster close coordination, especially in the aspect of coastal pollution, between the two neighboring countries.

In order to give focus to the urgent need to control marine pollution and conserve the remaining coral and other marine life in the coastal waters of Singapore, it was proposed that a marine conservation area or marine park

28. The *Showa Maru*, a Japanese-owned tanker, hit Buffalo Rock to the south of the Singapore Strait and spilled an estimated 5,000 mt of oil.

29. The ASEAN Oil Spill Contingency Plan was approved in 1972. It calls for oil combat equipment and materials to be held in readiness and for cooperation among the PSA, several government agencies, and the oil companies.

be established in the waters of the Southern Islands.[30] To put such a scheme into effect will require the cooperation of a number of agencies to regulate shipping, fishing (commercial and pleasure), scuba diving, spear fishing, recreational/tourism facilities, and solid-waste disposal within the designated area. In an area of existing intensive use of sea space and marine resources, prospects of creating such a scheme are not good. As in other spheres of endeavor, government initiatives are called for, to formulate an appropriate scheme and provide the financial and other necessary resources. However, the possibility of involving private participation in the scheme was also suggested.

TOWARD A MANAGEMENT PLAN

Singapore does not as yet have a coastal zone plan similar to that of many states in the United States or the European coastal countries. Sri Lanka has already developed a coastal management plan,[31] and under the ASEAN-USAID CRM Project, all of the participating countries are finalizing their coastal management plans. Thus far, there has been no official recognition of the need for such a plan in Singapore. However, it is simply a matter of time before this takes place, judging by the rising consciousness of having achieved the status of a developed nation and the desire for a better quality of living and hence an even more aesthetically satisfying environment.

The move toward conserving the "old Singapore" within the city in the 1980s is an important step in the right direction. There have also been indications of stronger interest and the realization of the value of a better-planned and -managed coastal zone. The designation of two coastal sites, one in Simpang to the northeast of the mainland and in Kampong Bugis on the Kallang River (fig. 1), for private planners to come up with waterfront development ideas has created exciting opportunities to enliven the built environment.

For official approval of a coastal zone plan, it will be necessary to undertake a thorough review of existing legislation related to the use and control of coastal land or sea space and resources. New legislation can then be put together, and existing agencies, whether the URA or the PSA, can be given the responsibility to manage the coastal zone on an integrated multisectoral basis.

30. Chia Lin Sien, "A Proposed Marine Conservation Area in the Southern Islands of Singapore," in *Technical Workshop on Integrated Coastal Area Management,* 26–30 November 1988, Singapore, ed. Chou Loke Ming and Chua Thia-Eng (Manila: ICLARM, in press).

31. Kem Lowry and H. J. M. Wickremeratne, "Coastal Area Management in Sri Lanka," *Ocean Yearbook 7,* ed. Elisabeth Mann Borgese, Norton Ginsburg, and Joseph R. Morgan (Chicago: University of Chicago Press, 1988), pp. 263–93.

CONCLUSION

The Singapore experience in urban coastal management has been valuable; the nation has gone through a 30-year period of rapid development and has emerged as an advanced developed nation. A great deal has been achieved, including a well-managed, clean, and green environment. The balance between development and environment has begun to tip in favor of the latter, but more needs to be done to improve the coastal environment in order to achieve an aesthetically pleasing total environment while retaining the ability to meet essential developmental needs.

It is clear that an integrated multisectoral management approach should be adopted. Such an approach will create greater opportunities for multiple uses of limited coastal resources and avoid conflicts among the different coastal sectors. The management framework is already in place, but higher priority should be given to the preservation and rehabilitation of the coastal environment and marine life. There are considerable opportunities for exploiting the land-sea linkages for a more ecologically sound coastal environment and for creating recreational, scientific, and educational facilities.

While the Singapore case may be unique in many ways, the creation of an urban coastal zone management plan for an intensely developed equatorial metropolitan area would serve as a model for other major coastal conurbations. These cities can derive considerable benefits and savings from the experience of similar plans in developed nations, including Singapore.

The Ocean in Japanese Literature: An Overview

Harvey A. Shapiro
Osaka Geijutsu University, Japan

INTRODUCTION

Because Japan is an island country on the western edge of the Pacific Rim, one would expect it to be ocean-oriented. Judging by the very small volume of Japanese literature about the ocean, however, one cannot help but wonder what the role of the sea is in Japanese civilization. This article reviews the Japanese literary tradition of ocean and coast.

THE CONDITIONS OF JAPAN'S COAST AND NEARSHORE SEA

The 2,500-km-long Japanese archipelago has a land area of 377,815 km^2.[1] This is only 0.25% of the world's total land area, but Japan's 200-mile EEZ amounts to some 3.86 million km^2, about half that of the United States.[2] Japan has one of the most intensively used coastlines in the world, with an average per capita length of only 30 cm.[3] Poor in mineral resources but with an economy heavily dependent on them, Japan participates in marine-related activities of all kinds and is one of the world's most extensive users of ocean space and resources.[4] For instance, Japan obtains nearly half of its annual protein intake from ocean fisheries, with coastal offshore areas now providing over 75% of that intake. It imports well over 50% of some necessary metal resources (for example, zinc) and often nearly all of its resources (for exam-

1. Bureau of Statistics, *Statistical Handbook of Japan* (Tokyo: Statistical Bureau's Management and Coordination Agency, 1988).
2. R. S. Lee, "Legislative Approaches to Coastal Area and Resource Management," in *Development and Management of Resources of Coastal Areas*, ed. K. Szekielda and B. Breuer (New York: German Foundation for International Development and United Nations, 1976), pp. 438–44.
3. T. Tanigawa et al., *New Marine Area Use Technical Manual* (Tokyo: Fuji Technosystems, 1978).
4. Tsuneo Akaha, *Japan in Global Ocean Politics* (Honolulu: University of Hawaii Press and Law of the Sea Institute, 1985).

ple, iron, nickel, and oil) via maritime transport.[5] Japan also imports much of its food (for example, soybeans [95.8%], corn [100%], and wheat [90.5%]).[6] In addition, there are no fewer than 50 marine parks, 14.5 million sports fishermen, some 150 million annual beach-goers, and a total of 4,000 ports, which suggests that Japan is indeed an ocean-oriented nation. However, when one considers the dreadful state of its coastal water quality, due in large part to economic growth based on coastal landfill and on fill-use industrial and related urban activities, I, for one, have considerable doubt about the appropriateness of calling Japan an "ocean-oriented" society.

The majority of Japan's landfill activities date from about 1956.[7] The continuing loss of valuable fishing grounds and natural coast (for example, the remaining natural coastline amounted to only 52% in 1975, 49% in 1980, and 46% in 1985), the infamous Minamata and other cases of heavy-metal pollution disease from eating poisoned fish, and the loss of coastal access and nearshore recreational opportunities[8] have been among the many impacts of reclamation and related activities. Such activities have not gone on without resistance from a vocal minority of concerned citizens and scientists.[9] Japan's present abusive approach to the use of its coastal waters and coastline is a result of its total commitment to "economic determinism," which tends to insist on construction and development at the expense of nature.[10] I am not convinced, however, that this is the only reason. Several years ago, I conducted a survey of citizen attitudes toward Osaka Bay (in western Japan). One memorable reply was that "the Bay is a toilet where we throw away what we don't want in our garden, the land." Such a view would seem to reflect a deep cultural reason for not being worthy of the description "ocean-

5. Ministry of International Trade and Industry (MITI), *White Paper on International Trade and Industry* (Tokyo: MITI, 1982).

6. Akaha (n. 4 above).

7. Harvey A. Shapiro, "Japan's Coastal Environment and Responses to Its Changes," in *The Coastal Zone: Man's Response to Change*, ed. Kenneth Ruddle (Geneva: Harwood Academic Publishers, 1988), pp. 491–519. Editors' note.—See also the article by Shapiro, "The Landfilled Coast of Japan's Inland Sea," in *Ocean Yearbook 7*, ed. Elisabeth Mann Borgese, Norton Ginsburg, and Joseph R. Morgan (Chicago: University of Chicago Press, 1988), pp. 294–316.

8. "565 km of Coast Lost to Development in 6 Years," *Mainichi Daily News* (12 November 1985): 8.

9. Harvey A. Shapiro and T. Zanic, "Coastal Citizens' Movements in Japan and California," in *Coastal Zone '83*, ed. Orville T. Magoon (New York: American Society of Civil Engineers, 1983), pp. 1803–22; Harvey A. Shapiro, "More on Japan's Inland Sea Coastal Citizens' Movements," in *Coastal Zone '85*, ed. Orville T. Magoon (New York: ASCE, 1985), pp. 2110–27; Harvey A. Shapiro, "Still More on Japan's Inland Sea Coastal Citizens' Movements," in *Coastal Zone '87*, ed. Orville T. Magoon (New York: ASCE, 1987), pp. 2197–2208.

10. I. L. McHarg, "The Place of Nature in the City," from a lecture presented at a meeting of the Japan Association for the International Garden and Flower Exposition (EXPO 90), Tokyo, 5 December 1988.

oriented." This article briefly reviews the Japanese attitude toward the sea as reflected in its literature.

THE SEA IN EARLY JAPANESE LITERATURE

Despite its development as an ancient rice-growing, metal-working civilization in East Asia,[11] Japan was still prehistoric before A.D. 400. The earliest record of the study of Chinese in Japan is dated 405, when a Korean was appointed to tutor a Japanese prince in Chinese. A few historical writings date from the early seventh century, but all are in Chinese and are therefore not considered to be "Japanese" literature.[12] The oldest pieces of true native Japanese literature are the songs in the ancient *Kojiki* (Record of ancient matters), in the *Nihongi* (Chronicles of Japan from earliest times to 697), and in the *Norito* (Liturgies of Shinto [Japan's indigenous belief system]). The *Kojiki*[13] was begun in 682 and completed in 712. It contains traditional myths of the Japanese people, beginning with those that form the basis of Shinto.[14] Volume 1 describes the myth of the genesis of Japan. The two original *kami* (deities) stood on the bridge joining heaven and earth, pushed down a jeweled spear and stirred up the brine (the sea) until it curdled; then they drew up the spear, from which the heavy brine dripped down to form an island. Descending to that island, they proceeded to give birth to the rest of the islands of the archipelago and then to various deities, including the sea deity Wadatsumi and the deities of the water gates (river mouths), estuaries, and ports. Similar accounts are recorded in the *Nihongi*, a contemporary work (in Chinese) completed in 720.

Unlike the verse of these two chronicles, the prose of this period (before 700) is best reflected in a series of *norito*, prayers to the deities of the Shinto faith.[15] Though the prayers are very old, they did not take their present form until the seventh century and were not written down until the early tenth century. The most famous prayer is the Oharai, the general purification ceremony. I witnessed this service in November 1988 at the Shinto funeral of my Japanese mother-in-law. Most Japanese witnessed it on television during the funeral of Emperor Hirohito (Showa) on 24 February 1989. The "vast ocean" is referred to as a place to be purged of sin: "To purge and purify them [all

11. R. H. P. Mason and J. C. Caiger, *A History of Japan* (Tokyo: Charles E. Tuttle Co., 1972).

12. W. G. Aston, *A History of Japanese Literature* (Tokyo: Charles E. Tuttle Co., 1972).

13. B. H. Chamberlain, *The Kojiki: Records of Ancient Matters* (Tokyo: Charles E. Tuttle Co., 1981).

14. Aston (n. 12 above).

15. T. Kimura, "Umi no Nihon bungakushi" (History of Japanese sea literature), *Meijyo Journal* no. 3 (September 1903): 56–63.

offenses], let the goddess Haya-akitsu-hime, who dwells in the . . . tides of the raging sea . . . swallow them up, and let the god Ibukido Nushi . . . spirit them out and away to the nether region . . . dissolve and destroy them."[16]

The eighth century, specifically 710–94, the so-called Nara period, has been called the "golden age of poetry" in Japan.[17] Unlike Europe, Japan produced no long poems, only short poetic expressions of emotion, love, and longing for home, as well as praise of love and wine, elegies for the dead, and the like. The beauties of nature, including the seasons, waves breaking on the shore, seaweed drifting, bird songs, and so forth, have a central place. The poetry of this and the following Heian period (794–1186) was written by and for members or officials of the imperial court. Probably the most famous collection is the 4,500 or so poems of the *Manyoshu* (Collection of a thousand leaves), completed early in the ninth century, but containing poems dating mainly from the last half of the seventh century and first half of the eighth. Only a few poems in it speak of the sea, and they are not strictly literature of the sea, as it is used only as a metaphor for love and emotions. For instance, one poem reads in part,

> I cherish you my darling, as the Sea God the pearls,
> . . . .
> Your lovely eyebrows, curved like the far-off waves,
> ever linger in my eyes,
> My heart unsteady as a rocking boat.[18]

One of the oldest and most popular Japanese legends, the legend of Urashima Taro, is included in this collection too, though the oldest record of it is found in the *Nihongi,* where it is said to have occurred in 477 to a fisherman in northern Kyoto prefecture. Briefly, the legend tells of a fisherman who was poor because he couldn't catch any fish. One day, while he was out on the sea, his boat sank to the bottom (the land of immortality and happiness), where he saw the palace of the sea god. He was invited in by the beautiful daughter of the sea god, and he stayed for 3 years (actually for centuries). The day he left to return home, the princess gave him a jewel box but warned him never to open it. When he got home, he couldn't find his house, his village, or anyone he knew. He thought that, if he opened the jewel box, everything would reappear. As he opened the box there was a puff of white smoke, and his youthful body shriveled up, his hair turned white, he turned into an old man, and discovering in a mirror in the box

16. Aston (n. 12 above), p. 12.
17. Ibid., p. 33.
18. William Theodore de Bary, ed., *Manyoshu: The Nippon Gakujutsu Shinkokai's Translation of One Thousand Poems* (New York: Columbia University Press, 1969), p. 128.

what had happened to him (mirrors are important in Shinto rituals), he died, done in by his own curiosity.[19]

Almost no prose remains today from the Heian period, as it was not until the early tenth century that Japanese writers began writing in the Japanese language. The famous Buddhist monk of the ninth century, Kukai, brought religion to Japan from China and founded many shrines around Japan. Folklore has it that in one night Kukai built a bridge to an island so that he could pray there at a shrine dedicated to people lost at sea. The island, off the coast north of Shionomisaki, is connected to the mainland by a row of pylon-shaped rocks, called the Hashikui-iwa (Bridge pier rocks).

A poet named Ki no Tsurayuki, a court noble, was the first to write prose. His most important work, related to the sea as a context or setting for life, was his *Tosa Diary*. It was written over 2 months in 935 during a perilous journey he made by sea back to the capital, Kyoto, after completing his 4-year term of service in Tosa (now called Kochi, on Shikoku Island). It describes the ordinary life of a traveler (by boat) in Japan at that time. He wrote of his seasickness, of a storm, and of his thoughts and recollections, in simple but elegant Japanese with occasional humorous episodes.[20]

A few sea-related poems of the period can also be found in the *Kokinshu* (Collection of ancient and modern poems), a collection of some 1,100 poems compiled in 905 by Ki no Tsurayuki and others, and completed about 922. Here are but two examples: first, a love poem by a military officer worried that his love was ignoring him.[21]

> Like the waves which dash to shore at Aumi Bay
> I would have you rush to me,
> Yet it seems you fear my eyes
> Even on the path of dreams.

Next is an example of a lovely sea-related seasonal poem:

> The colored leaves of autumn
> floating on the white crests of rising waves,
> Look!
> Are they not small boats set sail
> by some fisherfolk.

Finally, there are the 31 volumes of the *Konjaku Monogatari* (Tales, an-

---

19. Ibid., pp. 216–18.
20. William N. Porter, trans., *Ki no Tsurayuki's* The Tosa Diary (Tokyo: Charles E. Tuttle Co., 1981).
21. L. R. Rodd and M. C. Henkenius, trans., *Kokinshu* (Tokyo: University of Tokyo Press and Princeton: Princeton University Press, 1984), pp. 210, 133.

cient and modern), also written during this period. It contains numerous Chinese, Japanese, and Indian legends and is an example of realism in early Japanese prose literature. One tale, "The Fisherman's Battle,"[22] is about seven fishermen who always carry weapons of war on their fishing boat. One day they are caught in a terrible storm, and, blown off course, they drift until they reach an island. They are asked by the inhabitants to help in a battle against those of a neighboring island. This turns out to be a battle between a giant centipede and a serpent-man. They help the serpent-man win and are invited to bring their families and live on this island paradise.

There is little else related to the sea during the rest of that period, and even less during the following so-called Kamakura period (1186–1332) as Japan moved into semi-isolation to protect itself from foreign invasion from the continent. There are occasional references to naval battles, as in the quasi-historical work called the *Gempei Seisuiki*, a history of the rise and fall of the Gen and Hei families of the 12th century. Similar references can be found in the famous 12th-century *Heike Monogatari* (Story of the Hei clan), at the end of which the 8-year-old mikado (emperor) is drowned, along with his nurse, in the sea at Dan-no-ura, near Shimonoseki, following the defeat of the Hei in a naval battle.[23]

Three centuries later, much was written about the pirates that ruled the waters and terrorized all who sailed the Inland Sea and lived on the islands within its boundaries. The pirates prayed at Oyamazumi Shrine on Oshima for successful forays. The merchants of the 16th century, tired of the plundering and loss of merchandise, paid off the pirates to ensure safe passage for the crews and the cargoes of the trading ships. Two of the 16th-century pirate barons, the Murakami brothers, are buried in imposing tombs on the hills above the port of Takehara, near Hiroshima.

About 350 years ago, the ancient fishing port at Mihonoseki, on the Shimane Peninsula, had five magnificent pine trees called the Seki no Gohon Matsu (Checkpoint's five pines). These pines were important navigational markers for the fishermen who sailed from the port to the distant Oki Islands. Aghast when they heard that their provincial lord proposed to cut the pines down, the villagers wrote a ballad in protest—an environmental protest song to stop the lord from his intended action.

Coastal net-whaling from small boats was developed in Taiji about 1606, and there were probably many folk legends about whale hunting. Whaling was a traditional activity for some coastal communities of Japan and a vital resource for their continued existence. Whale-meat recipes are even recorded in a 1489 Japanese cookbook. Detailed drawings and descriptions of whales are known from at least as early as the 18th century. One, a watercolor scroll, depicts 23 varieties of whales. Between the 18th and early 20th centuries

22. H. Naito, *Legends of Japan* (Tokyo: Charles E. Tuttle Co., 1972), pp. 11–16.
23. Aston (n. 12 above).

Woodcut of small Japanese whaling vessels netting and harpooning whales. (By permission of the Kendall Whaling Museum, Sharon, Massachusetts.)

there were several very important whaling ports around Japan. Whalers from Britain, America, and Hawaii settled in Omura port, on Chichijima, in the Ogasawaras as early as 1830. Here is a tale from a much earlier time, about the great Buddha, or Daibutsu, of Kamakura and a great whale.[24] Cast in bronze in 1252, the approximately 50-foot high Daibutsu has withstood tsunami and storms that destroyed the building it is housed in.

Buddha and the Whale

Folklore has it that a large whale heard persistent rumours of the large size of the Daibutsu. The whale was jealous. He persuaded a friendly shark to measure the Buddha. The shark did his best but had to convince a rat to do the actual measuring. The rat took 5,000 steps to get all the way around the Buddha, which translates to a circumference of about 97 ft. This was really too much for the whale, who still did not believe that anything on Earth could rival his bulk, so he himself went to investigate. When he reached the shallow water near shore, he put on magic boots and walked up to the temple. He tried to get inside, but was too big for the entrance. A priest came out and asked the whale why he had come.

The whale said, "The little animals that live in the sea and on dry

24. T. M. Hawley, "The Whale: A Large Figure in the Collective Unconscious; or, A Freudian Field Day," *Oceanus* 32, no. 1 (spring 1989): 117–18.

land insist on telling preposterous stories of a Daibutsu so large that it surpasses even myself in size. I know that there is nothing on earth that can match my bulk, and so have come to prove that these little animals are liars, and are merely jealous of my great size."

We can imagine how taken aback the priest must have been, but before he could stammer a reply to the whale, who stepped out through the doorway—stooping as he came—but the Daibutsu himself! In fact, the Buddha was surprised to see a creature so large as the whale, but calmly allowed the priest to measure them both with his rosary. The whale was able to return home happily, as his length was two inches beyond the Daibutsu's height. The Buddha, being perfect and not afflicted with undue pride, returned to his temple and reassumed his lotus position as he remains today.

The tsunami was known in Japan as something that caused great devastation. These waves have long been depicted in woodcuts and paintings, and probably in literature. Here is a folktale about an old man who remembered what his grandfather told him about tsunami. The folktale, translated from the original Japanese, provides insights into the nature of the tsunami hazard in ancient Japan and the perception of the hazard among certain elements of the Japanese population. It is quoted here in abridged form.[25]

A Living God

From immemorial time the shores of Japan have been swept, at irregular intervals of centuries, by enormous tidal waves,—tidal waves caused by earthquakes or by submarine volcanic action. These awful sudden risings of the sea are called by the Japanese *tsunami*. . . . The story of Hamaguchi Gohei is the story of a [tsunami] calamity which happened long [ago].

He [Hamaguchi] was an old man at the time of the occurrence that made him famous. He was the most influential resident of the village to which he belonged: he had been for many years its *muraosa*, or headman; and he was not less liked than respected. The people usually called him *Ojiisan*, which means Grandfather; but, being the richest member of the community, he was sometimes officially referred to as the Choja. . . .

Hamaguchi's big thatched farmhouse stood at the verge of a small plateau overlooking a bay. The plateau, mostly devoted to rice culture, was hemmed in on three sides by thickly wooded summits. From its outer verge the land sloped down in a huge green concavity, as if scooped out, to the edge of the water. . . . Ninety thatched dwellings and a Shinto

25. Lafcadio Hearn, *Gleanings in Buddha-Fields: Studies of Hand and Soul in the Far East* (Boston: Houghton Mifflin and Co., 1897), pp. 16–27.

temple, composing the village proper, stood along the curve of the bay; and other houses climbed straggling up the slope for some distance on either side of the narrow road leading to the Choja's home.

One autumn evening Hamaguchi Gohei was looking down from the balcony of his house at some preparations for a merry-making in the village below. There had been a very fine rice-crop, and the peasants were going to celebrate their harvest by a dance in the court of the *ujigami* [Shinto parish temple] . . . .

The day had been oppressive; and in spite of a rising breeze there was still in the air that sort of heavy heat which, according to the experience of the Japanese peasant, at certain seasons precedes an earthquake. And presently an earthquake came. It was not strong enough to frighten anybody; but Hamaguchi, who had felt hundreds of shocks in his time, thought it was queer,—a long, slow, spongy motion. Probably it was but the after-tremor of some immense seismic action very far away. The house crackled and rocked gently several times; then all became still again.

As the quaking ceased Hamaguchi's keen old eyes were anxiously turned toward the village. . . . [He] became aware of something unusual in the offing. He rose to his feet, and looked at the sea. It had darkened quite suddenly, and it was acting strangely. It seemed to be moving against the wind. It was running away from the land.

Within a very little time the whole village had noticed the phenomenon. Apparently no one had felt the previous motion of the ground, but all were evidently astounded by the movement of the water. They were running to the beach, and even beyond the beach, to watch it. No such ebb had been witnessed on that coast within the memory of living man. . . . And none of the people below appeared to guess what that monstrous ebb signified.

Hamaguchi Gohei himself had never seen such a thing before; but he remembered things told him in his childhood by his father's father, and he knew all the traditions of the coast. He understood what the sea was going to do. . . . He simply called to his grandson:—

"Tada!—quick,—very quick! . . . Light me a torch."

Taimatsu, or pine-torches, are kept in many coast dwellings for use on stormy nights, and also for use at certain Shinto festivals. The child kindled a torch at once; and the old man hurried with it to the fields, where hundreds of rice-stacks, representing most of his invested capital, stood awaiting transportation. . . . He began to apply the torch to them, hurrying from one to another as quickly as his aged limbs could carry him. The sun-dried stalks caught like tinder; the strengthening seabreeze blew the blaze landward; and presently, rank behind rank, the stacks burst into flame, sending skyward columns of smoke that met and mingled into one enormous cloudy whirl. Tada, astonished and terrified, ran after his grandfather, crying,—

"Ojiisan! why? Ojiisan! why?—why?"

But Hamaguchi did not answer: he had no time to explain; he was thinking only of the four hundred lives in peril. . . . Hamaguchi went on firing stack after stack, till he had reached the limit of his field; then he threw down his torch and waited. . . . Hamaguchi watched the people hurrying in from the sands and over the beach and up from the village, like a swarming of ants, and, to his anxious eyes, scarcely faster; for the moments seemed terribly long to him. The sun was going down; the wrinkled bed of the bay, and a vast shallow speckled expanse beyond it, lay naked to the last orange glow; and still the sea was fleeing toward the horizon.

. . . a score of agile young peasants [arrived], who wanted to attack the fire at once. But the Choja, holding out both arms stopped them.

"Let it burn, lads!" he commanded,—"let it be! I want the whole *mura* here. There is a great danger,—*taihen da!*"

The whole village was coming; and Hamaguchi counted. . . . The growing multitude, still knowing nothing, looked alternately, in sorrowful wonder, at the flaming fields and at the impassive face of their Choja. And the sun went down.

. . . [sobbed Tada, "Grandfather] is mad. He set fire to the rice on purpose: I saw him do it!"

"As for the rice," cried Hamaguchi, "the child tells the truth. I set fire to the rice. . . . Are all the people here?"

The Kumi-cho and the heads of families looked about them, and down the hill, and made reply: "All are here, or very soon will be. . . . We cannot understand this thing."

"*Kita!*" shouted the old man at the top of his voice, pointing to the open sea. "Say now if I be mad!"

Through the twilight eastward all looked, and saw at the edge of the dusky horizon a long, lean, dim line like the shadowing of a coast where no coast ever was,—a line that thickened as they gazed, that broadened as a coastline broadens to the eyes of one approaching it, yet incomparably more quickly. For that long darkness was the returning sea, towering like a cliff, and coursing more swiftly than the kite flies.

"*Tsunami!*" shrieked the people; and then all shrieks and all sounds and all power to hear sounds were annihilated by a nameless shock heavier than any thunder, as the colossal swell smote the shore with a weight that sent a shudder through the hills, and with a foam-burst like a blaze of sheet-lightning. Then for an instant nothing was visible but a storm of spray rushing up the slope like a cloud; and the people scattered back in panic from the mere menace of it. When they looked again, they saw a white horror of sea raging over the place of their homes. It drew back roaring, and tearing out the bowels of the land as it went. Twice, thrice, five times the sea struck and ebbed, but each time with lesser surges:

then it returned to its ancient bed and stayed,—still raging, as after a typhoon.

On the plateau for a time there was no word spoken. All stared speechlessly at the desolation beneath,—the ghastliness of hurled rock and naked riven cliff, the bewilderment of scooped-up deep-sea wrack and shingle shot over the empty site of dwelling and temple. The village was not; the greater part of the fields was not; even the terraces had ceased to exist; and of all the homes that had been about the bay there remained nothing recognizable except two straw roofs tossing madly in the offing. The after-terror of the death escaped and the stupefaction of the general loss kept all lips dumb, until the voice of Hamaguchi was heard again, observing gently. . . .

"That was why I set fire to the rice." He, their Choja, now stood among them almost as poor as the poorest; for his wealth was gone—but he had saved four hundred lives by the sacrifice. Little Tada ran to him, and caught his hand, and asked forgiveness for having said naughty things. Whereupon the people woke up to the knowledge of why they were alive, and began to wonder at the simple, unselfish foresight that had saved them; and the headmen prostrated themselves in the dust before Hamaguchi Gohei, and the people after them. . . . gifts could never have sufficed as an expression of their reverential feeling towards him; for they believed that the ghost within him was divine. So they declared him a god, and thereafter called him Hamaguchi Daimyojin, thinking they could give him no greater honor;—and truly no greater honor in any country could be given to mortal man.

One has to wait until the 17th century to find much in the way of sea literature. One begins to wonder why a country like Japan, with such a highly developed culture and heavy dependence on the sea, had not developed likewise in the field of maritime literature.

Occasional gems that may be considered sea literature have survived, such as the following haiku by the poet Basho (1644–94):[26]

A lobster in a pot
Dreaming awhile
Under the summer moon.

Cool seascape with cranes
Wading long-legged in the pools
Mid the tideway dunes.

26. Mason and Caiger (n. 11 above), p. 198.

The seashell-bringing wind
Has wrought its magic on the shore
At Waka-no-ura.

Basho wrote the lobster poem during an evening boat trip on Akashi Bay, in western Japan. He saw a lobster in a container still filled with warm seawater. The lobster looked quite content despite its inevitable fate, but the poet was not so sure about the fate of humanity.[27] The "shell-bringing wind" is the *kaiyose,* which blows on the 20th day of the second month of the lunar calendar, according to ancient lore.

The poet and painter Bason (1716–83) wrote this haiku.[28]

The sea at spring time,
All day long it rises and falls,
Yes, rises and falls.

The legend of Urashima Taro, mentioned earlier, was revived during the 17th century, but all in all, this was a tiny output on such a geograpically relevant topic from a country with such a great culture otherwise. A millennium is a remarkably long time for an island country to be without a history or tradition of sea literature.

SOME POSSIBLE REASONS FOR JAPAN'S LACK OF SEA ORIENTATION

Two possible reasons for Japan's apparent lack of sea orientation in its literature have been suggested by knowledgeable Japanese literary experts. First, in 663 the Japanese navy of some 27,000 men was utterly defeated while trying to invade the Korean peninsula in the so-called Pekson-kang War.[29] In an attempt to keep the humiliating news from the people, as well as to discourage the possible discovery of the truth, the government began a deliberate program of destroying all written records pertaining to the event, as well as sea-related literature for centuries to come.[30] Thus, anything in this genre that exists today is likely but a trickle of what might have been, or possibly was, produced.

27. Ibid., p. 199.
28. D. C. Buchanan, *One Hundred Famous Haiku* (Tokyo: Japan Publishing Co., 1973), p. 23.
29. Y. Miyamoto, *Umi to Nihonjin* (The sea and the Japanese) (Tokyo: Yasaka Shobo, 1987).
30. Two interviews with Professor Shunitsu Nakakoji, scholar of Chinese and Japanese literature, March and August 1986, Takarazuka, Osaka, Japan.

The second reason is that the only people who could read and write until the 17th century were the privileged few of the nobility. Since they neither fished nor farmed, it is only natural that they would write practically nothing on those themes. Also, since they lived relatively far from the sea and seldom visited it, any use of the sea in their writings was only as a background or metaphor, not as the subject of their work.[31]

Either or both of these possible reasons combined with over 1,000 years of semi-isolation (794–1582) and total isolation (1619–1867) served to reinforce this general lack of interest in or concern about the sea in the general public. When the total isolation commenced, Dutch and Portuguese traders were in Japan as visitors and residents. In 1634, the Dutch traders were allowed to stay in Japan but were ordered to move to Degima. Degima is a 13,000 m^2 fan-shaped, artificial island in Nagasaki Bay. Now reclaimed, only brass studs mark the outline of Degima, and a narrow canal still separates it from the mainland. At the time of the total seclusion, however, it was Japan's window to the Western world. During the period of isolation, the sea was a means of escape and thus legally accessible only to fishermen and seamen. It also was a route for possible invasion from abroad, to be protected against at all cost. This lack of familiarity bred a sense of fear and mystery about the sea, which, even to those who had access to it, spread no further than the Sea of Japan, the Yellow Sea, and the East China Sea.[32] In fact, it was not until the mid–19th century that the Pacific Ocean even had a name in Japan; it was merely called "the sea."[33]

The Edo period (1600–1867) witnessed a popularization of literature, which accompanied the spread of education in Japan to the lower classes, who became better educated than ever before, more prosperous, and able to read and even buy books. An extensive literature appeared on a broad range of topics, but there was practically nothing about the sea, from which most people were, as before, isolated.

THE SEA IN MORE RECENT JAPANESE LITERATURE

With the appearance of Commodore Perry's naval force, the so-called Black Ships, in Tokyo Bay in July 1853, carrying a letter from President Fillmore

---

31. Interview with Ms. Michio Yamashita, poet and writer, in Yokohama, Japan, December 1988.

32. N. Ishimo, "Gaikoku to Nihonjin" (Other countries and the Japanese), in *Nihon no Kaiyomin* (The sea race of Japan), ed. J. Miyamoto et al. (Tokyo: Miraisha, 1974), pp. 73–81.

33. T. Haga, "Umi to Nihonjin" (The sea and the Japanese), in *Umi to bunmei* (Sea and civilization), ed. T. Hamada (Tokyo: Tokyo University Press, 1987), pp. 3–37.

to the emperor requesting trade and diplomatic relations,[34] things began slowly to change. Though strong on land, Japan's relative ignorance of, and lack of concern about, the sea made its coastal cities vulnerable to bombardment by those from overseas who would otherwise be ignored or refused trade and landing rights. One after another, European nations sought and obtained trade and diplomatic privileges, as the Americans had earlier. Foreign ships increasingly appeared in Japanese ports, and foreigners were gradually allowed to live permanently in several port cities. It was not until 1857 that Japan built a ship large and strong enough to cross the Pacific Ocean, and in 1860 it sailed for the first time to San Francisco.[35] Thus, the floodgates of cultural inflow had begun to open. All of this resulted in a rising national interest in the sea, an interest that would take a century to take root, finally doing so, it has been said, as recently as 1965–70,[36] when Japan reached its peak of economic growth and environmental pollution, especially of the sea and coastal zone.

With the growing interest in the sea during the Meiji period (1868–1910) and the Taisho period (1910–25), people in various fields began taking an active interest in the sea. The famed ethnologist Kuni Yanagita (1875–1962) searched for the origin of the Japanese and found that they had come by sea from the southeast via the Black (Japan or Kuroshio) Current. His famous study of the sea and related customs was the first major attempt by a Japanese scholar to write about the various elements of Japan's "sea culture," up to then virtually unrecorded.[37]

An equally famous poet and a friend of Yanagita, Toso Shimazaki wrote poetry on this same topic.[38]

From a far away unknown island,
A coconut comes floating by

This refers to the origin of the race in the islands and lands of the south and west Pacific.

A more popular and well-known form of sea literature can be found in the folk songs taught in Japanese schools. Here is a stanza from a song about a boy who was born and raised by the sea.[39]

34. E. O. Reischauer, *Japan: The Story of a Nation* (Tokyo: Charles E. Tuttle Co., 1974).

35. Miyamoto (n. 29 above).

36. Ibid.

37. K. Yanagita, *Kaiyo no Michi* (Road on the sea), new edition, vol. 12 (Tokyo: Chikuma Shobo, 1972).

38. Haga (n. 33 above), p. 5 (author's translation).

39. I. Nakano, trans., *101 Favorite Songs Taught in Japanese Schools* (Tokyo: Japan Times, 1983), p. 68.

Son of the Sea

I used to bathe in the wide open sea!
Just like a lullaby I heard the waves go wild.
How I liked to breathe in the grand space so free!
This is the way that I grew as a child.

That poem was published in 1910. The following one, which simply describes the sea and shore during the day and night, was published 3 years later:[40]

Sea

Where rows of pine trees vanish in the distance
There comes a white sail floating in sight,
High up are fishing nets drying on the shore,
Seagulls glide silently in their flight,
Look, the sea in the daytime!

Mountains and islands veiled in darkness,
Dim are the fishing lights far from the land,
Sea waves break, sea waves gleam softly in moonlight,
How lightly sea winds blow o'er the beach sand,
Look, the sea at nightime!

A song of more recent origin (1941) is about the dynamic movement of the sea and children's desire to see foreign countries that lie beyond it, at a time when such things were forbidden due to World War II.[41]

Sea

How wide and how endless, oh mighty sea,
Moon will rise, sun will set, wind will blow free.
Rolling on, rolling back, breaking in foam.
Sea waves roll, sea waves break, where is their home?
Sailing a ship on the wide open sea,
To lands I've never seen, will you take me?

The government often referred to Japan as a "country of the sea,"[42]

---

40. Ibid., p. 117.
41. Ibid., p. 226.
42. Interview with Ms. Makiko Habu, poet and writer, in Yokohama, Japan, December 1988.

and there were several militaristic (naval) songs on this theme, such as "Umi Yukuba" (When we go to sea), which can still be heard in Japan on the loudspeaker cars driven by Japanese ultranationalists.

The postwar period brought a flourish of activity regarding sea literature. For instance, Kurahashi's 1970 short story, "To Die at the Estuary," associates the sea with death and the impermanence of life against the background of modern industrial Japan.[43] Then there is the doomsday association with the sea as reflected in the 1973 science fiction *Nippon Chimbotsu* (Japan sinks), which was made into an exciting film that shows how a sudden progression in the earth's plate tectonics along the Japan Trench would pull Japan beneath the surface of the sea in a matter of months. An equally shocking but nonfiction book on the life and suffering of the victims of Minamata mercury food poisoning is Ishimure's 1972 masterpiece *Kukaijodo* (Suffering sea, pure land).[44] This is just one of many of her works on the theme of antipollution. Michiko Ishimure (b. 1935), who has lived all her life in Minamata City on Kyushu Island in western Japan, uses her unique writing skills to get her readers to feel what the victims feel and thus understand them and their suffering, an accomplishment that won her a literary prize for nonfiction writing. Her writings contributed greatly to the effort of raising public concern about pollution and exposing the polluters. They have also helped to create doubts in many Japanese people about the correctness of Japan's destructive approach to the environment in general, and to the sea in particular, in its pursuit of economic growth, superiority, and affluence.[45]

Another writer who is now contributing to a growing sea literature is Makiko Habu, born in 1930 in Imabari City, west Shikoku Island, on the Inland Sea. She is now a resident of Yokohama. She has written prolifically, especially poetry, about the sea and seacoast for nearly two decades.[46] To her, the sea is alive; it has a face with deep blue eyes. She says she feels a deep sense of wonder and the supernatural about the sea, but at the same time she says she also feels fear, probably because she cannot swim.[47] All of her sea poems are memories of the sea and coast of her birthplace. Two examples are included here, as translated for the first time by me. In an excerpt of the first poem, she relates how to her the sea is alive.[48]

43. Y. Kurahashi, "To Die at the Estuary," translated by Dennis Keene, in *Contemporary Japanese Literature*, ed. H. Hibbert (Tokyo: Charles E. Tuttle Co., 1977), pp. 248–81.

44. M. Ishimure, *Kukaijodo* (Tokyo: Kodansha, 1972).

45. K. Habu, *Kindai e no majutsushi* (Wizard of our times, Michiko Ishimure) (Tokyo: Ozankyaku, 1982).

46. Makiko Habu, *Shishu: No to hana no hibi* (Collection of poems: days of fields and flowers) (Yokohama: Habu, 1972); Makiko Habu, *Shishu: Umi to Kijibato* (Poems: the sea and pheasant) (Yokohama: Habu, 1974); Makiko Habu, *Shishu: Watashi no minzokushi* (Poems: my ethnology record) (Yokohama: Sozo-hakkosho, 1982).

47. Interview with Habu (n. 42 above).

48. Habu, *Shishu: Watashi no minzokushi* (n. 46 above), pp. 78–79.

The Sea Returns to Its Former Color

When I swim,
the current flows,
the waves are turbid.
In a moment,
I am engulfed
in micro-organisms . . .
hundreds of tiny living things.
. . . The current flows,
the sea returns to its former color,
I swim again.

In the second example, excerpted here, she expresses the feelings of fish, not unlike the lobster of Basho.[49]

        The Great Tide

Memories of the unending sea
like a wave on the river's bottom,
rushing toward me!
The river bottom
with gentle ripples.
The dune
with warm tide pools remaining;
Fins stand upright,
the bewildered goby
of sand color.
Trapped there,
cold in a dream,
awaiting the high tide;
I also wait
fervently,
the new, deep, blue,
strong wave and spray.

Habu's steadfast literary efforts have given those fighting to save her beloved Odagahama beach and the adjacent sea in her hometown moral support in the face of a continuing series of legal setbacks.[50] She has also

---

49. Habu, *Shishu: No to hana no hibi* (n. 46 above), pp. 32–33.
50. "Court Okays Reclamation Plan: Within Mayor's Power, Illegal or Not," *Mainichi Daily News* 3 November 1988), p. 8.

been helping to increase public concern about the problems of the sea and coast, not unlike Ishimure, but in Habu's case in an effort to save a beach and its nearshore sea *before* it is lost. She appeals with the beauty of the sea and coast in her poetry to the hearts of her countrymen.[51] Unfortunately, their hearts are as yet insufficiently attuned to her gentle yet firm message, the message of the literary harbingers of what I see as Japan's new culture, now in what appears to be the early stages of its formation.

A most recent poem was published in a brochure advertising a 1990 competition to design the coastal zone area at Sagami Bay, Kanagawa prefecture, to achieve harmony between humanity and the sea. The *nagisa* is an old Japanese term used in literature and poety to describe the coastal zone.

We need the "Nagisa"[52]

Every day, at the "nagisa"
The tides flow and ebb,
Bringing morning and evening alternately.
As the season changes
Wind and light,
Everything changes at the "nagisa."

Everything drifts into the "nagisa,"
From far away in the distant ocean.
There at the "nagisa,"
Man and the divine,
This world and the other,
Life and death,
Mingle with each other.

Man was born in the sea,
The sea was his mother.
As long as man hears the call of nature
From the sea,
We need the "nagisa,"
Filled with the sound of waves,
And the tang of the salt sea air.

51. Makiko Habu, "Hama ga ari, hito ga iru" (Where there's a beach, there are people), in *Sekai Journal* (January 1989): 340–46.
52. "We Need the 'Nagisa,'" Surf '90 "Nagisa" Design Idea Contest, conducted by the Sagamiwan Urban Resort Festival in 1990 Association, Kanagawa Prefectural Government, Japan, p. 3.

CONCLUSION

In this article, an effort has been made to try to show why I have doubts about the appropriateness of calling Japan a sea-oriented nation despite its present dependence on and relationship to the sea. A broad, rather cursory, look at Japanese literature on the sea reveals a gap of nearly 1,000 years during which most Japanese apparently lost their original sense of sea and to some extent the coast. It is only recently that this sense is beginning to be revived, or perhaps we should say initiated, by Japan's growing relationships with the rest of the world, as well as by awareness of pollution and the continuing destruction of the environment of the nation's coastal zone. One can only hope that this vital, basic sense of the sea and coast will take root and at last turn the Japanese heart again toward the sea while there is still some unpolluted sea and natural coastline. Such a change of heart would be most welcome and is worth anticipating, eventually, from one of the world's most influential economic powers. As a model that others increasingly emulate, Japan, by becoming truly "sea-oriented, sea-loving," could have far-reaching positive impacts on the preservation of the world's nearshore seas and coastal environments.

Arms Control and Disarmament at Sea: What Are the Prospects?

Peter Haydon
Centre for Foreign Policy Studies, Dalhousie University

INTRODUCTION

Worldwide interest in naval arms control and disarmament has risen to a level not seen since the end of the First World War, when the remaining "great powers" tried to control the growth of each other's navies. The circumstances that are driving the present situation are very different from those of 70 years ago. At that time, naval power was concentrated in the hands of relatively few states, fleet structures were essentially compatible, and naval weaponry was generally less destructive. Today's world, on the other hand, is marked by a destabilizing proliferation of naval power and widespread access to highly destructive weapons. This does not mean, however, that we cannot learn from the experience of the interwar years.

Unfortunately, those lessons alone will not provide a complete solution for controlling the present proliferation of naval arms. The complexity of today's international system and the way in which the oceans have been militarized since the 1960s requires a broadly based approach to future naval arms control initiatives if they are to be more than just tokenism. Also, because so many states and organizations now have genuine concerns in the outcome of new naval arms control regimes, any initiative has to be based, in part, on an aggregation of international interests rather than on East-West considerations alone.[1]

The heart of the present problem is the almost uncontrolled growth of naval power in many parts of the world. This is not restricted to the navies of the superpowers and their allies, but includes many of the nonaligned and newly industrialized countries. In this respect, it is fair to say that many states now have far more naval capability than they need to protect their maritime

---

1. This was a basic premise of the Results and Conclusions portion of the First International Seminar on Naval Arms Limitations and Naval Activities Reduction in the World Ocean, sponsored by the government of the Soviet Union and held in Moscow, 6–10 February 1990.

interests.[2] Thus, a key assumption of this paper is that global stability could be improved by halting the present naval arms race and by reducing the level of armament at sea throughout the world. Accordingly, the question to be addressed is, How can this be accomplished?

Naval arms control has become an imprecise term and is now used to cover a multitude of interrelated issues. It is important, therefore, to clarify some of the basic definitions. At the risk of oversimplification, naval arms control initiatives can be categorized under one of four basic objectives.

1. reducing the numbers of sea-based strategic weapons, which today comprise only intercontinental ballistic missiles but which may be expanded to include other missile systems;[3]
2. banning or reducing the numbers of naval tactical nuclear weapons;
3. imposing numerical and/or geographic limitations on weapons and naval vessels; and
4. developing martime confidence-building measures.[4]

The interrelationship is complex. One way that it can be explained is by using the modern naval vessel as an example. On the one hand, a warship can carry strategic missiles and tactical nuclear weapons, it can sail freely around the world, and its presence in certain areas could be considered destabilizing. Those are the negative aspects. On the other hand, a warship can serve as a deterrent to aggression, and under some circumstances its presence in an area can be a stabilizing influence. It can also provide humanitarian assistance in a time of disaster. This leads to another assumption, that naval vessels in themselves are not destabilizing or threatening. Rather, it is the way in which they are used that can create uncertainty. It follows, therefore, that naval arms control is, inter alia, a political subject. So, a logical place to begin this discussion is by looking at the geopolitical issues that provide the impetus for the present level of interest in naval arms control and limitations.

2. The reasons for the present proliferation of naval armaments are complex. One review of the issue is the report of the UN Department of Disarmament Affairs, *The Naval Arms Race* (New York: United Nations, 1986). Editors' note.—See Frank Barnaby's article, "The Role of the Submarine," in this volume, in particular table 1 for a summary of the major warships of the navies of the world.

3. There is widespread and growing belief that the family of sea-launched cruise missiles (SLCMs) should be brought into the present arms control regime as strategic rather than tactical weapons.

4. The tendency is to expand the categories into those that might more closely correspond to negotiating themes, as was done with the Results and Conclusions part of the February 1990 Moscow seminar (n. 1 above), where 13 proposals were put forward.

GEOPOLITICAL ISSUES

East-West Relations

Walking back from the Cold War essentially requires that both superpowers relinquish the basic ideologies that brought the world to its present situation. The United States needs to abandon its strategy of containment and embark on a new concept of internationalism.[5] The Soviet Union must depart from basic Marxist/Leninist ideology that a final war between capitalism and communism is inevitable. Although *perestroika* has brought change to the Soviet Union, those reforms have, for the most part, been theoretical and political.[6] And there is still no sign that the Soviet Union intends to relinquish its status as a military superpower.[7] As a result, belief in the sincerity of the Soviet intentions requires an enormous act of faith on the part of Western leaders. This is a commitment that many are still unwilling to make without reservation, believing that the internal political struggle in the Soviet Union has not yet been resolved and, therefore, visible evidence of reform is needed to be certain that the Soviet leaders are sincere.[8] Such pessimistic views are not universally shared. Many scholars and some politicians are convinced that the Soviets are sincere and that their aim of removing the danger of world war must be given an early and positive response by the West.[9] But contradictory opinions are not restricted to the West. Consensus on many aspects of reform has yet to be achieved within the Soviet Union.

Recent improvements in East-West relations and the dramatic events of the 1989–90 Eastern European revolution have added new dimensions and urgency to the rather cautious process of European arms control and disarmament. It now seems likely that a significant conventional arms control agreement on the sizes of foreign forces stationed in Europe will be signed. Although the focus of such an agreement will be on the land and air forces, the need to include naval units was mentioned by Mikhail Gorbachev in his speech at the London Guildhall on 7 April 1989. The need for naval arms control was also mentioned during the December 1989 meeting between Pres-

---

5. The call for a new U.S. foreign policy and a global strategy is now widespread. Among the stronger arguments are those made by Zbignew Brzezinski in "America's New Geostrategy," *Foreign Affairs* (Spring 1988): 680–99.

6. Michael MccGwire, "A Mutual Security Regime for Europe?" *International Affairs* 64, no. 3 (Summer 1988): 361–79. Editors' note.—For another opinion see Elisabeth Mann Borgese, "Perestroika and the Law of the Sea" in this volume.

7. Henry Trofimenko, "The End of the Cold War, Not History," *Washington Quarterly* (Spring 1990): 21–35. Editors' note.—For another opinion, see Borgese (n. 6 above).

8. C. A. H. Trost, "The Morning of the Empty Trenches: Soviet Politics of Maneuver and the U.S. Response," *Proceedings* (August 1988): 13–16.

9. MccGwire (n. 6 above), pp. 377–78.

idents Bush and Gorbachev. It is realistic to assume, therefore, that the present round of Euopean arms control discussions could eventually include attempts to impose limitations on the naval weapons and capabilities of the superpowers and their alliances.

Theoretically, adding naval forces to the present arms control process should not be difficult, but many factors complicate the process. For instance, if some of the technical details are not addressed beforehand, future discussions on naval arms limitations could flounder in much the same way that the initial momentum of the recent "Open Skies" conference could not be sustained when the details of the concept came to be worked out.[10] It is not surprising, therefore, that the prospect of naval arms limitations is viewed with concern in a number of quarters. One such concern is that the asymmetrical naval structures of the superpowers and their allies have become so complex that the process of naval arms limitations can be successful only if addressed within a global framework.[11] Any future naval arms limitation discussions would have to be preceded by changes in the naval strategies of the superpowers.

Gorbachev has chosen to use arms control as the means of reducing the perceived threat to the Soviet Union. He is doing this so he can adopt a new defensive doctrine, within a concept of international mutual security, as the first step in switching production from military to consumer items.[12] So far, the Soviets have been unwilling to make more than token reductions in their force levels and, in fact, are continuing to modernize their forces.[13] The U.S. government, on the other hand, is under considerable pressure to reduce defense spending as a means of attacking the crippling deficit but is not willing to make significant changes in force structures without comparable Soviet reductions and look for meaningful Soviet force reductions as signs that Soviet military doctrine has changed.[14]

Although arms reductions are clearly in the interests of both superpowers, mutual distrust is hampering progress. In the Soviet Union, the doctrinal

10. *Globe and Mail* (28 February 1990), Sec. A, p. 1.

11. The difficulties created by multiple asymmetries within and between the military structures of continental and maritime states are explained by Rear-Admiral James A. Winnefeld, U.S. Navy (Retired), in "Avoiding the Conventional: Arms Control Battle," *Proceedings* (April 1989): 30–36.

12. William H. J. Manthorpe, Jr., "What Is Pushing Gorbachev into Arms Control?" *Proceedings* (December 1989): 43.

13. For instance, the recent announcement that some ballistic missile submarines would be taken out of service happened at the same time as the latest Typhoon-class SSBN was put into service, more than compensating for the combined capability of the old Golf-class SSBs being scrapped.

14. See Paul Dibb, *Is Soviet Military Strategy Changing?* Adelphi paper 235 (London: International Institute for Strategic Studies, 1989), pp. 35–46; and James J. Tritten, *Naval Arms Control: An Idea Whose Time Has Yet to Come* (Monterey, California: U.S. Naval Postgraduate School, 1989).

debate is over the question of how much defensive capability is enough,[15] or, in Soviet terms, of what constitutes reasonable sufficiency. The United States is addressing a similar question, but from a different perspective. Because the superpower military structures were built to oppose one another on a global basis rather than just in Europe, reductions need to be made on a correspondingly wide basis. The problem does not exist in Europe alone; superpower confrontation occurs on a worldwide basis. Herein lies one of the major problems. The Soviet Union and its European allies are continental states and essentially capable of self-sufficiency, while the United States and its allies are oceanic states, and their stability depends heavily on extensive sea lines of communication. This fundamental difference, which is acknowledged by the Soviets,[16] has created extensive asymmetries in force structures that are reflected in the respective maritime strategies.

Both the United States and the Soviet Union have based their maritime strategies on a concept of defense in depth through the projection of sea power. The United States acknowledges that it has a global maritime strategy, but the Soviet Union masks the true intent of its naval forces by claiming that its military doctrine calls for a defensive navy to conduct "retaliatory action against aggression from naval and oceanic directions if this arises."[17] The Soviet navy is also required to support state interests abroad, but to a lesser extent than the U.S. Navy. As a result, the structure of the two navies are very different, and each perceives the other differently. The U.S. naval forces that the Soviets see as threatening are, by American reckoning, counterbalances to the distinct Soviet military advantage on land. The balancing of these asymmetries within any future arms limitations regime will be difficult, but not impossible.

North-South Concerns

Within both the Soviet "new thinking" and the growing recognition that the United States needs to develop a new view of its place in the world lies the understanding that a new world order is emerging.[18] In Soviet thinking, this view has been expressed in the concept of the future Europe, the "common European home," as well as in the stated intention of playing a greater role in the United Nations. But a reduction in the likelihood of superpower con-

15. Dibb (n. 14 above), p. 46.
16. This is covered in the report of the comments made by Marshall Sergei Akhromeyev during his 1989 visit to the United States, in "Naval Arms Control: Where Do We Go from Here?" *Naval Forces* 10, no. 4 (1989): 59.
17. Admiral Vladimir Chernavin, Commander of the Soviet Navy, in an interview with *Danas* (Zagarev), 30 May 1989. Quoted in Foreign Broadcast Information Service Daily Report FBIS-SOV-89-112A, 13 June 1989.
18. MccGwire (n. 6 above), p. 378; and Brzezinski (n. 5 above), p. 682.

frontation will not remove all risks of regional instability.[19] Even without the risk of confrontation, the conditions that cause regional instability will remain, including those areas where the "iron hand of Soviet control" has been relaxed.[20] Furthermore, it is quite probable that as Western Europe advances toward a higher degree of unification and begins the rehabilitation of the nation-states of Eastern Europe, its focus will become much narrower (Eurocentric), with the result that the Third World could become a secondary economic interest. The consequences of this trend are self-evident. The Third World countries would again be relegated to a "second class" status with a corresponding likelihood that the cycles of poverty and civil war will increase rather than decline.[21]

As many of the states that emerged during the postcolonial period struggle to improve their standards of living and begin to exploit the potential of their ocean space, the likelihood of interstate conflict at sea will increase.[22] These trends have been summarized by Michael Morris:

> Third-World navies have been expanding along three interrelated dimensions: expansion of the national ocean zones, naval roles, and naval capabilities. In the case of some of the larger developing states, the expansion of naval roles and capabilities has even projected beyond the extensive national ocean zones. Though naval expansion has been uneven in the Third World it has generally transformed what was often a neglected armed service, the navy, and a neglected aspect of national development, ocean resources, into key dimensions of domestic and foreign policy.[23]

The recent Iran-Iraq war and the present situation in the Persian Gulf are clear examples of how easy it is for regional conflict to draw in other states,

19. Shahram Chubin, *The Super-powers, Regional Conflicts, and World Order*, Adelphi paper 237 (London: International Institute of Strategic Studies, 1989), p. 75. Editors' note.—See article by Joseph R. Morgan, "Naval Operations in Korean Waters," in this volume.

20. Richard Sharpe, "The Foreword to *Jane's Fighting Ships 1989–90*," *Sea Power* (July 1989): 38.

21. The cycles of poverty, famine, and civil war, compounded by the trend to fundamentalism in many regions, are difficult to break. Two short essays of particular note are "The World's Wars: Turn South for the Killing Fields," *Economist* (12 March 1988): pp. 19–22; and Charles Krauthammer, "How to Deal with Countries Gone Mad," *Time* (21 September 1987).

22. See David L. Larson, "Naval Weaponry and the Law of the Sea," *Ocean Development and International Law* 18, no. 2 (1987): 133; Michael A. Morris, *Expansion of Third-World Navies* (London: Macmillan Press, 1987); and Chubin (n. 19 above), p. 75. Editors' note.—See Michael A. Morris, "Comparing Third World Navies," in this volume, an article developed from his book *Expansion of Third-World Navies*. See also the article in this volume by Frank Barnaby, "The Role of the Submarine."

23. Morris (n. 22 above), p. 1.

particularly where international economic interests are at stake. The former is an even better example of the misuse of naval weaponry, in clearly showing that the problem of excessive naval power is not restricted to the industrialized states.[24]

TECHNOLOGY

The series of Middle East wars provided several examples of trends in naval warfare. First, the advent of the antiship missile has changed the basic nature of warfare at sea. The big ship is no longer the ultimate instrument of sea power; the antiship missile has effectively become an equalizing factor. Fired from shore, from small vessels, or from aircraft, these missiles have the capability to give a tactical advantage to the less powerful state. Moreover, they are now being produced and sold by many countries, to the extent that the spread of this capability has been likened to an epidemic.[25] Second, the effectiveness of mine warfare and the difficulty in countering it were brought home by the Iranian mining of the Persian Gulf. Third, there has been an increase in the number of states capable of producing nuclear weapons. Despite the extensive precautions taken by the nuclear powers to prevent such proliferation, the necessary components and technology can be acquired at a price. In addition to these trends, the evolution of the modern submarine and the development of new propulsion systems, which increases its ability to conduct covert operations, also serve to reduce the supremacy of the big warship.

Today, it is relatively easy for small states to acquire sufficient naval capability not only to defend the waters rightfully under their jurisdiction but also to project power in such a way as to take control of or deny other states the use of certain bodies of water or international straits.[26] The misuse of such capabilities is destabilizing. For instance, had the Iranians made greater use of their Chinese Silkworm antiship missiles, particularly against merchant ships or the nonbelligerent warships, the Gulf war might easily

---

24. Editors' note.—Both Iran and Iraq used naval weapons to harass neutral shipping, particularly oil tankers. The technology of the naval power used by the combatants ranged from sophisticated guided missiles (one of which struck a U.S. naval vessel and resulted in 29 deaths) to mines of World War II vintage and small "gunboats" of an archaic, almost primitive, design, manned by Iranian volunteer forces operating as guerilla forces at sea.

25. James T. Hacket, "The Ballistic Missile Epidemic," *Global Affairs* (Winter 1990): 38–57.

26. Editors' note.—See Lewis M. Alexander and Joseph R. Morgan, "Choke Points of the World Ocean: A Geographic and Military Assessment," *Ocean Yearbook* 7, ed. Elisabeth Mann Borgese, Norton Ginsburg, and Joseph R. Morgan (Chicago: University of Chicago Press, 1988), pp. 340–55.

have escalated. In such circumstances, the problem is twofold: the availability of the high-technology weapons, and their use.

ECONOMIC CONSIDERATIONS

Few states or economic unions enjoy the luxury of industrial or agricultural self-sufficiency; as a result, they must rely upon international trade for their economic stability. In turn, those dependencies translate into the building blocks of the international economy as we now know it. For instance, most industrialized states would cease to function without substantial imports of raw materials and, in some cases, energy and food staples. On the other hand, those states whose economies are mainly resource-based need to import technology and manufactured goods, while underdeveloped states need to import a wide range of commodities with only a limited export capability to offset those demands. In this way the various components of the global economy become interdependent, and the ties that bind them together are the sea lines of communication. One weakness of this economic structure is that it breeds specialization, which in turn increases dependence on international trade. The situation is not helped by the imperfect distribution of a number of key resources.[27]

To some extent, many of these dependencies are self-induced and merely support the states' choices of certain types of commodities over others. Moreover, the industrial cycle is still based on the use of new raw materials rather than reprocessed or recycled material. The global dependence on fossil fuels, which accounts for about one-third of all ocean transportation, also creates strategic concerns for many states. In time, environmental concerns and the recognition that many of the key resources are not infinite will eventually force states into energy conservation programs, and make them shift to less damaging energy sources and into widespread metal reprocessing. But this is unlikely to happen until political and/or economic pressures force the release of the large amounts of capital that such programs require. It must be accepted, therefore, that for the foreseeable future international trade in raw materials and energy products, with the associated dependencies, will continue. And therefore competition and, quite possibly, conflict over the rights to supply and move those commodities will also continue.[28]

27. See Hanns Maull, *Raw Materials, Energy, and Western Security* (London: Macmillan/International Institute of Strategic Studies, 1984).

28. The strategic importance of the main sea routes has been addressed by many analysts; for instance, Robert J. Hanks, *The Unnoticed Challenge: Soviet Maritime Strategy and the Global Choke Points* (Cambridge, Mass.: Institute of Foreign Policy Analysis, 1980); Roger Villar, "Vital Sea Routes," *Navy International* (August 1982): 1260–63; and Alexander and Morgan (n. 26 above). For a detail of the implications of the oil industry, see Peter R. Odell, *Oil and World Power* (London: Penguin Books, 1986).

THE LAW OF THE SEA

The 1982 UN Convention on the Law of the Sea (LOS Convention) was not specifically intended to be an arms control agreement; nevertheless, it has had a significant impact on the naval arms control process. Not only does the LOS Convention establish the foundations of an orderly regime for the peaceful use of the world's oceans, but it also has the potential to be an important conflict-resolution device. These impressive accomplishments are not without criticism, however. For instance, Ambassador Arvid Pardo has written that "not only does the 1982 Law of the Sea Convention contribute nothing substantial to arms control or disarmament in the marine environment but, on the contrary, by the vagueness or ambiguity of many of its key provisions it could foster controversy and contribute to increasing international tension."[29] Specifically, Pardo's concerns were that limitations had not been placed on the military use of the oceans and the seabed.[30] In this respect, of the specific issues he has cited at various times, six stand out:

1. the extent of the High Seas and seabed that naval forces could use for peaceful purposes;
2. the rights of warships to unimpeded passage through straits;
3. the rights of foreign warships in EEZs;
4. the extent of action that foreign warships could take on the Continental Shelf;
5. the right of foreign warships to entry into Territorial Seas without consent; and
6. the use of force to remove nonfriendly, foreign warships from the Territorial Sea.[31]

These apprehensions were well founded. UNCLOS III did nothing to constrain the use of force at sea, because such initiatives were beyond its mandate. Even if naval arms control had been an aim of the convention, it is doubtful if anything could have been accomplished under the circumstances, particularly in view of the difficulty experienced in dealing with other contentious issues. Because the major naval powers did not press for such initiatives, any attempt to expand the provisions dealing with the military use of the

29. Arvid Pardo's foreword in *The Denuclearization of the Oceans*, ed. R. B. Byers (London: Croom Helm, 1986), p. iii.
30. The problems of the military use of the oceans and the perceived shortcomings of the LOS Convention have been addressed by many scholars and lawyers. One of the more interesting examinations is that by Elmar Rauch, "Military Uses of the Ocean," in *The German Yearbook of International Law* (Kiel: Kiel University, 1985), pp. 229–67.
31. Pardo (n. 29 above), p. iii.

oceans would have had little or no chance of success. In this respect, as Ken Booth has pointed out, the military provisions of the convention essentially met the requirements of the major naval powers.[32]

Pardo's concerns deal with how states employ their warships in foreign waters and on the High Seas, especially where the actions of those vessels might appear threatening. If warships had been constrained to their own waters, the LOS Convention would have been a very rigid arms limitation agreement.[33] But to seek such sweeping constraints within the framework of the convention could have jeopardized any hope of its general acceptance and might have undermined the valuable contribution the convention has made to international security. Thus, it may be more appropriate to look at Pardo's concerns in terms of a "code of conduct" for warships in foreign waters and on the High Seas. In the form of legal obligations, perhaps in the manner of the International Rules for the Prevention of Collisions at Sea or as an extension of the present bilateral agreements between some navies for the prevention of incidents at sea, this notion becomes a more realistic concept of naval arms limitations, and one that has sound historic precedents.

HISTORIC ASPECTS

Naval arms control is not a new idea. It effectively started in 1817 with the Rush-Bagot Treaty, which was designed to prevent a naval arms race on the Great Lakes and whose provisions are still observed.[34] Since then, several attempts have been made to develop meaningful naval arms limitations agreements. As a result, conventional (nonnuclear) naval arms control has considerable historical experience upon which to draw.

As a result of widespread concern over the deterioration of the traditional balance of power in Europe, especially the Anglo-German naval race, two arms limitation conferences were held in The Hague, in 1899 and 1907. Although these conferences were not able to slow the naval arms race, several positive results were accomplished, including the establishment of an international court and the development of an international agreement on the conduct of modern warfare.[35] The First World War prevented further progress and, at the same time, created a number of new problems.

The very complicated period of history that links the two world wars of

32. Ken Booth, *Law, Force, and Diplomacy at Sea* (London: George Allen and Unwin, 1985).

33. Ibid., p. 75.

34. John H. Barton and Lawrence D. Weiler, *International Arms Control: Issues and Agreements* (Stanford, California: Stanford University Press, 1977), p. 32.

35. Ibid., pp. 32–37, 15–18.

this century saw considerable naval arms control activity.[36] Those initiatives, in fact, were technically naval arms limitations because their intent was to reduce and limit, both numerically and geographically, the sizes of the three large navies in the Pacific Ocean. The intention was to create spheres of influence throughout the Pacific, including a three-power agreement, the Washington Agreement, to freeze the strategic naval forces of the United States, Japan, and Great Britain. By a mathematical process, the capabilities of strategic warships (battleships and battle cruisers, aircraft carriers, cruisers, destroyers, and submarines) were restricted on the basis of numbers, displacement, armament, and foreign base support. The three powers initially bound themselves to abide by the provisions of the agreement and agreed that their respective naval intelligence staffs would carry out the necessary verification. France and Italy were later added, albeit reluctantly, to the signatories to create a secure strategic environment in the Pacific. The first London Treaty (1930) extended and amplified the provisions of the Washington Treaty but without the full support of France and Italy. The second London Treaty (1936) tried to make further adjustments, but by then world stability was beginning to crumble under the weight of rearmament. In hindsight, the original Washington and London treaties were well intentioned and comprehensive attempts to establish a more secure environment. But unfortunately they failed to foresee the possible return of militarization in the Pacific or to prevent the European naval rearmament of the late 1930s.

Naval arms limitations were not considered after the Second World War. Instead, a complicated and largely involuntary formula of economics, decolonization, and technology determined the manner and speed by which states restructured their navies. This process was further complicated by the advent of the Cold War, the spread of nuclear weapons, and by the evolution of strategies based on nuclear deterrence. The consequences of this rearmament, which are part of the reason for the present proliferation of naval power, can be summarized as follows:

1. In response to Soviet interventions in Eastern Europe, the United States took the central position in the politics of containment, which relied on extensive use of the world's oceans, and also became the "champion" of the freedom of the seas, as much in its own interest as for the benefit of others, when Britain abrogated that function in 1947.

2. The Soviet Union overreacted to the threat posed by the U.S. Navy and precipitated a naval arms race which expanded in the 1970s into all

36. This section draws on ibid.; Stephen Roskill, *Naval Policy between the Wars* (New York: Walker and Company, 1968); Robert A. Hoover, *Arms Control: The Interwar Naval Limitation Agreements* (Denver: University of Denver, 1980); and Robin Ranger, "Learning from the Naval Arms Control Experience," *Washington Quarterly* (Summer 1987): 47–58.

oceans. Most analysts now accept that Khrushchev's decision to rely on nuclear submarines and missiles to counter the U.S. Navy was the turning point in this sequence of events, which had been triggered by the U.S. deployment of Polaris intercontinental ballistic missiles.

3. The spread of nuclear weapons technology to France, Britain, China, India, and many other states has vastly complicated U.S.-Soviet attempts to reach agreement on halting or even slowing the use of those weapons.

4. Arms control initiatives have, until recently, concentrated on nuclear weapons, with the result that the capabilities of nonnuclear forces have continued to grow, unchecked, at frightening rates. The widespread application of modern technology to military purposes has been one of the contributing factors in the worldwide growth of military structures.

5. The pursuit of expansionist foreign policies, for ideological and strategic reasons, by the superpowers and their allies led initially to a polarized international system. This subsequently diffused, but into a world order that could not provide an effective counterbalance to the military supremacy of the superpowers.

Thus, since the end of the Second World War a hierarchy of sea power has evolved.

1. Two superpowers, with a wide range of naval capabilities
2. Major naval powers who either have a power projection capability and/or have or are thought to have nuclear weapons (Great Britain, China, France, India, Israel, North Korea, South Africa)
3. Industrialized maritime states who have smaller navies, without nuclear weapons, for collective and self-defense, some of whom have more naval capability than they need for these purposes
4. Less developed states, several of whom have acquired considerable naval capability, such as North Korea, Libya, and Cuba

The common denominator within most navies in this hierarchy is the possession of sophisticated and highly destructive weapons. What this means is that the traditional restraints on sea power, where only the former great powers had large and powerful navies, have vanished. Instead, there is now a proliferation of virtually uncontrolled naval weaponry.

MARITIME INTERESTS AND SEA POWER

There is a consistent correlation between economic maritime interests and the acquisition of naval power. Moreover, historic trends have been perpetuated by new nations as they grow toward industrial maturity. These relationships can be seen by comparing the four levels of naval power with five levels of maritime economic interest, which are

1. industrialized states with diversified economies and a wide range of maritime interests and responsibilities,
2. geographically smaller but widely diversified industrialized states,
3. states with less diversified or developed industrial bases but with extensive maritime responsibilities,
4. smaller, less developed states, and
5. underdeveloped states.

Table 1 shows a ranking of the states that make the most use of the oceans, and table 2 shows the detailed ranking of states by their military and paramilitary capability. However, a general comparison between economic and military maritime power can be made from table 1 by using the naval tonnage column.

NAVAL POWER

One of the factors that complicates any comparative examination of relative levels of maritime interest and naval power is the rapid growth in the number of sovereign states with maritime interests. Because the international system has not yet reached the level of maturity that makes the need for a legal right of self-defense unnecessary, maritime states naturally want to protect their maritime interests and to ensure that the waters under their jurisdiction remain free for lawful use. This requires that they maintain some form of constabulary capability within those ocean areas. The way in which that capability is provided is a matter of choice. In this respect, many states also follow tradition by using their navies as extensions of the sovereign state in diplomatic roles. Unfortunately, some states have confused the reasonable concepts of self-defense and the diplomatic use of warships with the right to embark on adventurism and as means of applying unwarranted coercion. Thus, one of the concerns in developing a more stable maritime environment is the relative power structure at sea.

Several attempts have been made to establish hierarchies of maritime states on the basis of their naval and other maritime capabilities, the best known probably being that done by Admiral Richard Hill in *Maritime Strategy for Medium Powers*.[37] The trouble with these taxonomies is that they do not pay enough attention to the political factors, particularly the perceived need by governments to be able to influence external incidents outside the framework of established alliances. For our purposes, the simple, four-level hierarchy of naval power is adequate.

37. J. R. Hill, *Sea Power for Medium Powers* (London: Croom Helm, 1986).

TABLE 1.—COMPARATIVE MARITIME INTERESTS RANKINGS (for top twenty states)

Length of Coastline	EEZ Area[a]	Off-shore Oil/Gas Production	Fish Catch Landed	Merchant Shipping Tonnage[b]	Exports[c]	Naval Tonnage[d]
Canada	U.S.A.	U.S.A.	Japan	Japan	Malaysia	U.S.A.
Indonesia	France	U.K.	USSR	USSR	Belgium	USSR
Denmark	Australia	Norway	U.S.A.	Greece	Ireland	China
USSR	Indonesia	Mexico	Chile	U.S.A.	Libya	U.K.
Australia	New Zealand	Venezuela	China	China	Netherlands	France
Philippines	U.K.	Australia	Peru	Philippines	Kuwait	Japan
U.S.A.	Canada	Netherlands	Norway	U.K.	Korea, Rep	India
Norway	USSR	USSR	Korea, Rep	Italy	Chile	Taiwan
U.K.	Japan	Malaysia	Denmark	Korea, Rep	Ivory Coast	Italy
New Zealand	Brazil	Indonesia	Thailand	India	Sweden	Turkey
China	Mexico	Saudi Arabia	India	Norway	Portugal	Peru
Greece	Denmark	India	Indonesia	Brazil	Trinidad and Tobago	Spain
Japan	Papua New Guinea	Nigeria	Korea, Dem P Rep	France	Saudi Arabia	Brazil
France	Chile	Brazil	Iceland	Spain	Germany, Fed Rep	Canada
Mexico		Denmark	Guatemala	Denmark	Denmark	Pakistan
Brazil		Thailand	Philippines	Taiwan	Tunisia	Netherlands
Turkey		Iran	Canada	Germany, Fed Rep	Indonesia	Greece
India		New Zealand	Spain	Iran	Canada	Australia
Chile		Ghana	Mexico	Netherlands	Israel	Argentina
Colombia		Italy	Ecuador	Poland	Ecuador	Germany, Fed Rep

SOURCES.—These rankings are based on data extracted from a wide range of sources, particularly International Institute for Strategic Studies, *The Military Balance 1989–1990* (London: IISS, 1989); Elisabeth Mann Borgese, Norton Ginsburg, and Joseph R. Morgan (eds.), *Ocean Yearbook 7* (Chicago: University of Chicago Press, 1988); Lloyd's *Register of Shipping 1987*; and *The Times Atlas and Encyclopaedia of the Sea*, 1990 edition.

[a] Taken from Clyde Sanger, *Ordering the Oceans* (Toronto: University of Toronto Press, 1987), p. 65. Only the largest 14 states were listed; the remaining order is controversial.

[b] Flags of convenience omitted. If they were to be added, Liberia, Panama, Cyprus, Bahamas, and Singapore would be in the top 10 states.

[c] Exports are ranked by percentage of GNP and are essentially a potential vulnerability rather than a measure of strength.

[d] Tonnages of major warships (PSC) only.

Superpowers

The huge navies of the superpowers form the first level of the sea-power hierarchy. Both are extensions of their respective political systems and are concerned with projecting power competitively around the world. It is the competitive nature of that interaction that is the greatest cause for concern. The naval role within Soviet military doctrine has been debated at length in many forums, particularly after Admiral S. G. Gorshkov wrote *The Sea Power of the State*.[38] From this and other Soviet writings, a number of important facts emerge: First, Soviet foreign policy and military doctrine are interrelated, and the Soviet navy is, therefore, an instrument of political power. Second, although the Soviets are adamant that the primary purpose of their enormous navy is defense of the homeland, they are careful to add that this defense will be conducted at considerable distance from the Soviet homeland and will include forces for "retaliatory" defense. Third, Soviet expansion into the Third World in the 1970s had strategic objectives, with the result that Soviet sea power evolved not only out of a need to influence world events but also in response to the American military posture. Last, for the Soviet Union, the competition with the United States is not just a question of trying to match the U.S. Navy in total capability; it is more the need to be able to do similar things.

Opponents of the present American concept of maritime strategy have claimed that it has no roots in foreign policy. This is simply not true. The U.S. Navy took on the role of the "guardian of the freedom of the seas" from the Royal Navy after the Second World War, when all the former great maritime powers had either been defeated or had been made bankrupt by the war. The U.S. maritime strategy, which has become highly controversial, has been described officially as a "concept of operations for the effective global employment of naval forces to protect the interests of the United States and our allies and support our national policy objectives. It is the same strategy that the United States has pursued in the name of peace for the past 40 years."[39] As mentioned earlier, one of the basic problems with the maritime strategies of the superpowers is mutual distrust based on entrenched misconceptions of each other's intentions. Thus, an objective in any future naval arms control discussion will be the resolution of those differences in perception.

38. S. G. Gorshkov, *The Sea Power of the State* (Annapolis, Maryland: Naval Institute Press, 1979).

39. Admiral Carlisle A. H. Trost, U.S. Navy, in a speech to the Leningrad Naval School on 12 October 1989, as reported in *Sea Power* (December 1989): 27–30.

Major Naval Powers

The second order of sea power includes several states with very diverse interests and allegiances. Some are regional powers with self-imposed mandates to maintain a commanding military presence in an area, such as India and China. Others are former "great" maritime and colonial states who still see the need to be able to take independent military action at sea, essentially Great Britain and France. For all these states, membership in an alliance presents a potential conflict of interests between requirements for collective action and requirements to take independent action. The common denominator is that these states all have large naval forces and/or have or are thought to have nuclear weapons.[40] Constraint in the use of that power is a major concern for all other states. For instance, one question that has bothered many analysts is, Would Britain have used tactical nuclear weapons in the Falklands had they been faced with a humiliating defeat?[41] Whereas the nuclear weapons of the United States and the Soviet Union exist under a condition of mutual deterrence, weapons of mass destruction in the hands of other states are less constrained and represent a distinct threat to global stability. Moreover, the ambiguous maritime strategies endorsed by these "secondary" nuclear-weapons states could be destabilizing under certain circumstances.

Industrialized Maritime States

Third in the hierarchy come the traditional maritime states who, with a few exceptions, represent the balance of the industrialized states. Generally, these states have little requirement to exert national influence at large in the world, outside of some form of collective action. Even though some may still have overseas territorial responsibilities, such as the Dutch, the likelihood of their having to resort to the use of force independently is very small. The majority of these states enter into security agreements with other states or with one of the superpowers, but not usually on a bilateral basis with any of the independent nuclear powers. Unfortunately, the industrialized maritime states tend to become the pawns in the balance of international naval power and could be committed to full-scale operations without the right or outright ability to prevent either superpower from taking unilateral action. Such states are

40. Editors' note.—See Frank Barnaby's article, "The Role of the Submarine," in this volume, particularly table 3, "Submarines in the World's Navies."

41. George H. Quester, "The Nuclear Implications of the South Atlantic War," in *The Denuclearization of the Oceans,* ed. R. B. Byers (London: Croom Helm, 1986), pp. 119–33.

faced with the unenviable prospect of having to maintain naval forces for collective defense that may have little or no direct value in furthering national interests outside alliance considerations. The relative strategic predictability of this group of states is generally a stabilizing factor in international conflict resolution. In terms of strategies, they probably do not need elaborate independent statements beyond those necessary to substantiate requirements for specific naval capabilities and for the management of their own waters.

Less Developed States

The final group is the most difficult to address, for it contains the coastal states of the Third World, many of whom might elect to use their military forces locally to further their own aims and to resolve conflicts. The fact that neutral shipping was deliberately attacked in the recent Iran/Iraq war is proof that in some cases it cannot be assumed that a belligerent will act rationally or will respect the traditional laws of warfare. In addition, concern has been expressed that, as some coastal states attempt to move the boundaries of their territorial waters further away from their shores, the potential for conflict at sea increases.[42] If the uncertainty of international terrorism is added to this situation, the potential for maritime instability in the Third World is high. When the economic interests of the industrialized states come under threat as a result of a Third World maritime dispute, it is very difficult to prevent them from taking action to protect their interests, as we are now witnessing in the Persian Gulf.

Global Distribution of Naval Power

Table 2 shows a ranking of states on the basis of various elements of naval power. Two facts stand out:

1. Although the capability to project significant military power, in terms of both amphibious forces and naval fleets, into other areas is concentrated in the hands of only a few states, many others have some power projection and sea-control capability.
2. The advantage now given to smaller states by the availability of modern weapons, mainly missiles, becomes both an equalizing factor (as seen in the Israeli, Cuban, North Korean, and South African fleets) and a destabilizing factor, as demonstrated in the Persian Gulf recently.

---

42. Larson (n. 22 above) discusses this likelihood at length.

TABLE 2.—COMPARATIVE NAVAL POWER RANKINGS (for top twenty states)

Submarines	Major Warships	Destroyers and Frigates	Minor Warships (PC/MW)	Coast Guard Vessels	DD/FF + PC/MW	DD/FF + PC/MW + CG	Amphibious Vessels
USSR	USSR	USSR	China	Japan	China	USSR	USSR
U.S.A.	U.S.A.	U.S.A.	USSR	USSR	USSR	China	U.S.A.
China	Japan	Japan	Korea, Dem P Rep	U.S.A.	Korea, Dem P Rep	Japan	China
Germany, Fed Rep	China	China	Romania	Sweden	U.S.A.	Korea, Dem P Rep	Brazil
Korea, Dem P Rep	U.K.	U.K.	Germany, Fed Rep	Greece	Romania	U.S.A.	Germany, Dem Rep
France	France	France	Mexico	Spain	U.K.	Sweden	Indonesia
India	Taiwan	Taiwan	Korea, Rep	Malaysia	Japan	Greece	Korea, Rep
Turkey	Italy	India	Turkey	Canada	Korea, Rep	Romania	Philippines
Japan	India	Korea, Rep	Yugoslavia	Saudi Arabia	Germany, Fed Rep	Spain	Poland
Norway	Korea, Rep	Italy	U.K.	Denmark	Taiwan	Turkey	Taiwan
Peru	Turkey	Turkey	Germany, Dem Rep	India	Turkey	Korea, Rep	Thailand
Sweden	Greece	Greece	Taiwan	Arab Emirates	Germany, Dem Rep	U.K.	Ecuador
Egypt	Spain	Canada	Sweden	Turkey	Mexico	Germany, Fed Rep	Egypt
Greece	Canada	Spain	Spain	Venezuela	Spain	Germany, Dem Rep	France
Italy	Germany	Germany, Dem Rep	Vietnam	Romania	Yugoslavia	India	Greece
Spain	Brazil	Brazil	Libya	Italy	France	Yugoslavia	Iran
Brazil	Pakistan	Pakistan	Israel	Peru	India	Mexico	Iraq
Australia	Netherlands	Netherlands	Japan	Egypt	Sweden	Malaysia	Italy
Libya	Indonesia	Indonesia	Poland	Poland	Cuba	France	Japan
	Germany, Fed Rep	Germany, Fed Rep	India	Yugoslavia	Greece	Denmark	Libya

Source.—This table is based on data in International Institute for Strategic Studies, *The Military Balance 1989–1990* (London: IISS, 1989).

Notes.—DD/FF Destroyers and frigates
PC/MW Patrol craft/mine warfare vessels
CG Coast guard vessels

Thus, we are left with the question, What constitutes excessive naval capability? It could be argued that it might be more appropriate to ask, What constitutes sufficient naval capability? Both questions are perfectly valid. However, the issue of *sufficiency* is essentially one that should be addressed by each state to determine its needs. The problem of *excess capability* should be addressed by the community of states.

NAVAL ARMS CONTROL AND LIMITATIONS

It can be assumed that, for the foreseeable future, maritime states will need to maintain some form of naval capability to meet national requirements in all or some of four basic areas.

1. To conduct constabulary duties in waters under national jurisdiction
2. To maintain commitments to collective security agreements and to support other international security commitments
3. To preserve national security from unexpected threats from the sea
4. For diplomatic purposes

Eventually, it may become possible to reduce or further limit some of those requirements, but until such time as the international system no longer contains irrational factions the right of self-defense must be upheld. The question is, therefore, How much naval capability is enough? Clearly, those states with greater maritime interests and responsibilities need more capability than those with only limited maritime interests. In this respect, one of the first requirements in developing a realistic naval arms control process would be to separate what might constitute reasonable naval capabilities for self-defense and constabulary tasks from other naval capabilities. This, of course, brings us back to the East-West balance of power and the present European arms control regime. Although some naval force capabilities can be linked to the European land situation, primarily amphibious and land attack capabilities, the nature of naval forces makes direct linkage to the land situation difficult. This is due, in part, to the fact that naval forces are seldom committed to any single theater of operations and are required to undertake a wide range of political and security tasks. Because of the prevailing inequalities in maritime power that evolved under the mantles of East-West collective defense, where immense military force on land was offset by naval power, any arms control process will have to address the complex structure of military asymmetries. The present round of discussions on European land-force limitations is an essential prerequisite to any naval arms limitation initiative, for it allows one tier of asymmetries to be removed, thereby making the naval asymmetries less complex to resolve, but only in that one region.

In adopting a global approach to naval arms control, the full implications of East-West initiatives need to be measured as part of the process. For instance, any reductions in naval tactical nuclear weapons would have to be looked at in terms of their overall impact on stability and include all weapons in specific categories rather than just those held by the superpowers. Similarly, concepts of limiting the sizes of naval forces would have to be considered on the basis of global stability. Failure to take a global approach to future naval arms control and limitation initiatives would run the risk of making the same mistake that caused the disarmament plans of the 1920s and 1930s to fail, when postwar euphoria did not recognize the possibility that another major arms race could be initiated. At the time, the innovators didn't extend their thinking sufficiently far ahead to realize that global stability was a prerequisite for an effective naval arms limitation regime. It would be shortsighted to try to reduce East-West naval capabilities to a point at which some other state would then hold the dominant naval power. In fact, naval arms control has to seek a true balance of power in all the world's oceans.

The questions that need to be addressed now are, By what means can the world's oceans be made more safe? and In which order should those measures be implemented?

Strategic Arms Limitations

It would be incorrect to assume that strategic weapons systems operate in isolation. The Soviets give the protection of their sea-borne strike capability a very high priority. And for good reason, for it is a declared principle of the U.S. maritime strategy to hold all Soviet naval units at risk in time of war. In this regard, the Soviets are not different; they also intend to attack any U.S. strategic missile submarines they encounter. As many authors have established, there is a clear linkage between the reliance on submarines as part of a strategic deterrence force and the level of naval activity in areas where they operate.[43] It follows, therefore, that an opportunity for tactical naval arms limitations exists within the framework of strategic arms limitations. Again, the linkages are complex, and some form of balancing formula would be required. For such initiatives to be meaningful, all states engaged in operating and prosecuting sea-borne strategic weapons systems would have to be included in the discussions. Although simple bilateral discussions between the superpowers would be beneficial, any reductions would have to ensure that vacuums, which could be exploited by third parties, were not created. In simple terms, the scope of strategic arms limitations talks must eventually be

---

43. For instance, see Mark Sakitt, *Submarine Warfare in the Arctic: Option or Illusion?* (Stanford, California: Stanford University Press, 1988). Editors' note.—See Frank Barnaby, "The Role of the Submarine" in this volume.

expanded to include the French, British, and Chinese, as well as any other states with strategic nuclear capabilities.

Tactical Nuclear Weapons

At one time tactical nuclear weapons were thought to have both a deterrent and an equalizing capability. In the early 1970s, the Soviets placed great store in their tactical nuclear weapons as the only way they could overcome their technical disadvantages. Despite revisions of the military doctrine in the late 1970s and the subsequent "new thinking," there still seems a reluctance on the part of the Soviet military to abandon tactical nuclear weapons completely. Instead, their initiatives to date have focused on the development of nuclear weapons–free zones (NWFZ). Despite the views of Ambassador Paul Nitze on the irrelevance of tactical nuclear weapons in the modern naval context, the United States is equally reluctant to put these weapons on the table. On balance, there is absolutely nothing to be lost by pressing for a reduction or, preferably, a ban on tactical nuclear weapons within the existing arms control regime. The logical, broader requirement to discuss all such weapons will be much harder to initiate, but if the initiative is taken within the European forum, it may be easier to extend the discussion to include other states rather than have to start afresh.

Although not specifically restricted to tactical nuclear weapons or the oceans, a number of nuclear weapons treaties have been ratified in the last 30 years. These include the Outer Space Treaty, the Treaty of Tlateloloco, the Non-Proliferation Treaty, and the Seabed Treaty.[44] Some are geographic, others are general. Those that relate to the banning of nuclear testing can be verified with ease, but those that merely seek to prevent the movement of nuclear weapons are difficult to enforce and thus must rely on the integrity of states to abide by the provisions. The weakness of such concepts has been proven on several occasions, and this tends to result in an overriding concern that unless compliance with such treaties can be verified and enforced in case of noncompliance, they will remain susceptible to violation. Thus, in establishing realistic naval arms control agreements the dual problems of compliance and verification must be incorporated into the final document, otherwise it becomes meaningless. The drawback is that verification can be almost prohibitively expensive and leads to questions on how to handle the information once it has been obtained. An interesting example of this came from the Canadian initiative for a nuclear weapons and submarine exclusion

44. Details of these treaties can be found in many documents. One of the best sources is the United Nations, *Status of Multilateral Arms Regulations and Disarmament Agreements,* which is issued periodically. Editors' note.—See "Disarmament: The Sea-Bed Treaty and Its Third Review Conference in 1989" in Appendix B of this volume.

zone in the Arctic Basin, which formed part of a paper presented at the October 1989 Canada-USSR Arctic Co-operation Conference.[45] The nature of the Arctic Basin and the types of submarines that were to be excluded called for a verification process based on a very sophisticated underwater surveillance system, which at the time was beyond the limit of known technology as well as being likely to cost several billion dollars. Ironically, the most effective verification system was another nuclear-powered submarine.

Confidence-Building Measures

The concept of an international code of conduct for naval forces has already been mentioned. By introducing a series of rules and procedures for the manner in which warships should conduct their business on the seas, such a code could have widespread application. Moreover, once such rules have been accepted, ambiguities could be resolved through a legal dispute settlement process, thereby preventing the structure from becoming moribund. More importantly, such a system could flow naturally from the growing number of bilateral agreements for the prevention of incidents at sea that many navies are now accepting.[46] These agreements, which are not yet all-encompassing, serve to prevent misunderstandings arising at sea in a manner that does not automatically require a political decision. In other words, they prevent the accidental escalation of incidents at sea through uncertainty of the other party's intentions. Admittedly, rules can be broken, but to do so would require a deliberate decision on the part of a unit or formation commander rather than a political act. By providing for safety distances and "back-down" procedures, these agreements do much to further the peaceful use of the oceans.

Other confidence-building measures that can be used on land and at sea are agreements between states to limit the movement of forces between theaters as well as the scope of exercises that their forces can conduct without formal notification. In the past, both the NATO and Warsaw Pact alliances have conducted large naval exercises in many parts of the world. Sometimes the initial deployment of forces to the exercise area was ambiguous and could have been read as a threatening gesture. Similarly, transfers of large numbers of ships from one area to another could give rise to concern on the part of

45. Sponsored by a number of Canadian nongovernment organizations and organized by the Canadian Centre for Arms Control and Disarmament. The conference was held in Ottawa, 24–26 October 1989.

46. For instance, the 1972 Agreement between the Government of the United States and the Government of the Union of Soviet Socialist Republics on the Prevention of Incidents on and over the High Seas and a similar agreement between Canada and the Soviet Union signed in November 1989.

other states. Both superpowers have used large naval and amphibious exercises as "demonstrations" of political concern and resolve.[47] It stands to reason that procedures to remove ambiguity and uncertainty from the movements of military forces by sea would enhance stability. Again, such measures are relatively easy to discuss and agree upon and would form a very valuable prelude to more fundamental arms control agreements. The interesting point about such initiatives is that they are the beginning of a system of controls on the peaceful use of the oceans by naval forces. As such, they are complementary to the principles embodied in the LOS Convention.[48]

One of the other measures that lends itself naturally to inclusion within a broad family of confidence-building measures is the creation of a UN naval force. This has been discussed at various times in the past, but the need for such a capability has not been universally supported, and recent papers on enhanced UN peacekeeping activities have tended to shy away from the naval dimension. The Soviet Union restated the need for a UN naval force in May 1988 as a means of responding to such incidents as the Persian Gulf war. Obviously such a force cannot be restricted to peacekeeping because the requirement to contain an interstate conflict could well require a UN force to interpose itself between the belligerents. Thus, it would be acting on behalf of the United Nations prior to the declaration of a truce or cease-fire. Essentially, the role would then become one of peacemaking rather than peacekeeping. In view of the experience of the Korean War, this may not yet be acceptable to the United Nations. Also, the Security Council may not be willing to create the necessary precedent. Such a concept would require extensive discussion and consultation to establish terms of reference whereby a UN naval force could be used to prevent the escalation of war. Because the involvement of the superpowers actually creates the greatest threat to escalation, any UN naval force might have to exclude the presence of either the United States or the Soviet Union. It is unlikely that this would be acceptable to the USSR. Nevertheless, it should not prevent the concept from being considered. Although a UN naval force would represent a substantial change in UN security policy, it might provide an incentive to bring a large number of naval powers to a negotiating table in order to discuss the need to impose limits on the existing hierarchy of naval power. In this respect, ratification of the 1982 LOS Convention is one of the various ancillary actions that would be universally beneficial.

47. For instance, the Soviet amphibious exercises held very close to the Polish border during the 1981 political crisis, and exercises held in the Caribbean by the Americans in October 1962, immediately prior to Kennedy's response to the Soviet deployment of missiles to Cuba.

48. There is now extensive literature on the subject of naval confidence-building measures. A recent UN seminar, held at Helsingør, Denmark, on 13–15 June 1990, which I attended, was devoted to that subject.

Naval Arms Limitations

Historic examples provide some of the alternatives to and also show some of the difficulties with naval arms limitations. It should be reasonably clear now that for limits on sizes, capabilities, or locations of naval forces to be realistic, a number of conditions must be satisfied. Foremost is the requirement that any agreement be verifiable. If it were not so, there would be little practical purpose in its existence. Also, any future agreement would have to be backed by a means of enforcement with some system of penalties in case of violation. These may seem frightening prerequisites, but in reality they need not be if the concept is based on a principle of international accountability. In other words, states that fail to abide by agreements should be brought before some international body to account for their actions.

Although an international approach to arms control is theoretically necessary because of the widespread proliferation of naval armament, this should not preclude narrower objectives. In fact, a number of preliminary agreements are almost certainly prerequisite to addressing the existing asymmetrical structure. Thus, a universal approach to naval arms control may have many interdependent layers. These could function on either numerical or geographic parameters. The experience with the Intermediate-Range Nuclear Forces (INF) Agreement tends to indicate that one of the more promising approaches to arms control is to address specific weapons systems or classes of weapons. Cruise missiles, tactical nuclear weapons, and a number of newer "high-tech" systems could be considered in this category.[49] There is also some support for addressing specific capabilities. The Soviets have proposed on several occasions that U.S. amphibious capability and aircraft carriers become the basis for naval arms control discussions.[50] Because of the counterbalancing requirement for those forces in the European balance of power, this, of course, would require that a number of other asymmetries be addressed. By the same token, the Soviet numerical supremacy in submarines should also be addressed. In the short term, these are unlikely prospects for arms limitations, and as discussed, there are a number of linkages to the European situation that should be dealt with first.

To support a more universal initiative, which might well address the deep-seated problem of excessive weaponry in the Third World and which could possibly be run in parallel with the series of East-West discussions, it might be possible to begin work on the basic question of what actually constitutes an excess of naval capability. Relatively simple formulas, such as those developed to support the Washington Treaty, would not work in the present situation where there are so few similarities in either weapon performance or

49. Thomas J. Welch, "Technology Change and Security," *Washington Quarterly* (Spring 1990): 111–20.
50. See "Naval Arms Control" (n. 16 above).

TABLE 3.—RELATIVE SIZES OF NAVIES

COUNTRIES WITH COASTLINES > 750 KM

	Argentina	Angola	Algeria
	Bahamas	Burma	China
	Brazil	Cuba	Ethiopia
	Cape Verde	Finland	
	Chile	Honduras	
	Colombia	India	
	Costa Rica	Iran	
	Dominican Rep	Italy	
	Ecuador	Japan	
	Fiji	Mauritania	
	France	Mexico	
	Gabon	Morocco	
	Greece	Mozambique	
	Ireland	Nicaragua	
	Jamaica	Portugal	
	Malaysia	Spain	
	Norway	Sweden	
	Oman	Tanzania	
	Peru	Thailand	
	Saudi Arabia		
	Somalia		
	South Africa		
Australia			

Under 0.1	0.1 to 0.25	0.25 to 1.0	1.0 to 2.5	2.5 to 5.0	5.0 to 10.0	Over 10.0
Canada	Denmark	Sudan	Tunisia	Korea, Rep	Germany, Fed Rep	Germany, Dem Rep
Iceland	Haiti	Suriname	Turkey	Libya	Israel	Korea, Dem P Rep
Indonesia	Madagascar	U.K.	USSR	Pakistan	Nigeria	Taiwan
New Zealand	Panama	Venezuela	U Arab Emirates	Sri Lanka	Yugoslavia	
	Philippines	Yemen, P Dem Rep	U.S.A.	Vietnam		

COUNTRIES WITH COASTLINES < 750 KM

Under 0.1	0.1 to 0.25	0.25 to 1.0	1.0 to 2.5	2.5 to 5.0	5.0 to 10.0	Over 10.0
		Belize	El Salvador	Brunei	Albania	Belgium
		Cameroon	Ghana	Guinea	Bahrain	Bulgaria
		Eq Guinea	Guatemala	Guinea-Bissau	Bangladesh	Iraq
		Liberia	Ivory Coast	Kuwait	Benin	Poland
		Sierra Leone	Kampuchea	Togo	Congo	Romania
			Kenya	Trinidad and Tobago	Gambia	Singapore
			Lebanon		Netherlands	Syria
			Qatar			Zaire
			Senegal			
			Seychelles			
			Uruguay			
			Yemen Arab Rep			

Sources.—Based on data from International Institute for Strategic Studies, *The Military Balance 1990–1991* (London: IISS, 1990); and *The Times Atlas and Encyclopaedia of the Sea* (New York: Harper and Row, 1989). App. 2.

Note.—Based on number of destroyers/frigates and patrol craft/mine warfare vessels per 100 km of coastline.

ship capability. Instead, a formula developed proportionally on the maritime responsibilities of states might prove useful. For instance, relating the numbers of smaller warships (destroyers, frigates, patrol vessels, etc.) to the length of a state's coastline, as shown in table 3, makes for an interesting comparison. It can be seen that some states have more relative naval power than others with greater maritime responsibilities. Although this example is imprecise, it shows that it is not difficult to develop a workable formula for determining acceptable levels of relative naval power. This is not an end in itself, but it is, perhaps, a useful part of a larger and more complex process that allows some correlation between regional concerns and numerical considerations. Imposing limitations on either vessels or weapons is bound to be unpopular, because eventually the international arms industry must come under scrutiny.

CONCLUSION

One view about naval arms control is that it is unlikely to succeed in the longer term if not conducted on a global basis. However, within that belief is a firm conviction that a wide range of subsidiary initiatives can and should be pursued, particularly those that will begin to address the asymmetries in the naval structures of the superpowers and their allies. Further, it is obvious that unless some restraint is put on the spread of military technology, those asymmetries will expand beyond their present bounds and create a universally unstable situation. But such steps cannot be taken outside the community of nations, for the problems lie in the use of naval power as much as in its existence. This political factor is, in fact, as destabilizing as the weaponry itself. Again, the Persian Gulf experience is a clear example. Ideally, both destabilizing factors should be removed, but this may not always be possible. Instead, other solutions will have to be sought. These require innovative thinking and, above all, a willingness to discuss the issues.

In this respect, the various proposals and recommendations made by the First International Seminar on Naval Arms Limitations and Naval Activities Reduction in the World Ocean, held in Moscow in February 1990, are interesting. They reflect a unique blend of traditional views, tempered by reality and by the pressing need to advance the naval arms control process, and some innovative concepts. With the exception of the call for the creation of "zones of peace," which still cannot be verified effectively, the proposals make an interesting and practical agenda.

1. Maintain the dialogue on naval arms control.
2. Make naval arms control a global rather than a superpower concern.
3. Press for a shift to defensive maritime strategies.
4. Take measures to reduce pollution and radioactivity caused by naval operations.

5. Ban tactical nuclear weapons.
6. Approach naval arms control from the perspective of a balance of interests rather than a balance of naval forces.
7. Seek a wide range of naval confidence-building measures, including the creation of a UN naval force.

Although these are realistic objectives, several are very complex and would be very difficult to implement without almost universal concurrence. Thus, it would appear that an important step in progressing in naval arms control and limitations is the direct involvement of the United Nations. Obviously, some initiatives can be started through the existing European arms control regime, but unless they are eventually expanded, the process will be incomplete. The last and probably the most important of the various initiatives is the need to further the entire spectrum of naval confidence-building measures. These stand to have the greatest effect on global security with the least resistance, either bilaterally or multilaterally. Moreover, as confidence grows, it will become very much less difficult to address the more complex issues.

Naval Operations in Korean Waters

Joseph R. Morgan
East-West Center

INTRODUCTION

Recent events indicate that the superpower military rivalry is far less serious now than at any time since the end of World War II. Mutual disarmament has been agreed on for a number of nuclear weapons, and conventional arms agreements are presumably forthcoming. Moreover, agreed reduction or even complete elimination of tactical nuclear weapons launched at sea is a distinct possibility.

Although ideological rivalry and the Cold War have apparently ended, at least for the foreseeable future, conflicts between smaller powers are still possible. Despite evidence that the prospect for conflict in East Asia is less now than at any time since the end of the Korean War in 1953, the possibility of conflict on the Korean peninsula should not be discounted. North and South Korea are still in an uneasy truce; technically the formal end of hostilities on 27 July 1953 involved North Korea and United Nations forces and was not agreed to by South Korea.

Another war on the Korean peninsula would almost certainly involve naval forces, and both Koreas maintain navies whose missions include defense of their country against attack by the other. Although the 1950–53 conflict was fought primarily on land, and naval forces of the United States were largely unopposed, the influence of sea power on the conduct of the war and the final result provides some lessons that might be applicable in the future. To be sure, war between the two Koreas is not inevitable, and there is evidence that both sides are attempting to avoid it, but at the same time both countries are prepared for it. These preparations include the maintenance of modern well-equipped navies, which would play an important role in determining the result of the hostilities.

THE NORTHWEST ISLANDS

The truce established at the end of the Korean War left some potentially conflict-prone divisions of territory between North and South Korea. In 1953,

the Armistice Agreement, which effectively ended the actual hostilities between the two Korean governments but not the state of war, provided for
five islands west of the Han River estuary and near the coast of North Korea
to remain under control of the United Nations Command. The islands,
Paengnyong-do, Teachong-do, Sochong-do, Yonpyong-yolto (which consists
of two small islands and various rocks and islets), and U-do, had only 13,000
people on them over an area of 5 square miles in 1978.[1]

The northern limit line, drawn by the commander of the United Nations
Forces 11 years after the armistice, lies north of the islands. Although North
Korea claims Territorial Seas around the islands, it has not taken action to
enforce its claims. In practice, South Korea normally insures that its fishing
vessels stay well south of the line to avoid provoking North Korea into any
action, but occasionally fish migrations tempt fishing boats to move close to
the line for short periods of time. Although the existence of these islands
under United Nations command is unprecedented, and the issue calls for an
agreement between the two Koreas, it is quite unlikely that any resolution of
the problem can be anticipated in the near future.[2]

South Korea considers the islands of great strategic value and has armed
forces stationed on them.[3] It is possible that an incursion into the area around
the islands, claimed as a North Korean Territorial Sea, could spark armed
conflict between the two nations, but it is more likely that any military activities in the vicinity would occur only if North Korea specifically chose to invade
the islands with a view to overwhelming the South Korean forces stationed
on them.

NORTH KOREA MILITARY ZONES

On 1 August 1977 North Korea established two military security zones. The
security zones were instituted by the following declaration: "The military
boundary is up to 50 miles from the starting line of the territorial waters in
the East Sea (Sea of Japan) and to the boundary line of the economic sea

---

1. J. R. V. Prescott, *The Maritime Political Boundaries of the World* (London and
New York: Methuen, 1985), pp. 48, 49.
2. Ibid., p. 49.
3. In a paper presented at the International Conference on the Regime of the
Yellow Sea: Issues and Policy Options for Cooperation in the Changing Environment,
19–20 June 1989, Seoul, Korea, Dr. Hyun-Ki Kim, a professor at the National Defense
College of the Republic of Korea, described the five islands as exceptionally strategic
due to their location as the most advanced South Korean naval base. He argued that
the probable intentions of North Korea are to occupy the Seoul metropolitan area,
including the five islands, by force and then to propose armistice talks immediately
after the occupation has become a fait accompli.

zone in the West Sea (Yellow Sea)."[4] Foreign ships, both military and civilian, were banned from entering the zone unless they had obtained appropriate prior agreement or approval.

Although the military security zones are supported neither by recognized principles of international law nor by the 1982 Convention on the Law of the Sea and, for that reason, have been objected to by a number of maritime countries as being restrictive of freedom of navigation, they do serve a useful purpose. Since both Koreas have navies whose avowed missions are the defense of the country against attack by a hostile force, and neither navy is configured for long-range deterrent operations, there is no reason for the South Koreans to operate naval vessels in waters surrounding North Korea. The military security zones thus help to preserve the uneasy peace.

SEAPOWER IN KOREA

Historical Notes

Koreans have a long history of the use of sea power to defend their territory against attack by an aggressor. One of Korea's (both North and South) great military heroes was Admiral Yi Sun-shin, who was instrumental in defeating Japanese naval forces in 1592–98. The Japanese had successfully invaded and overrun the Korean peninsula on their way to northern China, when Admiral Yi developed a maritime strategy to prevent resupply of the Japanese forces and to achieve subsequent defeat of the Japanese navy, resulting in command of the Sea of Japan by Korea. The admiral's strategy and tactics were based on the development of a new ship type: the *Kokukson* (turtle warship), reputedly the first ironclad warship. The Korean fleet defeated several Japanese fleets during the 6-year naval war, with the last resounding Korean victory taking place in 1598. During the final decisive battle Admiral Yi was mortally wounded. Numerous statues of the admiral can be found in South Korea, particularly in Seoul and at Chinhae, an important South Korean naval base and home of the Republic of Korea Naval Academy.[5]

Seapower in Korea 1950–53

Although the Korean War was a classic land war, with well-defined front lines, armies pitted against each other, and traditional concepts of strategy

4. Quoted in Bruce D. Larkin, "East Asian Ocean Security Zones," *Ocean Yearbook 2*, ed. Elisabeth Mann Borgese and Norton Ginsburg (Chicago: University of Chicago Press, 1980), p. 289.

5. Yushin Yoo, *Korea the Beautiful: Treasures of the Hermit Kingdom* (Los Angeles and Louisville: Golden Ponds Press, 1987), pp. 13, 76, 78.

employed, American sea power did play a part. Sea power supported land forces, brought supplies into South Korea, and prevented the resupply of North Korea by sea. In addition, there are several examples of the use of naval forces in support of ground units: by amphibious assaults, shore bombardment, and the routine transportation of army and marine corps units to the Korean peninsula.

Any attempt to evaluate the effectiveness of sea power in bringing the war to a successful conclusion is hampered by the undeniable fact that the sea power was almost entirely provided by the U.S. Navy, which was for the most part unopposed by North Korea or its Chinese ally. Nevertheless, some valuable lessons can be learned from a brief analysis of the sea war in Korea.[6]

UN ground forces were steadily on the defensive for the first 82 days of the war and had retreated from the thirty-eighth parallel to a small perimeter around the port of Pusan.[7] The tide of the war changed, however, on 15 September 1950 with a dramatic example of the use of sea power. A successful amphibious landing at the port of Inchon, far to the north of the Pusan perimeter, put the North Korean army on the defensive and permitted the breakout of South Korean forces from Pusan, as troops were withdrawn by North Korea for combat operations to the north. Although there could be no opposition to the Inchon landing by the North Korean navy, which was nonexistent at the time, the choice of Inchon as a landing site was controversial. The tidal range on the west coast of Korea is as much as 33 feet, and there were no beaches in the vicinity on which a conventional landing could be made. Environmental conditions were far more important as obstacles to a successful operation than were the North Korean armed forces. Nevertheless, General MacArthur's selection of Inchon had undeniable advantages, in the form of strategic surprise, psychological effects on North Korea, and reduction of U.S. casualties.[8]

Other amphibious operations conducted by the United States were not as successful. A planned invasion of the important North Korean port of Wonsan was delayed by the need to clear the landing approaches of mines. In the early stages of the war, the U.S. Navy was ill-prepared for mine-clearance operations; few minesweepers were available, most having been mothballed at the conclusion of World War II. The extent of the North Korean mine-laying efforts came as a surprise to the United States. The result was not only the delay of landing operations but the loss of two minesweepers, with considerable personnel casualties. The geography of the Korean peninsula lends itself to defensive mine-laying operations. A definitive work on the use of sea power in the Korean War of 1950–53 has this to say:

6. Malcolm W. Cagle and Frank A. Manson, *The Sea War in Korea* (Annapolis: U.S. Naval Institute, 1957), p. 3.
7. Ibid., p. 75.
8. Ibid., p. 77.

Actually the Korean peninsula was almost ideally suited for an experiment in defensive mine warfare. After the UN's entry into the war, the Communists could foresee that U.S. naval forces would take every advantage of the amphibious warfare specialty to move northward. The landings at Pohang and Inchon were eloquent testimony to this special skill. Moreover, the Communists recognized the vulnerability of Korea's eastern coast to amphibious assault, and also to bombardment from the sea. The waters off the east coast were deep and the coastal plain narrow. The coastline was reasonably straight, and the 100-fathom curve lay fairly close to shore. Off the good harbors of Wonsan and Hungnam, there was a large shelf of shallow water which made mine planting exceptionally effective. On the opposite shore, Korea's western coastline was a honeycomb of shallows, with the Korean rivers emptying into the Yellow Sea. Nowhere in the Yellow Sea was the water more than sixty fathoms deep; mean tidal range was twenty-one feet. While not ideal, the west coast was certainly mineable.[9]

The mines launched at Wonsan were from junks and sampans, and many were "drifters," which when launched from a North Korean port might travel the full length of the peninsula in fifteen days, carried south by the prevailing ocean currents. Initial sweeping operations were for moored contact mines, and U.S. minesweepers were generally successful in their mine-clearing efforts. When the principal channel into Wonsan had been swept of contact mines, and the U.S. forces believed the menace from mines was over, magnetic mines were discovered. These mines lie on the bottom and explode when a ship with the necessary magnetic signature passes over them. Although wooden-hull minesweepers are less susceptible to magnetic mines than larger, steel vessels, they are not immune. The first casualty to a magnetic mine was the South Korean minesweeper YMS-516, a sister ship of the small minesweepers being used by the American navy. An explosion under the keel "blasted that small ship into bits and pieces."[10]

There was little doubt that a virtually nonexistent navy could have a considerable influence on the employment of sea power by a greatly superior force, by the use of mines. Mines are an inexpensive weapon, easy to lay, and difficult to sweep or clear. According to U.S. Navy Admiral Joy, then Commander Naval Forces Far East, "The main lesson of the Wonsan operations is that no so-called subsidiary branch of the naval service, such as mine warfare, should ever be neglected or relegated to a minor role in the future. Wonsan also taught us that we can be denied freedom of movement to an enemy objective through the intelligent use of mines by an alert foe."[11]

9. Ibid., pp. 121, 122.
10. Ibid., pp. 130, 143.
11. Quoted in ibid., p. 151.

Wonsan was not the only North Korean port to be mined. Chinnampo (now Nampo), the principal port for the capital at Pyongyang, was also heavily mined, and U.S. use of the port after Pyongyang had already been captured was delayed by the need for mine-clearing operations. The problem at Chinnampo differed from that at Wonsan primarily because of very different hydrographic conditions. Chinnampo, on the west coast of North Korea, has a minimum tidal range of 12 feet, and tidal currents are as swift as 5 knots. Moreover, the water is muddy, a consequence of silt carried down the Taedong River into the Daido-Ko estuary. Both channels into the port were mined. The lengthy approaches to the port had to be swept; the minesweeping operations started 69 miles from the port and proceeded through a delta that was 33 miles from the dock facilities. Fortunately, there was no enemy resistance, but the use of mines clearly delayed military operations by the U.S. forces in the area. When an attack by Chinese forces changed the character of the war completely some time later, Chinnampo became an important evacuation port for UN forces. The previously thorough mine-clearing efforts made the subsequent evacuation a complete success; remarkably, there were no casualties.[12]

Minesweeping efforts were needed throughout the remainder of the war. Although the operations were more routine and less urgent than those earlier in the war, which were in direct support of amphibious operations or needed evacuations, mine clearance efforts remained an important, dangerous aspect of allied naval combat operations. Five minesweepers were sunk during the war,[13] a heavy cost in material and personnel casualties enacted by an enemy with only the most rudimentary of navies.

In addition to amphibious assaults and mine warfare, sea power played other roles in the 1950–53 war, principally in direct support of land forces. Shore bombardment was carried out frequently, sea lines of communication were maintained to South Korean ports, North Korean ports were blockaded, and in the case of the important port of Wonsan an actual siege took place.[14]

The siege began on 16 February 1951 and lasted until the last day of the war, 27 July 1953. During this 892-day period, American sea power was completely dominant; it was opposed only by shore batteries in the harbor itself and by remining efforts carried out by small sampans usually operating at night. American destroyers operated in the harbor, supported by offshore air power from aircraft carriers when needed. Occasional bombardments by battleships and cruisers were scheduled to add to the gunfire efforts of the destroyers. Minesweeping was carried out by South Korean and U.S. minesweepers, on a generally routine basis, with only occasional interference by North Korean shore batteries.

12. Ibid., pp. 153, 163.
13. Ibid., p. 221.
14. Ibid., pp. 398–439.

How influential was sea power in bringing the war to a successful conclusion, assuming that a truce after 3 years of fighting that left both sides in control of about the same territories as they occupied at the beginning of the war can be considered a success? It is difficult to answer the question for a number of reasons. First, the United States was unopposed by either a North Korean or Chinese navy; hence, the ability of the American forces to control the strategic lines of communication that supplied the South Korean and U.S. armies might have taken place without benefit of naval forces. Second, the initial failures of the U.S. forces to invade North Korean ports due to the presence of mines can be considered a victory by the North Koreans, which was achieved without a genuine navy. Finally, the siege of Wonsan, which completely cut off supplies of arms, ammunition, and food to North Korea's important port, might have had a very limited effect on the conduct of the war, since resupply took place overland from China throughout most of the war. On the other hand, when North Korea invaded the South on 25 June 1950, there were no U.S. troops in Korea. "With so few combat forces initially available, control of the seas (taken for granted as is too often the case) was a prerequisite in implementing the United Nations decision to resist aggression against the Republic of Korea. Without the capability to use the seas, the decision to intervene on a rocky peninsula half a world away would have been meaningless and unenforceable. With control of the seas the decision was sound and reasonable."[15]

IS ANOTHER KOREAN WAR POSSIBLE?

A Scenario

There are hopeful signs that ideological struggles between countries espousing communism and those following a free-enterprise, democratic system of economy and government will not take place in the future. Relations between the Soviet Union and the United States have improved dramatically in recent years, and despite events in China during June 1989, Chinese-American relations are still relatively good. However, the possibility of peaceful reunification on the Korean peninsula remains uncertain.

Armed conflict in Korea is unlikely in the near future. There are still 43,000 U.S. troops in Korea, many on the "front line," and an attack by North Korea would almost certainly involve U.S. forces, which would not be limited to army units but would involve air and sea power as well. A sensible strategy for North Korea would be to wait for the withdrawal of U.S. forces from South Korea, which will likely occur in the next few years if East-West rela-

15. Arleigh Burke, then U.S. Chief of Naval Operations, 1 May 1957. Quoted in ibid., preface.

tions continue to be relatively cordial. Moreover, it does not seem probable that the Soviet Union or China would support North Korean efforts to unify the Koreas by force at a time when there is genuine effort to improve relations with both western Europe and the United States. The South Korean navy no doubt appreciates the combination of air and sea power, such as was demonstrated during the 1950–53 Korean War, when U.S. and British aircraft carriers were "decisive in defeating the North Korean aggressor." According to at least one South Korean scholar, however, "the most likely scenario in resumption of the Korean War is that both U.S. and USSR navies might stay out of Korean waters to avoid any risky direct confrontation."[16]

If any fighting were to take place it would be between the two Koreas "going it alone," without superpower allies. It may be, of course, that North Korea would still receive supplies and continue trade with both the USSR and China, and trade between South Korea and Japan, the United States, and other trading partners would continue. Continuing trade would provide armaments, food, and other necessities; hence, armed conflict between the two Koreas might last a relatively long time. Fighting would be carried out primarily on land, with sea power serving to support ground forces as well as to carry out the classic functions of navies: to defend coasts against attack, project force from sea to land, control trade routes, blockade ports, bombard key enemy installations, and provide gunfire support to army units engaged with the enemy.

The war will be "conventional," since it is reasonable to assume that the superpowers would intervene to prevent a serious escalation of the conflict that would involve nuclear weapons.

A war between North and South Korea might begin with a deliberate invasion across the demilitarized zone (DMZ) between the two countries. A decision by North Korea to invade South Korea appears to be unlikely as long as U.S. Army troops continue to occupy frontline positions in support of the South Korean army. An attack across the DMZ by South Korea is equally unlikely in the near future, since consultation with the United States would be required under the terms of the defense agreement between the two countries. But war might begin with seemingly unimportant provocations at sea.

South Korean fishing boats in hot pursuit of fish might stray into the waters surrounding the northwest islands and be attacked by patrol vessels of the North Korean navy. Paengnyong-do lies much closer to North Korea than to South Korea; it is 176 km from Inchon and only 11 km from Wolnae-do in North Korea. Hence, intrusions into its waters might be particularly

---

16. Kang Young-O, "Surface Force Structure—Need for Korea," paper presented at the International Conference on the Regime of the Yellow Sea: Issues and Policy Options for Cooperation in the Changing Environment, SLOC Study Group—Korea, 19–20 June 1989, Seoul, Korea.

upsetting to North Korean forces if tensions were high for other reasons. The northwest islands have South Korean military garrisons on them, which must be supplied by South Korean ships. An attack by the North on one of these ships could escalate into a full-scale war.

The first military objective in such a case might be the capture of the five northwest islands by the North. The islands have long been a thorn in the side of North Korea. They are far from the mainland of South Korea and are thus ideal for South Korean surveillance of North Korean military activities. Moreover, they can be used as bases by the South Korean navy to interdict traffic from North Korea across the Yellow Sea to China. The islands themselves are small and of no great value, but they are symbolically important to South Korea as the northernmost outpost of national territory. Loss of them would be a serious blow to military morale as well as to the South Korean civilian populace living on these minuscule bits of land.

North Korea's military zones in both the Yellow Sea and the Sea of Japan can be viewed as a factor in maintaining stability on the Korean peninsula, since they reinforce the notion that the defensively oriented South Korean navy has no reason to operate outside of the country's own waters. But an inadvertent entry into one of the zones without prior approval could result in an attack on the offending vessel by one or more of the numerous fast attack craft and patrol vessels in the North Korean navy. Such an attack would certainly be protested verbally by the South Korean government and, if relations between the two countries were in a particularly sensitive state, might be countered by the South Korean navy.

Strategic Geography

The Korean peninsula separates the Yellow Sea from the Sea of Japan. Consequently, both North and South Korea have coasts on two bodies of water, and their coastal defense problems are considerable. South Korea has the less favored geography, since it is an "insular" nation from the standpoint of its dependence on the sea for the carriage of ammunition, supplies of all kinds that would be needed for the conduct of a war, and most of its food supplies. Its only land border is with North Korea, with which it is technically still at war and with which it has no diplomatic or trade relations. North Korea is somewhat better off, as the experience of the 1950–53 war demonstrated. Supplies could be transported over the common border between China and over a short riverine boundary with the USSR; both countries might be allies, albeit noncombatant, in the event of armed conflict or, at the least, might be sympathetic with the North.

A blockade of key North Korean ports, such as Nampo on the Yellow Sea and Wonsan on the Sea of Japan, could make the situation more difficult for the North, but an effective blockade of South Korean ports would make the conduct of a long war virtually impossible for the South.

Both Koreas require a "two-ocean navy" for optimum use of sea power, but neither country has enough ships to defend two coasts and project power in two seas at the same time. South Korea is better off, since the transfer of ships from the Yellow Sea to the Sea of Japan and vice versa can probably take place without much interference from North Korean surface forces. On the other hand, attempts by North Korea to move ships from one sea to the other would require that they run a gauntlet and be subject to attack by the South Korean navy along a long sea route. Both North and South Korea can overcome the obstacles of unfavorable geography by building larger, more capable navies. But neither country shows signs of becoming a sea power with more than regional force projection capabilities.

The Korean Navies

The naval forces of North and South Korea are compared in table 1. It is obvious from the numbers and types of ships available to the two countries that neither aspires to be a true sea power, capable of projecting force oceanwide. Neither fleet has aircraft carriers, battleships, cruisers, or nuclear submarines. The 21 submarines in the North Korean navy and the 5 available to South Korea are diesel-powered types of modest capabilities. In a hierarchy of Third World navies both navies qualify in the Adjacent Force Projection category, defined as having "impressive territorial defense capabilities and some ability to project force well offshore (beyond the EEZ)."[17] The capability of defending their own territories and at the same time projecting force outside the EEZs makes both navies forces to be reckoned with. Each navy seems to be configured to combat the other, and both are designed for the primary mission of fighting another war on the Korean peninsula.

The strength of the North Korean navy is clearly in its submarine and fast attack craft.[18] The submarines could be used defensively to prevent South Korean amphibious forces from invading North Korean territory. More likely, however, they would be employed to interdict shipping approaching

17. See Michael A. Morris, *Expansion of Third-World Navies* (New York: St. Martin's Press, 1987), p. 25.

Editors' Note.—See also Joseph R. Morgan, "Small Navies," *Ocean Yearbook 6,* ed. Elisabeth Mann Borgese and Norton Ginsburg (Chicago: University of Chicago Press, 1986), pp. 362–89; and two articles in this volume, Michael A. Morris, "Comparing Third World Navies," and Frank Barnaby, "The Role of the Submarine."

18. Editors' Note.—See also the article by Frank Barnaby, "The Role of the Submarine," in this volume. Note that the figures for submarines in the navies of North and South Korea are slightly different. This is because the sources used for the navy inventories, though both highly reputable, cannot provide completely current information due to the necessary production time for each publication. In a time of great flux, when navies are both increasing and decreasing rapidly, published tables cannot reflect sudden changes in numbers of vessels and produce one definitive number, although both these sources are accurate within a few months.

TABLE 1.—NAVAL FORCES OF NORTH AND SOUTH KOREA

	North Korea	South Korea
Submarines	21	5 (1)
Destroyers	—	9 (8)
Frigates	3	7
Corvettes	—	24 (2)
Fast attack craft (M)	32	11
Fast attack craft (T)	168	—
Fast attack craft (G)	157	—
Fast attack craft (P)	—	68
Patrol craft	40	—
Landing ships (tank)	—	8
Amphibious craft	141	—
Minesweepers	23	8

SOURCE.—*Jane's Fighting Ships 1989–90* (London: Jane's, 1989).
NOTE.—(M) missile; (T) torpedo; (G) gun; (P) patrol; number in parentheses indicates planned or under construction.
EDITOR'S NOTE.—Note that the figures given in this table are not exactly the same as those given in tables 1 and 2 in the article by Frank Barnaby, "The Role of the Submarine." Minor discrepancies such as these occur due to the time lapse between compilation and publication of the source of the figures. Such delays cannot be avoided, but the figures provided by both sources (*Jane's Fighting Ships 1989–90* and *The Military Balance 1989–90*) are the leading sources of such information and are both equally correct.

South Korean ports, thereby preventing the South from resupplying needed armaments, equipment, fuel, and food. A dozen or more submarines—allowing for a number to be in maintenance or repair status—stationed around the coast of South Korea could seriously interfere with strategic lines of communication, vital to the South because of its islandlike character.

Fast attack craft, with their limited effective range and sea-keeping qualities, are generally considered defensive in nature. They would be used as coastal defense forces, ready to attack approaching enemy ships. They are also ideal for short-range raids on South Korean territory near the DMZ, which serves as the border between the two countries.

South Korea emphasizes surface forces of moderate size—destroyers, frigates, and corvettes. The nine destroyers (with eight more under construction or planned) can be used as antisubmarine vessels, to support amphibious forces, and to bombard an enemy coast. Some might also be used to escort merchant ships, but more likely the smaller, less powerful frigates and corvettes would be used for that purpose.

Comparing the amphibious warfare capability of the two navies indicates that North Korea, with large numbers of amphibious craft, might intend to use them on frequent small-scale raids on the South, which although largely of a nuisance character, might nevertheless serve to wear down the South Korean army and tie it down in defensive actions. The South, on the other hand, with eight large landing ships (LSTs, etc.), might mount one or more amphibious landings of fairly large scale. These would be supported by a preinvasion bombardment by the destroyers.

The minesweeping capability of the two navies can be compared directly. The fact that the North Korean minesweepers outnumber the South's by 23 to 8 indicates both a clear superiority in strength and the increased importance North Korea attaches to the threat of mines. South Korea seems to have forgotten the mine warfare lessons of the 1950–53 war, while the North remembers how effective their own defensive mining campaign was in preventing or delaying invasion of their territory.

Only a few comments concerning the actual combat capabilities of the two navies can be made, since factors such as morale, training, and maintenance are largely unknown. Some things are clear, however. Most of the North Korean submarines are of the Soviet Romeo class, and a new design is needed.[19] Only 32 of the very numerous fast attack craft consist of missile-equipped vessels. The great majority are armed with torpedoes and guns and can be judged to be less effective.

South Korea's destroyers are mostly of the obsolete U.S. Gearing and Summer classes and, despite having been backfitted with Harpoon guided missiles, are reaching the end of their useful life. Hence, the eight destroyers proposed for construction will not increase the combat capability of the navy greatly, since they will serve merely to replace the older vessels, albeit with improved capability. Of the seven frigates, three are of the modern Korean Ulsan class; the others are older U.S. designs of presumably lesser effectiveness. On the other hand, three of the five South Korean submarines are recent additions to the fleet and are of a modern design. Nevertheless, the small number of submarines, no matter how capable each one is, is a limiting factor in the effectiveness of the combat fleet.

On balance, it appears that North Korea has naval superiority over the South, at least in quantitative terms. There is no doubt that South Korea has the more robust economy, however, and its increasing national wealth enables it to build a bigger, more sophisticated navy. The relative poverty of North Korea makes a buildup of naval force without the assistance of China or the Soviet Union very improbable. The recent increase in numbers of submarines in the South Korean fleet and the planned replacement of the aging destroyer force with new designs are indicative of the trend toward naval force parity. There is no indication of a naval arms race between the two Koreas, since available information does not indicate any plans of the North for either quantitative or qualitative improvement of its navy.

Twenty-one diesel submarines, many of obsolescent design, and nine old destroyers, albeit modernized to some extent with guided missiles, compose the bulk of the sea power available to North and South Korea, respectively. Both navies may serve an essential coastal defense function, however, and in the event of a stalemate between opposing air and ground forces, sea power could be a deciding factor.

19. P. D. Jones and J. V. P. Goldrick, "Far Eastern Navies," *Proceedings* (U.S. Naval Institute) 113/3/1009 (March 1987): 66.

CONCLUSION

The following statement sums up a prevailing view concerning the unification of Korea. It was written by a Korean, but non-Koreans might very well agree with it.

> A divided Korea is an economic and political anomaly which cannot happily survive. The Korean people are one people, historically, ethnically and culturally. The present two halves of the divided nation complement each other, and Koreans of both halves would like to share a unified government. The division of Korea stands in the way of a truly permanent peace in the Far East. Almost all Koreans understand these basic facts, and whatever their other differences, all are united in the firm conviction that Korea should be unified. In an era of negotiation and political elasticity, the United States and the Soviet Union ought to be working toward overdue talks on Korean unification. However, history has taught the Koreans that it must be largely up to their own people, of both North and South, to achieve the goal of uniting their fatherland.[20]

Recent events and changes in thinking in both the United States and the Soviet Union might foretell the end of the long-standing communist versus capitalist rivalry and, with it, the end of ideological confrontations that have spread outside the specific concerns of the superpowers to a number of other countries. If East and West Germany can reunite, why not North and South Korea?

Peaceful reunification during the 1990s is entirely possible. At present, however, neither Korea has sufficient trust in the motives of the other to believe that the reunification will take place without some form of military confrontation. Hence, although both Koreas desire peace, both are maintaining a state of preparedness.

Any future armed confrontation would almost certainly be fought by opposing Korean forces with no direct support from powerful allies. Sea power would be an important factor, and the two Korean navies would weigh heavily in determining the outcome.

Once reunification has been achieved, hopefully by peaceful means, the united Korean navy will be in a position to undertake other missions. A navy configured for defense, monitoring and surveillance of the EEZ, disaster relief, search and rescue, piracy and drug-traffic abatement, and marine scientific research would be appropriate.

20. Yushin Yoo (n. 5 above), p. 99.

Comparing Third World Navies

Michael A. Morris*
Clemson University

INTRODUCTION

Third World navies have been expanding in three interrelated dimensions: national ocean zones, naval roles, and naval capabilities. In the case of some of the larger developing states, the expansion of naval roles and capabilities has even projected beyond the extensive national ocean zones. Though naval expansion has been uneven in the Third World, it has generally transformed what was often a neglected armed service, the navy, and a neglected aspect of national development, ocean resources, into key dimensions of domestic and foreign policy. Third World navies have accordingly tended to gain in importance in recent years, domestically with respect to national security and resource policies and internationally with respect to neighbors and to the traditional maritime powers.[1]

HISTORICAL BACKGROUND

The disintegration of colonial empires and the resultant decolonization after the Second World War produced a stream of newly independent Third World states over the ensuing decades. At the end of the Second World War

---

*EDITORS' NOTE.—Michael Morris has previously written articles for *Ocean Yearbook* on the ocean policies and activities of the South American states and an article on the military aspects of the exclusive economic zone.

1. In a recent book, I developed and applied a methodology—the Third World naval hierarchy—for analyzing the implications of the spread of maritime weapons systems in the Third World (*Expansion of Third-World Navies* [London: Macmillan and New York: St. Martin's Press, 1987]). This article presents an overview of some prominent trends affecting Third World navies, in order to assess the continuing viability of the Third World naval hierarchy. Editors' note.—See also Michael Mac-Gwire, "The Horizontal Proliferation of Maritime Weapon Systems," in *Navies and Arms Control*, ed. G. H. Quester (New York: Praeger, 1980), pp. 155–68; Joseph R. Morgan, "Small Navies," in *Ocean Yearbook 6*, ed. Elisabeth Mann Borgese and Norton Ginsburg (Chicago: University of Chicago Press, 1986), pp. 362–89.

the independent Third World consisted of about 20 Latin American states and a smattering of other states in Africa, the Middle East, and Asia. Figure 1 documents the increase in independent Third World states and relates it to the increase in Third World states with navies. It is clear from figure 1 that most Third World states have regarded a navy as a necessary concomitant of national sovereignty, though not quite as urgent a task as the establishment of national militia. The gap between the two lines is explained largely by the absence of navies in landlocked and/or very small Third World states. Some landlocked states do have navies for patrolling lakes and rivers, and many very small countries support a navy, though it may be only token in size and capability. The remaining dependent territories are likely to establish navies on gaining independence, though they are all weak and most have very small land areas. A few of them would encompass large ocean zones.

Though numerous "token" Third World navies have not grown, expansion has been the general rule. The drive to establish navies and then to expand them appears to be universal, so that even small states, when they are able, support naval expansion. This applies particularly to the oil-rich states.

FACTORS CONTRIBUTING TO NAVAL EXPANSION

Historical trends and the extension of national zones seaward ("national enclosure") have contributed to the increase in number and maturity of Third World navies. Institutional growth of Third World navies has been an integral part of national growth, but national enclosure dramatized the need to defend offshore areas and thus spurred naval development. Third World countries that were poor in land-based resources naturally realized the potential importance of maritime resources, particularly as marine-resource technology improved. For smaller navies, the development of cheap yet potent weaponry reinforced the determination to build up naval power to protect extended national zones and resources.

Maritime rivalries and disputes with neighbors have also contributed to naval growth. Particularly in the Third World, the extension of national ocean zones has tended to compound traditional rivalries and led to disagreements over boundaries and resource allocation. The greatly expanded ocean zones have brought more states into maritime contiguity in more places than before. Extended zones need up to three kinds of boundary delimitations: the Territorial Sea, the exclusive economic zone (EEZ), and the Continental Shelf.

Third World naval expansion has been closely associated with relations with developed states. These states have been the major suppliers for Third World navies and have had a variety of motives of their own for stimulating arms sales. Sometimes the prime motive has been purely commercial, particularly in the case of developed states whose domestic markets are not suffi-

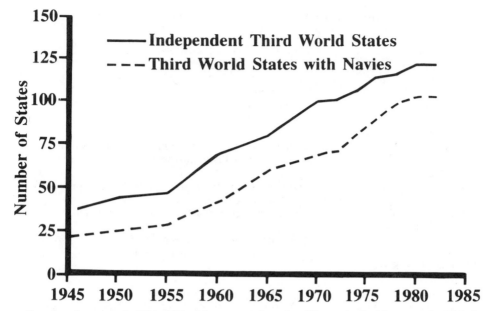

FIG. 1.—Increase in Third World states and navies. Chronological increase in Third World independent states was derived from successive issues of *Europa Yearbook*. Chronological increase in Third World states with navies was derived from successive issues of *Jane's Fighting Ships*. Third World states have qualified for listing in *Jane's* when some kind of formal naval organizational structure is established and when at least a modicum of naval vessels is acquired, such as several patrol craft. (Michael A. Morris, *Expansion of Third-World Navies* [London: Macmillan and New York: St. Martin's Press, 1987], p. 7.)

ciently large to support a viable arms industry. In other cases, political motives lead developed states to manipulate arms transfers either to assist development or to stunt it. In either case there is a tendency to accentuate the dependency of Third World navies, which are often already very vulnerable. In some important cases, Israel and Argentina for example, great-power pressure or arms embargoes have triggered the development of a domestic arms industry.

Domestic, regional, and extraregional factors have all contributed to Third World naval expansion. Nationalism has helped to fuse external and internal factors by propelling Third World states to claim large offshore zones as part of the national patrimony and to build up navies to protect these zones.

CHARACTERISTICS OF NAVAL EXPANSION

A standard scheme for classifying warships has been adopted in order to make it possible to describe Third World naval expansion through weaponry

TABLE 1.—WARSHIP CLASSIFICATION SYSTEM

	Characteristics
Major surface warships	
Aircraft carrier	
Cruiser	> 6,000 tons
Destroyer	3,000–6,000 tons
Frigate	1,100–3,000 tons
Corvette	500–1,100 tons
Submarines	Includes only conventionally powered submarines without ballistic missile capability
Light forces	
Fast attack craft	Includes only torpedo- or missile-equipped craft
Fast (or large) patrol craft	> 20-mm gun
Coastal patrol craft	< 20-mm gun, machine gun or unarmed
Amphibious forces	
Landing ship	According to *Jane's* designation LS
Landing craft	According to *Jane's* designation LC
Mine-warfare forces	
Minelayer	
Minesweeper, ocean	According to *Jane's* designation MSO
Minesweeper, coastal	According to *Jane's* designation MSC, includes inshore minesweepers
Supply ships	Includes support, oiler, repair, depot, collier, and tender ships, > 100 tons
Surveying vessels	Includes oceanographic research vessels
Other ships and vessels	Includes transport, tug, salvage, rescue ships, etc., > 100 tons

SOURCE.—Michael A. Morris, *Expansion of Third-World Navies* (London: Macmillan and New York: St. Martin's Press, 1987), p. 14.

comparisons (see table 1). While this classification system can be applied to all navies, it is especially appropriate to those of the Third World. It does not, for example, include nuclear submarines, which no Third World navy possessed until 1988, but does distinguish between different kinds of light forces, which are a major constituent of Third World navies.[2] In particular, it differentiates between fast attack craft (FAC), which though relatively small are quite potent, being armed with torpedoes or missiles, and other small craft such as fast patrol craft (PC) or coastal patrol craft ([PC]), which are much less potent. Figures 2 and 3 use this classification in showing the aggregate expansion of Third World navies by weaponry categories.

Several conclusions can be drawn from the data on aggregate growth of Third World navies. There is an overall pattern of growth over the entire

2. In early 1988 the Soviet Union transferred a nuclear submarine to India.

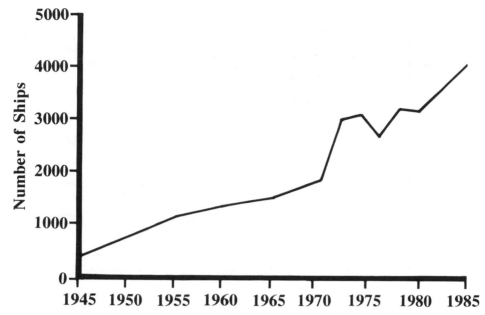

Fig. 2.—Expansion of Third World navies, all kinds of warship categories aggregated (excluding coastal patrol craft). (Michael A. Morris, *Expansion of Third-World Navies* [London: Macmillan and New York: St. Martin's Press, 1987], Appendix 2A and p. 15.)

postwar period, as newly independent states established navies and began to build them up. Growth was noticeably rapid during the early 1970s but declined briefly after the 1973–74 oil crisis, and expansion renewed in the 1980s (see fig. 2). The current level is quite high, and the rapidity of growth is noteworthy in view of the fact that small craft have been excluded from the measurements. Figure 3 elaborates on growth trends. It shows that an increase in the more expensive major warships has been less dramatic than an increase in FACs. All four categories listed have expanded considerably over the postwar period, though there has been a tendency for the number of major surface warships and submarines to stabilize in recent years.[3]

In addition to the generally adverse global economic conditions over the last decade, the escalating cost of modern warships has tended to constrain naval growth. In this, the naval development of the Third World parallels that of the developed states. All navies have been adversely affected by escalating warship costs, and desired growth targets have time and again had to be cut back. Continuing Third World naval expansion is particularly impres-

3. Editors' note.—For more information about this subject, see Frank Barnaby, "The Role of the Submarine," in this volume.

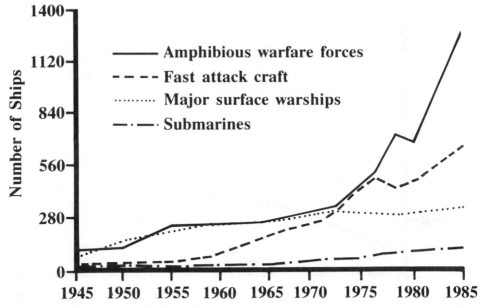

Fig. 3.—Expansion of Third World navies by selected weaponry categories. (Michael A. Morris, *Expansion of Third-World Navies* [London: Macmillan and New York: St. Martin's Press, 1987], Appendix 2A and p. 15.)

sive in light of these economic constraints, which have required considerable sacrifice from relatively poor developing states. By the late 1980s, there was an aggregate Third World tendency for naval expansion to level off, but, as will be shown, this has not led to a contraction or compacting of the Third World naval hierarchy.

Both horizontal and vertical proliferation are features of Third World naval expansion.[4] Vertical proliferation refers to the diffusion of advanced weapons technology, including improvements in such fundamental weapon characteristics as range, accuracy, payload, and systems reliability. It also covers enhanced capabilities of sensor and surveillance systems. Third World navies have generally benefited from these technological advances through the acquisition of new warships and associated weapon systems and the modernization of older vessels. Vertical proliferation brings its own problems in the form of rapidly escalating costs. Third World navies have participated in improvements in weapon technology pertaining to light forces, major warships, submarines, and naval aviation but have not acquired nuclear submarines or other nuclear-powered vessels, with the exception of the 1988 USSR nuclear submarine transferred to India.

Horizontal proliferation refers to the spread of new weapon technology to more and more states. Figure 1 shows that the rapid increase in the number

4. See Morris (n. 1 above).

of independent Third World states has been closely paralleled by the increase in Third World navies. The spread of new technologies to these navies has, not unnaturally, been uneven. Token navies have shared only minimally in the naval buildup, while other Third World navies have developed into respectable fighting forces, at least as regards their contiguous coastal areas.

THE ROLES OF THIRD WORLD NAVAL FORCES

Third World naval forces have three key roles: constabulary/regulatory, territorial (coastal) defense, and force projection at sea. Navies of developed states normally perform these roles too, but the Third World context gives them distinctive features. Third World navies are often much more involved in constabulary/regulatory duties than are the navies of developed countries. In the absence of a well-developed infrastructure, Third World navies often support communications, security, and coastal development, and perform a variety of riverine services and police functions as well. More conventional constabulary roles are also performed through patrolling offshore areas to assure observance of coastal-state regulations by nationals and foreigners. With national enclosure, this includes regulation and enforcement out to the 200-mile limit.

Constabulary responsibilities in the EEZ derive from the new resource and resource-related rights and include surveillance and enforcement for fisheries, offshore oil, and pollution control. Resource protection can pose difficult enforcement problems, for example when fishing grounds lie well offshore or in isolated areas. Offshore resources, however, tend to be concentrated geographically so that their protection is easier than enforcement of the whole EEZ. Moreover, offshore resource protection is greatly facilitated by the law. The EEZ consensus legitimizes the coastal state as the offshore policeman, while violators of EEZ resource rights risk being revealed as poachers. Accordingly, light forces often suffice for EEZ constabulary responsibilities although both maritime and air surveillance are required in appropriate combinations.

Inshore and offshore territorial (or coastal) defense is a more ambitious undertaking and requires more potent combat capabilities. Territorial defense functions may vary from fairly modest efforts to prevent use of the sea in inshore areas to more ambitious attempts to control the use of the sea in offshore areas. Requirements for major surface warships, submarines, and naval air capabilities will escalate accordingly.[5]

With the advent of the EEZ in particular, constabulary and inshore/

5. "Inshore and offshore territorial defense" is used rather than the more conventional term "coastal defense" because of the Third World tendency to territorialize enclosed areas or to assimilate the defense of such areas with that of domestic land areas.

offshore territorial defense roles often overlap. Offshore resource control is usually of some importance to the domestic economy and is also the most obvious symbol of intent to command foreign respect for new coastal states' EEZ rights. Many coastal states define resource protection more broadly than mere surveillance of valuable offshore oil wells or fishing grounds using light constabulary forces. Resource protection tends instead to be broadly defined as defense of EEZ resource and resource-related rights and is seen as involving substantial enforcement activity throughout the EEZ to the extent that national capabilities permit. EEZ control of resource rights is expansive and not easily distinguishable from overall EEZ control. Similarly, inshore/offshore territorial defense can become involved in EEZ constabulary duties. In the vast spaces of the EEZ, naval force may be required to back up light constabulary forces or to punish recurrent violators when the relatively limited coercive capabilities of patrol forces do not suffice. Since coastal state EEZ rights often overlap the rights of foreign states, naval power may be required to enforce ambitious attempts to control the entire EEZ.

The national enclosure movement, as advocated by Third World states, emphasizes coastal-state discretion in extended offshore zones when controlling hostile or prejudicial foreign activities. Third World nationalism has given added impetus to this drive to control ocean space out to the 200-mile limit. It is this expansive thrust that tends to fuse constabulary roles in the EEZ with inshore/offshore territorial defense activities.

The constabulary and territorial defense functions in turn tend to merge into the third naval role: force projection at sea. Since constabulary responsibilities are recognized in law out to the 200-mile limit and since these legitimize some coastal defense functions, force projection at sea in this new context refers most appropriately to naval operations beyond the EEZ. Naval forces tasked with naval defense out to 200 miles have at least some potential for deep-water roles beyond the 200-mile limit. Any Third World navy that aspires to national control of the EEZ will be inclined to establish a secure perimeter or buffer beyond the EEZ in order to improve control of the national zone. Thus the roles within and beyond the EEZ are linked in Third World security planning. Recurring proposals for peace zones in Third World ocean basins have received widespread endorsement from coastal states.[6] Such zones would exclude great-power navies while permitting littoral navies to carry out operations in adjacent national and international zones.

In areas adjacent to national zones, force projection at sea may therefore be considered as a related, additional step in the building up of a navy that purports to control national ocean space out to 200 miles. Third World navies

6. Editors' note.—See article by Stanley D. Brunn and Gerald L. Ingalls, "Voting Patterns in the UN General Assembly on Uses of the Seas," in *Ocean Yearbook 7*, ed. Elisabeth Mann Borgese, Norton Ginsburg, and Joseph R. Morgan (Chicago: University of Chicago Press, 1988), pp. 42–64, which reviews the UN General Assembly's voting actions regarding, inter alia, establishing the Indian Ocean as a zone of peace.

do not yet aspire to distant force projection at sea. The progressively more demanding roles for Third World navies that have been outlined encourage naval expansion as each new role makes new demands in terms of weaponry requirements. Each successive role establishes the national presence at sea in more impressive, and geographically more extensive, terms.

THE IMPACT OF THIRD WORLD NAVAL EXPANSION

The expansion of Third World navies has had a significant impact on national military establishments, on management of offshore resources and defense of national ocean zones, on neighbors, and on maritime powers. Attention has tended to focus on one dimension or another of this multifaceted expansion and to derive implications from it. Such limited perspectives have tended to lead to incomplete, and at times contradictory, assessments of implications based on overly selective evidence. Broadly viewed, however, two schools emerge—one of which attributes to naval growth very much greater regional and global significance than does the other.

Some commentators have dramatized the impact of the Third World naval build-up. They maintain that technological innovations have called the whole concept of naval superiority into question. The ship-launched missile, in particular, will be the great equalizer by dramatically increasing the power of small states. Already the "explosive growth of Third World naval power" has produced "subtle shifts in the world's balance of naval power," and "as small, affordable anti-aircraft and anti-submarine systems join these navies, the world's balance of naval power may rest in the Third World.[7] On the basis of such alleged trends, others have concluded that the proliferation of modern weaponry points toward the demise of naval diplomacy, with presence forces becoming a "mere folkloristic manifestation."[8]

Contrasting conclusions have been argued with equal force. For example, numerous commentators have stressed that modern weaponry alone will not suffice to produce effective Third World navies. Even though some modern naval weaponry is potentially impressive, Third World states are still plagued by pervasive poverty and limited expertise. Such disadvantages of underdevelopment will severely handicap their use of complex weaponry in demanding combat conditions.[9] The implications for the great powers vary accordingly. Extended national zones and modern weaponry in Third World hands will allegedly lead the maritime powers to select times and places for

7. R. L. Chambers, "Third World Navies: New Weight in the Balance of Power," *U.S. Naval Institute Proceedings* (August 1979): 117–19.

8. E. Young, "New Laws for Old Navies: Military Implications of the Law of the Sea," *Survival* (November/December 1984): 262.

9. L. Dowdy, "Third World Navies: New Responsibilities, Old Problems," *Marine Policy* (April 1981): 147–48.

intervention with greater care. However, more selective and decisive interventions will be more salient and will lead to a "revival of naval diplomacy."[10]

From a historical perspective, even technological change may not be as great as it might at first seem. A recent survey of modern warship innovations concluded that "probably the greatest surprise of naval development since 1946 is the relative absence of radical change."[11] The buildup of Third World navies with modern weaponry is a new development, but from this perspective it is not likely to alter fundamental relationships among people, weapons, and the sea, or among large and small powers.

On some basic matters there is agreement between the two schools of thought, and certain differences can be attributed to a difference of emphasis in interpreting the same facts. All agree that the acquisition of modern weaponry by Third World navies has enhanced defensive capabilities and made maritime powers more hesitant to interfere or intervene. All agree that the efficiency of Third World navies will be impaired by domestic problems related to underdevelopment. There is also agreement that the maritime powers will continue to exert naval coercion on Third World states, as in the past, when they regard vital interests as threatened.

A basic difference in orientation does, nevertheless, separate the two schools. The first basically regards the expansion of Third World navies as heralding a fundamental shift in international relations. The second regards this naval expansion as a significant development, but a development unlikely to alter fundamental power relationships. The debate, conducted in generalities with reliance on highly selective evidence, appears to have reached an impasse.

This study aspires to a more specific and systematic assessment of the impact of Third World naval expansion. The military capabilities of individual Third World navies, in particular, need to be related to one another and then to be considered in specific contexts. Systematic comparison of Third World navies in specific contexts in turn provides a firm basis for deriving discriminating conclusions and generalizations. The Third World naval hierarchy offers just such a balance of specificity and generality.

THE HIERARCHY OF NAVAL CAPABILITY

General patterns in the structure and setting of more than 100 Third World navies may be studied by ranking these navies within a hierarchy divided

10. Ken Booth, "The Military Implications of the Changing Law of the Sea," in *Law of the Sea: Neglected Issues,* ed. J. K. Gamble, Jr. (Honolulu: Law of the Sea Institute, University of Hawaii, 1979), pp. 372–76.

11. N. Friedman, *Modern Warship: Design and Development* (London: Conway Maritime Press, 1979), p. 187.

into six groups. Each Third World navy has, of course, its peculiarities, but important patterns nevertheless emerge from the application of the hierarchical model.

This study has had four successive classification stages, taking into account both quantitative and qualitative considerations in order to develop a naval hierarchy for the Third World. An initial ranking was made at the first stage. There were some reclassifications of individual navies at each of the subsequent stages. The initial classification was made on the basis of quantitative criteria relating to weapons, and this was refined at the second stage by the use of qualitative criteria. The two final stages introduced additional criteria related to the land-based and sea-based support of the fleet. The third stage relied on broad indicators of naval power, and the fourth dealt with national power-base indicators. For a navy to merit high rank, both the quantity and quality of weaponry must be impressive (stages 1 and 2). Naval power and national power (stages 3 and 4) are also required to sustain an effective fleet.

Each stage thus includes distinctive yet complementary criteria, and the overall classification process provides a dynamic, multifaceted profile of Third World navies. The reliance on multiple criteria in successive stages of classification helps to correct possible distortions in the model. Navies cannot achieve a high ranking merely by excelling in a single criterion. Neither can an "off-the-shelf" navy with impressive weaponry yet little expertise or backup. Such navies that are weak in one particular area would not, on the other hand, be unreasonably handicapped.

The initial comparison (stage 1) was crudely quantitative and was based on the number of different key kinds of fighting ships and supply vessels. It included all naval vessels except patrol boats, surveying vessels, and such vessels as were included within a classification of "other ships and vessels." Worksheets were sorted in accordance with these criteria in order to distinguish stronger from weaker navies. This initial comparison revealed various patterns or clusters of warship distribution in terms of sea-keeping and geographical reach. Simple but systematic quantitative criteria characteristic of each cluster were derived and are shown under "Naval/Naval Aviation Structure" in table 2. The table expresses these criteria in terms of naval capabilities. These quantitative criteria refined the initial sorting of the worksheets and led to a tentative ranking of all 104 Third World navies.

The functionally oriented names designating each rank synthesize the respective equipment characteristics. The functions characteristic of navies in each rank broadly correspond to the three roles of Third World navies: constabulary roles, territorial defense (inshore and offshore), and force projection at sea (adjacent and distant). It was shown above that these roles are progressively more ambitious. Navies in successively higher ranks are able to undertake not only more ambitious roles but also a greater number of increasingly ambitious roles.

TABLE 2.—EQUIPMENT CHARACTERISTICS BY NAVAL RANK (1980)

Rank	Category of Third World Navies	Naval/Naval Aviation Structure	Naval Capabilities	States in Each Rank (Alphabetical Order)
6	Regional force projection navies	All Third World naval and naval aviation equipment categories strongly represented. More than 15 major warships and/or submarines.	Impressive territorial defense capabilities and some ability to project force in the adjoining ocean basin.	Argentina, Brazil, India
5	Adjacent force projection navies	Most Third World naval and naval aviation equipment categories well represented. More than 15 major warships and/or submarines.	Impressive territorial defense capabilities and some ability to project force well offshore (beyond the EEZ).	Chile; Iran; Korea, Dem P Rep; Korea, Rep; Peru
4	Offshore territorial defense navies	Quite a few Third World naval and naval aviation equipment categories well represented, including some larger units at upper levels. 6–15 major warships and/or submarines.	Considerable offshore territorial defense capabilities up to EEZ limits.	Colombia, Egypt, Indonesia, Libya, Mexico, Pakistan, Philippines, Taiwan, Thailand, Venezuela

3	Inshore territorial defense navies	Third World naval and naval aviation equipment categories moderately represented at lower levels and only sparsely represented at upper levels, if at all. 1–5 major warships and/or submarines.	Primarily inshore territorial defense with limited offshore defense capability.	Bangladesh, Cuba, Dominican Rep, Ecuador, Ethiopia, Ghana, Malaysia, Myanmar, Nigeria, Syria, Uruguay, Vietnam
2	Constabulary navies	Sparse representation of Third World naval equipment categories at lower levels only. Naval aviation minimal or nonexistent. No major warships, but fast attack craft.	Some ability to prevent use of coastal waters, with concentration on constabulary functions.	Algeria; Gabon; Guinea; Guinea-Bissau; Iraq; Oman; Saudi Arabia; Singapore; Somalia; Tanzania; Yemen Arab Rep; Yemen; P Dem Rep
1	Token navies	Only minimal representation at lower levels of Third World naval equipment categories. No fast attack craft; only patrol craft and/or landing craft. Naval aviation nonexistent.	Unable even to patrol national Territorial Seas effectively. Impotent in the EEZ.	62 navies, listed in table 3

Source.—Michael A. Morris, *Expansion of Third-World Navies* (London: Macmillan and New York: St. Martin's Press, 1987), pp. 25–26.

Third World states with rank 1 navies are unable to perform any significant naval roles and are consequently denominated as "token navies."[12] Rank 2 navies, the "constabulary navies," are able to perform constabulary missions and little else. Rank 3 navies have some impressive weaponry but are largely limited to inshore defense, the "inshore territorial defense navies." Rank 4 navies are better equipped for a defense role well offshore and are the "offshore territorial defense navies." Rank 5 and 6 navies, in addition to having constabulary and territorial defense capabilities, are able to project force at sea in varying degrees. Rank 5 navies are able to project force beyond the EEZ, the "adjacent force projection navies," while rank 6 navies have greater ability to project force in the adjoining ocean basin, the "regional force projection navies."

Table 3 presents the final results of the four-stage naval classification. It shows the hierarchy for 1980 of 104 Third World navies in six ranks according to geographical region. It will be noted that there are specific weaponry increments for each ascending rank (table 2). Each new rank embodies the strengths of the rank below plus at least one important weaponry increment. Rank 1 is an exception in that the navies so classified are only "token" navies. Table 4 updates the hierarchy to 1986.

Several studies have carried the Third World naval hierarchy up to the present. For example, the South American naval hierarchy identified in the book *Expansion of Third-World Navies* has remained stable up to 1990, according to a former director and researcher at the University of Chile's graduate international relations faculty (Instituto de Estudios Internacionales).[13] I have just completed a reassessment of the earlier Caribbean/Central American naval hierarchy from the same book and have reached a similar conclusion about the continuing stability of its structure.[14]

OVERVIEW OF THE THIRD WORLD NAVAL HIERARCHY

The development and implementation of the Third World naval hierarchy makes possible certain broad conclusions. It becomes evident that Third World navies vary greatly in capabilities and potential. Naval expansion in the Third World has been very uneven—62 navies in the bottom rank with mere token capabilities and only 3 navies in the top rank. Thirty-nine navies fill the other four ranks.

12. Third World states with no navies and hence no naval functions were given a zero rating and don't appear in the tables.
13. Pilar Armanet, *Políticas Marítimas Sudamericanas*, Documento de Estudio no. 6 (Santiago, Chile: Comisión Sudamericana de Paz, 1990).
14. Michael A. Morris, "Caribbean Navies and Coast Guards Compared," a paper presented at the 12th Annual Conference of the Association of Caribbean Studies, Santo Domingo, Dominican Republic, 29–31 July 1990.

Variations between individual navies within the same rank are also considerable. The position of a navy within a given rank helps to indicate whether it is rising or stable. Algeria, Iraq, and Saudi Arabia, for example, were at the top of rank 2 in 1980, and all of them have undertaken significant programs of naval expansion, which have propelled them to a higher rank. There are other indicators that help to identify those navies that are likely to rise in rank. A rising navy will share many of the characteristics of its present rank but will have a stronger, more diversified performance in a number of indicators than is characteristic of other navies in the same rank. The Nigerian navy, for example, appears likely to rise above its current rank 4 status, and this is reflected in a growing budgetary commitment, which is now larger than that characteristic of its current rank. A relatively high level of Nigerian naval expenditure is reflected in a variety of other indicators such as number, kind, and age/modernization of vessels, and this will probably result in upward mobility in due course.

Each rank is distinct, and variations within ranks are sufficient to allow hierarchical ordering within the rank. Nevertheless, there are some important overlaps between ranks. Navies in ranks 3 and 4, for example, emphasize territorial defense in varying degrees, while navies in ranks 5 and 6 have force-projection capabilities, also in varying degrees. Navies in ranks 3 to 6 all possess conventional naval combat capabilities and may be designated "upper-rank navies" to distinguish them from the lower-rank navies, which do not have these combat capabilities (ranks 1 and 2). Even among the lower-rank navies, rank 1 navies stand apart in only having token naval capabilities, while rank 2 navies are able to perform constabulary functions.

It is not very difficult to qualify as a rank 1 or rank 2 navy in view of the limited budgetary and weaponry requirements. There was a rapid increase in lower-rank navies during the 1970s so that in 1980 there were 74 navies in ranks 1 and 2, contrasting with the 30 upper-rank navies. Rank 1 navies are more symbolic than indicative of a commitment to naval development. There is, however, a key threshold separating the navies of rank 2 from those of the bottom rank. Rank 1 navies are usually navies only in name, and states with rank 1 navies generally either do not aspire to or are so poor as to be unable to achieve a rise in naval status. To qualify for rank 2 status would entail a far-reaching political decision to transform a token naval establishment into a real navy with appropriate equipment, training, and infrastructure. The current rank 2 navies, in contrast, can already perform constabulary functions. Many aspire to higher status, and some can probably achieve this in due course.

The threshold between ranks 2 and 3 is again substantial. For a rank 2 navy to rise to rank 3 there must be considerably greater budgetary and weaponry commitments, since the higher rank requires some inshore territorial defense capabilities. Between 1978 and 1980 there were no new entrants into the upper ranks, and by 1986 only three additional navies managed to

TABLE 3.—THIRD WORLD NAVAL HIERARCHY BY REGION (1980)

Rank	Category of Navies	South America	Central America and Caribbean	South Atlantic	Africa	Mediterranean	Middle East	Indian Ocean	Far East	Totals by Category	
										Number	Percentage of Countries
6	Regional force projection navies	Brazil Argentina		(Brazil) (Argentina)				[Australia] India	[Australia]	3	2.9
5	Adjacent force projection navies	Peru Chile					Iran	(Iran)	Korea, Dem P Rep Korea, Rep.	5	4.8
4	Offshore territorial defense navies	Venezuela Colombia	(Venezuela) (Colombia) Mexico	(Venezuela)	Egypt Libya	(Egypt) (Libya)	(Egypt) (Libya)	Pakistan (Indonesia) (Thailand)	Indonesia Thailand Taiwan Philippines	10	9.6
3	Inshore territorial defense navies	Ecuador Uruguay	Cuba Dominican Rep	[South Africa] (Nigeria) (Uruguay) (Ghana)	[South Africa] Nigeria Ethiopia Ghana	[Israel] (Syria)	[Israel] Syria	(Malaysia) (Ethiopia) Myanmar Bangladesh	[New Zealand] Vietnam Malaysia (Myanmar)	12	11.5
2	Constabulary navies			(Guinea-Bissau) (Gabon) (Guinea)	Algeria Somalia Guinea-Bissau Gabon Guinea Tanzania	(Algeria)	Iraq Saudi Arabia Yemen, P Dem Rep Oman Yemen Arab Rep	(Iraq) (Saudi Arabia) (Somalia) (Yemen, P Dem Rep) (Oman) (Yemen Arab Rep) (Tanzania)	Singapore	12	11.5

		Region columns								Total	%
1	Token navies	Suriname Guyana Bolivia[a] Paraguay[a]	Guatemala (Suriname) Bahamas Costa Rica Trinidad and Tobago Haiti El Salvador Panama Nicaragua Honduras Barbados (Guyana) Jamaica St Vincent St Lucia Grenada Belize St Kitts	(Mauritania) (Senegal) (Zaire) (Angola) (Suriname) (Ivory Coast) (Congo) (Cameroon) (Liberia) (Benin) (Guyana) (Cape Verde) (Sierra Leone) (Gambia) (Eq Guinea)	Morocco Mauritania Senegal Tunisia Zaire Angola Sudan Ivory Coast Congo Mozambique Kenya Cameroon Liberia Madagascar Benin Zanzibar Cape Verde Sierra Leone Gambia Togo Mauritius Djibouti Eq Guinea Comoros Malawi[a] Mali[a]	(Morocco) (Tunisia) (Lebanon)	Kuwait U Arab Emirates Qatar Lebanon Jordan Bahrain	(Kuwait) Sri Lanka (U Arab Emirates) (Qatar) (Bahrain) Mozambique) (Kenya) (Madagascar) (Mauritius) (Djibouti) Maldives Seychelles (Comoros)	Kampuchea Brunei Papua New Guinea Fiji Tonga Solomon Is Laos[a]	62	59.6
	Total	12	19 (23)	(24)	37	(7)	13 (15)	7 (29)	16 (17)	104	99.9

SOURCE.—Michael A. Morris, *Expansion of Third-World Navies* (London: Macmillan and New York: St. Martin's Press, 1987), pp. 34–35.

NOTE.—Both quantitative and qualitative considerations were taken into account in developing the hierarchy and are described in detail in this article. States are ranked hierarchically within as well as between ranks. States are listed without parentheses for their main region and within *parentheses* for regions they overlap. Regional totals are for regions they overlap, without parentheses for states in the key region and within parentheses to include overlapping states. Navies of four enclave developed states (Australia, Israel, New Zealand, and South Africa) are included in the hierarchy for illustrative purposes, because they are physically located amid developing states. While these enclave navies are positioned approximately with respect to the ranks of the hierarchy, they are not included in the regional totals. Their distinctive position is emphasized by placing them in *brackets*.

[a] Landlocked.

TABLE 4.—CHANGES IN RANK IN THE THIRD-WORLD NAVAL
HIERARCHY, 1980–85

Rank	Category of Navies	Rise in Rank	Decline in Rank
6	Regional force projection navies		
5	Adjacent force projection navies		
4	Offshore territorial defense navies		Iran
3	Inshore territorial defense navies	Algeria, Iraq Saudi Arabia	
2	Constabulary navies	Angola, Bahrain, Brunei, Ivory Coast, Kenya, Kuwait, Morocco, Nicaragua, Qatar, Tunisia, U Arab Emirates	
1	Token navies		

Source.—Michael A. Morris, *Expansion of Third-World Navies* (London: Macmillan and New York: St. Martin's Press, 1987), p. 37.

Note.—See table 3 for 1980 rank. In all cases, substantial changes from 1980 to 1985 have either raised or lowered the navy in question by at least a full rank.

qualify. With foreign assistance a number of lower-rank navies could probably rise fairly quickly.

Greater stability and exclusivity are characteristics of the upper ranks. Entry to these ranks is difficult, and upward mobility within the upper ranks requires considerable effort, since the budgetary and weaponry increments for each ascending rank are very substantial. The cost of warships is high and rising. The lead time for acquiring major warships can stretch over a number of years, and the life cycle of major warships is several decades, during which time heavy maintenance costs must be borne. Greater diversity as well as a lengthy and intense commitment to naval power is necessary if status is to be maintained in the higher ranks, particularly ranks 5 and 6. Between 1978 and 1980 only Ecuador and Thailand among upper-rank navies changed status, and only by one rank in each case.

In spite of the sustained commitment of resources required, there are some 30 Third World states that maintain upper-rank navies. This figure is all the more impressive when it is remembered that many of these navies are of relatively recent origin. None of the upper-rank navies has suffered significant decline because of inability to maintain inventory levels previously achieved. Rare cases of naval decline, such as post-Sukarno Indonesia and post-Shah Iran, have been linked to domestic and foreign policy changes and do not seem likely to be protracted. Most states with upper-rank navies are in fact fairly large and viable entities, and most could probably sustain a rise in rank were a political decision made to this effect.

There are some significant regional variations, with the Far East and South America particularly well represented in the upper ranks and Africa and the Caribbean/Central America especially heavily represented in the lower ranks. Most regions are characterized by the existence of a small number of upper-rank navies and a much greater number of lower-rank navies.

Though the relative positions of individual navies are continually changing, such mobility tends to be within a particular rank. In the lower ranks, relative positions shift fairly readily, since only small increments of naval power are needed for a boost in status. Since moving from one rank to another requires a large and sustained commitment, the overall naval hierarchy can be considered fairly stable.

It lies in the interest of all Third World navies to reinforce this stability and to prevent unbridled expansion. The naval hierarchy, as delineated here, emphasizes the benefits of stability and restraint, and the elusiveness of an unending search for security through higher naval rank. To the extent that these insights are accepted, self-interest may slow the continuing expansion of Third World navies.

The Role of the Submarine

Frank Barnaby*
Military Technology Consultant

INTRODUCTION

The submarine has become the crucial element in many of the major navies of the world. Operated by 43 navies, it is the most numerous of the major warships (table 1). Strategic nuclear submarines are becoming the mainstay of the superpowers' strategic nuclear arsenals. For example, about 70% of the warheads carried by American strategic ballistic missiles are on submarine-launched ballistic missiles[1] (table 2). Consequently, much effort is being put into antisubmarine warfare activities. This paper will discuss the role of the submarine and its effects on nuclear policies.

EDITOR'S NOTE.—This article is the third by Frank Barnaby for *Ocean Yearbook* on the subject of naval activities. His earlier articles are "Superpower Military Activities in the World's Oceans," *Ocean Yearbook 5*, ed. Elisabeth Mann Borgese and Norton Ginsburg (Chicago: University of Chicago Press, 1985), pp. 223–39; and "Strategic Submarines and Antisubmarine Warfare," *Ocean Yearbook 1*, ed. Elisabeth Mann Borgese and Norton Ginsburg (Chicago: University of Chicago Press, 1978), pp. 376–79. For further information on naval forces, in particular submarine activities, see in this volume Michael A. Morris, "Comparing Third World Navies"; Joseph R. Morgan, "Naval Operations in Korean Waters"; and Peter Haydon, "Arms Control and Disarmament at Sea: What Are the Prospects?" See also Joseph R. Morgan, "Small Navies," *Ocean Yearbook 6*, ed. Elisabeth Mann Borgese and Norton Ginsburg (Chicago: University of Chicago Press, 1986), pp. 362–89; Kevin McCann, "The Soviet Navy: Structure and Purposes," *Ocean Yearbook 6*, ed. Elisabeth Mann Borgese and Norton Ginsburg (Chicago: University of Chicago Press, 1986), pp. 346–61; Andrzej Karkoszka, "Naval Forces," *Ocean Yearbook 2*, ed. Elisabeth Mann Borgese and Norton Ginsburg (Chicago: University of Chicago Press, 1980), pp. 199–225; Owen Wilkes, "Ocean-based Nuclear Deterrent Forces and Antisubmarine Warfare," *Ocean Yearbook 2*, ed. Elisabeth Mann Borgese and Norton Ginsburg (Chicago: University of Chicago Press, 1980), pp. 226–49; Ronald Huisken, "Naval Forces," *Ocean Yearbook 1*, ed. Elisabeth Mann Borgese and Norton Ginsburg (Chicago: University of Chicago Press, 1978), pp. 412–35; and SIPRI, "The ASW Problem: ASW Detection and Weapons Systems," *Ocean Yearbook 1*, ed. Elisabeth Mann Borgese and Norton Ginsburg (Chicago: University of Chicago Press, 1978), pp. 380–85.

1. International Institute of Strategic Studies, *The Military Balance 1989–1990* (London: International Institute of Strategic Studies [IISS], 1989).

MAJOR WARSHIPS IN THE WORLD'S NAVIES

The world's navies operate a total of about 2,100 major warships, including aircraft carriers, battleships, cruisers, destroyers, frigates, and submarines. According to figures published by the International Institute for Strategic Studies (IISS) for mid-1989, 28 of these warships are aircraft carriers; 4 are battleships; 86 are cruisers; 298 are destroyers; 755 are frigates; and 908 are submarines[2] (tables 1 and 3).

SUBMARINES IN THE WORLD'S NAVIES

The navies of 43 countries operate 908 submarines. Six countries—the United States, the USSR, the United Kingdom, France, China, and India— operate nuclear-powered submarines. These statistics bring home the importance of the submarine to the world's navies.

NUCLEAR POLICIES

Strategic nuclear submarines, with their load of ballistic missiles, play a crucial role in the policies of the nuclear-weapon powers. Improvements in the accuracy of delivery of warheads carried by submarine-launched ballistic missiles may dramatically change nuclear strategies. Nuclear strategies will also have to be reconsidered should strategic nuclear submarines become vulnerable. Both changes in nuclear strategies will increase the risk of nuclear world war.

The nuclear policies of the advanced nuclear-weapon powers, particularly the United States and the USSR, are already changing because of technological advances in land-based strategic nuclear weapons and their supporting technologies. The targets at which nuclear weapons are aimed are generally determined by the accuracy with which the weapons can be delivered. Inaccurate nuclear weapons are seen to be useful for nuclear deterrence, by threatening an enemy with unacceptable death and destruction; accurate nuclear weapons are seen as more useful for fighting a nuclear war than deterring one by assured destruction.

Nuclear war–fighting policies are, in turn, likely to evolve into nuclear war–winning policies. Each of these changes is likely to increase the risk of nuclear war even in periods of East-West détente.

Generally speaking, nuclear deterrence depends on the theory that, if one's potential enemy knows that they will suffer unacceptable damage if they attack you or that the attack is most likely to fail, they won't attack in

2. R. G. Purver, *Arms Control in the North,* National Security Series no. 5 (Kingston, Ontario: Center for International Relations, Queen's University, 1981).

TABLE 1.—MAJOR WARSHIPS IN THE WORLD'S NAVIES

Country	Aircraft Carriers	Battleships	Cruisers	Destroyers	Frigates	Submarines
Albania						2
Algeria					3	4
Argentina	1			6	7	4
Australia				3	9	6
Bangladesh					3	
Belgium					4	
Brazil	1			9	7	7
Bulgaria					3	4
Canada					19	3
Chile			1	8	2	4
China				19	37	93
Colombia					4	2
Cuba					3	3
Denmark						3
Ecuador				1	1	2
Egypt				1	5	10
Ethiopia					2	
France	2		2	4	36	21
German Dem Rep					19	
Germany, Fed Rep				7	7	24
Greece				14	7	10
India	2			5	21	17
Indonesia					15	2
Iran				1	5	
Iraq					5	
Israel						3
Italy	1		3	4	22	10
Japan				6	57	15
Korea, Dem P Rep					2	21
Korea, Rep				11	7	3
Libya					3	6

Malaysia				4	
Mexico			3	1	
Morocco					6
Netherlands			4	11	
New Zealand				4	
Nigeria				2	
Norway				5	12
Pakistan			7	10	6
Peru		2	8	4	11
Philippines				3	
Poland			1	1	4
Portugal				14	3
Romania				4	1
Saudi Arabia			1	8	
South Africa					3
Spain			4	14	
Sweden					8
Syria				3	11
Taiwan			26	10	3
Thailand				5	4
Tunisia				1	
Turkey			12	10	15
U.K.	2		13	34	31
Uruguay	4			2	
U.S.A.	14[a]	41[b]	68	102	133
USSR	4	37[c]	52	171	368
Venezuela				6	3
Vietnam				7	
Yugoslavia				4	5

SOURCE.—International Institute of Strategic Studies, *The Military Balance 1989–1990* (London: International Institute of Strategic Studies, 1989).

EDITORS' NOTE.—Note that the figures given in this table for the North and South Korean navies are not the same as those cited in this volume in Joseph Morgan's article, "Naval Operations in Korean Waters." Minor discrepancies such as these occur due to the time lapse between compilation and publication of the source of the figures. Such delays are unavoidable, but as both sources (*Jane's Fighting Ships 1989–1990* and *The Military Balance 1989–1990*) are equally authoritative, the figures provided are correct.

[a]Five U.S. Navy aircraft carriers are nuclear-powered: 4 Nimitz class and 1 Enterprise class.

[b]Nine U.S. Navy cruisers are nuclear-powered: 4 Virginia class, 2 California class, 1 Truxton class, 1 Long Beach class, and 1 Bainbridge class.

[c]Three USSR Navy cruisers are nuclear-powered: Kirov class.

TABLE 2.—WEAPONS AND DELIVERY SYSTEMS: SUBMARINE-LAUNCHED BALLISTIC MISSILES

Country	Type		Year First Deployed	Missiles	Missiles Deployed	Total Warheads	Yield per Warhead	Circular Error Probability
U.S.A.	Poseidon C-3		1971	10 MIRV	208	2080	50 kt	450
	Trident C-4		1979	8 MIRV	384	3072	100 kt	450
	Trident D-5		1990				475 kt	c. 120
Total				18	592	5152		
USSR	SS-N-6 Serb	mod 3	1973	2 MIRV	192	192	1 MT	1300
	SS-N-8 Sawfly	mod 1	1973	1	286	286	1.5 MT	1500
		mod 2	1973	1				900
	SS-N-17 Snipe		1980	1	12	12	1 MT	1400
	SS-N-18 Stingray	mod 1	1978	7 MIRV	224	1568	500 kt	1400
		mod 2	1978	1			1 MT	900
		mod 3	1978	7 MIRV			500 kt	900
	SS-N-20 Sturgeon		1983	10 MIRV	120	1200	200 kt	500
	SS-N-23 Skiff		1986	4 MIRV	96	384	100 kt	c. 900
Total				34	930	3642		
U.K.	Polaris A-3 TK		1967	3 MRV	64	64	200 kt	900
France	M-4		1985	6 MIRV	48	144	150 kt	n/a
	M-20		1977	1	48	48	1 MT	n/a
Total				7	96	192		
China	CSS-N-3		1984	1	12	12	c. 2 MT	n/a

Source.—International Institute of Strategic Studies, *The Military Balance 1989–1990* (London: International Institute of Strategic Studies, 1989).
Notes.— MRV = re-entry vehicle; MIRV = multiple independently-targetable re-entry vehicle; n/a = not available.

the first place. The policy of nuclear deterrence by assured destruction rests on four tenets. First, the nuclear forces of the deterrer must be fashioned exclusively for retaliation in response to an attack by the other side's weapons of mass destruction or in response to a threat of annihilation. Second, the nuclear forces—including their command and control systems—must be capable of prompt action. Third, the threat on which the deterrence rests must be the killing of a large fraction of the enemy population and the destruction of much of their economy. Fourth, the enemy must be aware of the threat in time to deter them from making the actions that will provoke the massive retaliation.

The commonly held view that the very destructiveness of nuclear weapons precludes the outbreak of nuclear war is false. Even if "rational" behavior is assumed, nuclear war is unlikely to occur only if it is believed that neither side can win. If one power perceives a chance of winning, then there is a risk that it will decide to strike while it has the advantage. Moreover, in a serious crisis, the side that perceives it is at a disadvantage may, if it believes the use of weapons of mass destruction is very likely, attack first and perhaps prematurely, in the hope of reducing the damage it thinks it is almost bound to suffer.

A paradox of the nuclear age is that nuclear deterrence based on mutual assured destruction (MAD), if it works at all, only does so with inaccurate nuclear weapons. As more-accurate nuclear weapons are deployed, the enemy may assume that your nuclear weapons are targeted on their nuclear forces and not on their cities. The cities then cease to be effective hostages. In other words, accurate nuclear weapons weaken and eventually kill nuclear deterrence based on assured destruction.

A relatively small number of nuclear weapons are needed for assured destruction: the number necessary to target the enemy's significant cities. In each of the two superpowers, for example, there are at most 200 cities with populations greater than about 100,000 people. If the relations between states, even hostile ones, are being determined rationally, a very small number of nuclear weapons that can be reliably delivered to their targets is enough for a minimum nuclear deterrent. For the superpowers, this number is certainly much less than 100.

Why is it, if 1 or 10, or maybe 100 or so nuclear weapons on target are all that are needed to deter, that the United States and the USSR have each deployed some 10,000 strategic nuclear weapons? A nuclear force of this size is likely to provoke a move away from nuclear deterrence based on assured destruction, particularly as more-accurate nuclear weapons are deployed.

IMPROVING WARHEAD ACCURACIES

The improvement of the accuracy of the delivery of nuclear warheads is a crucial qualitative technological advance in the nuclear arms race. Develop-

TABLE 3.—SUBMARINES IN THE WORLD'S NAVIES

Country	Total Submarines	Conventional Submarines					Nuclear-Powered Submarines		
		Strategic	Tactical	Coastal	Other[a]	Total	Strategic	Tactical	Total
Albania	2				2	2			
Algeria	4				4	4			
Argentina	4				4	4			
Australia	6				6	6			
Brazil	7				7	7			
Bulgaria	4				4	4			
Canada	3				3	3			
Chile	4				4	4			
China	93		88		88	88	1 Xia	4 Han	5
Colombia	2				2	2			
Cuba	3			3	3	3			
Denmark	3			3		3			
Ecuador	2				2	2			
Egypt	10				10	10			
France	20		10		10	10	6 total 1 L'Inflexible 5 Le Redoubtable	4 Rubis	10
Germany, Fed Rep	24			24		24			
Greece	10		10		10	10			
India	17		16		16	16		1 Chakra	1
Indonesia	2				2	2			
Israel	3			3		3			
Italy	10				10	10			
Japan	15				15	15			
Korea, Dem P Rep	21				21	21			
Korea, Rep	3				3	3			
Libya	6				6	6			
Netherlands	6				6	6			
Norway	12				12	12			
Pakistan	6				6	6			
Peru	11				11	11			
Poland	4				4	4			
Portugal	3				3	3			
Romania	1				1	1			
South Africa	3				3	3			
Spain	8				8	8			
Sweden	11				11	11			

Country	Total				Nuclear	SSBN	SSN
Syria	3		3	3			
Taiwan	4		4	4			
Turkey	15		15	15			
U.K.	31	10	15	10	21	4 Resolution	17 total: 6 Trafalgar, 6 Swiftsure, 2 Valiant, 3 Churchill
U.S.A	133	3	2	5	128	35 total: 9 Ohio, 12 Franklin, 8 Madison, 6 Lafayette	93 total: 41 Los Angeles, 37 Sturgeon, 10 Permit, 1 Lipscomb, 1 Narwal, 3 Shipjack
USSR	368	129	19	154	214	63 total: 5 Typhoon, 18 Delta-I, 4 Delta-II, 14 Delta-III, 5 Delta-IV, 15 Yankee-I, 1 Yankee-II, 1 Hotel-III	151 total: 12 Yankee-I, 3 Hotel-II, 5 Oscar, 1 Papa, 10 Charlie-I, 6 Charlie-II, 5 Echo-I, 29 Echo-II, 3 Yankee, 5 Hotel + Yankee, 4 Akula, 3 Sierra, 5 Alfa, 16 Victor-I, 7 Victor-II, 23 Victor-III, 12 November, 1 Uniform, 1 X-Ray
Venezuela	3	3	3	3			
Yugoslavia	5	5	5	5			

Source.—International Institute of Strategic Studies, *The Military Balance: 1989–1990* (London: International Institute of Strategic Studies, 1989).

Editors' Note.—Note that the figures given in this table for the North and South Korean navies are not the same as those cited in this volume in Joseph Morgan's article, "Naval Operations in Korean Waters." Minor discrepancies such as these occur due to the time lapse between compilation and publication of the source of the figures. Such delays are unavoidable, but as both sources (*Jane's Fighting Ships 1989–1990* and *The Military Balance 1989–1990*) are equally authoritative, the figures provided are correct.

[a]Or not known.

ments in American nuclear weapons illustrate what improvements in accuracy are possible.

The accuracy of a nuclear warhead is normally measured by its circular error probability (CEP), defined as the radius of the circle, centered on the target, within which half of a large number of warheads of the same type fired at the target will fall. The Americans have continually improved the guidance system of their intercontinental ballistic missiles so that the CEP has been considerably reduced. For example, the CEP of a Minuteman II warhead, first deployed in 1966, is about 370 m; the new American intercontinental ballistic missile—the MX—has a CEP of about 100 m. The latest Soviet intercontinental ballistic missile, the SS-25 Sickle, has a CEP of about 200 m.

Similar developments are taking place in submarine-launched ballistic missiles. The new American Trident D-5 submarine-launched ballistic missile, for example, has a CEP of about 120 m, whereas the CEP of the older Trident C-4 is 450 m. Warheads with CEPs of about 100 m or less are war-fighting weapons (table 2).

Trident-2 and MX warheads may eventually be fitted with terminal guidance, in which a laser or radar set in the nose of the warhead scans the ground around the target as the warhead travels toward it through the earth's atmosphere. The laser or radar locks onto a distinctive feature in the area, such as a tall building or hill, and guides the warhead with great accuracy to its target. With terminal guidance, Trident-2 and MX warheads will have CEPs of 40 m or so.

NUCLEAR WAR–FIGHTING AND WAR-WINNING

The move from MAD to nuclear war–fighting is virtually certain if large numbers of accurate nuclear weapons are deployed. These are, in military jargon, "counter-force" rather than "counter-city." With nuclear weapons capable of destroying even very hardened military targets, nuclear war–fighting based on the destruction of hostile military forces becomes the preferred policy. Accurate nuclear weapons are, in other words, most likely to change the nuclear policy from nuclear deterrence to nuclear war–fighting. This change is likely to occur whether or not the political leadership wants to make the change. It happens because of technological developments.

A nuclear war–fighting policy can justify the deployment of a large number of nuclear weapons. For example, U.S. military officers, responsible for America's strategic nuclear targeting plans, have identified as many as 40,000 military targets in the USSR that are reckoned to be suitable for nuclear attack!

As tactical nuclear war–fighting weapons are deployed, they will be integrated into military tactics at relatively low levels of military command. Then the military will more easily come to believe that, if a war occurs, nuclear weapons will be used. Nuclear war becomes "fightable." And of course the military will believe that, if they have to fight a war, it is winnable.

The belief in the "fightability and winnability" of nuclear war will make such a war more likely. The deployment of nuclear war–fighting weapons also leads to perceptions that "limited nuclear wars" and "protracted nuclear wars" are possible. These also increase the probability that a deliberate nuclear war will occur. The more sophisticated nuclear weapon systems become and the more complex nuclear strategies become, the greater the danger that nuclear war will break out by accident, madness, or miscalculation.

"War-fighting deterrence," as the present policy has been called, will give way to war-winning strategies, in which it is argued that victory is possible in a nuclear war. A range of military technologies is being developed that will strengthen military and political perceptions about the possibility of fighting and winning a nuclear war. The most important of these technologies are those related to antisubmarine warfare, anti–ballistic missile systems, and antisatellite warfare systems.

If one side could severely limit the damage that the other side's strategic nuclear submarines could create in a retaliatory strike, and it believed it could destroy, by anti–ballistic missiles, the enemy missile warheads that survived a surprise attack, then the temptation to make an all-out first strike may become virtually irresistible, particularly during a period of international crisis.

Developments in antisubmarine warfare are particularly disturbing. Now that land-based ballistic missiles are vulnerable to a first (preemptive) nuclear strike by hostile land-based missiles, East-West nuclear deterrence depends mainly on the continuing invulnerability of strategic nuclear submarines.

If strategic nuclear submarines do become vulnerable, a first nuclear strike may be seen as desirable and even essential to prevent the enemy from acquiring a first-strike capability. Hence the importance of developments in antisubmarine warfare.

ANTISUBMARINE WARFARE (ASW) SYSTEMS

Antisubmarine warfare (ASW) systems are designed to detect, identify, and destroy enemy submarines.[3] Modern submarines are so effective once they get within range of enemy warships that the best way of dealing with them is to engage them before that.

3. Editors' note.—For further information, see Frank Barnaby, "Strategic Submarines and Antisubmarine Warfare," and SIPRI, "The ASW Problem: ASW Detection and Weapons Systems," in *Ocean Yearbook 1,* ed. Elisabeth Mann Borgese and Norton Ginsburg (Chicago: University of Chicago Press, 1978), pp. 376–79 and pp. 380–85; and Owens Wilkes, "Ocean-based Nuclear Deterrent Forces and Antisubmarine Warfare," *Ocean Yearbook 2,* ed. Elisabeth Mann Borgese and Norton Ginsburg (Chicago: University of Chicago Press, 1980), pp. 226–49.

To attack enemy submarines at long range, the American navy, for example, relies mainly on long-range maritime patrol aircraft, particularly the P-3 Orion, and on its own attack submarines. The Long Range ASW Aircraft (LRAACA), a land-based four-engine ASW patrol aircraft, a derivative of the P-3 design, is under development to supplement the P-3.

Any enemy submarines that evade detection at long range will be attacked by the surface warships. An American carrier battle group, for example, will use formations of surface ships carrying passive and active sonar systems to detect hostile submarines and torpedo-armed helicopters to attack them.

In ASW, detection is the critical element. Detection methods are being improved by increased sensitivity of detectors, better integration between various sensing systems, and better computer processing of the data collected by sensors.

The main categories of ASW sensors are electronic, based on radar, infrared, or lasers; optical; acoustic, particularly active and passive sonar; and magnetic, particularly the magnetic anomaly detector, which measures the disturbance to the earth's magnetic field caused by the presence of the submarine. Sensors may be carried on aircraft and ships, deployed on satellites in space, or fixed to the bottom of the ocean.

Sonar devices rely on sound to detect objects in the ocean. Although light doesn't travel well through water, sound does. During the Second World War, the development of underwater acoustic technology accelerated rapidly, spurred on by the needs of ASW. The technology led to the development of sound navigation and ranging (sonar).

In an active sonar device, a pulse of sound is transmitted from the sonar transmitter. If, in its passage through the ocean, it hits an object like a submarine, some of the sound will be reflected back, and some of the "echo" will be collected by the sonar receiver. The time taken for the sound to travel to the object and back to the sonar receiver is measured. From this time and the known velocity of sound, the distance to the submarine is calculated. If a number of sonars are used, the direction the echoes come from gives the position of the submarine. A typical sonar system, which may be towed behind a ship, is an array of acoustic transducers. A transducer acts as both a transmitter and a receiver, emitting short pulses (bursts) of sound waves at regular intervals and listening for echoes between the pulses.

The ASW activities of the United States and the USSR are global and continuous, involving a total system of great technological complexity, including the use of a network of foreign bases. In American ASW activities, fixed undersea surveillance systems, based on arrays of hydrophones and monitoring a large area of ocean, play a key role. (A hydrophone is an electroacoustic transducer used to detect sound transmitted through water.) Mobile and air-dropped systems supplement the fixed sea-bottom sensors.

The U.S. Navy operates special ships, called Tagos Ocean Surveillance Ships, which are platforms for the Surveillance Tower-Array Sensor System

(SURTASS), with long-range surveillance capabilities to extend ASW coverage to those parts of the world ocean not covered by the fixed, ocean-bottom systems. P-3 Orion aircraft, which operate from a number of bases throughout the world, are provided with information about the general location of Soviet submarines. An aircraft then uses large numbers of sonobuoys and sophisticated computers to process the data and pinpoint the submarines. A sonobuoy is dropped and floats on the sea to pick up noise from any submarine and transmit a bearing on it to the aircraft. Three such bearings enable the aircraft to fix the position of the submarine.

ASW systems are also carried on surface warships, particularly cruisers and destroyers. American warships, for example, carry Tactical Towed Array Sonar (TACTAS), in which a network of sonobuoys is towed behind a ship to detect any submarine in the vicinity. ASW helicopters on the ships attack any submarines with torpedoes or depth charges. An example of such a helicopter is the American SH-60B Seahawk Light Airborne Multipurpose System (LAMPS) Mk III, a computer-integrated ship helicopter system that deploys sonobuoys and processes information from them. It is also a platform for radar and electronic warfare support measures. The SH-60F is a derivation of the SH-60B, providing quick reaction inner-zone protection for a carrier battle group using an improved tethered sonar.

The most effective weapon system for detecting and attacking enemy submarines is another submarine—the hunter-killer submarine. A hunter-killer is usually a nuclear-powered submarine equipped with ASW sensors, underwater communications equipment, computers for data analysis, and computers to fire and control ASW weapons, particularly torpedoes and ASW missiles. The United States operates a fleet of 96 nuclear attack submarines; the USSR operates 147 nuclear attack submarines (out of a total tactical submarine fleet of 280). The U.S. Navy is constructing a new nuclear attack submarine—the Seawolf class—which will be quieter, faster, and more heavily armed than earlier classes.

Once detected, enemy submarines can be destroyed with torpedoes, missiles, or depth charges. The Americans are developing two new long-range ASW missiles that will be able to attack enemy submarines at distances beyond torpedo range. The ASW Stand-off Weapon (ASWSOW) is a new version of the currently deployed Submarine Rocket (SUBROC). The Vertical Launch ASROC (VLA) is to replace the existing Antisubmarine Rocket (ASROC). The ASWSOW will be a dual-capability (conventional and nuclear), all-digital, quick reaction missile, using the most sophisticated targeting capabilities. It will be able to attack enemy submarines at the maximum range of future sensors, including over the horizon.

Soviet ASW weapons include the SS-N-14 Silex (ASROC type), the SS-N-15 Starfish (SUBROC type), the SUW-N-1 (FRAS-1) (ASROC type), the 53-68 nuclear-armed torpedo, and the 65 nuclear-armed torpedo. Other ASW missiles include the Australian Ikara and the French Malafon.

Existing ASW missiles are not autonomous. They rely on receiving guidance data during their flights. Normally, the warhead homes acoustically on the sound waves emitted by the enemy submarine. An alternative is a nuclear warhead that is so destructive that it does not need accurate delivery. The SUBROC nuclear warhead, for example, has an explosive yield of between 1 and 5 kt and can destroy submarines within a radius of a few kilometers from the point of explosion.

ASW SANCTUARIES

Because strategic ASW developments are so destabilizing, it has been suggested that, to preserve the invulnerability of the sea-based nuclear deterrent, certain areas of the oceans should be declared ASW sanctuaries, so that strategic nuclear submarines could operate in them without fear of detection. By hindering ASW activities this would considerably reduce the risk of nuclear war by miscalculation and accident, as well as by intent.

Purver, for example, has suggested a Polar Basin treaty.[4] The treaty would first be limited to the surface and seabed areas of the Polar Basin, thereby prohibiting such ASW systems as surface warships. This would severely hamper superpower ASW activities. If a Polar Basin treaty could be negotiated, other areas of the oceans could be made into ASW sanctuaries.

4. Purver (n. 2 above).

Problems concerning Conservation of Wildlife in the North Sea*

Patricia W. Birnie
IMO International Maritime Law Institute, Malta

INTRODUCTION

Absence of Wildlife Dimension to North Sea Ministerial Conferences[1]

The North Sea Ministerial Conferences had until 1990 concentrated their attention on control and prevention of pollution of the North Sea, virtually ignoring both the impact of this on living resources in this area and the numerous problems facing their conservation and management, to the extent that this takes place. It appeared, however, that the United Kingdom was prepared to see some extension of the conferences' scope, at least for cetaceans. The question has thus arisen whether the scope of future conference declarations should be extended to include wildlife conservation or whether these problems can adequately be handled within the existing legal regime. This regime is based on numerous international conventions, some of which establish commissions specific to their purposes, others of which are operated and administered through existing organizations such as the European Communities (EC) or the Council of Europe, which themselves generate conventions or rules and recommendations. The drawback of this system is that it is not "networked" in the ordinary meaning of that term; that is, the threads of the conventions are not arranged in the form of a net encompassing all the problems within a single fabric.[2] They are generally addressed pragmati-

---

*EDITORS' NOTE.—A shorter version of this article was published as "The North Sea: Perspectives on Regional Environmental Cooperation," *International Journal of Estuarine and Coastal Law Special Issue* 5 (1990): 252–70.

1. The problems of the North Sea environmental regime and the role of the North Sea Ministerial Conferences were highlighted in Steinar Andresen, "The Environmental North Sea Regime: A Successful Regional Approach?" in *Ocean Yearbook 8*, ed. Elisabeth Mann Borgese, Norton Ginsburg, and Joseph R. Morgan (Chicago: University of Chicago Press, 1989), pp. 378–401.

2. "Work in which threads, wires, or the like, are arranged in the form of a net" (*The Shorter Oxford Dictionary* [Oxford: Oxford University Press, 1978]), s.v. "network."

cally, as single-species problems, despite the increased calls for an ecological approach to marine management, within which the interrelationship of species themselves and with the marine environment would be taken into consideration.[3]

Principles of Living Resource Conservation

The threefold objectives of conservation of marine living resources have been laid down in the World Conservation Strategy:[4] maintenance of ecological processes and life-support systems, the preservation of genetic diversity, and the sustainable utilization of species and ecosystems. The first is a condition precedent for the other two. It takes account of the fact that each species has unique characteristics and its own geographic distribution, resulting from evolutionary processes spanning thousands of years. During this process each species has adapted to the specific conditions of its habitat, including water temperature and chemistry, the interrelationships within its food chain, and the conditions of the ocean floor and the species and plants inhabiting it, as well as those of the superjacent water column.

Only a few of the marine species in the North Sea are currently harvested. These include various kinds of fish for human consumption, such as North Atlantic salmon, cod, haddock, mackerel, and herring, and for industrial purposes (processed into meal or oil), such as sand eel and Norway pout.[5] Although some marine mammals, such as various large cetaceans (whales) and pinnipeds (seals, Phocidae), have been exploited from time to time in this area, none is currently taken in directed fisheries by North Sea states other than Norway, which continues to catch some minke whales. Some smaller cetaceans, such as pilot whales, however, are still harvested by Faroe Islanders and others. Dolphins and porpoises are also killed incidentally in the nets used for other fisheries. Catching of small cetaceans is not regulated by the International Whaling Commission (IWC), though this might change following decisions taken at the 1990 IWC meeting. Shellfish such as mollusks and crustaceans, including mussels, scallops, crabs, and lobsters, are heavily

---

3. See, for example, C. de Klemm, "Living Resources of the Oceans," in *The Environmental Law of the Sea,* ed. D. M. Johnston, IUCN Environmental Policy and Law Paper no. 18 (Gland, Switzerland: International Union for the Conservation of Nature, 1981), pp. 71–192, especially pp. 71, 79–84.

4. International Union for the Conservation of Nature, *World Conservation Strategy* (Gland, Switzerland: IUCN, 1980); see also the "World Charter for Nature," GAR 37/7, 9 November 1982, reprinted in *International Legal Materials* 23 (1983): 455–60.

5. Editors' note.—For further information about commercial fishing in the North Sea, see James R. Coull, "The North Sea Herring Fishery in the Twentieth Century," in *Ocean Yearbook 7,* ed. Elisabeth Mann Borgese, Norton Ginsburg, and Joseph R. Morgan (Chicago: University of Chicago Press, 1988), pp. 115–31.

exploited commercially and destroyed also by other activities related to the seabed, such as bottom trawling and commercial oil and gas exploration and seabed mining. Birds, which are also part of the North Sea wildlife, are not harvested, though they too are killed accidentally by fisherman using gill nets and by oil pollution. But all marine species, whether harvested or not, need to be preserved for the role they play in maintaining the complex relationships of all species and the ecosystems of the commercially harvested species. They are valuable, to an extent as yet unknown and largely unexploited, as genetic resources that might provide new sources of food, industrial, and pharmaceutical products. Yet in this century, exploitation of many species of fish, shellfish, and marine mammals, especially since World War II, has increased to such an extent in response to both local and global demands, that many have been overexploited and some have drastically declined as a result of this and other factors. Increasing pollution and use of the sea for navigation, mineral extraction, and recreation, for example, probably affect marine wildlife and contribute to the decline of many species. Most of these activities take place on or over the Continental Shelf. This area also contains most fisheries, especially near the coast. The North Sea consists entirely of the Continental Shelves of its seven surrounding states and thus presents a large number of management problems.

Threats to the North Sea Wildlife[6]

Many scientists and involved organizations are especially concerned by the lack of protection afforded to small cetaceans and the lack of research into and monitoring of the problems threatening their survival. Four factors deriving from human activities threaten dolphins, porpoises, and small whales in the North Sea, as well as other species: deliberate kills and accidental catching in fishing nets, overfishing, contamination by pollutants, and habitat disturbance. Information is incomplete and thus contentious, but the suggested threats to these and other species of wildlife are outlined here.

Overexploitation
Some overexploited fisheries, such as shellfish, demersal, and pelagic fisheries,[7] are not threatened with extinction because their reproductive capacity

---

6. This section is based on a presentation, "Cetaceans in U.K. Waters," given by P. E. Evans to a meeting of the Marine Forum, 13 June 1989. Copies are available from Marine Forum, 80 York Way, London N1 9AG, England. For a more extended account see P. E. Evans, "Ecological Effects of Man's Activities on Cetaceans," obtainable from P. E. Evans, Department of Zoology, University of Oxford, Oxford, England.
7. Editors' note.—See, for example, Coull (n. 5 above).

allows stock levels to recover fairly quickly. This is not the case with seals, whose breeding habits make them vulnerable to capture, and cetaceans, whose slow rate of reproduction may inhibit recovery if stock levels become too low. However, when fisheries do collapse to such an extent that they fail to recover their place in the ecosystem, this may result in serious ecological and economic consequences. By-catches (incidental catches) exacerbate the problem for all wildlife.

In the North Sea it is thought possible that the overexploitation and collapse of the herring fishery may have caused the decline in harbor porpoises, the seasonal and geographical distribution of which closely follows the herring spawning. The harbor porpoise and the white-beaked and white-sided dolphins may also be threatened by overfishing of sand eel, sprat, and Norway pout for industrial purposes. Severe breeding failures in puffins and arctic terns in the Shetland Islands are already the visible result of this overexploitation.[8]

Pollution

Much of the waste of modern industrial society is discharged into the sea, inter alia, from ships, platforms, aircraft, and land-based outfalls. The details of the pollutants concerned and their regulation by the relevant organizations, commissions, and treaties (e.g., the London Dumping Convention, the Oslo and Paris Commissions, and EC) are dealt with in Andresen's article and references therein.[9] The pollutants include solid substances, such as plastic, tar balls, and other toxic wastes; sewage; effluents; and runoffs. Some substances whose discharge is banned under existing conventions could, if discharged, harm or kill not only fish but also birds and marine mammals, especially if discharged regularly in coastal waters and estuaries, where waters are warm and the pollutants are slow to disperse. Over a period of years, marine mammals can accumulate toxic chemical residues from pesticides, fertilizers, and dumped wastes. Other substances currently not subjected to controls and conditions concerning the time and place of dumping and quantity dumped may be potentially lethal. Spillage of massive quantities of oil or chemicals can result from ship strandings or collisions, as well as from deliberate discharges, and such activities wreak great havoc, particularly on seabirds.[10]

High levels of PCBs (polychlorinated biphenyls), organochlorine pesticides such as dieldrin, DDT, lindane, and heptachlor, and heavy metals such

8. A. Webb, "Ecological Effects of Man's Activities on Seabirds," in *North Sea Forum: Report* (March 1987), pp. 101–19, at p. 103 and passim.

9. See Andresen (n. 1 above).

10. Advisory Committee on Pollution of the Sea (ACOPS) annual reports detail such incidents and provide maps and charts of the locus of oil spills and damaged seabirds. Editors' note.—The ACOPS report for 1989–90 is found in App. A in this volume.

as mercury and cadmium (regulated by the London and Oslo Dumping and Paris Land-Based Pollution Conventions) have been detected in dolphins, porpoises, and pilot whales in the North Sea, especially those found along the Dutch and German Wadden Zee coast. The pollutants are absorbed by marine organisms and move up the food chain. As they are cumulative and persistent, the effects may be long-term and progressive. There are as yet unconfirmable suggestions that these pollutants weaken the immune systems of marine mammals and birds, lowering their resistance to disease, and affect the reproductive capacity of fish, crustaceans, birds, and mammals and the hatching of bird and fish eggs. It is possible that this factor contributed to the huge number (over 17,000) of common seals that died in the North and Baltic Seas in 1988 of a viral infection, but this has yet to be confirmed.[11] Small cetaceans are among the most vulnerable to such effects. The huge algae bloom that recently occurred along the Swedish and Danish coasts is bound to have affected dolphins and porpoises, as well as killing fish, but figures are not yet available. It is notable that bottlenose dolphins that used to appear regularly in the Thames, Humber, and other major English estuaries have virtually disappeared and are found only in the Moray Firth, the least polluted of the main Scottish estuaries. The harbor porpoise has also vanished from estuaries adjacent to industry (Thames, Firth of Forth, Firth of Tay, Clyde, Tyne and Tees, Merseyside, and the Bristol Channel).

Destruction of Habitat

Habitat destruction is produced not only physically, by trawling, dredging, in-filling, or draining of spawning, breeding, or nursery grounds of birds and mammals, but also ecologically by the introduction of changes that inhibit a species' continued existence. These can be brought about by pollution, warming of the water, increasing salinity, decreasing nutrients, increased eutrophication, and algae blooms that absorb all the oxygen from the water. Until recently, little research had been done on the effects of these activities on habitats, and estuaries still continue to be dammed or dredged, barrages contemplated, marshes drained, and harbors and piers extended on beaches around the North Sea. In particular, very little research has been done on pollution levels in cetaceans in U.K. waters; data available were obtained 15–20 years ago. There is no information on current levels in porpoises on British North Sea coasts and thus on regional contaminant levels, and little knowledge of their biological effects on cetaceans. It is suggested also that the life cycle of dolphins and porpoises in the North Sea is disturbed by shipping, which includes merchant, fishing, naval, recreational, and passenger ferry vessels. Some deaths are caused by collisions, and the echolocation of small cetaceans may be interfered with by sidescan sonar and echo sounders. There has been much exploitation of the North Sea floor for oil, gas,

11. See infra, Sec. "Pinnipeds (Seals) in U.K. Waters" and footnotes 51 and 52.

sand, and gravel. This has increased traffic, oil spills, oil/metal and other waste dumping and discharges, loss of hazardous cargos, and extensive use of seismic surveys, which generate high-energy and low-frequency sound waves that impinge on the habitat of small cetaceans and may interfere with their echolocation and communication.

Introduction of exotic (alien) species may also challenge the habitat of local species. Exotic species may be introduced deliberately or accidentally. Since their natural predators are not simultaneously introduced, the new species may expand dramatically, replace some local species, and perhaps spread diseases from the parasites and diseases they bring with them.

Directed and Incidental Catches

Deliberate kills include the catching of a few bottlenose whales in the Faroe Islands and the islanders' annual kill of about 2,000 pilot whales,[12] as well as white-sided and white-beaked dolphins and orca. Norwegian whalers have killed orca; Iceland captures some live orca (many of which die) for sale to aquaria; French and Spanish fishermen kill several thousand common dolphins and other species, including the bottlenose dolphin and harbor porpoise.

The rapidly increased use of long-lasting monofilament nets has resulted in the drowning of many dolphins, porpoises, and whales.[13] Danish fisheries are thought to snare thousands; the Greenland salmon gill net fishing is estimated to entrap about 1,500. British drift-netters take a large by-catch of white-beaked and common dolphins off the Scilly Isles, and Dutch fishermen have similar by-catches, but these are poorly documented. It is suspected that incidental catches are far greater than hitherto supposed because data have been so lacking.

Possibilities of a Multispecies Approach to Management

Is it desirable and possible to manage the diversity of species living in the North Sea on the basis of the ecological approach to conservation, as now advocated by some states and environmentalists? The question has been ad-

12. The Faroe Islanders have a long tradition of hunting the pilot whale, originally for subsistence purposes, but more recently as a sport, since the products are no longer needed. It is now widely considered that this killing is both cruel in technique and unnecessary. The most extensive report and review of this hunt is contained in "Pilot Whaling in the Faroe Islands: A Second Report," by the Environmental Investigation Agency (London: Crusade against All Cruelty to Animals, Humane Education Centre, Avenue Lodge, Bounds Green Road, London N22 4EU, 1986).

13. Editors' note.—See article by Paul G. Sneed, "Controlling the 'Curtains of Death': Present and Potential Ocean Management Methods for Regulating the Pacific Driftnet Fisheries," in this volume.

dressed in several papers, and various solutions and formulas have been put forward for this purpose. These were reviewed in a recent article by Gulland.[14] He pointed out that the North Sea is not only one of the world's most important fishing grounds, it is also a complex marine ecosystem, bordered by seven countries, one of which (Norway) is not a member of the EC. In the North Sea at least 20 species of commercial fish are found, some of which are fished by Norway, many of which interact, and some of which are caught incidentally. Yet the North Sea is still mainly managed as though each species existed in isolation, by means of setting quotas for each species without regard to the effect on others. There is also long-standing overcapacity in fishing fleets, and fishing mortality greatly exceeds theoretically optimum values; little effective action has been taken on this problem. Quotas maintain the status quo as far as possible rather than endeavoring to restructure fishing practice in accordance with new restrictive management objectives, and they are founded on the vast expansion in catches from the 1960s onward. States are reluctant to introduce new policies or measures without clear scientific evidence that they require or will produce benefits, and this is hard to obtain despite the coordinating and evaluative work of the regional scientific body, the International Council for the Exploration of the Sea (ICES), of which the EC is not a member. The EC lays down a Common Fisheries Policy (excluding cetaceans and seals) for its members on a tactical basis; there are few data on its effect on other species. Both ICES and the EC, moreover, have ignored economic issues and confined themselves to biological ones, without much involvement of the industries concerned; they have not provided a forum for discussion of multispecies management or for identification of strategic issues and goals. Questions concerning which fish should be taken and whether greater weight should be accorded to industrial value, human consumption, or particular species have not been asked, let alone answered. It is true that large whales and seals are now specifically protected (at least for the time being) but not in the context of their role in the ecosystem alongside other species. Norway and Iceland, Canada, and the USSR have held some discussions on such an approach, however. Introducing a comprehensive multispecies approach to include marine mammals, birds, fin fish, and shellfish would make a complex situation even more complex.

Gulland's conclusion is that North Sea policies are in a vicious circle; until there is pressure to reduce fishing effort, there will be no substantive discussion of the forms of benefit resulting, but this pressure will not arise until the benefits have been identified and in themselves exert the catalytic

14. J. A. Gulland, "The Management of North Sea Fisheries: Looking towards the 21st Century," *Marine Policy* 11 (1987): 259–72. See also David Symes, "North Sea Fisheries: Trends and Management Issues," in *International Journal of Estuarine and Coastal Law Special Issue* 5 (1990): 271–87; he considers that the North Sea presents a challenge of "management under extreme uncertainty."

pressure. A start could be made by dealing with the scientific uncertainty—but where to begin? Gulland suggests that a new forum is required for this discussion, either by extending ICES's scope or establishing regional management councils of the kind instituted under the U.S. Fishery Conservation and Management Act of 1976.[15] He holds back from proposing a better way of managing North Sea fisheries, since that enormous task can be tackled only when it is accepted that changes in policy are necessary; as challenges to present government policies would be involved, it would be better, in his view, to encourage more independent research in universities and institutes.

Gulland advocates institution of more forums, though he notes that adding another bureaucratic layer may further complicate the issue. Iljstra, in a 1988 article,[16] suggested that, given the growing interest in environmental issues, there may now be an opportunity to reconsider a proposal made in 1974 by a David Davies Study Group, that there should be a standing conference of all North Sea states, involving the EC and the Council of Europe as independent members, to coordinate all aspects of the North Sea regime through various issue-specific committees.[17] Ijlstra suggests that the present ministerial conferences could be expanded to include fisheries, conservation and management of seabed and other marine resources, and sea and air transport, in addition to the present pollution issues, but doubts, given the widespread means for regional cooperation that now exist, that this will prove politically acceptable, however useful it may be.

The wide range of treaties and established forums relevant to North Sea environmental issues has been outlined elsewhere.[18] What will be addressed here are those aspects of the regime that relate to conservation of marine wildlife in the North Sea—fisheries, marine mammals, birds—and the regime's adequacy and aptness for resolving existing problems in relation to these species.

LEGAL FRAMEWORK FOR CONSERVATION AND PROTECTION OF NORTH SEA WILDLIFE

As the North Sea is only a small part of the world's oceans, we must first identify the relevant international regime. The starting point must, therefore,

---

15. Pub. Law L94-265, 16 *United States Code,* Sec. 181 as amended.

16. Ton Ijlstra, "Regional Co-operation in the North Sea: An Inquiry," *International Journal of Estuarine and Coastal Law* 3 (1988): 181.

17. M. Sibthorp, ed., *North Sea: Challenge and Opportunity* (London: Europa, 1974). See also R. B. Clark, ed., *The Waters around the British Isles: Their Conflicting Uses* (Oxford: Clarendon Press, 1987), in which the David Davies Institute revised these conclusions.

18. See Patricia W. Birnie, "The North Sea: A Challenge of Disorganised Opportunities," in *The North Sea: A New International Regime,* ed. D. C. Watt (Guildford, United Kingdom: Westbury House, 1980), pp. 3–29; Clark (n. 17 above).

be the United Nations Conferences and Conventions on the Law of the Sea (UNCLOS).

UNCLOS I, II, and III

UNCLOS I produced four conventions in 1958: that on the High Seas[19] confirmed freedom of fishing on the High Seas beyond the Territorial Sea but made no provision for conservation and offered no specific protection to marine mammals; that on Conservation of Fisheries,[20] necessarily based on freedom of fishing on the High Seas, could only urge cooperation and participation in conservatory regimes aimed at maintaining the optimum sustainable yield of fisheries; it too made no specific reference to marine mammals and, in any event, was not widely ratified. UNCLOS II failed, so it was left to UNCLOS III to address the problem of conservation of marine mammals. This it did but not without compromise and, thus, resort to some ambiguous provisions.

Article 56 of the 1982 LOS Convention[21] provides that coastal states can exercise sovereign rights, inter alia, to explore and exploit, conserve, and manage the living resources of the seabed and the water column in 200-mile exclusive economic zones (EEZs) measured from their Territorial Sea baselines. Coastal states also have jurisdiction in these zones in regard to protection and preservation of the marine environment and marine scientific research; the conditions under which these jurisdictions are exercised are laid down in Parts XII and XIII, respectively. Coastal states' exclusive rights over fisheries are set out in Article 62 and are subject to various conservatory duties under Article 61.

All North Sea states have extended their fisheries jurisdiction to 200 miles either as exclusive fisheries zones (EFZs) or EEZs,[22] and five of the seven North Sea states have signed but not ratified the LOS Convention.[23] Its provisions relating to fisheries were negotiated by consensus and are generally considered to have entered into customary international law through state practice (though it is not uniform in its details). The LOS Convention requires the coastal state to determine the catch and its own capacity to harvest it; only

19. Geneva Convention on the High Seas, done at Geneva, 29 April 1958, in force 30 September 1962, *United Nations Treaty Series* 450: 11.

20. Convention on Fishing and the Conservation of the Living Resources of the Sea, done at Geneva, 29 April 1958, *United Nations Treaty Series* 559: 285.

21. *The Law of the Sea: Official Text of the United Nations Convention on the Law of the Sea with Annexes and Index* (New York: United Nations, 1983), sales no. E.83.V.5 (hereinafter cited as LOS Convention).

22. The European Communities (EC) coordinated the adoption of these by its North Sea states in 1976 and 1977.

23. The signatories are Belgium, Denmark, France, the Netherlands, and Norway.

the surplus to this must be made available to certain other states. Taking account of the best scientific advice, the coastal state must ensure by conservation and management measures that living resources are not endangered by overexploitation, and must cooperate with subregional, regional, and global international organizations to this end. The measures must aim at keeping harvested species at levels of maximum sustainable yield (MSY) "as qualified by relevant environmental and economic factors, including the economic needs of coastal fishing communities and taking into account fishing patterns, the interdependence of stocks and any generally recommended international minimum standards, whether subregional, regional or global."[24] In taking these measures the states must also "take into consideration the effects on species associated with or dependent upon harvested species with a view to maintaining or restoring populations of such associated or dependent species above levels at which their reproduction may become seriously threatened."[25] Relevant scientific information must be regularly exchanged through competent international organizations (subregional, regional, or global) with all concerned states participating.

Coastal states are required to promote optimum utilization of the living resources in the EEZ. The convention adopts a species approach, however, under which special conditions are laid down for highly migratory species (Article 64); marine mammals (Article 65); anadromous (e.g., salmon) and catadromous (e.g., eels) species (Articles 66 and 67, respectively). Article 68 places sedentary species under the Continental Shelf regime (Part VI). It also requires the states concerned, when stocks cross national boundaries or boundaries between national and international zones, to try to agree on coordinated conservatory measures through appropriate subregional or regional organizations.

LOS Convention Provisions on Highly Migratory Species

It is important in the case of marine mammals, which are usually also highly migratory species, to distinguish between the requirements of Articles 64 and 65, since some marine mammals (cetaceans: families Physeteridae, Balaenopteridae, Balaenidae, Eschrichtiidae, Monodontidae, Ziphiidae, and Delphinidae) appear in Annex I, which lists the species to which Article 64 applies. This is because in early UNCLOS drafts there was no separate article protecting marine mammals, only Article 64 on highly migratory species. When the more conservatory Article 65 was added, it was decided to leave the above cetacean families in the Article 64 Annex, even though Article 64 requires states to promote optimum utilization of the species listed. Cetaceans are taken incidentally in catches directed at some of the other species listed in the Annex to Article 64 (mackerel, pomfrets, marlins, sharks, swordfish, sailfish,

24. LOS Convention, Art. 61(3).
25. Ibid., Art. 61(4).

sauries), and it is therefore appropriate to leave the relevant regional organizations (for tuna, for example) with primary responsibility to deal with this problem. States fishing the Annex I species in a region are required to cooperate directly or through such organizations to ensure both conservation and optimum utilization within and beyond the EEZ.

Provisions on Marine Mammals
Marine mammals are removed from all requirements of optimum utilization and subject to the following requirement of Article 65:

> Nothing in this Part restricts the right of a coastal state or the competence of an international organization, as appropriate, to prohibit, limit or regulate the exploitation of marine mammals more strictly than provided for in this Part. States shall co-operate with a view to the conservation of marine mammals and in the case of cetaceans shall in particular work through the appropriate international organizations for their conservation, management and study.

North Sea states can thus prohibit the taking of cetaceans in their EEZs and EFZs and must "work through" the "appropriate" organizations for their conservation. The ambiguity of these phrases should be noted. Most states accept regulation of cetaceans by some international body. All North Sea states except Belgium are members of the IWC but do not agree on whether it is "appropriate" for all, or only some, cetaceans. Is there a choice of "appropriate" bodies in this region for management of large or small cetaceans?

Available Management Bodies for Conservation of Marine Mammals: Their Limitations

For our purposes the relevant bodies are the EC, Northeast Atlantic Fisheries Commission (NEAFC), ICES, and IWC.

European Communities
Between 1976 and 1977 the EC concerted the adoption of 200-mile fisheries zones by its North Sea members. A Common Fisheries Policy (CFP) for fishing of certain species in these zones (excluding cetaceans and seals) was concluded in 1983. Since then the EC itself has been the fishing management body. The EC, and not its individual members, is now a member in its own right of the relevant fisheries commissions (excluding the IWC and sealing bodies).[26]

---

26. For an excellent and exhaustive account of the EC role, see R. R. Churchill, *EEC Fisheries Law* (Dordrecht, Netherlands: Martinus Nijhoff, 1987). See also S. Johnson and G. Corcelle, *The Environmental Policy of the European Communities* (London: Graham and Trotman, 1989), pp. 237–47, especially pp. 239–42.

The scope of the EC's competence in relation to fish is related to the list of agricultural products outlined in Annex II to the EC treaty, which lists them according to the usage of the 1950 Brussels Convention on Customs Nomenclature. It includes "fish, crustaceans and molluscs" and "preparations" thereof, animal products not elsewhere specified or included; dead animals covered by Chapter I or Chapter III; animals unfit for human consumption; and fats and oils of fish and marine mammals, whether or not refined. The position of marine mammals (especially whales and seals) is thus somewhat problematic; their meat and their fats and oils are covered, but some products otherwise dealt with in the Brussels Convention are excluded: whalebone, walrus and narwhal tusks and other teeth/tusks of whales and seals, ambergris and other animal products used in pharmaceutical products, spermaceti, skins, and furs.[27] If animals are caught only for purposes of obtaining one of these products (e.g., killing seals for pelts), the catching stage (but not the marketing and trading stage) would fall outside the CFP's scope. Some animals, as in whaling, are caught in order to get a number of products, some of which are in, and some outside, Annex II. It is thus not clear whether whaling comes within the CFP; the IWC considered that it did and in 1979 proposed that the EC should try to become party to the International Convention for Regulation of Whaling (ICRW).[28] Some EC Council members disputed this, and the matter was left unresolved. Thus the EC has only observer status at the IWC meetings but has adopted a Regulation on Common Rules for Imports of Whales and Other Cetacean Products[29] (subsequently revised) recognizing that conservation of cetacean species calls for measures that will restrict international trade. These measures are to be taken at EC level while conforming to the EC's international obligations. The regulation banned import into the EC of the main whale and other cetacean products, listed in an annex to the regulation. Cetaceans for this purpose include whales, porpoises, and dolphins. There is some concern that the removal of customs barriers in 1992 may undermine the undoubted effectiveness of this regulation.

The EC has also, by directives, banned the commercial import of the

27. Churchill (n. 26 above), pp. 54–56; Johnson and Corcelle (n. 26 above), pp. 239–42.
28. International Convention for the Regulation of Whaling, 2 December 1946, printed by direction of the International Whaling Commission (IWC), *United Nations Treaty Series* 161: 72.
29. Council Regulations (EEZ) 348/81, 20 January 1981, O.J. L 39/1, 12 January 1981, reproduced and explained in Patricia W. Birnie, *Legal Measures for Conservation of Marine Mammals*, IUCN Environmental Policy and Law Paper no. 19 (Gland, Switzerland: IUCN, 1982), pp. 85–88. This has since been revised and extended; see Council Regulations 3786/81 (correction in L131/30) and L.377/42. See also N. Haigh, *EEC Environmental Policy and Britain*, 2d ed. (Harlow, United Kingdom: Longman Group, 1987, reprinted 1990), pp. 300–303.

skins of harp and hooded seals.[30] This was followed by a dramatic decline in the harvest of these species in the Canadian Northwest Atlantic, but it has played no role in regulating North Sea seals. Whether the EC, if its council were so minded, would be the most appropriate body to manage marine mammals in the North Sea remains controversial. It has been pointed out that "the Community as a fisheries regulatory body is diffuse and multipartite,"[31] which makes it difficult to communicate with, and is a very bureaucratic body, which probably also reduces its efficiency since it has a relatively small and overburdened staff. These weaknesses could, of course, be remedied were EC member states prepared to address them.

Northeast Atlantic Fisheries Commission[32]
The original 1959 NEAF Convention was renegotiated in 1980 to take account of the subsuming of a large part of the High Seas area within its member states' 200-mile zones from 1977 onward. Problems arose from the EC's insistence (as required under the Rome Treaty and EC practice) on adhering to the new convention in its own right, replacing its individual member states, but eventually the Eastern European partners agreed to accept the EC as a party, and a new convention was concluded in 1980, entering into force in 1982. It establishes a new commission charged with the task of taking account of the best scientific advice available (to be supplied by ICES) and of furthering both conservation and optimum utilization of fisheries. But under the new convention, it can make recommendations only for fisheries found beyond its parties' 200-mile zones, in order to ensure consistency among those recommendations applying to stocks occurring both in 200-mile zones and beyond, those recommendations affecting these stocks through interspecies relationships, and those measures taken by parties within their zones. Recommendations become binding if not objected to within a specific period. Within the zones the NEAFC can make recommendations to parties only at their initiative and only if the recommendation made is voted for by the party concerned.

The NEAFC provides a forum for consultation and exchange of information on the state of fisheries and on management policies, including examination of the effect of the policies on the resources. It can also cooperate with member states concerning management of stocks that migrate between their 200-mile zones and the High Seas and can coordinate scientific research

30. See European Communities Dir. 83/129, O.J. 1983 L 91/30, as extended by Dir. 85/444, O.J. 1985 L 259/70; see also Johnson and Corcelle (n. 26 above), pp. 240–42.
31. Churchill (n. 26 above), p. 283.
32. Convention on Future Multilateral Co-operation in the Northeast Atlantic Fisheries, text in O.J. 1981 1 227/22, Cmnd. 8474; its 13 parties include all North Sea states, that is, EC and Norway.

in cooperation with ICES. Enforcement is by national (or EC) means. NEAFC has adopted only one measure to date, concerning minimum mesh sizes for capelin beyond 200 miles. The EC implements its measures by adoption of regulations. Like its predecessor, the new NEAFC has given no advice on marine mammals or pinnipeds, which it leaves to the IWC and ICES, respectively.

Its potential as a management body for these species in the North Sea seems weak, but it could be given such a role if its parties so agree. The total allowable catch limits for the Canadian harvest of hooded and harp seals, for example, are now recommended by the Scientific Council of the Northwest Atlantic Fisheries Organization (NAFO), a body similar to NEAFC.

International Council for the Exploration of the Sea[33]

The ICES includes all North Sea states among its members. It is not a regulatory body but a long-standing forum for scientific collaboration and debate at the regional level. Its aims are to promote research into the study of the sea, including living resources and pollution in the North Sea and North Atlantic, and to draw up programs for this, organize the necessary research, and publish the results. It has no managerial role and cannot make decisions, though it can make recommendations on problems researched and can highlight issues. Fishing rights related to the North Sea are regulated by the EC (see above) and Norway.

ICES has a small secretariat but works through a committee system, including about 20 fishery groups reporting to an Advisory Committee on Fishery Management. There is a similar committee on pollution and many working groups on specific questions. Many scientific papers are read at meetings. Independent scientific arguments are thus well aired and criticized, but states are not always diligent in supplying the required data and other information. ICES early established a subcommittee on whales that led to the conclusion of the first whaling convention in 1931.[34] It has studied, evaluated, and reported on the biological effects of new uses of the sea such as oil exploitation, sand and gravel extraction, and waste disposal (including dumping and effluent discharges), and has reported on the accumulation of oceanic contaminants in marine mammals[35] and on the status of some seal stocks. It has, on request, become the scientific advisory body to NEAFC and the Oslo

---

33. For its history, see A. Went, "70 Years a Growing," International Council for the Exploration of the Sea 1902–72, Rapp. P.-V. Réun. Cons. Int. Explor. Mer. 165, ICES, Copenhagen, Denmark (1972); for recent developments, see B. Parrish, "The Future Role of ICES in the Light of Changes in Fisheries Jurisdiction," *Marine Policy* 3 (1979): 232–38.

34. Convention for the Regulation of Whaling, signed at Geneva, 20 September 1931, *League of Nations Treaty Series*, CLU no. 3586.

35. Clark (n. 17 above), pp. 339–43; see also ICES Res. C. Res 1979/4:19, inviting members to study the effects of pollutants on marine mammals.

and Paris Commissions and has collaborated with bodies such as the International Oceanographic Commission (IOC), FAO, WHO, IMO, and EC. It presents the sole means in the North Sea of integrating the comprehensive advice on all fish stocks (and marine mammals) on which regulations are based and of evaluating the effect on fisheries of other activities in that area (e.g., sand and gravel pollution). Obviously it can and does play a very useful role in coordinating the research necessary for management of North Sea wildlife, if its members allow it. Some are reluctant to invite it to address such sensitive issues as small cetaceans and seals, but ICES has the potential to do so and is studying these questions. It already cooperates with NEAFC, FAO, Unesco, IOC, and to a lesser extent with United Nations Environment Programme (UNEP). It is uniquely equipped to facilitate scientific cooperation on *all* North Sea issues but is a link in the chain, not a comprehensive body.

Scientific advice can, of course, be developed and coordinated through other international bodies not specifically focused on the North Sea, such as UNEP, the International Union for Conservation of Nature (IUCN), IOC, and the Group of Experts on the Scientific Aspects of Marine Pollution (GESAMP).

International Whaling Commission[36]

The IWC was established in 1946 and held its first meeting in 1959. Its original purpose was primarily to regulate the catching of the great whales and to enable the orderly development of the whaling industry (especially in Antarctica) by the ICRW. The ICRW consists of substantive articles and a schedule of regulations, amendable annually by a three-quarters majority vote of the IWC. The IWC consists of one commissioner from each contracting government, each with one vote. It has 36 member governments, which include all North Sea states except Belgium, but the majority are nonwhaling developing states, some of which exploit small cetaceans.

Because of the severe depletion of all stocks as advised by its Scientific Committee, in 1982 the IWC, by a majority vote, using its so-called New Management Procedures, imposed zero quotas on all stocks of whales harvested commercially. This moratorium became operative from 1985 for coastal whaling and 1985/86 for pelagic operations. By 1990 the whale stocks were to have been comprehensively assessed, but at its 1990 meeting the IWC

---

36. Established by the International Convention for Regulation of Whaling (n. 28 above). For a full description of the history of international whaling and its regulations, see Patricia W. Birnie, *International Regulation of Whaling*, 2 vols. (Dobbs Ferry, New York: Oceana, 1985); J. E. Scarff, "The International Management of Whales, Dolphins, and Porpoises: An Interdisciplinary Assessment," *Ecology Law Quarterly* 6 (1977): 323–638; J. Tonnessen and A. Johnsen, *The History of Modern Whaling* (London: C. Hurst, 1982); Ray Gambell, "The International Whaling Commission: Quo Vadis?" *Mammal Review* 20 (1990): 31–43.

postponed this review to 1991.[37] Some whaling IWC members (including Norway, Iceland, and Denmark [Faroes]) are likely to press for a resumption of whaling. Some of the states still whaling in 1982 lodged formal objections, as is permitted under the ICRW's objection procedures; all except Norway have withdrawn them under various forms of pressure from other members and nongovernmental organizations. The United States threatened to impose economic sanctions, as permitted under its relevant national laws relating to conservation agreements to which the United States is a party[38] (a ban on fish imports and prohibition of access to fisheries in the U.S. EEZ), against states certified as undermining the ICRW, since the United States now regards the ICRW as a conservation treaty. Nongovernmental organizations also have initiated consumer boycotts of fisheries products, and some, such as Greenpeace, have actively intervened at sea in the North Sea, as elsewhere, to prevent ongoing whaling operations.

The United Kingdom ceased whaling in 1963, the Netherlands in 1964. France and Germany have not been active in whaling for many years. Denmark is not engaged in harvesting the great whales, but its Faroe Islanders still have a pilot whale hunt and take some white-sided dolphins. Norway continued to take minke whales until 1990, having initially lodged on objection to the moratorium. It now takes a few minke whales under special permits for scientific research, which may be issued under Article VIII of the ICRW at the discretion of the states concerned, even on stocks otherwise protected, and this is very controversial. The IWC has established guidelines for the issue of such permits, indicating, inter alia, the questions to which the research should provide answers. In 1988 Norway instituted a 5-year program to study and monitor minke whales in the Northeast Atlantic, including food selection and intake, digestion, and body composition. Up to 30 were to be taken in 1988, and 5 were to be tagged and released as part of an ecological study designed to provide information for future multispecies management of the Barents Sea.[39]

It is possible that the multispecies approach might ultimately be introduced in the North Sea and its environs. Meetings have taken place among some whaling states to discuss the adoption of a new convention based on this approach. It is supported by some states (Iceland, Norway, USSR, and

37. "Breakaway Threat as Challenges on Whaling Ban Fail," *London Times* (7 July 1990): 6.

38. Pelly Amendment to the Fishermen's Act, 1967, 22 *United States Code*, Sec. 1978 as amended; Packwood-Magnuson Amendment to the Magnuson Fishery Conservation and Management Act, 1976, Pub. Law 94-265, 16 *United States Code*, Sec. 181 as amended.

39. For these and other events in the IWC after 1985, see annual reports on IWC meetings in *Marine Policy,* and the IWC chairman's "Report of the Annual Meetings" available from the IWC, The Red House, Station Road, Histon, Cambridge CB4 4NP, England.

Japan) interested in continuing whaling, and environmentalists suggest that the intention is, by this means, to justify the culling of whales or seals, allegedly to facilitate the growth in populations of other species.

There is concern that, though the large whales are now protected from capture, small cetaceans are not.[40] What animals are included in "small cetaceans" is a matter of some confusion and disagreement. Whales, porpoises, and dolphins are separate species but are all part of the order of Cetacea.[41] The term "small cetaceans" is quite arbitrary but is often conveniently used (as it will be here) to cover all cetaceans not yet regulated by the IWC.[42] The ICRW could be applied to them ad hoc if they require regulation for purposes of conservation (Article V), since the substantive convention does not define "whale" or "whaling," and whales, under Article V, can be regulated by including the appropriate requirements in the Schedule of Regulations. The whales so regulated are identified by their scientific nomenclature in the schedule's list of "interpretations." This would accord with the objectives of the ICRW, expressed in its Preamble, of protecting *all* whales from overfishing, but its adoption would require a three-quarters majority in the IWC. Nine of the IWC member states, mostly those exploiting small cetaceans, including Denmark, whose Faroe Islands engage in the controversial annual pilot whale hunt, have opposed this development. They and "like-minded states" in the IWC contend that such extension would be a violation of the ICRW on the grounds that the Final Act of the 1946 Washington Whaling Conference, at which the ICRW was concluded, recommended to the governments attending, "as a guide" (to secure harmonization in seven languages of the colloquial terminology relating to species then exploited), a Chart of Nomenclature of Whales that was annexed to the Final Act.[43] These states argue that the ICRW was intended to apply only to the whales listed on the chart. The only small cetaceans listed were two species of Ziphiidae, the Arctic and Antarctic bottlenose whales. The minke (regarded as the smallest of the large or the largest of the small whales), widely exploited since the 1970s, especially by Norway (following the industry's pattern of moving on from overexploited stocks to as yet unexploited ones), and the orca were not listed but were later added to the schedule.

40. For details of this problem, see Birnie (n. 36 above); Patricia W. Birnie, "Small Cetaceans," *Marine Policy* 5 (1981): 277–81; IWC annual reports (n. 39 above).

41. See Nikki Meith, "Saving the Small Cetaceans," *Ambio* 13 (1984): 1–13; Nikki Meith, "Small Cetaceans: The Forgotten Whales," *International Whale Bulletin* 4 (1989), special edition.

42. Meith, "Saving the Small Cetaceans" (n. 41 above), p. 3; Meith, "Small Cetaceans" (n. 41 above), p. 1; S. G. Brown, "Research on Large and Small Cetaceans: Conservation and Management," *Ambio* 15 (1986): 171.

43. Birnie (n. 36 above), vol 2, p. 703; International Whaling Conference, Annex 4: Nomenclature of Whales, Washington, D.C., 20 November–2 December 1946, Final Act.

The IWC, therefore, currently regulates only these small cetacean species; the bottlenose whale has for some time been a protected stock in the North Atlantic, and the minke, along with all other species commercially exploited, has since 1985 been subject to zero catch limits, which must be reviewed upon the basis of the best scientific advice. By 1990 the IWC was required by the moratorium amendment to undertake a comprehensive assessment of the effects of the moratorium on whale stocks and consider whether or not to modify the ban. In the event, as the necessary data were not available, the IWC in 1990 agreed to postpone this review to its 1991 meeting. The IWC's Scientific Committee began to examine the status of small cetaceans from 1971 onward. It established a Sub-committee on Small Cetaceans to keep a watch on those taken in direct catches. Member states were then required by the schedule to keep records, not on small cetaceans as such, but merely on whales taken only in so-called small-type whaling operations, defined in the schedule as "catching operations using powered vessels with mounted harpoon guns hunting extensively for minke, bottlenose, pilot or killer whales." The IWC approved, for administrative purposes only, a list of the small cetaceans of the world. Scientific papers identifying those needing protection were frequently submitted to the Scientific Committee; these included the northern bottlenose whale and the harbor porpoise in the North Atlantic.

The Scientific Committee indicated that some international body was required to manage stocks of all cetaceans not yet covered by the IWC. No progress was made on this until the 1990 IWC meeting, when it was areed that cautious consideration should be given to the issue. At meetings held in 1978 (Copenhagen), 1979 (Estoril), and 1981 (Reykjavik), the IWC considered revision of the ICRW and its possible replacement by an international cetacean convention that would clearly apply to all species of cetaceans. A new convention's extension to small cetaceans, especially those (the great majority) found in the EFZs and EEZs, proved the most controversial issue of all. Some states (led by Canada before it withdrew from the IWC) proposed that, in the 200-mile zone, coastal states should have the prime responsibility for managing all cetaceans not yet regulated by the IWC and that the appropriate organizations to work through, as required in LOS Convention Article 65, would be the regional fishing organization (e.g., the NEAFC or NAFO). For fisheries within the 200-mile zone the organization would advise only if requested to do so by the coastal state concerned, and its advice would not be binding, though it should be taken into account in formulating management measures. In the North Sea this approach would be unsatisfactory, given the limitations on the EC, its membership in NEAFC, and the restricted role of the new NEAFC within the 200-mile zones. Although concern has been raised before, the IWC has never taken action on issues of protecting whales from harassment and pollution and of preserving their habitats, especially their breeding grounds.

The recent proposals to reconsider revision of the ICRW (initiated in the IWC in 1987 by the USSR) are unlikely to meet with any greater success, whether or not the moratorium on whaling is lifted. Thus there is no immediate prospect of any further international protection of small cetaceans by the IWC, despite its agreement to look at the issues, and there is no other organization exclusively concerned with this problem, though it should be noted that UNEP has planned meetings on small cetaceans, the first of which took place early in 1990.

Should a convention be developed exclusively for the North Sea?[44] Or should efforts be made instead to improve the ratification and use of other wildlife conservation conventions, which while not specifically or exclusively focused on protection of small cetaceans (or seals, or birds) can be extended for this purpose?

Current Status of Wildlife in the North Sea

Cetaceans in U.K. Waters[45]

Twenty-four species of whales and dolphins have been recorded in British and Irish waters; 12 are seen regularly, the others rarely. The harbor porpoise and bottlenose dolphin are most frequently sighted because of their inshore distribution, but minke whales and white-beaked, common, and Risso's dolphins are often seen, usually off the west coast. Originally most information on cetaceans (though on large whales only) came from the whaling industry; this had ceased in U.K. waters by 1930. From 1913, however, reporting of stranded cetaceans was required. Since 1973 this has been coupled with the Marine Mammal Society's scheme for whale sighting and reporting (on status, distribution, and seasonal movements concerning feeding and reproduction) by a network of 750 marine biologists and amateur volunteer observers and a few dedicated cruisers, to enable examination of relevant ecological factors. This now provides the main information on whale activities and the status of stocks in the area. The resulting information is patchy, of course, and interpretation of the data presents problems. Although there is a harbor porpoise project in Shetland, no resources are forthcoming to enhance any of these studies. Much is therefore unknown or speculative. Since 1985 one scientist has been surveying small cetacean by-catches off British coasts, but these are difficult to determine; action by the appropriate fishery departments is required, especially if plans for a coordinated European sightings scheme are to go ahead. Other individuals and institutes are work-

44. This could be developed through ministerial conferences. It appears that the United Kingdom is at least favorable to such an idea.

45. This section is based on the papers by Evans (n. 6 above).

ing on particular problems.[46] Even with these inadequacies of data, changes in status of populations have been noticed, such as widespread declines in harbor porpoises in the southernmost North Sea and the English Channel, with declines recently indicated in the northern North Sea, and a scarcity of bottlenose dolphins along the Channel coast. On the other hand, more pelagic species, such as the long-finned pilot whale, sperm whale, and some beaked whales, have been recorded more frequently.

Many of the problems besetting cetaceans in the North Sea are international, and solutions require collaboration among European states for their study and for effective countrols. A European Cetacean Society was established in 1987 to bring cetologists together for these purposes. EC support is now required for the society's activities. The issue of how best to protect small cetaceans internationally needs reexamination since, as will be seen below, present conventions and bodies are not ideally suited to this task.

Pinnipeds (seals) in U.K. Waters
Common (harbor) seals and gray seals are those most often found in the North Sea. Common seals are found throughout the Northern Hemisphere, but gray seals are found only in the North Atlantic and Baltic Sea. In the United Kingdom, the Natural Environment and Research Council (NERC), through its Sea Mammal Research Unit (SMRU) formed in 1977, is required by statute[47] to provide scientific advice on matters related to the management of seal populations; the SMRU surveys annually all major gray seal breeding colonies in Britain, the smaller ones every 3–6 years, and common seal stocks whenever possible. As seals are migratory but cyclically return to the same land sites to pup, they are at that time particularly vulnerable to capture. Although seals disperse widely in the North Sea and are impossible to count while they breed colonially on certain uninhabited islands, it is possible to use aerial surveys to count or otherwise estimate pups, and these figures can be used as the basis for estimating population size (the maximum likelihood or other techniques). Inevitably there is some margin of error, which gives rise to controversy.[48]

In 1982 the total size of the British gray seal population was estimated to be 84,000 animals, compared to 82,500 for 1981, with the English population constant and the Scottish increasing. Common seals (generally surveyed by boat every 5–6 years but annually by air in The Wash) were estimated (at a minimum) to be 20,500 animals in 1982.[49] By 1985 the total British gray seal

---

46. For example, M. Klinowska, University of Cambridge, Department of Psychology, who inter alia has produced a review of U.K. dolphinaria.

47. Conservation of Seals Act, 1980, C. 30.

48. See "Seal Stocks in Great Britain: Survey Conducted in 1982," *U.K. Natural Environment and Research Council News Journal* no. 3 (1984): 7–9.

49. Ibid., p. 9.

population was estimated to be 92,000 animals and the common seal to be 24,700.[50] By 1987 the figures for common seals off the United Kingdom were estimated at 25,700 (18,900 off the North Sea coast; 6,800 off the Atlantic coast); in northwest Europe outside the United Kingdom there were thought to be 11,000, in the Wadden Zee 8,650. With the addition of 30,000 off Iceland, the total for the Northeast Atlantic was put at 75,350.

A few stocks of gray seals have never been hunted, but some were hunted until 1980 or 1982. Since then a few licenses have been issued to take a few gray seals around salmon farms (four taken in 1985) and off the Faroe Islands (37 taken in 1985), where the stocks were respectively increasing (Outer Hebrides and North Sea) or stable. The status of some stocks is unknown. Common seals were commercially hunted in the Inner and Outer Hebrides and off the west coast of Scotland until 1981, off Orkney until 1982, and in The Wash and off Shetland until 1973. The Wash and Shetland stocks have since increased; the status of other stocks is unknown. Licenses have been granted to fish farmers to take seals off the Hebrides and west coast, but only one was killed there in 1985. They have also been granted to fish farmers and salmon netsmen on the east coast of Scotland, but none was taken in 1985. A few scientific permits have also been issued, but the total number of gray seals killed in 1985 was only 43, and only one common seal was killed. The controversial culling has ceased since the 1970s, when it proved impossible scientifically to verify the need for and effects of culls.

Unfortunately, however, at least 17,230 common seals in Northwest Europe died in 1988 (including off the United Kingdom, the Netherlands, Lower Saxony, and Norway, and in the Danish Wadden Zee, Linfjorden, Kattegat, and Skagerrak).[51] The estimated die-off in the United Kingdom (2,690) is probably an underestimate. The seals died in an epidemic at first believed to be caused by a canine distemper,[52] but no conclusive evidence has yet been identified as to the actual cause of the disease. Its causes and effects

---

50. "Seal Stocks in Great Britain: Surveys Conducted in 1985," *Natural Environment and Research Council News* 1 (March 1987): 11–13; figures for 1987 forthcoming in the same publication.

51. The mysterious plague of 1988 killed more than 17,000 harbor (common) seals along the entire North Sea coastline. In some places the death rate was 80% of the estimated seal population.

52. The cause of the mass die-off was a previously unrecorded virus related to canine distemper. Scientists have named the disease phocine distemper; its symptoms include severe respiratory and nervous system problems, lethargy, and death. The virus was first isolated by Albert Osterhaus, a Dutch veterinarian and virologist. It is thought that the seals, already known to be weakened by living in a polluted environment, were unable to cope with the virus, and its effects on the population were devastating. See "Seals Deaths: Sea Mammal Research Unit (SMRU) Coordinates U.K. Monitoring," *Natural Environment Research Council News* 7 (October 1988): 6–8; D. Eis, "Seal Deaths in Europe," *Ambio* 18 (1980): 144; Bob Krist, "Foster Mother to Holland's Seals," *International Wildlife* (July/August 1989): 21–24.

have been studied by the SMRU, among others, including whether it is being transmitted to coastal otters in Europe (including the Shetlands) and whether there is any link with pollution, to which marine mammals are particularly vulnerable as top predators and because toxic compounds collect in their blubber.[53] The link has not yet been established, nor do gray seals seem so vulnerable to the disease, but the SMRU advised that the closed season for both common and gray seals should be extended as a precautionary measure. So far the U.K. Parliament has not approved the necessary amendment to the 1971 Conservation of Seals Act to enable this. Many complex questions remain unresolved, especially concerning the long-term impacts of the disease on the stocks and the effects on it of pollutants, on which information is lacking. DDT and PCB compounds have been found in gray seals, especially off some parts of the United Kingdom's North Sea coasts, but there is no evidence of any effect of these on their survival. Higher levels have been found in seals from the Dutch Wadden Zee, and a decline there in harbor seals may have been contributed to by the effect of this on reproduction rates: seals in the Frisian Wadden Zee may be the most affected; harbor seals have vanished in the West Scheldt. Swedish and Danish harbor seals, on the other hand, have increased since hunting was banned in the mid-1960s. A European Seal Group is studying these problems, as is ICES; they cannot be resolved without collaboration at the international level. At least the epidemic has provided an opportunity to monitor the whole North Sea harbor seal population.

Birds and Seabirds[54]

Many species of birds feed in estuaries around the North Sea, including seabirds, which rely entirely or mainly on the sea for their food. Birds from the many seabird colonies on the North Sea coastline and from elsewhere feed in the North Sea; these colonies are of international importance. Some seabirds (divers, grebes, sea ducks, shags, cormorants, black guillemots) feed close inshore, some offshore (gulls, kittiwakes, auks, fulmars, gannets), but all dive below the surface for their prey. Pollution can thus affect birds on the surface and below it, and through the food chains, especially birds that feed on shellfish or crustaceans. Some feed on small fish, including sand eels, increasingly taken in industrial fisheries; overfishing of these thus reduces the birds' diet and affects breeding. All birds have to come to land to breed;

53. See "Seal Deaths" (n. 52 above).
54. See for further details M. L. Tasker and Michael Pichkowski, *Vulnerable Concentrations of Birds in the North Sea* (Peterborough, United Kingdom: Nature Conservancy Council, 1988), produced by the Seabirds at Sea Team (Mark Tasker, Michael Pienkowski), Chief Scientist Directorate, Nature Conservancy Council; Tasker and Pienkowski, *The Conservation of Seabirds in the North Sea* (Peterborough, United Kingdom: Nature Conservancy Council, 1989).

destruction or degradation of land habitat can profoundly disturb breeding as well as feeding patterns.

It is only fairly recently that information has been forthcoming on bird concentrations in the North Sea and on the threats to them, both natural (e.g., temperature and current changes, fish distribution) and artificial (pollution, overfishing, drainage of land, and other forms of habitat degradation). Pollution threats include oil spills, pesticides, chemicals from fertilizers and dumping, plastics, and net entanglement;[55] egg collection has also posed a hazard to survival of some species.

Relevant Conventions[56]

Convention on Trade in Endangered Species of Wild Fauna and Flora, 1973 (CITES)[57]

The CITES Convention, generally regarded as one of the most effective for its purposes, protects species threatened by international trade in them or their products (referred to as "specimens" thereof). The EC is a party to CITES and has implemented it. The species are listed in three appendices according to the degree of threat. Appendix I species (the most threatened) require special permits from both exporting and importing countries, and trade in these species cannot be carried out for purely commercial reasons (including trade in specimens taken beyond national jurisdiction, e.g., from the High Seas or airspace). Appendix II lists species that are threatened only if trade is not regulated and for which only import permits are required. Appendix III lists species regulated by states to prevent overexploitation and for which effective prevention requires the cooperation of other states. All the large cetaceans are now listed on Appendix I or II, but it is possible to make reservations to CITES; Japan, for example, has entered a reservation on all types of whales. CITES also covers seals and birds, some of which have been listed.

CITES is administered by an efficient Secretariat (despite some controversies), which has become a repository of information on trade. It remains

55. For a full account, see Webb, (n. 8 above), pp. 101–19.
56. For the text and discussion of all these conventions, inter alia, see Simon Lyster, *International Wildlife Law* (Cambridge: Grotius Publications, 1985): texts are in the appendices at pp. 307–461. See also Birnie (n. 36 above), pp. 391–97, 509–18; Birnie, (n. 29 above), pp. 29–49; Meith, "Saving the Small Cetaceans" (n. 41 above), pp. 8–9; Meith, "Small Cetaceans" (n. 41 above), p. 4; Patricia W. Birnie, "The Role of Law in Protecting Marine Mammals," *Ambio* 15 (1986): 137–43.
57. Convention on Trade in Endangered Species of Wild Fauna and Flora, 1973 (CITES), *International Legal Materials* 12 (1973): 1085–1104; Lyster (n. 56 above), pp. 384–406. Editors' note.—Lists of parties to CITES and marine species protected by CITES are found in App. G of this volume.

one of the most powerful instruments to protect species that are traded in, and could be used to protect any commercially valuable small cetaceans. At present, the few large whales straying into the North Sea are already listed, and no small cetaceans are commercially exploited by North Sea states. The Faroes pilot whales are consumed or stored locally.

Convention on Wetlands of International Importance Especially as Waterfowl Habitat, 1971 (Ramsar)[58]

The Ramsar Convention relates to habitat protection and defines wetlands in very broad terms. Wetlands, many of which are found alongside tidal estuaries, are very productive and provide essential habitat for many waterfowl, fish, amphibians, reptiles, mammals (including otters), and plants. Wetlands in states surrounding the North Sea have been destroyed at an unprecedented and alarming rate in recent years for land reclamation, for agriculture (including afforestation), and for industrial purposes. The purpose of the Ramsar Convention is to stem this tide of losses. Its parties include Denmark, West Germany, the Netherlands, Norway, and the United Kingdom, but not France or Belgium. Wide European participation has ensured protection of most of the Palearctic flyway. The convention is particularly important for birds dependent on these habitats; thus it is not enough for only North Sea states to adhere to it; the greatest possible number of non-European parties is also required to ensure that all areas over which North Sea birds migrate are covered. The Wadden Zee, for example, is now Ramsar-protected by the Netherlands, West Germany, and Denmark, despite its use for oil and gas extraction, fisheries, tourism, and military exercises. It is important both to waterfowl and seals.

A List of Wetlands of International Importance is maintained by the IUCN under the convention; each party must list suitable wetlands within its territory, including at least one when it joins the convention. There are now over 400 wetlands on the list. North Sea state parties have listed 20 North Sea sites: Belgium has listed 2, Denmark 5, West Germany 4, Norway 2, the Netherlands 4, the United Kingdom 3. The criteria for inclusion, laid down in the convention and subsequently elaborated, relate to the significance of the wetlands in terms of ecology, botany, zoology, limnology, or hydrology. Initially wetlands of international importance to waterfowl were given priority. Some states (e.g., the United Kingdom and the Netherlands) argue that only sites already protected by national legislation should be listed. The international listing ensures the site's full protection; for example, a leaking foreign-flag oil tanker had to be removed from a U.K. North Sea site. Listing does not, however, entail many obligations. Parties must formulate and imple-

58. Convention on Wetlands of International Importance Especially as Waterfowl Habitat, 1971 (Ramsar), *International Legal Materials* 11 (1972): 968–76; Lyster (n. 56 above), pp. 183–207.

ment plans to "promote" conservation by "wise use," establish wardens, and inform the Bureau (provided by the IUCN) of any changes in ecological character (the United Kingdom and Denmark have acted on this). Parties can consider this information, along with submitted reports, at their triannual conferences.

Despite its potential weaknesses, which include a lack of provisions for financial aid and of amendment procedures (necessitating conclusion of protocols if changes are required), the Ramsar Convention has progressed. Amendments were made in 1987 (relating to financial support and establishment of a permanent Bureau and Standing Committee) and are being applied even before their entry into force, with beneficial effect on North Sea sites. Plans to develop some listed wetlands have been contested and have had to be abandoned (e.g., the Ouse Washes, Rhine sites, the Norfolk Broads). Ramsar provides a valuable tool for "watchdog" nongovernmental organizations to monitor the sites, aided by the Bureau, according to procedures laid down by the parties. The majority of sites listed are within nature reserves; most parties have made inventories of all their wetlands, whether or not listed. Some parties, encouraged by conference guidelines, require execution of environmental impact assessments before development takes place. The convention encourages research and exchange of data on wetlands. This has had results, and priority "Action Points" have been laid down by the Conference of the Parties.[59] The fourth meeting was held in 1990.

Convention concerning the Protection of the World Cultural and Natural Heritage, 1972 (World Heritage)[60]
The main aim of the World Heritage convention is to protect the world's most prized cultural and natural areas of "outstanding universal value," which could include maritime sites. Its value in wildlife protection is limited; although natural habitats do contain animals and birds of interest, they can be listed only if of exceptional significance. So far no European sites, let alone North Sea ones, have been listed, and it does not seem likely that the latter will ever qualify. The convention does have potential for wildlife habitat protection, and several areas of natural value have been listed, but it can apply only to physical areas and not to individual species such as whales. Once listed, the area must be protected according to standards laid down in the convention. Denmark, France, West Germany, Norway, and the United Kingdom are parties to it.

59. For current progress, see *Proceedings of the Third Meeting of the Contracting Parties*, Regina, Saskatchewan, Canada, 27 May–5 June 1987.
60. Convention concerning the Protection of the World Cultural and Natural Heritage, 1972, *International Legal Materials* 11 (1972): 1358; Lyster (n. 56 above), pp. 208–38.

Convention on the Conservation of Migratory Species of Wild Animals, 1979 (Bonn Convention)[61]

The Bonn Convention, which currently has 29 parties, aims to protect species "a significant proportion of whose members cyclically and predictably cross one or more jurisdictional boundaries." It does this by listing species in appendices and requiring or encouraging "range states" (i.e., states exercising jurisdiction over any part of the range of a particular migratory species, or states whose flag vessels are engaged in catching such species beyond national jurisdiction) to take measures according to the listing. Species listed in Appendix I are those in danger of extinction throughout all or a significant portion of their range. The convention imposes strict conservation obligations on range states. Appendix II lists species that either have an "unfavourable conservation status" as defined in the convention and require conclusion of international agreements for their effective protection, or have a conservation status that would significantly benefit from international cooperation. The convention does not directly require states to take measures for these species but requires them to conclude agreements to protect species on the basis of guidelines laid down in the convention. Parties are also required to promote cooperation in, or support research on, migratory species.

The wide ratification essential to the success of the convention has not happened, and there are many ambiguities in the terms and definitions, which require interpretation through state practice. Unfortunately there is very little such practice; despite strenuous efforts by the convention's one-person Secretariat and concerned states and groups, not a single agreement has yet been concluded, whether relating to the North Sea or elsewhere.

The first conference of the parties in 1985 resolved that the Secretariat should take appropriate action to develop a number of agreements on specific species, including the North Sea and Baltic Sea populations of small cetaceans, but numerous problems have been encountered on these,[62] though work has been undertaken on a North Sea small cetacean agreement.[63]

A Small Cetacean Working Group was established, drawn from the convention's Scientific Council and outside governmental and nongovernmental organization experts. It produced a draft regional agreement for conservation of these species in the North Sea and Baltic Sea areas. The working group and the Secretariat also began work on a proposal to list small cetaceans

61. Convention on the Conservation of Migratory Species of Wild Animals, 1979 (Bonn Convention), *International Legal Materials* 19 (1980): 15, 411–27; Lyster (n. 56 above), pp. 278–98. Editor's note.—Lists of parties to the Bonn Convention and the marine species protected by the convention are found in App. G of this volume.

62. Other failed agreements are on Palearctic waterfowl, bats, and storks; some overlap the activities of other bodies (Lyster [n. 56 above], pp. 289–91, 297–98).

63. See M. Gibson, "The Bonn Convention and Its Role in Protecting Small Cetaceans," paper presented at Marine Forum Workshop. Appendix D may be obtained from the Marine Forum (n. 6 above).

in Appendix II, which was considered by the Scientific Council in 1988; irreconcilable differences of opinion were revealed on the scope and content of the agreement, and no state was willing to sponsor it. The Scientific Council decided to set up its own working group to look at small cetaceans globally and to continue the work of identifying species that should be added to the appendices. As agreements can be concluded only if species are listed in Appendix II, the Scientific Council discussed a Netherlands proposal for including eight species found in the North and Baltic seas. It approved seven—common (harbor) porpoise (*Phocoena phocoena*), bottlenose dolphin (*Tursiops truncatus*), common dolphin (*Delphinus delphis*), Risso's dolphin (*Grampus griseus*), long-finned pilot whale (*Globicephala melaena*), white-beaked dolphin (*Lagenorhynchus albirostis*), and white-sided dolphin (*Lagenorhynchus acutus*)—but deleted minke whale (*Balaenoptera acutorostrata*), currently protected under the ICRW. The Second Conference of the Parties approved the addition of the above seven species to Appendix II but disbanded the original regional working group and asked the Scientific Council to give priority to its global review, with a view to adding further species to Appendix II at the third conference in 1991. A globally based Scientific Council working group on small cetaceans has been established, including a broad range of scientific expertise (both the secretary of the IWC and the convener of its Scientific Committee's Subcommittee on Small Cetaceans are included as key participants).

The secretary of the Bonn Convention and the chairman of its Scientific Council attended the 41st IWC Meeting in San Diego in June 1989 to further the work of reviewing small cetaceans and updating the range state and common names lists for the newly listed species. Further support for this study was given at the 1990 IWC meeting.[64]

The Bonn Convention is also applicable to birds and seals, but they have fared no better under it than the small cetaceans. No agreements have been concluded on either group, although a draft agreement has been prepared for Palearctic waterfowl, which cross the North Sea and other areas.

Convention on the Conservation of European Wildlife and Natural Habitats, 1979 (Berne Convention).[65]

The Berne Convention, concluded under the auspices of the Council of Europe, with both the EC and individual states now party to it, aims to conserve wild flora and fauna and their natural habitats, promote cooperation between countries in their conservation efforts, and particularly to protect endangered and vulnerable species, including migratory species. The convention provides in general for conservation of wildlife and habitats and for special protection

64. "Breakaway Threat" (n. 37 above).
65. Convention on the Conservation of European Wildlife and Natural Habitats, 1979, *European Treaty Series* no. 104: 428–41; Lyster (n. 56 above).

for species listed in appendices, on criteria laid down in the convention. Appendix I covers strictly protected plants, Appendix II strictly protected animals, and Appendix III protected animals, including all birds not listed in Appendix II. Hundreds of species are listed. Parties are obliged to protect all *important* breeding and resting areas; deliberate change affecting Appendix II species is prohibited; directed catches are prohibited, as is possession and internal trade in Appendix II species, as necessary for protection. Final decisions on amendments to the convention and invitations to become parties require the approval of the Committee of Ministers of the Council of Europe.

The convention requires cooperation and research. Parties must take the measures necessary to maintain populations of all species of animals and plants at levels corresponding to ecological, scientific, and cultural requirements; some limited exceptions are permitted but not to the extent that the result will be detrimental to the survival of the species concerned. Unlike the Bonn Convention, the Berne Convention requirements are binding, not hortatory. It is also backed by a substantial administration—a Standing Committee of the Parties and the Secretariat facilities of the Council of Europe itself. The convention is not without a number of ambiguities that need to be resolved in state practice. For example, it is not clear whether the use of drift nets and gill nets that cause large incidental mortality to Appendix II marine species is prohibited under Appendix II, which prohibits *deliberate* capture.

Several small cetaceans are listed in Appendix II (including common, bottlenose, white-beaked, white-sided, Risso's striped, and roughtooth dolphins; harbor porpoises; long-finned pilot, killer, and false killer whales), but for the convention to be effective, breeding grounds must be protected and therefore identified. The resting and breeding grounds of small cetaceans are unknown. A solution would be to designate as resting and breeding grounds coastal areas where these cetaceans are known to occur. Like the ICRW, the Berne Convention does not address other threats such as interaction with fisheries, incidental catch, or food depletion. Pollution prevention could be covered to the extent that it falls within the requirement that parties "have regard" for conservation implications of their activities, but there is no direct linkage to the conventions dealing with pollution from various sources, for example, the Oslo (dumping), Paris (land-based), and IMO MARPOL (vessel-source) conventions. The Berne Convention has been partly implemented, but no party has legislation automatically protecting habitat (except for bats in the United Kingdom); existing protection of habitat provides for significant exceptions for agriculture and industrial development for example, often on coasts and estuaries. Exceptions made must be reported biannually to the Standing Committee. No party (except in relation to bats) has provided total protection of breeding or resting sites for Appendix II species. IUCN reports illustrate many infractions in this respect. The United Kingdom is endeavoring to make an inventory of all important breeding and resting sites, but protection of these is imperfect, although the Wildlife and Countryside Act of 1971 was designed to ensure protection through a system of national nature reserves, marine nature reserves, and sites of special scientific interest. The last in particular do not always receive strict protection.

CONCLUSIONS

While there is a framework for protection of wildlife in the North Sea, both at the international and regional level, it is not an integrated system or a holistic regime for protection. Conventions have been concluded, commissions established, and responsibilities allocated to existing bodies on a pragmatic basis. There is considerable overlap between bodies and conventions, with no one body responsible for taking the lead. The EC is a useful but not wholly appropriate body for this. It now has a clearer legal basis for environmental action under the provisions introduced in its 1987 Single Act, though these contain some ambiguities and preserve the EC's primacy of economic considerations.[66] Further development of its role in the environmental field is contemplated, but its concerns remain primarily those of economic development, within the goal of economic integration. Moreover, Norway is not a member, while several Mediterranean states, whose concerns are different from those of North Sea states, are. Better participation in and enforcement of the existing conventions are required, as well as development of further measures, such as a regional convention on small cetaceans, whatever the institutional developments.

It has been reported that new life is being breathed into the UNEP/FAO Marine Mammals Action Plan;[67] its Planning and Coordinating Committee has been institutionalized, and it held its first meeting in 1988; real coordination of its various parts is now contemplated; it is intended that better communication beween all concerned commissions and secretariats should be established and that action will be initiated on the priorities set. UNEP in its Regional Seas Programme has also begun to add conservation protocols to its framework conventions, supplementing existing protocols on pollution prevention.[68] It is still possible for the North Sea to be subjected to such an approach: that is, a framework convention establishing a North Sea Committee, with existing conventions and bodies linked to it.

66. See Johnson and Corcelle (n. 26 above), pp. 342–45.
67. Greenpeace statement to the Second Meeting of the Parties to the Bonn Convention, Geneva, 11–14 October 1988, "Legal Protection of Small Cetaceans in the North and Baltic Seas," p. 1; "New Life for the Action Plan," *Pilot* 3 (April 1989): 1, 16.
68. Editors' note.—See article by Peter M. Haas, "Save the Seas: UNEP's Regional Seas Programme and the Coordination of Regional Pollution Control Efforts," and "UNEP Regional Seas Programmes Agreements" (App. G), both in this volume.

Just before the 1988 UNEP meeting, the Director-General of IUCN convened a meeting of the secretariats and standing committee members of the Ramsar Convention, CITES, Bonn Convention, and Berne Convention. Representatives of the World Heritage convention and ICRW were also asked to attend but could not; the World Wildlife Fund sent an observer.[69] The participants agreed that considerable benefit would ensue from closer cooperation in program development and activities, joint approaches to problem solving and data matters, dovetailing meetings, and developing common formats for reports. It was agreed to institutionalize meetings of the various secretariats, which IUCN would convene annually. It might be possible also to colocate the various secretariats in the new building to be provided for IUCN by the Swiss government. The need for greater interaction among secretariats was also stressed at a meeting of experts on biological diversity convened by UNEP in 1988.

If such moves are contemplated at the international level, should not such rationalization also begin at the regional (or subregional) level of the North Sea? This was a question that some thought should urgently be addressed by the Third North Sea Ministerial Conference.

In the event, the conference agreed to adopt a memorandum of understanding (an informal agreement between administrations that does not have the status of a treaty) on the protection of small cetaceans in the North Sea, "as an interim step" toward the conclusion of a regional agreement between North Sea and Baltic Sea coastal states under the Bonn Convention.[70] This would aim to protect habitat and species. It was agreed that the protection of marine wildlife in general should be improved, and that a common and coordinated approach should be adopted for developing species and habitat protection and appropriate conservation measures for the North Sea, especially with regard to seals, sea and coastal birds, habitat, and site protection. It was also decided to consider the impact of fisheries on the ecosystem and the impact of the marine environment itself on fisheries, taking into account their socioeconomic values. Proposals for such studies do not, of course, require any measures to be taken at this stage.

Though limited, the first step to better protection of North Sea wildlife has been taken. This is progress indeed, albeit hesitant and slow. As such it is to be welcomed, and a firmer march forward is to be encouraged by all the organizations and forums discussed in this article.

69. "Cooperation among Nature Conservation Conventions," *Ramsar* 3 (April 1989): 1. See Draft Report on the Implementation of the Berne and Bonn Conventions; European Parliament Committee on the Environment, Public Health, and Consumer Protection, E.P. Doc. 2-536/84 (1988); *Implementation of the Berne Convention* (Gland, Switzerland: IUCN, 1986), which reviews relevant national legislation; Simon Lyster, *Problems of U.K. Implementation of Berne,* Report for Wildlife Link (London: Wildlife Link, 1985).

70. Ton Ijlstra, "Laws of the Sea: The Third International North Sea Conference," *Marine Pollution Bulletin* 21 (1990): 223–26.

Conservation and the Antarctic Minerals Regime

Francis Auburn*
Centre for Commercial and Resources Law, University of Western Australia

The Convention on the Regulation of Antarctic Mineral Resource Activities (CRAMRA)[1] was finalized in 1988 by the Antarctic Treaty Consultative Parties. The convention was the result of lengthy negotiations and attempts to satisfy a number of conflicting national interests including those of claimant/nonclaimant, mining/environmental, and developed/developing states. It is generally accepted that the most likely possibility for Antarctic mineral activities in the future is for offshore oil and gas.[2]

A crucial provision of the convention prohibits Antarctic mineral resource activities except in accordance with the convention.[3] For exploration and development a management scheme must be approved.[4] The commission established under CRAMRA must identify an area before exploration and

---

*EDITORS' NOTE.—This is the second article on Antarctic resources that Professor Auburn has written for the *Ocean Yearbook* series. His first article was "Legal Implications of Petroleum Resources of the Antarctic Continental Shelf," *Ocean Yearbook 1*, ed. Elisabeth Mann Borgese and Norton Ginsburg (Chicago: University of Chicago Press, 1978), pp. 500–515. Several other articles on resources of the Antarctic are to be found in the *Ocean Yearbook* series. They are G. L. Kesteven, "The Southern Ocean," *Ocean Yearbook 1*, pp. 467–99; Barbara Mitchell, "The Southern Ocean in the 1980s," *Ocean Yearbook 3*, ed. Elisabeth Mann Borgese and Norton Ginsburg (Chicago: University of Chicago Press, 1982), pp. 349–85; and John Bardach, "Fish Far Away: Comments on Antarctic Fisheries," *Ocean Yearbook 6*, ed. Elisabeth Mann Borgese and Norton Ginsburg (Chicago: University of Chicago Press, 1986), pp. 38–54. See also the report on CRAMRA, "Antarctic Treaty Special Consultative Meeting on Antarctic Mineral Resources: Final Act and Convention on the Regulation of Antarctic Mineral Resource Activities," in Appendix B of this volume.

1. For full text of Convention on the Regulation of Antarctic Mineral Resource Activities, see *International Legal Materials* 27 (1988): 859. For text of Final Act and an abridged version of CRAMRA see Appendix B of this volume.

2. F. Bastianelli, "Le potentiel miner de l'Antarctique: Conditions opérationelles et régime juridique," in *International Law for Antarctica*, ed. F. Francioni and T. Scovasi (Milan: Giuffre Edizore, 1987), p. 472.

3. CRAMRA (n. 1 above), Art. 3.

4. Ibid., Art. 45.

development can be considered. This decision is made by consensus.[5] A detailed series of environmental principles must be complied with before decisions on mineral resource activities are taken.[6]

CRAMRA provides for the most elaborate set of institutions to date within the Antarctic Treaty System. The commission is to consist of all the CRAMRA parties that were treaty consultative parties on 25 November 1988 and every other CRAMRA party satisfying an Antarctic mineral resources research test.[7] The commission's functions include the adoption of measures to protect the Antarctic environment, deciding whether to open up an area, and approving the budget.[8]

However, the central institution for most effective purposes is the regulatory committee established for each area identified by the commission.[9] The regulatory committee must approve a management scheme for each application for exploration and development of a block. The management scheme covers a wide range of vital issues, including measures for environmental protection, liability, resource conservation, financial guarantees, insurance, and enforcement.[10] The 10-member committee is carefully balanced to include the United States, the Soviet Union, four claimants (including the claimants for the area in question), and four nonclaimants.[11] At least three members must be developing countries.

Voting majorities in the regulatory committee are central to CRAMRA, as they represent the balancing of the various interests of the consultative parties. Substantive decisions need a two-thirds majority.[12] Authorization of exploration, essentially constituted by approval of a management scheme, and approval of development pose the additional requirements of a simple majority of the four claimants and a simple majority of the other six members. The decision-making procedures of the regulatory committee force claimants to support each other and also pressure other like-minded groups of nations (such as those favoring mining) to form alliances.

The Scientific, Technical, and Environmental Advisory Committee[13] is open to all parties to CRAMRA. Its function is to advise the commission and the regulatory committees on the scientific, technical, and environmental aspects of Antarctic mineral resource activities. In particular it provides advice on the crucial decisions needed to identify an area for exploration and

5. Ibid., Art. 41(2).
6. Ibid., Art. 4.
7. Ibid., Art. 18(2).
8. Ibid., Art. 21(1).
9. Ibid., Art. 29(2).
10. Ibid., Art. 47.
11. Ibid., Art. 29(2).
12. Ibid., Art. 32(3).
13. Ibid., Art. 23.

development, to prepare guidelines, and to approve applications.[14] In these cases the Advisory Committee undertakes a comprehensive environmental and technical assessment.[15] In each decision, the commission or regulatory committee is to take full account of the views of the Advisory Committee.

A central and innovative feature of CRAMRA is that it contains detailed and obligatory environmental principles. Article 2 states that the convention provides means for assessing the possible environmental impact of resource activities, determining whether such activities are acceptable, governing the conduct of these activities, and ensuring they are undertaken in strict conformity with the convention.

These principles may be compared with Article X of the Antarctic Treaty, requiring each party to exert appropriate efforts to the end that no one engages in any activity in Antarctica contrary to the principles or purposes of the treaty. There has been controversy over the effect of this article. In particular it is not clear whether conservation of the environment, which is not mentioned in the treaty, can be regarded as a principle or purpose of the treaty.[16]

The CRAMRA objectives and general principles may also be seen in the light of the conservation principles of the Convention for the Conservation of Antarctic Marine Living Resources.[17] Article II of CCAMLR provides for broad conservation principles for ecosystem management. However, other provisions of CCAMLR contradict the principles.[18] Conservation measures

14. Ibid., Arts. 41, 43, 45, 54.

15. Ibid., Art. 26(4).

16. F. M. Auburn, *Antarctic Law and Politics* (Bloomington: Indiana University Press, 1982), p. 120. Editors' note.—The Antarctic Treaty was signed on 1 December 1959. The text of the treaty may be found in *United States Treaty Series* 12:794; *United States Treaties and Other International Agreements*, no. 4780; *United Nations Treaty Series* 402:71; and as an appendix to Kesteven, pp. 493–99.

17. The full text of CCAMLR may be found in *International Legal Materials* 19:841; *United States Treaties and Other International Agreements*, no. 10240; and is republished in *Ocean Yearbook 3*, ed. Elisabeth Mann Borgese and Norton Ginsburg (Chicago: University of Chicago Press, 1982), pp. 497–512. For more information about CCAMLR, write to Dr. Darry L. Powell, Executive Secretary, CCAMLR, 25 Old Wharf, Hobart, Tasmania 7000, Australia. Editors' note.—For information about CCAMLR and Antarctic ecosystem management, see Kenneth Sherman, "Biomass Yields of Large Marine Ecosystems," *Ocean Yearbook 8*, ed. Elisabeth Mann Borgese, Norton Ginsburg, and Joseph R. Morgan (Chicago: University of Chicago Press, 1989), pp. 132–37.

18. For example, the nonderogation from the whaling and sealing conventions in Article VI, "Nothing in this Convention shall derogate from the rights and obligations of Contracting Parties under the International Convention for the Regulation of Whaling and the Convention for the Conservation of Antarctic Seals" (International Convention for the Regulation of Whaling, signed 2 December 1946 and in force 10 November 1948 [*United Nations Treaty Series* 161:72; *United States Treaties and Other International Agreements*, no. 1849]; Convention for the Conservation of Antarctic Seals,

adopted by the CCAMLR Commission have stressed the single-species rather than the ecosystem approach.[19] At the most recent meeting of the commission at the time of writing of this paper,[20] the commission's failure, even for single species, was emphasized by the conflict between the Soviet Union, calling for more data from fishing vessels, and other nations calling for catch restrictions for some depleted species.[21]

The crucial difference between the objective and general principles of CRAMRA, on the one hand, and those of the Antarctic Treaty and CCAMLR, on the other hand, is that CRAMRA contains specific and obligatory provisions for the enforcement of the general principles. Article 4 prohibits Antarctic mineral resource activities unless certain key principles are observed. Decisions must be based upon information adequate to enable informed judgments about possible impacts. No activities shall take place unless such information is available.

No resource activities shall take place until it is judged that they will not cause significant adverse effects on water or air quality; significant changes in the atmospheric, terrestrial, or marine environments; significant changes in populations of species of fauna or flora; further jeopardy to endangered species; or substantial risk of degradation to areas of special biological, scientific, historic, aesthetic, or wilderness significance.

No resource activity shall take place until it is judged that safe technology and procedures are available to comply with these principles and that the capacity exists to monitor key environmental parameters and respond effectively to accidents.[22]

Critics have pointed out that the crucial environmental standards of Article 4 rely largely on undefined terms such as "significant adverse effects."[23] Assessments of environmental practices have revealed widespread failure to comply with national standards and the measures prescribed under the Antarctic Treaty.[24] These assessments have been reinforced by on-site inspec-

---

concluded February 1972, in force 11 March 1978 [*International Legal Materials* 11:251; *United States Treaties and Other International Agreements*, no. 8826; and *United States Treaty Series* 29:441]).

19. F. M. Auburn, "Uses and Exploitation of Antarctica," *Journal of Polar Studies* 1 & 2 (1987): 49–59.

20. July 1990.

21. Commission for the Conservation of Antarctic Marine Living Resources (CCAMLR), *Report of the Eighth Meeting of the Commission* (1989), p. 29.

22. CRAMRA (n. 1 above), Art. 4.

23. J. R. Burgess, "Comprehensive Environmental Protection of the Antarctic: New Approaches for New Times," paper given at the Conference on Antarctica: An Exploitable Resource or Too Important to Develop, at the University of London, 21 February 1990, p. 11.

24. For example, B. S. Manheim, Jr., *On Thin Ice* (Washington, D.C.: Environmental Defense Fund, 1988).

tions, particularly those carried out by Greenpeace.[25] However, some national authorities consider that the impacts generated by past scientific research activities are insignificant, taking into account Antarctica's vast size.[26] On this interpretation it could be argued that local impacts are not significant adverse effects. In that case, they would only be covered by the obligatory principles if they constituted a substantial risk to areas of special significance for biological or other reasons.[27] Thus the protection afforded to the local environment under CRAMRA is questionable. Having regard to the work of consultative meetings, it may also be doubted whether the protection for areas of special significance would actually be enforced in case of a conflict between perceived national interests and environmental requirements.

CRAMRA prohibits mineral resource activities in any specially protected area (SPA) or site of special scientific interest (SSSI) or any other areas protected under the Antarctic Treaty. The CRAMRA Commission may also designate protected areas for historic, ecological, environmental, scientific, or other reasons.[28] Decisions about resource activities shall take into account the need to respect other established uses of Antarctica.[29] The latter provision helps to ensure that resource activities should not indirectly impact on an SPA. For example, flying helicopters or other aircraft in a manner that would unnecessarily disturb bird and seal concentrations would have to be minimized.[30] Other protected areas include new types of treaty-protected areas, such as the multiple-use planning area (MPA) introduced by the 1989 consultative meeting.[31]

The MPA is of particular significance in relation to CRAMRA. Any mineral resource activities would be likely to utilize shore facilities near stations or areas of other use and interest because of the sparsity of suitable coastal sites and the absence of services generally available elsewhere. The MPA is

25. For example, Greenpeace International, *Report on the Visit to Dumont d'Urville* (Amsterdam: Greenpeace International, 1989).

26. U.S. Congress, Office of Technology Assessment, *Polar Prospects: A Minerals Treaty for Antarctica* (Washington, D.C.: GPO, 1989), p. 144.

27. CRAMRA (n. 1 above), Art. 4(2)(e).

28. Ibid., Art. 13(2).

29. Ibid., Art. 15(1). "Other" suggests that Antarctic mineral resource activities are an "established use" even though the consultative parties deny that resource activities have actually taken place yet.

30. Agreed Measures for the Conservation of Antarctic Fauna and Flora, Art. VIII. The agreed measures were adopted by the Third Consultative Meeting of the Consultative Parties to the Antarctic Treaty in 1964, reprinted in *Treaties and Other International Agreements on Fisheries, Oceanographic Resources, and Wildlife Involving the United States* (Washington, D.C.: GPO, 1977), pp. 28–34; and reprinted in James N. Barnes, *Let's Save Antarctica!* (Richmond, Victoria, Australia: Greenhouse Publications, 1982), pp. 44–45.

31. 1989 Antarctic Treaty Consultative Meeting, Recommendation XV-11 (hereafter cited as ATCM).

intended to assist in coordinating human activities in those areas where such activities pose identified risks of mutual interference or cumulative environmental impacts. Each MPA shall be subject to a management plan developed through consultation between interested parties to the Antarctic Treaty.[32] Management plans already exist for SSSIs. Weaknesses of the SSSI management plans include the failure to assess site values in a regional or continental context, inadequacies in the guidelines for drafting plans, and inadequate protection of peripheral areas.[33]

MPAs may contain SPAs, SSSIs, and historic monuments. The management plan may provide for specific measures applicable to the building and operation of stations and related logistic support activities. It may also apply to vessel and aircraft operations, scientific research, and visitors.[34]

Discussion of the MPA concept at the 1989 consultative meeting indicated the problems foreseen by some consultative parties. The original term "Antarctic protected area" (APA) was rejected because of fears of restricting access to areas and impeding the freedom of scientific research. Concern was expressed as to who would propose designation of such areas and their management plans and who would implement them. Stress was placed on cooperation in the planning and coordination of activities and minimizing cumulative environmental impacts. The number and size of MPAs should be kept to a minimum.[35]

The United States has prepared an illustrative draft management plan for the area around Palmer Station on Anvers Island in the Antarctic Peninsula.[36] In January 1989 the Argentine ship *Bahia Paraiso,* carrying tourists and fuel, ran aground 2 km from Palmer Station, where biological projects being carried out are susceptible to oil spills.[37]

The Environmental Defense Fund alleged that the wreck could have been avoided if the Argentines had taken greater precautions.[38] In particular, the fund argued that the channel that the *Bahia Paraiso* sought to negotiate is marked on U.S. and British charts as containing dangerous ledges and pinnacles.[39] Although the accident happened in close proximity to the U.S.

32. Ibid.

33. P. L. Keage, P. R. Hay, and J. A. Russell, "Improving Antarctic Management Plans," *Polar Record* 25, no. 155 (1989): 312.

34. 1989 ATCM (n. 31 above), Rec. XV-11.

35. Final Report of the Fifteenth ATCP Consultative Meeting, 1989, pp. 28–29.

36. Ibid., p. 27.

37. "Oil Slick Could Scar Antarctic 'for a Century,' " *New Scientist* (11 February 1989): 31.

38. Environmental Defense Fund, "*Bahia Paraiso* Oil Slick Only the Tip of the Iceberg," press release, 16 February 1989, p. 1.

39. B. S. Manheim, Jr., "Antarctica: The Last Fragile Frontier," *Christian Science Monitor* (23 March 1989): 6.

station and in the easily accessible Antarctic Peninsula, it was 9 days before a containment boom was placed around the wreck.[40]

The short-term implication of the wreck was to lend substantial support to the opponents of CRAMRA.[41] In the longer term the *Bahia Paraiso* raises the question of the effectiveness of future MPA designations. This is important to CRAMRA because the MPA is the sole existing mechanism to manage multi-use areas in Antarctica. An MPA will be integrated into CRAMRA under Article 13(1) of the convention.

The *Bahia Paraiso* wreck and other environmental problems in the Antarctic Treaty System, such as the building of the French airstrip at Pointe Géologie,[42] underline the need for the MPA concept, particularly with regard to activities under CRAMRA. However, the MPA, when integrated with CRAMRA, will raise new problems. CRAMRA states the need to respect other established uses of Antarctica. These include the operation of stations and their associated installations, support facilities and equipment, scientific investigation and cooperation, navigation, and aviation.[43] The management scheme for a block under CRAMRA must make provision to avoid and resolve conflict with other legitimate uses of Antarctica.[44] So CRAMRA would seem to indicate that an existing MPA would either prevail over possible conflicting uses under CRAMRA or a process of conflict resolution would have to be adopted.

The MPA itself is a novel concept for Antarctica. For a coastal station the MPA would be most suitably applied to all the current uses of the area. It would cover not only the station itself, but also a wide surrounding area, including any SPA or SSSI close to the station, and could also extend to the navigational channels used by supply vessels. To be effective it would have to be enforced.

During the negotiation of the agreed measures, the problem of jurisdiction with regard to permits to enter SPAs raised difficulties.[45] Similar concerns have already arisen regarding the MPA.[46] Jurisdiction is so closely allied to sovereignty that the consultative meetings have been unable to tackle it

---

40. "Sunken Argentinian Supply Vessel: A Continued Concern," *Antarctic* 11, no. 11 (1989): 442.

41. Australian Minister for Foreign Affairs and Trade, "Protecting Antarctica," news release, 19 March 1990, p. 3.

42. F. M. Auburn, "Aspects of the Antarctic Treaty System" *Archiv des Volkerrechts* 26, no. 2 (1988): 207.

43. CRAMRA (n. 1 above), Arts. 15(1)(a), (b), (f).

44. Ibid., Art. 47(s). It is noteworthy that Article 15(1) refers to established uses. It would appear that the intention is that all established uses are to be regarded as legitimate uses, since established uses have to be respected under Article 15(1).

45. Auburn (n. 16 above), pp. 271–72.

46. See n. 35 above, and accompanying text.

effectively.[47] However, the MPA will require the exercise of jurisdiction because of the likelihood of conflict between the various uses within the MPA. The recommendation specifically refers to the possibility of "mutual interference" between different activities.[48] The most likely solution to the problem for an MPA involving some CRAMRA activities would be for the MPA to be integrated into the management scheme. This would require some fine distinctions to be made between inspections under the Antarctic Treaty (specifically safeguarded under CRAMRA)[49] and inspections under CRAMRA itself.[50]

The Antarctic Treaty consultative meetings have, to date, shown themselves to be unable to deal with serious environmental problems such as Pointe Géologie and the *Bahia Paraiso*.[51] Lack of international cooperation in environmental enforcement is a major defect in the Antarctic Treaty. In theory it could be remedied by the MPA on a local level, and this would be applicable to CRAMRA. However, the past record of the consultative parties does not give ground for optimism that the MPA concept will indeed prove effective.

An apparent effect of the criticism of the consultative meetings' environmental record has been the laying down of guidelines for environmental impact assessment. The procedure requires an initial environmental evaluation (IEE) to determine whether the activity might reasonably be expected to have a significant impact. If such an impact is foreseen, then a comprehensive environmental evaluation (CEE) is to be prepared. Inter alia, unavoidable impacts will be identified, and the significance of the predicted effects will be evaluated in relation to the advantages of the proposed activity. Key environmental indicators are to be monitored. On the basis of the CEE the appro-

47. "In any case of dispute with regard to the exercise of jurisdiction in Antarctica the Contracting Parties concerned shall immediately consult together with a view to reaching a mutually acceptable solution" (Antarctic Treaty, Art. VII[2]).

48. 1989 ATCM (n. 31 above), Rec. XV-11.

49. CRAMRA (n. 1 above), Art. 11.

50. Ibid., Art. 12. CRAMRA inspections are subject to a number of restrictions not applicable to treaty inspections. For instance, under CRAMRA, observers carrying out an inspection are under a duty to protect confidentiality of data and information. Inspections shall not impose an undue burden on the operations of stations, installations, and equipment visited. CRAMRA observers shall avoid interference with the normal operations of stations and installations.

51. Pointe Géologie was not even publicly discussed by a consultative meeting, despite the worldwide criticism of the impact of the construction operations on fauna. One response of the 1989 consultative meeting to the *Bahia Paraiso* wreck was to recommend increased mutual cooperation in the hydrographic survey and charting of Antarctic waters in order to contribute to the safety of navigation and the protection of the environment (1989 ATCM [n. 31 above], Rec. XV-19). If it is indeed the case that the *Bahia Paraiso* hit dangerous ledges marked on U.S. and British charts, then the recommendation seems to miss one of the major lessons to be learned from the accident.

priate national authority decides whether the activity should proceed in its original or in a modified form. In preparing a CEE, consultative parties are to be informed and given the opportunity to comment.[52]

The new procedure is a major improvement on the code of conduct for Antarctic expeditions and station activities, which requires an evaluation of the environmental impact of major operations in the Antarctic and a consideration of alternative actions.[53] The code of conduct has been repeatedly ignored, especially in the planning of new stations.[54] The new CEE procedure has been applied by the United Kingdom to the Rothera Point airstrip.[55] However, the CEE falls short of the most effective national standards. It is to some extent a self-judging procedure. "Appropriate national authority" may mean the authority that prepared the original plan. Although the United Kingdom did circulate the Rothera evaluation to private environmental organizations, this goes beyond the requirements of Rec. XIV-2. A CEE is only required where the activity might reasonably be expected to have a "significant" impact. This leaves considerable possibilities for interpretation. However, the recommendation specifically refers to the "planning process leading to decisions about scientific research programmes and their associated logistic support facilities." It will therefore be difficult to argue a new scientific station does not require a CEE. This was specifically recognized by the 1989 consultative meeting.[56]

The new CEE provisions established by the 1987 and 1989 consultative meetings have direct effects on the interpretation of CRAMRA. They set a minimum standard. The activities being considered will, in some cases, be the same. An example would be the planning of a new airstrip or a shore-based support facility. CRAMRA requires that Antarctic mineral resource activities take place in a manner consistent with the components of the Antarctic Treaty System.[57] This is reinforced by the requirement that a state must be a party to the Antarctic Treaty in order to become a party to CRAMRA.[58] Therefore a CEE under the Antarctic Treaty can be seen as binding on parties to CRAMRA. So, for example, a decision following a CEE that an airstrip should not be built on a particular site because of significant specified

52. 1988 Antarctic Treaty Consultative Party, Rec. XIV-2 (hereafter cited as ATCP).

53. 1982 ATCP (n. 52 above), Rec. VIII-11.

54. Auburn (n. 19 above), pp. 84–88.

55. British Antarctic Survey, "Proposed Construction of Airstrip at Rothera Point, Antarctica: Final Comprehensive Environmental Evaluation," Report of the British Antarctic Survey (Cambridge: IAS, 1989).

56. "Before establishing a new station or facility, Contracting Parties should prepare a Comprehensive Environmental Evaluation in accordance with Recommendation XIV-1" (1989 ATCM [n. 31 above], Rec. XV-17).

57. CRAMRA (n. 1 above), Art. 10(1).

58. Ibid., Art. 61(2).

environmental impacts would be a strong argument against a finding under CRAMRA that the effects were insignificant.

CRAMRA requires the Advisory Committee to undertake a comprehensive environmental assessment at the critical stages of deciding to open up an area, the preparation of guidelines by the regulatory committee, deciding on exploration applications, and permitting development.[59] The assessment shall be based on all information available to the Advisory Committee, including the information supplied by the party to CRAMRA on behalf of an operator.[60] The assessment shall include a consideration, where appropriate, of a number of factors. These include the adequacy of existing information to enable informed judgments to be made, the extent and intensity of likely direct environmental impacts, possible indirect impacts, cumulative effects, and the environmental consequences of the alternative of not proceeding.[61]

Although the Advisory Committee is a major improvement on current international environmental institutions under the Antarctic Treaty System, the procedure for a comprehensive environmental assessment under CRAMRA still suffers from major drawbacks. The regulatory committee shall refer to the Advisory Committee all parts of the exploration application that are necessary for it to provide advice, together with any other relevant information.[62] Therefore the regulatory committee decides which parts of the exploration application shall be referred to the Advisory Committee. This power to set the agenda is only one of the examples in CRAMRA of the dominant role of the regulatory committee and the subsidiary position of the Advisory Committee.

In preparing its advice the Advisory Committee may seek information and advice from other scientists and experts or scientific organizations as may be required on an ad hoc basis.[63] Some of the most effective assessments of the Antarctic Treaty System's environmental practices have come from private environmental organizations, including the Antarctic and Southern Ocean Coalition (ASOC), the Antarctica Project, the Environmental Defense Fund, and the World Resources Institute. Greenpeace has established an Antarctic station and conducts regular inspections. However, there has been strong resistance within the Antarctic Treaty System to the granting of observer status to such well-qualified organizations.[64] It will be pointed out that some national delegations have specifically included representatives of these

59. Ibid., Arts. 41, 43, 45, 54.
60. For example, ibid., Art. 44(2)(b)(iii).
61. Ibid., Art. 26(4).
62. Ibid., Art. 45(3). Here the application for an exploration permit is taken as an example. Similar comments apply to the other stages at which the Advisory Committee has to provide a comprehensive environmental assessment.
63. Ibid., Art. 26(5).
64. Auburn (n. 19 above), pp. 133–35.

nongovernmental organizations.[65] However, the organizations that these persons represented were not themselves given observer status at the consultative meeting, and the representatives were bound by the applicable rules of secrecy.

A good indication of the attitude of the Antarctic Treaty System to such knowledgeable and effective organizations is the position taken by the commission of CCAMLR to repeated requests by ASOC and Greenpeace for observer status. ASOC was given this status in 1988. But there are still reservations and possible limitations on the status of ASOC as an observer, and it has not been given observer status on the CCAMLR Scientific Committee.[66]

Greenpeace is a particularly interesting illustration of the attitude of the Antarctic Treaty System to well-informed, but unwelcome, critics. Since Greenpeace has established an all-year station, it would have been entitled to consultative status had it been a country and a party to the Antarctic Treaty. In 1989 the CCAMLR Commission once more rejected the application of Greenpeace for observer status.[67] The peculiar nature of this refusal is shown by the fact that Janet Dalziell, listed on the New Zealand delegation to the CCAMLR Commission as the representative of nongovernmental organizations,[68] was also on the 1989 delegation of New Zealand to the CCAMLR Scientific Committee, but listed as "Greenpeace NZ."[69]

CRAMRA may be seen as having taken a retrograde step in only providing for observer status on the Advisory Committee for treaty and CCAMLR parties who are not parties to CRAMRA.[70] In the light of the treatment of nongovernmental organizations to date by the Antarctic Treaty System, CRAMRA would appear to exclude such bodies from any meaningful recognized role. This is underlined by the Advisory Committee's power to seek advice from "scientific organisations."[71] So even ASOC, which has been given observer status in the CCAMLR Commission, would not be entitled to apply for that status as regards the CRAMRA Advisory Committee.

This defect in CRAMRA goes to the heart of the practical effect of environmental assessment. The exclusion of the most effective environmental

65. At the 1989 consultative meeting, Ms. Lyn Goldsworthy was the representative of nongovernmental environmental organizations on the Australian delegation. Mr. Charrier of the Cousteau Foundation was on the French delegation. Mr. N. Webber and Mr. A. Hemmings represented nongovernmental organizations on the New Zealand delegation. Ms. Lee Kimball of the World Resources Institute was on the U.S. delegation (Final Report [n. 35 above], pp. 266, 271, 274, 279).

66. CCAMLR (n. 21 above), p. 37.

67. Ibid., p. 38.

68. Ibid., p. 39.

69. Scientific Committee for the Conservation of Antarctic Marine Living Resources, *Report of the Eighth Meeting of the Scientific Committee* (1989), p. 60.

70. CRAMRA (n. 1 above), Art. 23(4).

71. Ibid., Art. 26(5).

critics of the Antarctic Treaty System from the CRAMRA institution responsible for environmental evaluation can only be seen as a strong indication of the attitude of the consultative parties to the crucial ecological principles of the convention.

The purely advisory role of the Advisory Committee is plain from the text of CRAMRA. However, there are further limitations that may be inferred in the light of the experience of the Antarctic System. The environmental assessment is to be "comprehensive" and is to be based on all information available to the Advisory Committee, including that provided by the operator, through the sponsor state, and any information requested on an ad hoc basis by the committee from scientists, experts, and scientific organizations.[72]

Taking the example of an application for an exploration permit, the regulatory committee shall meet "as soon as possible" after an application has been lodged to elaborate the management scheme,[73] which also constitutes authorization for the subsequent issue of the exploration permit "without delay."[74] The time scale is not stated, but it can be assumed that the regulatory committee will require the response of the Advisory Committee within a matter of months. It is unlikely that a regulatory committee would be prepared to wait for a year or more, because then it would not be meeting "as soon as possible after an application has been lodged."

The application for an exploration permit must include a detailed environmental assessment, specifically taking into account the factors to be considered by the Advisory Committee.[75] The time scale would appear to suggest that the Advisory Committee will have to carry out an evaluation based upon the operator's assessment and any additional available information.

However, the basic principles require that no Antarctic mineral resources activities shall take place unless the information available is adequate to enable informed judgments to be made upon their possible impacts.[76] It has been stressed that this is not merely a vague general objective but a legally binding rule. It is generally recognized that current scientific knowledge of offshore Antarctica is very sketchy, and there are no known mineral deposits of present commercial interest.[77] After years of heavy exploitation of Antarctic finfish, the Soviet Union could still say that, in the absence of more detailed historical and current biological data from fishing vessels, management proce-

72. Ibid., Arts. 26(4), (5).
73. Ibid., Art. 45(1).
74. Ibid., Art. 48.
75. Ibid., Art. 44(2)(b)(iii).
76. Ibid., Art. 4(1). The 1989 consultative meeting issued a declaration of intent in similar terms (1989 ATCM [n. 31 above], Rec. XV-14) but without binding legal effect.
77. J. E. Miekle and M. A. Browne, *Antarctic Mineral Resources Regime: Diplomacy and Development* (Washington, D.C.: Congressional Research Service, 1989), p. 1.

dures should not be enacted.[78] It is highly likely that the Advisory Committee would, on a purely scientific basis, have to advise that there were insufficient data in some instances to provide the requisite guidance to the regulatory committee.

The conventional means for obtaining sufficient information would be scientific research in Antarctica. It is common for programs of research, especially those involving complex issues, such as possible future environmental effects of offshore hydrocarbon exploration and development, to take several years. It is unlikely that a regulatory committee would be prepared to accept the views of the Advisory Committee that the decision on an application for an exploration permit should be deferred for 3 or 4 years.

If extensive scientific research is required, this will be extremely costly, especially as it would require prolonged use of vessels under the difficult conditions of the Southern Ocean. There would be no incentive for the operator to provide the funds, since the research could only hold up its application and the results might well prevent the permit being approved. On past experience with the Antarctic Treaty System, bodies having functions with some similarity to the Advisory Committee have not been given funds for extensive independent research, and have been restricted to the function of coordination on a limited budget.

For example, the Scientific Committee of CCAMLR gives advice on a full-scale resource regime that has been in operation for 8 years, with continuing overexploitation of several major finfish stocks. Yet its projected budget for 1991 was only A\$125,000.[79] This would barely suffice to cover the costs of a single research vessel for a few days, if the budget were allowed to be used for such a purpose. The scientific advisor to the Antarctic Treaty consultative meetings is the Scientific Committee on Antarctic Research. SCAR has repeatedly complained that it cannot meet all the demands placed upon it, especially those posed by requests from consultative meetings, without substantially increased funds.[80]

The unwillingness of the consultative meetings and CCAMLR Commission to provide adequate funds to their scientific advisory bodies for even the limited function of coordination of national research efforts is not accidental. The consultative meetings and commission wish to maintain the primacy of the political forum. Scientific research in Antarctica is still based on national efforts. The major international investigations, such as BIOMASS, involved

78. CCAMLR (n. 21 above), p. 29. It will be noted that the fishing vessels referred to were predominantly from the Soviet Union itself.

79. Scientific Committee for the Conservation of Antarctic Marine Living Resources (n. 69 above), p. 354.

80. Meeting of Scientific Committee on Antarctic Research (SCAR) Executive Committee, *SCAR Bulletin* 95 (1989): 6. The ATCM 4 months later continued to make requests to SCAR (for example, 1989 ATCM [n. 31 above], Recs. XV-5, XV-9, XV-10, XV-11) involving the committee in further expenditure.

the international coordination of whatever states decided to perform as part of their national research programs. It may be concluded that the fundamental binding principle of CRAMRA that no mineral resource activities will take place unless information adequate for informed judgments is available will not be observed in practice.

Monitoring can be seen as of special importance to CRAMRA, to ensure that the environmental principles are actually carried out. The Antarctic Treaty has a unique inspection system, giving observers nominated by the consultative parties full access to all parts of the continent at all times.[81] Until recently the right to inspect was exercised extensively only by the United States and was not based upon any anticipation of violations of the treaty or measures under it.[82] However, there has been an upsurge in the number of inspections and the countries carrying them out. A further recent development has been that some inspections have specifically examined environmental issues.[83] In part, countries may have been influenced by the Greenpeace inspection reports, which have provided a considerable body of specific information on environmental practices.[84] After considerable delay the CCAMLR Commission introduced a system of observation and inspection in 1988, but it is of a limited nature, being restricted to verifying compliance with measures in force in relation to the flag state concerned.[85] Even then there were demands that the inspections should be carried out equitably, as quickly as possible, and with a minimum-size inspection team.[86]

Neither the Antarctic Treaty nor CCAMLR directly provides for an effective long-term method of ensuring that the objectives and principles of the treaties are carried out in practice. In 1987 the consultative meeting guidelines for a CEE provided that key indicators of the environmental effects of the activity should be monitored and, where possible, environmental impacts should, as in all Antarctic activities, be minimized or mitigated.[87]

The 1989 consultative meeting specifically dealt with environmental monitoring. Inter alia, governments should undertake to establish environmental monitoring programs to verify the predicted effects and to detect the possible unforeseen effects on the Antarctic environment of waste disposal, oil contamination, station construction and operation, logistic support, and

81. Antarctic Treaty, Art. VII.

82. Auburn (n. 16 above), pp. 110–12.

83. Delegation of the United Kingdom, "Inspection Report on Specially Protected Areas in the Antarctic Peninsula Region," 1989 ATCM (n. 31 above), XV/ATCM/INF/35.

84. Greenpeace International, *1989/1990 Expedition Report* (Amsterdam: Greenpeace International, 1989), pp. 12–14.

85. CCAMLR, *Report of the Seventh Meeting of the Commission* (Hobart: CCAMLR, 1988), pp. 31, 33, 35.

86. CCAMLR (n. 21 above), p. 105.

87. 1988 ATCP (n. 52 above), Rec. XIV-2.

the conduct of science programs. Accurate records of national activities should be maintained.[88] The need for environmental monitoring on a periodic basis is now being recognized by national Antarctic programs.[89] Ecosystem monitoring is integral to the CCAMLR approach of conservation and regulation of the entire marine living ecosystem. Although considerable progress has been made,[90] the actual conservation measures agreed to by the commission still center on a single-species approach.

CRAMRA provides for inspection to be carried out by designated observers and also allows for a system of inspection under the management scheme,[91] but, as with CCAMLR, inspection under an Antarctic resources regime is to be narrower than that under the treaty. So inspections must be compatible, must reinforce each other, and shall not impose an undue burden on the stations and installations to be visited.[92] Data and information of commercial value are to be protected.[93] Thus, considerable constraints are imposed on inspection under CRAMRA, and there are gray areas affording operators and sponsor states the opportunity to prevent inspections or limit them considerably. The most vital element of monitoring under CRAMRA relates to the operator's compliance with the management scheme. However, the conditions for monitoring are to be left to the regulatory committee.[94] The Advisory Committee has no prescribed role in monitoring and, as previously suggested, is unlikely to be funded for such an active and expensive role.

It is not specifically stated who will perform the monitoring. Since the management scheme is drafted by the regulatory committee and the sponsoring state is a member of that committee and will have allies on it (e.g., other claimants), it is likely that monitoring will be left to the sponsoring state. If this were to happen, there would be a conflict of interests. This would be especially apparent if the operator were an instrumentality of the sponsoring state. It would then be doubtful whether there would be compliance with the basic principle that no resource activity shall take place until it is judged that the capacity exists to monitor key environmental parameters so as to identify adverse effects.[95]

It is not clear whether that principle would ever be effectively complied with. The *Bahia Paraiso* wreck suggests that monitoring of such effects will be difficult, even in the relatively favorable conditions of the Antarctic Peninsula.

88. 1989 ATCM (n. 31 above), Rec. XV-5.

89. U.S. Antarctic Program Safety Review Panel, *Safety in Antarctic* (Washington, D.C.: National Science Foundation, 1988), sec. 8, p. 5.

90. CCAMLR (n. 21 above), pp. 14–15.

91. CRAMRA (n. 1 above), Art. 12(1), 47(g).

92. Ibid., Art. 12(7).

93. Ibid., Art. 16.

94. Ibid., Art. 47(g).

95. Ibid., Art. 4(4)(b).

Periodic and effective monitoring of Antarctic oil wells will be expensive. There appears to be no real incentive for the sponsoring state or any other member of a regulatory committee to go to that considerable expense. Since the Advisory Committee will not be allowed to monitor compliance with the management scheme, it appears that the monitoring provisions of CRAMRA are of limited utility in protecting the environment.

Implementation of recommendations in force has been a persistent problem of the Antarctic Treaty, especially in environmental matters.[96] This may be linked to a wider pattern of failure of the treaty system to take adequate and immediate action on safety issues, such as the lessons to be learned from the crash of Air New Zealand flight TE901 in 1979 on Mount Erebus, in which all 257 people on board were killed.[97] Implementation has been a less serious problem with CCAMLR, because the relatively limited conservation measures that have been brought into force already take into account the views of fishing states.

Compliance with CRAMRA would raise new types of issues. The principles in Article 4 are directly binding and have to be observed by operators. The whole concept of a mining regime requires enforcement of rules at a number of levels. However, CRAMRA is very sketchy on the matter. Enforcement of the management scheme will be subject to rules to be devised by the regulatory committee in each case.[98] This raises problems similar to those involved with monitoring.

The regulatory committee will decide whether an operator has complied with CRAMRA, measures under it, and the management scheme. The committee has the power to modify, suspend, or cancel the scheme, cancel the exploration or development permit, or impose a fine.[99] Since the decision involves a matter of substance, it requires a two-thirds majority of the 10 members.[100] Assuming that the sponsor state supports the operator, it should be comparatively easy for that state to persuade three other states on the regulatory committee to support its view, by invoking either claimant solidarity, mutual mining interests, or other joint concerns. The sponsor state could also obtain the support of other states by pledging its own vote on another regulatory committee. If even this failed, then considerations of world political alliances and trade-offs in other international institutions could well suf-

---

96. F. M. Auburn, "The International Court for the Environment and Antarctica," in *Per un tribunale internazionale dell'ambiente*, ed. A. Postiglione (Milan: Giuffre Edizore, 1989), pp. 78–81.

97. F. M. Auburn, "The Erebus Disaster," *German Yearbook of International Law* 32 (1989): 194. It was only in 1989 that some substantive measures were taken by the consultative meeting on air safety (1989 ATCM [n. 31 above], Rec. XV-20).

98. CRAMRA (n. 1 above), Art. 47(p).

99. Ibid., Art. 51(3).

100. Ibid., Art. 32(3).

fice. It may be concluded that sanctions on an operator may be likely only if the operator is at odds with its own sponsor state.

If there are problems of compliance by parties to the convention with their obligations under it, the commission shall draw the attention of the party to the matter.[101] The past record of the consultative parties with environmental problems, such as the Pointe Géologie airstrip, clearly shows that they are not willing to publicly take to task a consultative party for possible breaches of environmental measures. The decision of the commission to take action is one of substance, so a three-quarters majority is needed, raising the same difficulties as with obtaining a qualified majority in the regulatory committee to discipline an operator. In any case, drawing the attention of the party to a breach of a rule is hardly an effective sanction. By the time the commission acts, the party's attention will have been drawn to the matter by press reports and other public statements of environmental groups.

For years, nongovernmental environmental organizations have been campaigning to have Antarctica proclaimed a global preserve, with complete protection from commercial exploration and development.[102] In 1987 Greenpeace established an all-year station on Ross Island and named it World Park Base.[103] The UN General Assembly repeated its call for an Antarctic prospecting and mining ban in 1989.[104] However, there was no real support for such proposals until a number of consultative parties, including Australia, France, New Zealand, and Belgium, indicated that they would not become parties to CRAMRA. If these countries adhere to their refusal, then the convention cannot come into force, because it must be ratified by all claimants, the Soviet Union, the United States, and four nonclaimants.[105]

At the 1989 consultative meeting, France and Australia put forward a proposal for a comprehensive convention for the preservation and protection of Antarctica. A convention would declare Antarctica a nature reserve. There would be comprehensive protection of the Antarctic environment. Activities covered would include shipping, port installations, air transport, airport in-

101. Ibid., Art. 7(7).

102. Barnes (n. 30 above), p. 10.

103. "Greenpeace Ship on Way to Antarctic Peninsula," *Antarctic* 11, no. 7 (1988): 296.

104. *Environmental Law and Policy* 20, nos. 1 & 2 (1990): 45.

105. "Being all States necessary in order to establish all the institutions of the Convention in respect of every area of Antarctica" (CRAMRA [n. 1 above], Art. 62[1]). It has been argued that "every area" means every area that is to be subject to resource activities as such, so claimants do not necessarily have a veto over entry into force (S. K. N. Blay and B. M. Tsamenyi, *The Convention on the Antarctic Mineral Resource Activities* [*CRAMRA*] [Hobart: Faculty of Law, University of Tasmania, 1989], p. 12). However, this interpretation is contrary to the Final Act and the intention of governments participating in the drafting of CRAMRA (F. M. Auburn, "Environmental Law and Practice," in *The Future of the Antarctic Regime: A World Order Perspective*, ed. R. Falk and S. Chopra [Princeton: Princeton University Press, forthcoming]).

frastructures, tourism, scientific stations and logistics, and waste discharges. General principles for environmental protection would be provided. There would be an environmental commission, scientific and technical committee, arbitration body, secretariat, and inspection and monitoring corps. The proposal is intended to complement the Antarctic Treaty System and be an integral part of it.[106]

Other papers on comprehensive conservation measures were put forward by New Zealand, Chile, the United States, and Sweden. In contrast with the Australia/France view, the U.S. working paper sought to fill in the gaps in the existing ecological measures by a vaguely worded suggestion for a recommendation to consider proposals for strengthening protection.[107]

The consultative meeting was clearly split on the various proposals. It decided to convene a special consultative meeting in 1990 to discuss all proposals.[108] It also fixed a meeting in 1990 to discuss the liability protocol required by Article 8(7) of CRAMRA.[109] The two meetings are closely related. The liability protocol must come into force for the party lodging the application before an application for an exploration permit is made.[110] The protocol must be adopted by the commission by consensus.[111] It is therefore an essential step in bringing CRAMRA into full effect. On the other hand, the Australia/France proposal was inseparably linked to those countries' rejections of CRAMRA.

That proposal has some useful and necessary innovations, including the inspection and monitoring corps, the environmental commission, the scientific and technical commission, and the secretariat. But it is still based firmly in the Antarctic Treaty System and is to be managed by the consultative parties, which include a number of countries whose past environmental record in the region has been strongly criticized on the basis of factual evidence in specific instances. For example, although France jointly sponsored the draft, it continued work on the Pointe Géologie airstrip, which has been the subject of a major environmental controversy.[112]

Opposition to the Australia/France proposal within the Antarctic Treaty System is based upon a number of arguments. They may be taken from the official views of the United States. The Australia/France proposal is seen by a number of consultative parties as aimed at replacing CRAMRA. The existing

106. Final Report (n. 35 above), pp. 202–13.
107. Ibid., p. 245.
108. 1989 ATCM (n. 31 above), Rec. XV-1.
109. Ibid., Rec. XV-2.
110. CRAMRA (n. 1 above), Art. 8(9).
111. Ibid., Art. 8(7).
112. It is noteworthy that the Australia/France proposal would specifically require that the comprehensive conservation convention apply to "air transport, the installation and maintenance of airport infrastructures" (Final Report [n. 35 above], p. 210).

system of environmental protection does need improvement but is basically sound.[113] It must be accepted that recent consultative meetings have taken significant steps to tackle environmental problems.[114]

The crucial feature of United States opposition to the Australia/France proposal, however, is that the United States perceives it as undercutting the Antarctic Treaty. Failure to agree on CRAMRA will, the United States believes, erode the cohesion of the treaty system itself.[115] It will be seen that the United States endorses the linkage of the issues of CRAMRA and the proposal for a comprehensive conservation convention. Having regard to the rejection of CRAMRA by Australia and France and their cooperation in the comprehensive conservation convention proposal, the conclusion is logical. This impression gains support from official statements emanating from the two countries.[116]

The Australia/France proposal is for a *comprehensive* treaty, laying down general principles for the protection of the Antarctic environment. It would have its own institutions which would, at a minimum, parallel those of CCAMLR and CRAMRA. Although specifically stated to be intended to reinforce the Antarctic Treaty System, the proposal only says that it will maintain the Antarctic Treaty in its entirety.[117] CRAMRA and even CCAMLR are not so mentioned.

The wider implications of the Australia/France proposal may be seen in the Draft Convention on Antarctic Conservation prepared by the Antarctica Project for submission to the consultative parties. This draft would prohibit any activity in Antarctica contravening the purpose of the convention.[118] Article 4 of this draft is modelled on Article 4 of CRAMRA and thus would forbid Antarctic activities unless adequate information is available to make an informed judgment that there would be no significant impacts on Antarctic ecosystems. Observer status would be given to "interested actors."[119]

There would be a commission composed of all parties to the convention.[120] An Environmental, Scientific, Safety, and Technical Committee would make recommendations that would bind the commission, unless there was a

---

113. Edward E. Wolfe, Deputy Assistant Secretary for Oceans and Fisheries Affairs, Department of State, in Hearings before the Subcommittee on Merchant Marine and Fisheries, U.S. House of Representatives (14 March 1990), pp. 11–12.

114. For example, 1989 ATCM (n. 31 above), Rec. XV-3 (Waste Disposal), Rec. XV-5 (Environmental Monitoring), and Rec. XV-10 (Specially Reserved Areas).

115. Wolfe (n. 113 above), p. 12.

116. See Australian Minister for Foreign Affairs and Trade (n. 41 above).

117. Final Report (n. 35 above), p. 213.

118. Antarctica Project, *Draft of the Convention on Antarctic Conservation* (12 May 1990), Art. 2. Editors' note.—For information about the Antarctica Project, ASOC, write to the Secretary, 1536 16th Street N.W., Washington, D.C. 20036.

119. Ibid., Art. 14, reflecting the painful experience of nongovernmental organizations requesting observer status in the Antarctic Treaty System.

120. Ibid., Art. 17(2).

consensus of the commission to the contrary.[121] An Antarctic inspectorate would review ongoing Antarctic activities to determine whether they were being conducted in accordance with the convention and measures in effect pursuant to it, certify responsible actors to undertake a specific activity, and undertake oversight of activities.[122] Very detailed provisions are made for IEE and CEE procedures. A monitoring system would be established.[123] Liability and compensation for damage to the Antarctic ecosystem would be dealt with specifically and in detail.[124]

The Antarctica Project draft is a carefully worked out amalgam of features of the Antarctic Treaty System (IEE and CEE), CRAMRA (general principles), and the Australia/France proposal (comprehensive conservation convention), considerably strengthened with innovative elements of environmental protection (observer status for interested actors). Perhaps the most interesting aspect of the Antarctica Project draft is that so much of it can be seen as a logical elaboration of the general ideas put forward by Australia and France.

Clearly such a treaty would be incompatible with CRAMRA and CCAMLR. This draft is not a proposal that might be accepted by the special consultative meeting on comprehensive measures. Renegotiation of both CCAMLR and CRAMRA would be most difficult, even if the consultative meeting had the political will to do so, which it probably has not. However, the draft is a valid model against which proposals for comprehensive conservation measures can be checked. The draft makes it clear that Australia and France have opened a Pandora's box that will be very hard to close.

The internal dispute between those consultative parties supporting CRAMRA and those opposed to it is the most serious public division the Antarctic Treaty System has had to face. This is underlined by the proposal for comprehensive conservation measures, which is directly linked to the dispute over CRAMRA. The natural tendency of the consultative meetings in the face of outside challenges, such as that from the General Assembly of the United Nations, has been to act as a group, whatever the internal differences may be. However, this dispute has been generated internally within the Antarctic Treaty System. It has been pointed out that the growing number and heterogeneity of consultative parties has largely insulated the system from effective outside assault, but at the same time it has rendered the system more open to major internal divisions.[125]

---

121. Ibid., Art. 39(3)(c).
122. Ibid., Art. 28(1).
123. Ibid., Art. 40.
124. Ibid., Chap. 5. Space allows only a selection of some of the major provisions of this draft for illustrative purposes.
125. Auburn (n. 19 above), pp. 293–95.

A novel element in this dispute is that some consultative parties that now oppose CRAMRA were previously strong supporters of the drafting of the convention. There is little doubt that the growing strength of environmental politics was a major consideration. The Australian Labour Party held on to power in a close election, and this was widely credited to its environmental policies, including its opposition to CRAMRA. However, this factor alone would not have contributed to such major policy reversals had it not been for the skilled and effective advocacy of nongovernmental environmental organizations.

The consultative meeting is partly responsible for its own difficulties. The Antarctic Treaty did not set up a legal regime. In default of any conventional basis for legitimacy, such as national sovereignty or a condominium, the consultative meeting has been driven to seek other justifications for its control of Antarctic affairs. Protection of the Antarctic environment is a particularly attractive theme[126] because everyone supports it. However, if the consultative meeting claims, in good part, to rest its exclusive management of Antarctica on its environmental record, this is an invitation to critics to press home past ecological problems and point out emerging ones, such as those to be found in CRAMRA.

126. For example, 1989 ATCM (n. 31 above), Rec. XV-22.

Reports from Organizations

Ocean Yearbook 10 will be a special anniversary issue, with major articles written by members of the *Ocean Yearbook* board of editors, International Ocean Institute (IOI) Board of Trustees, and IOI Planning Council. A major article will describe and discuss the IOI, including its genesis, aims, accomplishments, and future. Hence, this *Ocean Yearbook* reports only briefly on the activities of IOI since the publication of *Ocean Yearbook 8*.

Pacem in Maribus (PIM) XVII, with a theme of "Peace in the Oceans: The New Era," was held in Moscow in June 1989. Rotterdam was the venue for PIM XVIII, which had as its major theme ports as nodal points in a global transportation system. Planning is well underway for PIM XIX, which will be held in Lisbon in September 1991. The conference theme will be "Ocean Governance: National, Regional, Global." International mechanisms of sustainable development in the oceans will be discussed, and a book with approximately 25 chapters is planned. The major sections will include "The Existing Framework," "Ocean Governance: National Level," "Ocean Governance: Regional Level," "Ocean Governance: Global Level," and "Conclusions and Recommendations."

The reports and abridged reports contained in this appendix represent reviews of only some of the activities by organizations which deal with ocean-related matters. They are intended to provide the reader with basic coverage of important programs and directions being pursued by selected major international organizations.

THE EDITORS

Advisory Committee on Pollution of the Sea (ACOPS) 1990*

ACOPS continues to expand its international activities, as described briefly below:

INTERNATIONAL CONFERENCE ON PROTECTING THE ENVIRONMENT FROM HAZARDOUS SUBSTANCES

The Basle Convention on Control of Transboundary Movements of Hazardous Wastes and their Disposal was adopted by more than 100 countries and the EEC in April 1989. Six months later, some of the world's most senior policy makers attended the International Conference on Toxic Waste, organised by ACOPS and held at the IMO headquarters, London, October 3–5, 1989. The Conference marks the beginning of ACOPS' public awareness campaign on this issue and its purpose was to urge countries to give effect to the Basle Convention as soon as possible because this umbrella agreement requires 20 ratifications. To date, there has been only one ratification, from Jordan.

ACOPS' three-day Conference was opened by its President, the former British Prime Minister, Lord Callaghan, and its Chairman, Lord Stanley Clinton Davis. It was addressed by several European environment ministers, the EEC Commissioner with special responsibility for environment, as well as leading figures from Europe, Africa and South America; the Secretary-General of IMO, Mr. Chandrika Srivastava; the Executive Director of UNEP, Dr. Mostafa Tolba; and the Director of Environment of OECD, Mr. Long. All papers presented were published in a special edition of the international journal *Marine Policy* in May 1990. Messages of support were received from the British Prime Minister, Mrs. Margaret Thatcher; the Secretary General of the UN, Senor Perez de Cuellar; and the Secretary General of the Commonwealth, Sir Shridath Ramphal.

ACOPS was greatly encouraged to note that the UK Government, whose minister addressed the Conference, signed the Basle Convention on 6 October 1989. It hopes that additional signatures and, more importantly, ratifications, will follow.

VICE PRESIDENTS

In the course of 1988–90, there were further additions to the ranks of Vice Presidents: Professor Elisabeth Mann Borgese (Canada), Professor Gennady Polikarpov (USSR), and Mr. Chandrika Prasad Srivastava (India).

---

*EDITORS' NOTE.—This report was supplied by Ms. Patricia Dent, Editor, *ACOPS Yearbook*, Advisory Committee on Pollution of the Sea (ACOPS), 57 Duke Street, Grosvenor Square, London WlM 5DH, United Kingdom. An earlier report can be found in *Ocean Yearbook 5*, ed. Elisabeth Mann Borgese and Norton Ginsburg (Chicago: University of Chicago Press, 1985), pp. 294–304.

We were delighted to note, that since the publication of our last *Yearbook*, Mr. Cissokho was appointed Minister for Rural Development in Senegal; Mr. Louis Le Pensec became Minister for Overseas Departments and Territories in France; and Professor Alexander Yankov was nominated Vice President of the Bulgarian Academy of Sciences.

Many ACOPS Vice Presidents attended their second meeting which was held in London on 4 October 1989, under the Chairmanship of ACOPS' President, Lord Callaghan. Vice Presidents continue to promote ACOPS' global policies in their respective countries and regions, through appointments and correspondence with senior government ministers and officials, and by drawing attention to their special, regional problems.

ACTIVITIES WITH INTERGOVERNMENT AGENCIES

ACOPS was granted consultative status in 1988, and has since participated at meetings of the Contracting Parties to the London Dumping Convention. In particular, ACOPS' former Chairman, Baroness White, addressed the meeting of Contracting Parties on 30 October 1987 when ACOPS was invited, together with a small number of governments and intergovernmental agencies, to present its views on the implementation of Articles 1 and 2 of the London Dumping Convention.

ACOPS has also been active in the work of IMO, including the diplomatic conference on Salvage (17–28 April 1989). It also intends to participate at the November 1990 diplomatic conference on International Cooperation on Oil Pollution Preparedness and Response.

As much of ACOPS' work in 1989 focused on the problem of Toxic Waste, the Committee was represented, through its Executive Secretary, Dr. Sebek, at the Euro-African Ministerial Conference on Environment and Development in January 1989.

The Chairman, Lord Clinton Davis, and Vice President from the Mediterranean region, Mr. Zorzetto, also lead ACOPS' delegation at the diplomatic conference which was held in Basle from 20–22 March 1989, and which adopted the Convention on the Control of Transboundary Movements of Hazardous Wastes and their Disposal.

Links were also maintained throughout 1989 and in 1990 with many intergovernmental agencies, such as the European Communities, OECD, UNEP and IOPC Fund; various ministers and Government officials in Europe, North and South America, Africa and Asia, especially during visits of Lord Clinton Davis to Switzerland, Belgium, France, and the United States; Rear Admiral Stacey to France and Dr. Sebek to the USA, India, Egypt, Italy, France, Switzerland, USSR, Yugoslavia, and Senegal.

Contacts were also maintained with research and academic institutions, as well as with industrial, wildlife, and environmental agencies.

EDUCATIONAL ACTIVITIES

As was specified in our last *Yearbook*, ACOPS has decided to give high priority to educational activities and exchanges of information with university and research establishments worldwide. Some examples of such activities include Dr. Sebek's lectures at the University of Parma (1988), the International Ocean Institute's course to government officials from the Indian Ocean area, held in Egypt in November 1989, and papers given to seminars organised by the European Parliament (November 1988), and the Italian Supreme Court (April 1989). Dr. Sebek also participated at *Pacem in Maribus XVII,* held in June 1989 in Moscow, and at the US Academy of Sciences workshop, held in Washington in February 1990.

Mr. Trevor Dixon, ACOPS' Environmental Advisor, presented papers at conferences on beach debris and pollution in Hawaii (April 1989), San Sebastian (June 1989), and Athens (July 1989). Mrs. Jennie Holloway, ACOPS' Legal Officer, took part at conferences and seminars in Rome (April 1989), Moscow (June 1989), Copenhagen (October 1989), and Vienna and Budapest (March 1990). On the home front, a seminar held in London (November 1989) for UK local authorities on problems arising from waste on UK beaches was well supported.

WORKING GROUP ON MARINE POLLUTION OF CEMR

ACOPS has continued to provide technical advice to the Working Group on Marine Pollution of the Council of the European Municipalities and Regions (CEMR). ACOPS' Vice President, Mr. Gaetano Zorzetto, continues as Chairman of the Working Group, whilst Dr. Sebek serves as the Rapporteur. The Working Group met in Rome in April 1989 and in London in October 1989.

Reports from Organizations

American Society of International Law: Maritime Boundary Project*

The American Society of International Law received funding from the Ford and Mellon Foundations to conduct a study of international maritime boundaries. The Project began in 1988 under the direction of Professor Jonathan Charney assisted by Dr Lewis Alexander. They were assisted in developing the initial plans for the Project by Mr David Colson, Mr Keith Highet, Mr Lawrence Hargrove, Dr Bob Smith and Professor Louis Sohn.

A meeting of these organisers and fifteen invited participants was held in Washington D.C. on 13 and 14 December 1988. At that meeting the Project's aims were determined and tasks were allocated to all participants.

The Project has two primary aims. The first is to make readily available to governments and interested parties on a continuing basis information on the established maritime boundaries between coastal states. The second is to analyze the practice of states with regard to the delimitation of maritime boundaries by agreement. The work proceeded in the following manner. Dr Alexander identified all known maritime boundaries established since 1940. They were then divided into geographical regions and participants were nominated to deal with particular regions. The following regions and participants were designated.

North America	Lewis Alexander and Bob Smith
Middle America & Caribbean	Kaldone Nweihed
South America	Eduardo Jimenez Arechaga
Africa	A. O. Adede
Central Pacific/East Asia	Choon-Ho Park
Indian Ocean	Victor Prescott
Persian Gulf	Robert Pietrowski
Mediterranean and Black Seas	Tullio Scovazzi
Northern and Western Europe	David Anderson
Baltic Sea	Eric Franckx

During the first half of 1989 these regional analysts produced reports on each of the boundary agreements in their region. Each report of 2–5 pages described the bases for delimitation under the following headings.

Political, strategic and historical considerations
Economic and environmental considerations
Geographic considerations
Islands, rocks, reefs and low-tide elevation considerations

*EDITORS' NOTE.—This report was supplied by Professor Victor Prescott, Department of Geography, University of Melbourne, Australia.

Baseline considerations
Geological and geomorphological considerations
Method of delimitation considerations
Technical considerations
Other considerations

The reports also included the text of the agreement and an appropriate map. As they were completed they were distributed to all participants for comments which would allow them to be improved. Once the regional reports had been prepared nine other participants looked at the various considerations from a global perspective.

These participants with their topics were as follows.

Bernard Oxman	Political, strategic, historical issues
David Colson	Legal regime issues
Barbara Kwiatkowska	Economic and environmental issues
Prosper Weil	Geographic issues
Derek Bowett	Islands, rocks, reefs and low-tide elevation issues
Louis Sohn	Baseline issues
Keith Highet	Geological and geomorphological issues
Leonard Le Gault assisted by Blair Hankey	Method of delimitation issues
Peter Beazley	Technical issues

The essays which these participants prepared were circulated to all other members of the Project.

A second and final meeting of the members of the Project was held 13–16 December 1989 at Airlie House in Virginia. At this meeting all the completed work was reviewed and final plans were made for the revision of all papers with a view to their publication as soon as possible.

Reports from Organizations

Intergovernmental Oceanographic Commission (of Unesco): Report of the Secretary on Intersessional Activities (1989) to the Twenty-third Session of the IOC Executive Council, Unesco, Paris, 7–14 March, 1990*

OCEAN SCIENCE

A. Ocean Dynamics and Climate

IOC Committee on Ocean Processes and Climate
The Third Session of the IOC Technical Committee on Ocean Processes and Climate (Paris, 27–29 June 1989) was attended by representatives of 25 Member States and seven international organizations and bodies. The meeting reviewed the implementation of TOGA, the planning of WOCE, and the related ocean observing and data management systems, and made recommendations on future activities.

In cooperation with WMO, ICSU and SCOR, the International TOGA Scientific Conference will be held in Honolulu, 16–20 July 1990. The Secretary ensured active involvement of IOC in the activities of the WMO-UNEP Intergovernmental Panel on Climate Change and in the Second World Climate Conference with the joint sponsorship of WMO, UNEP, Unesco and ICSU, by being a member of the Organizing Committee. The WMO will co-sponsor the Intergovernmental WOCE Panel (IWP).

IOC Involvement in Global Change Studies
The Fifteenth Session of the Assembly, Paris, 4–19 July 1989, "emphasized the need for an active and strengthened involvement of IOC in these activities (IPCC; SWCC; and expected consideration by the UN General Assembly of a Convention on Global Change) and requested the Chairman and Secretary of IOC to take appropriate steps to ensure the recognition of the need for ocean monitoring as an integral part of the global environment studies" (document SC/MD/91, para. 69). Accordingly, during 1989 and 1990 participation increased over as broad and complex a range of activities as possible within the constraints of the limited resources of the IOC. IOC provided input to the following activities: the Second World Climate Conference, SWCC, Geneva, 29 October–7 November 1990; the three Working Groups of the WMO-UNEP Intergovernmental Panel on Climate Change (IPCC), namely WG-I on Scientific Assessment of Climate Change, WG-II on Climate Change Impacts and WG-III

---

*EDITORS' NOTE.—This report is excerpted from the Intergovernmental Oceanographic Commission (IOC), "Report of the Secretary on Intersessional Activities (1 January–31 December 1989)," to the Twenty-third Session of the IOC Executive Council, Unesco, Paris, 7–14 March, 1990, and was supplied by Mr. Gunnar Kullenberg, Secretary IOC, Unesco, 7, Place de Fontenoy, 75700 Paris, France.

on Response Strategies; and many meetings of international organizations and working groups. In preparation for the 1992 UN Conference on Environment and Development focus could be on the development of appropriate ocean and coastal zone case studies with specific propositions for action by governments, including the establishment of required ocean and coastal zone operational observations, and monitoring activities.

World Climate Program—Data, Applications and Research Components

WMO is responsible for the overall coordination of the World Climate Program, and for the planning and implementation of the World Climate Data Program and the World Climate Applications Program. The World Climate Research Program is being jointly implemented by the Intergovernmental Council of Scientific Unions and WMO. Efforts are aimed at rescuing and preserving old records, automating climate data management at the national level, improving data management procedures, compiling information on station networks, data sets and sources, and disseminating information on significant climatic events of regional and global consequence. Climate projects include:

(1) The Global Energy and Water Cycle Experiment (GEWEX)—to observe, understand and model the global atmospheric hydrological cycle, energy budget and exchanges of moisture and energy with the underlying surface;

(2) The World Ocean Circulation Experiment (WOCE)—to provide, for the first time, almost simultaneous observations of all oceans as a basis for determining the global ocean circulation and heat transport; and

(3) The development of numerical models of the climate system—exploiting the scientific advances in individual disciplines and integrating the observations from diverse sources into a consistent global picture, so as to provide the basis for exploring the range of possible future climate variations.

The Joint SCOR-IOC Committee on Climatic Changes and the Ocean

At its Tenth Session (Halifax, Nova Scotia, 14–20 June 1989), the Chairman of CCCO reviewed the role of the Committee since the creation of TOGA and WOCE. It is increasingly important that scientific advice in a usable and appropriate format should be available when needed. Clearly, CCCO has a major role in meeting this requirement. The Committee agreed that the terms of reference of the three CCCO Tropical Ocean Climate Studies Panels (Tropical Atlantic Ocean Climate Studies Panel; Indian Ocean Panel; and Tropical Pacific Ocean Climate Study Panel) should all enlarge their perspectives beyond the tropics and beyond TOGA and WOCE.

Tropical Oceans and Global Atmosphere (TOGA)

The TOGA Scientific Steering Group (SSG) and several associated panels met in Hamburg, 18–22 September 1989, as guests of the Meteorological Department, Max Planck Institute, Hamburg University. The Intergovernmental TOGA Board endorsed the proposed Coupled Ocean Atmosphere Response Experiment (COARE). The TOGA Implementation Plan and extensive model intercomparison and validation studies are underway. All TOGA Data Management Centres are fully operational.

World Ocean Circulation Experiment (WOCE)

The WOCE Scientific Steering Group (SSG) met at Wormley, 24–26 October 1989, at the offices of the Institute of Oceanographic Sciences as guests of the WOCE

International Project Office. The Group reviewed the progress towards implementation of WOCE and noted the evolution from "scientific" towards "national" responsibility. Concern was expressed at further slippage of satellite programs, and at the uncertainty of arrangements for real-time access to some remote sensor data.

B. Ocean Science in Relation to Living Resources (OSLR)

Sardine/Anchovy Recruitment Project (SARP)
SARP is presently implemented in six regions: North Sea, Iberian Upwelling, SW-Atlantic, SE-Pacific, Gulf of California, and Japanese waters. The scientific and logistic details of the multi-national, cooperative METEOR cruise to investigate recruitment variability of the Southwest Atlantic anchovy off the coasts of Argentina, Brazil, and Uruguay, were prepared during the IOC Workshop on SARP in the Southwest Atlantic in Montevideo, 21–23 August 1989. Sprat recruitment in the North Sea was studied on three research vessels of Denmark, Federal Republic of Germany, and the United Kingdom. The *ad hoc* Expert Consultation on SARP (La Jolla, California, 30 October–1 November 1989) served (1) to promote interregional exchange of data and experience, (2) to broaden the SARP concept, and (3) to include new technology, such as remote sensing in SARP studies. The POLARMAR-IOC Course on SARP Methods (Bremerhaven, Federal Republic of Germany, 2–15 September 1989) was funded by W. Germany and IOC and attended mainly by Latin-American scientists.

Tropical Demersal Recruitment Project (TRODERP)
Participation in TRODERP has grown considerably, particularly within the PREP (Penaeid Prawn Recruitment Project) in the WESTPAC Region. The main recommendations of the Second IOC-FAO Workshop on the Recruitment of Penaeid Prawns in the Indo–Western Pacific Region (PREP) in Phuket, Thailand, 25–30 September 1989, were to strengthen the PREP network in the WESTPAC Region by obtaining international funding and to promote the Rainfall Emigration Experiment (REX). The IOC Workshop to define IOCARIBE-TRODERP Proposals in Caracas, Venezuela, 12–16 September, outlined four proposals for funding in the IOCARIBE Region: PREP (Prawn Recruitment Project), FEDERP (Fish Estuarine Demersal Recruitment Project), CORDERP (Coral Reef Recruitment Project) and SOAR (Satellite Ocean Analysis for Recruitment).

C. Ocean Science in Relation to Non-living Resources (OSNLR)

The Third Session of IOC-UN/OALOS Guiding Group of Experts on the OSNLR Program was held in Bordeaux, France, from 21 to 25 February 1989. The Guiding Group of Experts endorsed project proposals developed for WESTPAC and IOCEA Regions. The group requested that the IOCARIBE regional group of experts formulate specific projects with a practical workplan for their region.

Implementation of OSNLR at the Regional Level
The IOCEA project, "Sediment Budget along the West African Coastline", was endorsed by Guiding Group. A regional cruise to study sedimentary regimes and budget

of a coastal region took place in October 1989. Nigeria provided a research vessel free of charge and regional participants' costs were covered by IOC.

Regional projects for WESTPAC, namely Palaeogeographic Mapping and Margins of Active Plates, were developed by the regional coordinator for OSNLR and circulated among Member States for comment. Workplans for both projects are being drafted.

An IOC Advanced Training Course in Continental Shelf Structures, Sediments and Resources was organized for 14 participants, with the financial assistance of Federal Republic of Germany. It was held at the Philippines Mines and Geosciences Bureau, Quezon City, the Philippines, from 1–15 October 1989 and consisted of two weeks of lectures and workshops at the laboratory, as well as a two-day cruise with the Philippines research vessel *R.P.S. Explorer* to demonstrate geophysical surveys, sampling and preliminary analysis on board.

The Third Meeting of the Regional Group of Experts for the southwest Atlantic was held in Buenos Aires, Argentina, from 24 to 25 August 1989. A bibliography for marine geosciences research for Argentina, Brazil, and Uruguay has been prepared by the Group and will be published in the IOC Technical Series.

Co-operation with Related Programs and Organizations
The Fifteenth Session of the Joint CCOP-IOC Working Group on Post-IODE Studies in East Asian Tectonics Resources (SEATAR) was convened in Bangkok, Thailand, on 30–31 October 1989, in conjunction with the Twenty-sixth Annual Session of CCOP, 25 October–3 November 1989. The Working Group reviewed the progress of sixteen years of research of the transect studies and concluded that three or four of the transect maps of high scientific quality should be published as early as possible, followed by a compendium volume with comprehensive interpretation of regional tectonics and mineral resource potential.

IOC co-sponsored a Joint CCOP/SOPAC-IOC Workshop on Geology, Geophysics and Mineral Resources of the South Pacific, Canberra, Australia, 24 September–1 October 1989. The Workshop received forty-five scientific papers and discussed future joint research opportunities.

The Secretary IOC approached the Directors of CCOP and CCOP/SOPAC in order to establish letters of understanding to improve cooperation in project implementation. At the meeting of experts for Cooperation on Oceanic Minerals in the Southeast Pacific (Quito, Ecuador, 12–16 June 1989) organized by CPPS, it was agreed to form a joint CPPS-IOC Regional Study Group of Experts on OSNLR. An *ad hoc* IOC-ICSEM meeting (Monaco, 14–16 December 1989) was organized to develop a Joint Mediterranean Component for the OSNLR Program within the framework of the IOC-ICSEM agreement.

D. Ocean Mapping

There are three ocean mapping programs of the Commission: General Bathymetric Chart of the Oceans (GEBCO), Geological/Geophysical Atlases of the Atlantic and the Pacific Oceans (GAPA), and a number of regional projects. All these programs have been progressing successfully. The IOC-IHO Guiding Committee for GEBCO decided to establish a Database for which the name "the GEBCO Digital Atlas" was accepted. This Database will be subject to continual updating and consist of digital contours of the GEBCO 5th Edition together with ship tracks and a Gazetteer of the undersea feature names. The GAPA project is close to completion. The Atlantic volume will be printed 1989/1990. The Pacific volume publication is planned for 1992.

Regional Ocean Mapping Projects
(1) International Bathymetric Chart of the Mediterranean (IBCM) and its Geological/ Geophysical Series;
   (2) International Bathymetric Chart of the Caribbean Sea and the Gulf of Mexico (IBCCA);
   (3) International Bathymetric Chart of the Central Eastern Atlantic (IBCEA); and
   (4) International Bathymetric Chart of the WESTPAC Area (IBCWA).

E. Marine Pollution Research and Monitoring

IOC-UNEP Group of Experts on Methods, Standards and Intercalibration (GEMSI)
A logistics-planning group meeting was held in Hamburg (Federal Republic of Germany, 11–13 July 1989). Manuals on the determination of petroleum hydrocarbons in sediments and of chlorinated biphenyls in open ocean waters were finalized by GEMSI. Two IOC/GEMSI experts participated in the Workshop on Validation/Quality Assurance in the Analysis of Trace Metals and Organics in the Marine Environment (Brisbane, Australia, 28 August–1 September 1989).

IOC-UNEP-IMO Group of Experts on Effects of Pollutions (GEEP)
The Fifth Session of GEEP (London, 17–20 April 1989) focused on the outcome of the Second Major IOC/GEEP Workshop on the Biological Effects of Pollutants, 10 September–2 October 1988, and the planning of the next major research and training workshop at Xiamen, People's Republic of China, in 1991.
   The Workshop on Biological Effects of Contaminants is co-sponsored by ICES and IOC and will be held at Bremerhaven, Federal Republic of Germany, 12–30 March 1990.
   Due to the success of the first FAO-IOC-UNEP/GEEP Workshop on Statistical Treatment and Interpretation of Marine Community Data (Piran, Yugoslavia, 15–24 June 1988), FAO, IOC and UNEP organized a second successful workshop at Athens, Greece, 18–29 September 1989. Three manuals describing procedures for measuring the biological effects of pollutants were finalized during 1989.

IOC-UNEP-IAEA Group of Experts on Standards and Reference Materials (GESREM)
The NOAA catalogue of Standards and Reference Materials for Marine Science has been updated and the UNEP-IAEA-IOC document "Reference Methods and Materials: A Programme of Comprehensive Support for Regional and Global Marine Pollution Assessments (Reference Method Zero)" has been published.

The Marine Pollution Monitoring System (MARPOLMON)
Operation of marine pollution monitoring projects has continued in the Mediterranean, Caribbean, and West and Central African Regions. The pilot project on monitoring litter pollution in the Mediterranean was completed.

The First Session of the IOCARIBE Group of Experts on Marine Pollution Research and Monitoring was held in San Jose, Costa Rica, 29 August 1989. The IOC Secretariat at Cartagena, Colombia, has set up a beach litter monitoring project and a litter monitoring data bank.

A project on marine pollution monitoring, research, control and prevention for the coastal and marine environment in IOCINDIO will be initiated in early 1990 by IOC in cooperation with national institutions.

A GIPME/MARPOLMON expert mission to Mauritius took place, 13–18 June 1989, to evaluate it as a potential site for marine pollution monitoring and training activities for IOCINCWIO. Following the recommendations of the mission, training workshop is planned in Mauritius for Spring 1990.

Co-operation on Technical Matters concerning GIPME/MARPOLMON with Other Bodies
(1) United Nations Environment Programme (UNEP).—In accordance with the 1987 Memorandum of Understanding between IOC and UNEP, several joint projects have been initiated or further developed during 1989. These are:

Mediterranean: IOC continues to support the implementation of MEDPOL. An IOC-UNEP Review Meeting on Oceanographic Processes of Transport and Distribution of Pollutants in the Sea was organized in Zagreb, Yugoslavia, 15–18 May 1989. The IOC-FAO-UNEP pilot project on monitoring of litter on beaches and in seawater is completed and the results were reviewed at a meeting in Haifa, Israel, 12–15 June 1989.

Caribbean: An IOC-UNEP Regional Workshop to review priorities for a joint IOC-UNEP Comprehensive Program (CEPPOL) was convened in San Jose, Costa Rica, 24–30 August 1989. Close cooperation continues between the Regional Coordinating Unit at Kingston, Jamaica, where an IOC program coordinator has been outposted since 1986, and the IOCARIBE Secretariat in Cartagena.

South East Pacific: IOC has been involved in the CONPACSE Program through regional components of GIPME/MARPOLMON. Experts attended the CPPS-UNEP-IOC Workshop on Monitoring Strategies for Eutrophication in South East Pacific Coastal Waters (Cartagena, Colombia, 7–11 August 1989) and the CPPS-UNEP-IOC Regional Seminar on Research and Monitoring of Marine Pollution in South East Pacific Countries (Cali, Colombia, 6–8 September 1989). A UNEP-CPPS-IOC-ECO(OPS) Workshop on Environmental Management of Coastal and Marine Areas of the South East Pacific was held in Lima, Peru, 23–27 October 1989. IOC also participated at the Sixth Interagency Consultation on Ocean and Coastal Areas Programme of UNEP, Geneva, 4–7 September 1989.

(2) International Maritime Organization (IMO).—IOC activities in support to the London Dumping Convention (LDC) were presented to the LDC 12th Consultative Meeting at IMO, London, 30 October–3 November 1989. The planned ICES-IOC Workshop on Biological Effects of Pollutants (Bremerhaven, 12–30 March 1990) was presented to the meeting by a GEEP expert.

Following a request by IMO's Marine Environment Protection Committee, GEEP considered the scientific basis for the identification of vulnerable marine areas, on the basis of a discussion paper prepared by a GEEP ad hoc group.

The IOC Course Module on the Scientific Background to the London Dumping and MARPOL 73/78 Conventions was presented to World Maritime University students in May 1989 for a third year. Five IOC lecturers met in Paris, 23–24 November 1989, and discussed the revision of the Course Module.

The Fourth Intersecretariat Consultation on IMO-IOC cooperation on matters related to marine pollution, research and monitoring took place in Paris, 23–24 November 1989.

(3) International Atomic Energy Agency (IAEA).—Co-operation with IAEA/ILMR has continued with emphasis on development of methods, quality assurance programs and standards, and reference materials. Implementation of the joint IAEA-IOC-UNEP project has continued in supporting work of the GIPME Groups of Experts.

(4) Joint Global Ocean Flux Studies (JGOFS).—The IOC was represented by the Chairman of GIPME at the Third Session of the Planning Committee for JGOFS (12–14 September 1989) and the JGOFS Pacific Planning Meeting (14–16 September 1989) both held at Honolulu, Hawaii.

(5) Training, Education and Mutual Assistance (TEMA) Components of GIPME.—Several GIPME training courses/workshops were held in different regions during 1989. With advice and input from GEEP, the INDERENA-PAC-UNEP-FAO-IOC Regional Training Course on Bioassays and Toxicity Tests for Caribbean countries was held in Cartagena, Colombia, 11–24 June 1989. Following its success, a Second Bioassays Training Workshop for CPPS countries will be held at Guayaquil, Ecuador, 22–25 January 1990.

In the WESTPAC Region, an IOC/GIPME expert conducted a training course on heavy metals analysis and effects assessment in four countries: Indonesia, Malaysia, Philippines, and Thailand (16 June–21 July 1989). During 1990 a second expert will lead a training course for the same countries on the effects of non-oil pollutants, particularly organochlorines. IOC supported the organization of a regional training course on the use of mesocosms in marine pollution studies for participants from WESTPAC and other developing regions (Xiamen, People's Republic of China, 10–16 December 1989). Through an outposted IOC Associate Expert in Montevideo, Uruguay, the training of Uruguayan scientists in heavy metal analysis has been initiated. A study grant was provided to one scientist from Uruguay.

(6) Joint IMO-FAO-Unesco-WMO-IAEA-UN-UNEP-WHO Group of Experts on the Scientific Aspects of Marine Pollution (GESAMP).—The IOC Secretary as Unesco Technical Secretary for GESAMP, participated in the 19th Session of the Group in Athens, 8–12 May 1989. IOC continued to provide support to several GESAMP Working Groups through participation in meetings and use of common experts from GEMSI and GEEP.

OCEAN SERVICES

A. Integrated Global Ocean Services System (IGOSS)

Pursuant to Recommendation XIV-10 "Marine Meteorological and Sea Ice Information Services for Navigation in the Treaty Area of the Southern Ocean" adopted by the Fourteenth Antarctic Treaty Consultative Meeting (ATCM) in 1987, a meeting of experts on the improvement of marine meteorological and sea ice services was jointly convened by SCAR, WMO, and IOC, in Leningrad, 20–24 February 1989. The aim of the meeting was to consider improving or developing operational marine meteorological and sea-ice information services in the Treaty Area of the Southern Oceans (viz. south of 60°S).

Pending the approval of IGOSS-V Recommendations by IOC and WMO governing bodies, early 1990 was devoted to implementing IGOSS-V decisions, viz.: to establish two EGOSS Groups of Experts and a Task Team; to accredit the Australian Specialized Oceanographic Centre; to forward the Guides finalized during the previous year and the EGOSS regular information service bulletins; and to request international participation in the preparation of Guides or revision of old ones.

The IGOSS Operations Coordinator attended the UN IGOSS-IODE Meeting (Washington, D.C., 3–6 April 1989) as IOC representative, with the aim of defining ways and means of moving toward the establishment of a Global Integrated Ocean Observing System through coordinated national and international efforts.

The Third Joint IOC-WMO Meeting for Implementation of IGOSS XBT Ship-of-Opportunity Programs was held in Hamburg, at the kind invitation of the Federal Republic of Germany, from 16 to 20 October 1989. The meeting agreed on a standardized line numbering scheme by combining the existing TOGA, WOCE and IGOSS schemes, and defined precise ways and means to manage and monitor the ship-of-opportunity program.

The Symposium on Operational Fisheries Oceanography was held at the initiative of Canada in St. John's, Newfoundland, from 23 to 27 October 1989. The Symposium was co-sponsored with Canada by NOAA (of USA) and IOC. The Symposium was attended by 250 participants from 30 countries. It represented a global forum for the presentation of state-of-the-art applications of real-time oceanography to fisheries.

B. New Ocean Observing Systems

Drifting Buoys
Items dealt with by DBCP-V included: its relationship with the World Climate Research Program, the World Weather Watch of WMO, the Integrated Global Ocean Services System (IGOSS) and the future global ocean-observing system; various coordination activities in the field of quality control of drifting-buoy data, code matters, procedures for submitting data onto the WMO Global Telecommunication System, etc.; drifting-buoy-related publications; and review of its workplan for the next intersessional period.

The Ninth Meeting on the Argos Joint Tariff Agreement (JTA) (Geneva, 23–25 October 1989) focused on technical and financial matters regarding the operation of the Argos System as a whole, including: report on the 1989 Global Agreement to secure a preferential tariff to "authorized users"; development of CLS/Service Argos; review of overall users' requirements; new proposed structure of the Global Agreement after 1990; and the terms and conditions of the 1990 Global Agreement.

The Commission was represented at the European Group on Ocean Stations (EGOS) Management Committee and Technical Sub-group Meetings held in Dublin, Ireland, on 20 and 21 June 1989, and, upon EGOS request, hosted the following meeting at Unesco headquarters on 6 and 7 December 1989. EGOS is an action group of the DBCP. The EGOS Management Committee established an EGOS Technical

Sub-group "to co-ordinate the technical and operational activities of contributions to EGOS programs to ensure their successful implementation".

Global Sea-Level Observing System (GLOSS)
In accordance with the Resolution EC-XXI.2, the First Session of the IOC Group of Experts on GLOSS was held at Proudman Oceanographic Laboratory (Bidston, UK, 19–23 June 1989). It reviewed the progress of GLOSS and determined implementation priorities for 1990–1991.

The Group considered the progress in planning the EGOSS Sea Level Pilot Project in the North and Tropical Atlantic and recommended initiation of this project as soon as possible. The Group noted the satisfactory progress in developing the IOCARIBE regional component of GLOSS and approved activities of India regarding the operation of its sea level stations in the framework of GLOSS and TOGA.

The Group also reviewed TEMA activities related to GLOSS and noted that assistance (provision and installation of instruments, delivery of spare parts, training of specialists, visits of consultations etc.), had been provided by individual Member States.

The Seventh Training Course on Mean Sea Level Measurement Techniques was hosted by Proudman Oceanographic Laboratory (Bidston, UK, 12–30 June 1989) for four participants from Nigeria, Costa Rica, Seychelles, and Tanzania.

C. International Oceanographic Data and Information Exchange (IODE)

The Sixth Session of the IODE Consultative Meeting (Paris, 13–15 March 1989) formulated the Provisional agenda for the Thirteenth Session of IODE and made proposals on actions to be taken for preparation of the session, scheduled for 17–24 January 1990 (UN, New York). The IOC Assembly was informed about the recommendation of the ICSU World Data Centre Panel (Moscow, August 1989) that a World Data Centre (Oceanography) be established in China.

Two *ad hoc* consultative meetings were organized (USA, January 1989, and Canada, July 1989) to initiate planning of the Global Temperature Salinity Pilot Project (GTSPP) that will unite historical IODE data holdings and quality-controlled IGOSS real-time data in an integrated database that will be continuously updated. The Workshop on GTSPP will consider the draft plan for GTSPP in New York, 15–16 January 1990, and at the Thirteenth Session of the Committee on IODE (New York, 17–24 January 1990).

Technical Aspects of Data Exchange and Data Monitoring
The ROSCOP form is used to record what types of oceanographic data have been collected on a research cruise or an analogous observational activity. This information is then available for international exchange and archiving in the IODE system. Work has continued during the year, in collaboration with ICES, to finalize a new edition of the form. The MEDI directory system contains short structured plain text entries describing data holding organizations and available data sets and data inventories. Entries have now been received from 2 international organizations and from 35 national organizations in 24 Member States describing a total of 207 data sets. MEDI information is available on diskette from the IOC Secretariat, and is being loaded

onto the NOSIE online information system operated by the US NODC. Consultations have been held with WEC-B, Oceanography, Obninsk, USSR, on their possible collaboration in the operation of the system.

Promotion of the GF3 Format for the exchange and archiving of data has continued. During the year IOC Manuals and Guides No. 17, Volume 4, User Guide to the GF3-Proc software, and Volume 6, Quick Reference Sheets for GF3 and GF3-Proc, have been published in English, and the French, Spanish and Russian versions are in progress.

IODE Training Activities

An IOC-Unesco Training Course on the Use of Microcomputers for Oceanographic Data Management was held at the Asian Institute of Technology, Bangkok, Thailand, 16 January–3 February 1989. Fifteen trainees from nine countries in the WESTPAC Region and four additional observers from the host country attended. The Japan Oceanographic Data Center, Tokyo, held its Eighth Oceanographic Data Management Course for the WESTPAC Region, 25 September–7 October 1989, attended by five specialists from China, Indonesia, Republic of Korea, Malaysia, and Thailand. Support was provided for a specialist from Ecuador to follow a training program on oceanographic data management from 23 October to 3 November 1989 at CEADO, Buenos Aires, the Argentinean NODC.

Marine Information Management (MIM)

During the year work has concentrated on strengthening IOC support to the FAO-IOC-UN Aquatic Sciences and Fisheries Information System (ASFIS) so as to improve the information services available to the marine science community. Close contacts with FAO and the UN have been maintained.

D. International Tsunami Warning System in the Pacific (ITSU)

The Twelfth Session of the International Co-ordination Group for the Tsunami Warning System in the Pacific (ICG/ITSU) took place in Novosibirsk, USSR, 7–10 August 1989. The representatives of 12 ITSU Member States and international bodies (IUGG, ITIC) participated in this session. The Session reviewed the development of the Tsunami Warning System in the Pacific since 1987 and noted that the national and regional Tsunami Warning Systems have improved. The Session recommended that efforts be continued by IOC in preparation of the project for Sub-regional Tsunami Warning Systems in the South West Pacific.

The Atlas of Tsunami Travel Time Charts, the Master Plan and the Communication Plan for the Tsunami Warning System have been published, as well as the Tsunami Brochure. The Session approved the immediate publication of the first edition of a Glossary of Tsunami.

Much attention was given to participation in the International Decade of Natural Disaster Reduction (IDNDR) and in this connection the Session recommended the preparation of two projects as a direct contribution to IDNDR: Program on Tsunami Public Awareness and Education, and the Project on Tsunami Disaster Mitigation.

The IOC Workshop on the Technical Aspects of the Tsunami Warning System (Novosibirsk, USSR, 4–5 August 1989) covered all the aspects of international cooperation pertaining to scientific and engineering research, and the coordination related to the mitigation of the tsunami hazard in the context of the International Decade of Natural Disaster Reduction. The report of the Workshop will be published in the IOC Workshop Reports Series.

REGIONAL ACTIVITIES

Caribbean and Adjacent Regions (IOCARIBE)

The IOCARIBE Newsletter (*IOCARIBE* News) was established in the second half of 1989, and 2 issues have been published. In order to review preparations for SC-IOCARIBE-III and decide on measures to increase interest and participation in IOCARIBE activities of Member States, IOC consultations were held in Cartagena, 9–12 October 1989, and followed by missions to some member states.

The Third Session of the Sub-commission for the Caribbean and Adjacent Regions was organized in Caracas, Venezuela, 4–8 December 1989, at the invitation of Venezuela, preceded by a Scientific Seminar, 28 November–2 December 1989. Eleven Member States and four cooperating organizations were represented at the Session and the draft IOCARIBE Medium-Term Plan 1990–95 was formulated.

Following the outlines established by the IOCARIBE Workshop on Physical Oceanography and Climate in 1986, a detailed project proposal "Ocean Circulation in the Caribbean Sea and Adjacent Regions: An Intergovernmental Oceanographic Commission Proposal to the European Space Agency for ERS-1 Science, Application and Validation" was approved by the European Space Agency (ESA) as part of the ERS-1 investigators plan. Central to this IOCARIBE project is an operational network of sea-level/weather stations throughout the region, a subset of which are GLOSS stations. Five tide gauges donated by NOAA to IOCARIBE Member States are being established. UNEP, in consultation with IOC, initiated regional studies within the framework of its Regional Seas Programme, on the possible impacts of sea-level and sea-surface temperature changes induced by climatic changes. Through consultations between UNEP/RCU/CEP and IOC/IOCARIBE, a Regional Task Team was established to prepare a study for the Wider Caribbean Region.

The IOCARIBE Secretariat has developed a major project proposal on associated effects of sea-level rise, sea-surface temperature change, ocean-transport processes, and other environmental changes. The proposal contains the following main elements: (1) physical oceanography and climate changes; (2) coastal sedimentation mechanisms and erosion problems; and (3) dispersion of pollutants in certain key areas.

The IOCARIBE regional component of the GIPME-MARPOLMON-CARIPOL Petroleum Pollution Monitoring Program has continued to be most successful and to date about 11,000 data points on beach tar, floating tar, dissolved and dispersed petroleum hydrocarbons (DDPH's) have been reported to the Regional Data Bank.

The Secretariats of UNEP and IOC convened a joint Regional Workshop to review priorities for Marine Pollution Monitoring, Research, Control, and Abatement in the Wider Caribbean Region in San Jose, Costa Rica, 24–30 August 1989. The Workshop was attended by 40 regional experts from 17 countries and representatives from international organizations. A framework was developed for a Regionally Coordinated Comprehensive Program for Marine Pollution Assessment and Control for the Caribbean (CEPPOL). The proposal for the CEPPOL Program was submitted to the Third Session of IOCARIBE and to the Fifth Intergovernmental Meeting of the Action Plan for the Caribbean Environmental Program and the Second Meeting of Contracting parties to the Cartagena Convention (Kingston, Jamaica, 10–18 January 1990).

The first Session of the IOCARIBE Group of Experts on Marine Pollution Research and Monitoring was convened on 29 August 1989 in San Jose, Costa Rica, in conjunction with the Workshop. The meeting discussed the ongoing CARIPOL Program and recommended that a monitoring program on marine debris should be included in CARIPOL-I.

Within the Ocean Science in Relation to Living Resources program, a workshop to further define IOCARIBE-TRODERP proposals was held in Caracas (Venezuela, 12–16 September 1989). The workshop finalized the three original Sub-project Proposals and in addition, a sub-project on Satellite Ocean Analysis for Recruitment (SOAR) was completed.

Western Pacific (WESTPAC)

Several activities have commenced or continued during 1989. Examples are: The IOC Assembly made a decision at its Fifteenth Session to establish the Sub-commission for the Western Pacific to replace the Regional Committee in order to strengthen regional activities. Progress has been made during the intersessional period towards the implementation of the cooperative research projects approved by the Fourth Session of WESTPAC (Bangkok, Thailand, 22–26 June 1987).

The IOC consultation meeting for the project on Toxic and Anoxic Phenomena Associated with Algal Blooms in the WESTPAC Region reviewed the red tides that have occured since the International Red Tide Symposium was held at Takamatsu, Japan, November 1987. Discussions were also held on the paralytic shellfish poisoning problems in the region and a workplan was formulated for this WESTPAC project. Support for the meeting was provided by Japan through its Funds-in-Trust in support of the WESTPAC Program.

A Management and Training Workshop on Pyrodium Red Tides was organized at Bandar Seri Begawan, Brunei Darussalam, 23–30 May 1989, by the Government of Brunei and ICLARM with the scientific and financial support from IOC as a part of the regional program.

The Second Workshop on Penaeid Prawn Recruitment in the WESTPAC Region was organized at Phuket, Thailand, 25–30 September 1989, and was funded by IOC and other organizations. It was attended by 16 scientists from Australia, China, Indonesia, Malaysia, Papua New Guinea, the Philippines, and Thailand. To strengthen interregional exchange of experience and data, two prawn researchers from Mexico and Spain were invited. The Project on Recruitment of Penaeid Prawns in the Indo-Western Pacific Region (PREP) uses a comparative geographic approach to enhance the understanding of the effects of fishing and environmental impact on penaeid resources to promote better management-oriented research on the prawn resources.

Based on the outcome of the Workshop on Marine Geology/Geophysics in the WESTPAC Region (Shanghai, China, 6–7 September 1988) and the Recommendations adopted by the Guiding Group of Experts on OSNLR at its Third Session (Paris, 2–5 March 1989), draft action plans are being prepared for the Project on a WEST-PAC Palaeogeographic Map and for the Project on Margins of Active Plates (MAP).

The IOC Workshop on a Co-operative Study of the Continental Shelf Circulation in the Western Pacific was organized at Chulalongkorn University, Bangkok, Thailand, 31 October–3 November 1989. Scientists from Australia, China, the Democratic People's Republic of Korea, Federal Republic of Germany, Indonesia, Japan, Malaysia, Thailand, and Viet Nam reviewed the status of physical oceanography studies being undertaken in their home countries with particular reference to continental shelf and coastal zone areas.

A Workshop on the Use of Sediments in Marine Pollution Research and Monitoring, originally planned for September 1989 in Dalian, China, has been postponed until Spring 1990. A consultation meeting for the project on the Assessment of River Inputs to the Seas in WESTPAC Region will follow the sediments workshop.

As part of WESTPAC program, the IOC-Unesco Training Course on the Use of Microcomputers for Oceanographic Data Management was organized at the Asian Institute of Technology, Bangkok, Thailand, 16 January–3 February 1989, with a view to increase the flow of oceanographic data and information in the region through IODE system.

An IOC Advanced Training Course on Continental Shelf Structures, Sediments and Resources was organized at the Mines and Geosciences Bureau, Manila, the Philippines, 1–15 October 1989.

North and Central Western Indian Ocean (IOCINCWIO)

The Editorial Board for the International Bathymetric Chart of the Western Indian Ocean (IBCWIO) has now been established, and its First Session was held in Antananarivo (Madagascar, April 1989).

In connection with the project on marine pollution monitoring and research, a GIPME/MARPOLMON expert mission to Mauritius took place, 13–18 June 1989, and following the recommendations of this mission, a Training Workshop is planned in Mauritius for Spring 1990.

RECOSCIX-WIO (Regional Co-operation in Scientific Information Exchange–Western Indian Ocean) is a marine science information network for the IOCINCWIO Region within the framework of ASFIS, building on the information component of the bilateral Kenya-Belgium Project in the Marine Sciences. Following the acceptance of the project by IOCINCWIO-II in December 1987, an IOC Associate Expert has established a Regional Dispatch Centre hosted by the Kenya Marine and Fisheries Research Institute, Mombasa. An ASFA CD-ROM system provided by the USA to IOC under VCP has been installed and is operational, as are the ASFA CD-ROM query and document delivery services. A library holdings database, WIOLIB, is being designed and will implement Unesco's mini-micro CDS/ISIS software at each participating library in the region. This will be a foundation for improved library management and regional sharing of scientific documentation. Initial funding has been provided by the Commission to start RECOSCIX-WIO on a pilot scale; extra-budgetary funds for full implementation are now being sought.

Central Eastern Atlantic (IOCEA)

A major event was the organization of the first regional oceanographic cruise as part of the implementation of the coastal erosion project, "Sediment Budget along the West African Coastline", within the framework of OSNLR. The cruise took place from 10 to 25 October 1989, on the R/V Sarkim Baka, which was provided by the Government of Nigeria. Analysis of the collected samples has started and a 3-day scientific meeting is scheduled to take place in Accra, Ghana, in January 1990, to prepare a final cruise report.

Preparations for the Second Session of the IOC Regional Committee for the Central Eastern Atlantic (Lagos, Nigeria, 5–9 February 1990) have been made and Nigeria will host the Session.

Central Indian Ocean (IOCINDIO)

A number of projects are progressively being implemented. The first meeting of the Steering Group on Coastal Water Dynamics was held in NIO, Karachi, 15–20 January 1989. The IOC-UNEP Task Team on Impacts of Climatic Changes on Coastal Zones and Areas in the South Asian Seas has collected various reports on the impact of sea level rise and case studies and are compiling a consolidated report. A Coastal Environment Management Program has been prepared under the guidance of ESCAP for Bangladesh, Sri Lanka and is being finalized for Pakistan. India has already prepared the Coastal Zone Management Program. These programs are useful for application in the IOCINDIO projects. A training workshop in marine pollution research and monitoring for the region is being organized with the National Institute of Oceanography, Goa, India, as the host institution.

Southern Ocean (IOCSOC)

IOC activities in the Southern Ocean are mainly involved in cooperative and coordinated investigations with other international organizations dealing with the problems of the marine environment in this area (the Antarctic Treaty, SCAR, SCOR, WMO, CCAMLR), planning of WOCE Core Project 2 "Southern Ocean" by WOCE SSG, and development of ocean observing systems through IGOSS and GLOSS.

The Consultative Meeting supported the proposal of SCAR to organize in 1991 the Antarctic Science Conference. The Secretary IOC approached the Chairman of SCAR with the offer to co-sponsor the Conference.

The joint SCAR-WMO-IOC scientific meeting on improving marine meteorological and ice information services for the Antarctic Treaty area of the Southern Ocean was held in Leningrad (USSR) in February 1989. The report and recommendations of the meeting were submitted to the Fifteenth Consultative Meeting of the Antarctic Treaty, Paris, 9–19 October, 1989.

The question on the improvement of the temperature-salinity observations in the Southern Ocean was discussed at the IGOSS Ship-of-Opportunity Meeting (Hamburg, FRG, 16–20 October 1989).

Development of sea-level observations in the Southern Ocean was considered by the IOC Group of Experts on GLOSS at its First Session (Bidston, UK, 19–23 June

1989) as well as by the WOCE SSG at its Thirteenth Session (Wormley, UK, 24–26 October 1989). Upon the recommendations of those groups, the *ad hoc* expert consultation on sea-level measurements in the Southern Ocean will be organized at the USSR Antarctic Research Institute in May 1990.

The WOCE SSG Working Group on WOCE Core Project 2 "Southern Ocean" continued the plan for the implementation of the Core Project 2 including the ocean observing system (moorings, float and drifter program, WOCE Hydrographic Program, surface fluxes and sea-level) and ACC modelling. The WOCE SSG at its Thirteenth Session (Wormley, UK, 24–26 October 1989) reviewed the status of the WOCE Core Project 2 planning and made recommendations on further actions needed.

Some IOC Member States are already responsible for the collection and processing of oceanographic data and preparation of oceanographic data products for the Southern Ocean Area. Within the IGOSS Program, Australia acts as IGOSS Specialized Oceanographic Centre to cover the Indian and South Pacific Oceans, including relevant sectors of the Southern Ocean, and within IODE, Argentina acts as Responsible National Oceanographic Centre for the Southern Ocean.

South Eastern Pacific (SEPAC)

The IOCARIBE Secretariat has participated in and closely followed the activities of the CPPS-UNEP Action Plan for Marine Pollution Research and Monitoring in the South-eastern Pacific, including the regional review on the Health of the Oceans and Impacts of Climatic and other Environmental Changes. The Permanent Commission for the South Pacific (CPPS) is the regional coordination unit for UNEP's South East Pacific Action Plan. The UNDP project on studies of "El Niño" will be jointly implemented by IOC and CPPS. At its Twentieth Session in Bogotá, Colombia, 28 August–1 September 1989, CPPS endorsed the establishment of a joint regional Group of Experts on OSNLR.

Southwest Atlantic

The IOC Associate Expert in Marine Pollution Research and Monitoring is supported by Sweden and stationed in Montevideo, and has implemented several activities. A meeting of experts working in relation to the OSNLR program was carried out in Argentina (24–25 August 1989). This meeting recognized that research in marine and coastal geology in the Southwest Atlantic is progressing in spite of the lack of funds. The meeting recommended greater regional communication and decided to hold a workshop in Maldonado, Uruguay, in 1990. The objective of this workshop will be to update the chart on the geology and geomorphology of the coast and the continental shelf of the Southwest Atlantic.

Mediterranean

The *ad hoc* meeting of experts, with IOC support, at Instituto de Ciencias del Mar (Barcelona, Spain, 9–11 March 1989) considered the provisional version of the International Research Program in the Western Mediterranean (PRIMO, in French) on physical oceanography and adopted the scientific aims of the proposed program. Working Groups were established to prepare files which will assist in drawing-up a Plan of Action for the proposed research.

The Assembly, at its Fifteenth Session (Paris, 4–19 July 1989), approved the proposal to set up this project (PRIMO) on the Western Mediterranean and ICSEM will co-sponsor it. IOC established an *ad hoc* group of experts to further develop, promote, and coordinate this project.

The Third Workshop on the Physical Oceanography of Eastern Mediterranean (POEM) was held at Harvard University, Cambridge, Massachusetts, 29 May–2 June 1989. The Workshop established a global scientific context for the emerging Mediterranean ocean science and focused on interdisciplinary scientific advances. Phase I of POEM ends on 31 December 1990. Phase II is planned from 1991 through 1997.

Other Areas of Cooperation

ICSPRO: The Inter-Secretariat Committee on Scientific Programs relating to Oceanography.

UN General Assembly and ECOSOC: IOC prepared a statement on "Marine Research and Ocean Services: Opportunities for Progress" for the 1989 Session.

ICSU: Consultations on interactions with IGBP were held at the First Interagency Co-ordinating Committee Meeting on IGBP with participation from ICSU, SC-IGBP, WMO, UNEP, Unesco and IOC (London, 11 January 1989); consultations on implementation strategy for ICSU-IGBP, and Second Interagency Co-ordinating Committee (Lisbon, 12–14 October 1989); consultations with SCOR on the intergovernmental structure for WOCE and matters related to CCCO (Washington, D.C., 1–3 May 1989); intersecretariat consultations on cooperation and matters related to the Fifteenth Assembly (Paris, 30 November 1989); consultations on the cooperation between Unesco and ICSU-IGBP (Paris, 26 February 1989), reaching an agreement of various specific tasks for different programs of Unesco, including those of IOC; and consultations on cooperation with the president of SCAR as regards Southern Oceans (24 October 1989).

IMO: Representation at the 12th Session of the London Dumping Convention Meeting (London, 30 October–3 November 1989), presenting a report on activities of relevance within GIPME; intersecretariat consultations (London, 12 January 1989, and Paris, 23–24 November 1989) reviewing all collaborative activities and preparing workplan for the future; intersecretariat consultations also in conjunction with GESAMP 19 (Athens, 8–12 May 1989); and presentation of the IOC Course Module on the Scientific Background for the Global Marine Pollution Control Conventions (MARPOL and LDC) at the World Maritime University (Malmo, 22–26 May 1989).

WMO: Consultations on collaborative ocean observations and climate-related research have been held on several occasions and IOC has also been participating in the International Organizing Committee for the Second World Climate Conference, and interagency discussions on climate-related programs (Paris, 20 January 1989; Geneva, 9 February 1989, and 30 October 1989; Paris, 30–31 October 1989; Geneva, 21 November 1989; Vienna, 12–13 December 1989; and Paris, 6–8 February 1989), in conjunction with the Tenth Session of the WMO Commission on Marine Meteorology, hosted by Unesco.

UNEP: Cooperative consultations have been held on several occasions. IOC also attended the Fifth Interagency Consultation arranged by UNEP, OCA/PAC (Geneva, 4–6 September 1989); GIPME consultations (Paris, 23 February 1989; Kingston, Jamaica, 8–11 March 1989; Athens, Greece, 8–12 May 1989; San Jose, Costa Rica, 23–30 August 1989; Geneva, 20–21 November 1989; Caracas, Venezuela, 4–7 December 1989); and the Secretary of IOC represented Unesco and IOC in the Session of the IPCC WG-II on Climate Impact Assessments (Geneva, 31 October–2 November 1989).

UN/OALOS: The Secretary of IOC participated part-time in the preparatory meeting of the UN Sea-Level Authority Preparatory Commission (Kingston, Jamaica, 8–11 March 1989), during which consultations on IOC-UN/OALOS cooperation were also held, and at the expert consultation arranged by UN/OALOS (New York, 6–7 September 1989). The IOC was represented at the *ad hoc* intersecretariat consultations arranged by UN/OALOS (Geneva, 12–14 July 1989).

CPPS: Considerable cooperation occurs with CPPS, over a widening spectrum, including marine pollution, non-living resources, El Niño, ocean processes, and climate. The IOC was represented at the Twentieth Session of CPPS (Bogota, Colombia, 28 August–1 September 1989).

ICES: The IOC was represented at the ICES Statutory Meeting (The Hague, the Netherlands, 6–11 September 1989), and in several ICES Working Group Meetings, including data exchange, marine pollution, recruitment; cooperative projects are being developed within IREP and within marine pollution research and monitoring through GIPME-GEEP.

ICSEM: Cooperation with ICSEM has been strengthened through the launching and joint sponsorship of the new research program (PRIMO) for the Western Mediterranean in the field of the physical oceanography presented to and adopted by IOC-XV. Following the recommendation of the Third Session of the Guiding Group Experts on OSNLR endorsed by the Assembly, IOC and ICSEM agreed to develop a Mediterranean component of the OSNLR Program focused on cooperative studies on sedimentary dynamics and coastal erosion. A preliminary *ad hoc* consultation of experts will take place at ICSEM Headquarters in Monte Carlo, 14–16 December 1989. It is also planned to hold an IOC-ICSEM Workshop on this subject in conjunction with the ICSEM Congress and Plenary Assembly in October 1990.

Other bodies: Joint activities with non-governmental organizations, especially the scientific advisory bodies of the Commission, have continued during the intersessional period. Specific joint programs are reported under relevant sections of this report.

Reports from Organizations

IMO International Maritime Law Institute (IMLI)*

The IMO International Maritime Law Institute (IMLI) has been established as an international center for the training of specialists in maritime law, including the international legal regime of merchant shipping and the law of the sea, with special emphasis on enhancing protection of the marine environment by facilitating implementation of the numerous international conventions, regulations, and procedures for furthering vessel safety and prevention of pollution concluded by the International Maritime Organization (IMO), a specialized agency of the United Nations, under the auspices of which the Institute has been established, as well as related conventions, such as those negotiated through UNEP (United Nations Environment Programme) and UNCTAD (UN Conference on Trade and Development), or *ad hoc*.

The Institute is situated on the island of Malta, which abounds in history and culture and has a rich maritime heritage and interest in the development of the maritime legal regime. The venue for the Institute thus is a most appropriate one. Under an Agreement concluded between IMO and the Government of Malta, the Government has provided attractive premises and facilities for the establishment and operation of the Institute. Accommodation for up to 20 students is available on campus, in handsomely appointed individual rooms, each with their own cooking and bathroom facilities.

The Institute opened on October 2, 1989. The official inauguration ceremony of the Institute, at which Dr. C. P. Srivastava, then the Secretary-General of the IMO, formally inaugurated the Institute, took place on November 4, 1989. The first meeting of its Governing Board took place thereafter in Malta.

Seventeen students have been in residence during 1989–90, each from a different developing country; there were also two Maltese students. All the students are graduate lawyers nominated by their government. They hailed from Sri Lanka, Somalia, Western Samoa, Seychelles, Panama, Philippines, Malawi, Ethiopia, Malaysia, Nigeria, Hong Kong, Bangladesh, Commonwealth of Dominica, Trinidad and Tobago, Papua New Guinea, Pakistan, and Bahrain, as well as Malta. The students ranged in age from 25 to 47 years and five of them were women. About half of the students had a maritime-related legal background, such as in a national shipping company; the rest were working in their governments' legal departments, such as the department of Justice or Attorney-General's office, since it is required that students are working or will work in official fields where they will be able to use the skills inculcated at IMLI.

The academic program for 1989/90 consisted of two main areas of study, namely, law of the sea, including environmental law aspects, and shipping law. The topics included public international law, international institutions, marine environmental law, international law of the sea, and commercial and regulatory shipping law. In addition, training was provided in the development and drafting of maritime legisla-

*EDITORS' NOTE.—This report was written by Professor Patricia Birnie, Director, IMO International Maritime Law Institute, P.O. Box 20, St. Julian's, Malta.

414

tion, in particular on the means of incorporating IMO and other maritime conventions into domestic legislation. Each student undertook, as part of the examination requirements, the drafting of a convention of his or her choice into national law. Each also had to write an 18,000-word essay on a maritime-related topic of some relevance to his or her country or region. The final evaluation of the students was based on two written examinations (on the Law of the Sea and on Shipping Law, respectively), the essay, and the drafting project. Successful candidates were awarded a Masters Degree in International Maritime Law.

The field of study is highly specialized, and the program as designed is intensive and demanding. It is expected that IMLI graduates, upon returning to their respective countries, will be able to facilitate the implementation of the many IMO and related international maritime and marine environment protection conventions, by assisting in their incorporation into their domestic legislation, and thus establishing sound maritime regimes comparable to those in developed countries, which nonetheless will be appropriately adapted to, and sensitive of, local needs.

The first Director of the Institute is Professor Patricia Birnie, formerly of the London School of Economics, a specialist in international law, especially law of the sea and international environmental law. The Senior Deputy Director is Professor P. K. Mikherjee of Canada, a Master Mariner and maritime lawyer, who is also experienced in the drafting of maritime legislation. Professor David J. Attard of the University of Malta is Special Adviser to the Director and a part-time member of the Faculty, teaching certain topics in international law and the law of the sea. In addition, there are three part-time lecturers who are Maltese maritime law practitioners with foreign postgraduate qualifications in maritime law. They teach selected topics in shipping law, and one assists with tutorials. Many distinguished visiting fellows drawn from both academic and maritime law practice and from international organizations and government legal offices in a variety of developed and developing countries have also given courses of lectures.

Needless to say, marine pollution is an important and major topic in the program. Several students have written their essays or based their drafting projects, or both, on the various aspects of this topic.

The Institute has had a successful start; its first graduation ceremony took place on June 23, 1990, at which the address to the new graduates was delivered by Judge Sir Robert Jennings, Q.C., a member of the International Court of Justice, who stressed the indispensable role of international law in the modern world and the importance of developing countries in its formulation and implementation. It was also attended by Dr. Srivastava, now Secretary-General Emeritus of the IMO, who is Vice-Chairman of the IMLI Governing Board. Three students graduated with distinction and five won prizes. The Malta Government Prize for best essay, presented by Dr. Fenech, the Parliamentary Secretary for Maritime Affairs, was awarded to Ms. Nkemdilim Nwandu (Nigeria) and Mr. Nimal Amaratunga (Sri Lanka) for their essays—respectively, "Shipbuilding and Shiprepair: Comparative Legal Perspectives" and "The Law of the Sea and Customary International Law." The IMO Secretary-General's Prize, presented by Dr. Srivastava on behalf of Mr. O'Neill (the IMO Secretary-General), for the best drafting project went to Ms. Gail Royer (Commonwealth of Dominica) and the IMLI prize, presented by the Director, for the best examination results was shared by Ms. Cathy Pui-ming Wong (Hong Kong) and Mr. John Chilundu Kondowe (Malawi).

The second course of IMLI will begin on October 1, 1990. A considerable number of applications have already been received. It is intended as far as possible, subject to availability of suitably qualified candidates, that the student body will be representative of developing countries different from those represented on the first course and that half (10) of those chosen should be women, since the primary aims of IMLI are to provide as many developing countries as possible with trained maritime lawyers competent to assist in the drafting of legislation to protect the marine environment, and to further the education of women from the developing world. All involved have been heartened by the cooperation and support received in launching IMLI on its successful maiden voyage.

Reports from Organizations

Maritime Report from Italy: Excerpts from *Italy and the Law of the Sea Newsletter*, no. 21 (February 1990)*

Italy and the Law of the Sea Newsletter (ILSNL) is published from time to time within the framework of research activities conducted in the University of Parma with the grant of the Italian Council for Scientific Research (CNR). It contains information in English about developments in Italy concerning the law of the sea, the protection of the marine environment, and Italian participation in the Antarctic Treaty system. The editors are Tullio Treves and Tullio Scovazzi. The assistant editors are Maria Clara Maffei and Laura Pineischi.

RESEARCH ACTIVITIES

The second edition of *Atlas of the Straight Baselines*, edited by T. Scovazzi, G. Francalanci, D. Romano and S. Mongardini, has been published. It contains 131 maps relating to bays and to coastlines deeply indented or fringed by islands.

A Conference on "International Responsibility for Environmental Harm Resulting from Industrial Activities" was held in Siena on 23–24 March 1990. It is organized as part of a research project financed by the CNR and carried out in the Universities of Siena and Parma under the direction of F. Francioni and T. Scovazzi.

MEETINGS

7th International Meeting "Sea and Territory", organized by the Italian Naval League and by the University of Palermo, was held in Palermo and Lampedusa on 5–8 June 1989. The meeting was coordinated by Prof. Guido Camarda.

ITALIAN TREATIES

Bilateral Treaties

Algeria—Agreement on transport and maritime navigation, signed in Algiers on 15 April 1986, in force from 1 October 1989 *Gazzetta Ufficial della Republica Italiana* (G.U. 203 of 31 August 1989).

France—Convention of the delimitation of the maritime boundaries in the area of the Mouths of Bonifacio, signed in Paris on 28 November 1986, in force from 15

*EDITORS' NOTE.—This report has been excerpted from *Italy and the Law of the Sea Newsletter*, no. 21 (February 1990). Copies of the *Newsletter* may be requested from the ILSNL headquarters, Istituto di Diritto Internazionale, Via Universita 12, 43100 Parma, Italy.

May 1989 (G.U. 131 of 7 June 1989). The Convention delimits the opposite 12-mile territorial sea of the parties in the Mouths of Bonifacio, between the islands of Corsica (France) and Sardinia (Italy). The delimitation line extends for a distance of about 40 n.m. and connects 6 points by 5 straight segments.

Multilateral Treaties

6 April 1974—Convention on a code of conduct for liner conferences (Geneva), in force for Italy from 30 November 1989 (G.U. 203 of 31 August 1989). Italy made a reservation and a declaration.

27 April 1979—International Convention on maritime search and rescue (Hamburg), in force for Italy from 2 July 1989 (G.U. 184 of 8 August 1989).

ITALIAN LEGISLATION

Ordinance for the Minister for the Coordination of Civil Protection of 19 April 1989 (G.U. 97 of 27 April 1989). Urgent measures for the collection and utilization of algae proliferated in the lagoon of Venice.

Decree of the Minister of the Environment of 26 April 1989 (G.U. 128 of 3 June 1989). Rules on the guarantee for transfrontier shipment of wastes.

Decree of the Minister of Merchant Marine of 3 May 1989 (G.U. 113 of 17 May 1989). Rules of fishing of bivalve molluscs.

Decree of the Minister of Merchant Marine of 3 May 1989 (G.U. 113 of 17 May 1989). Prohibition of fishing, catching, transporting and trading of cetaceans, turtles and sturgeons.

Decree of the Minister of Merchant Marine of 4 May 1989 (G.U. 113 of 17 May 1989). Rules of the temporary stop of fishing vessels.

Law No. 171 of 5 May 1989 (G.U. 109 of 12 May 1989). New regulations on pleasure craft navigation.

Law No. 183 of 18 May 1989 (G.U. Suppl. 120 of 25 May 1989). Rules for the protection of the ground. It relates also to the public intervention and planning for the protection of rivers, watercourses, wetlands and for the protection of the coasts from the invasion and erosion by maritime waters.

Decree of the Minister of Merchant Marine of 8 June 1989 (G.U. 146 of 24 June 1989). Institution of the Italian coast guard.

Law-decree No. 227 of 13 June 1989, converted with modifications into Law No. 283 of 4 August 1989 (G.U. 185 of 9 August 1989). Urgent measures against eutrophication of the coastal waters of the Adriatic Sea.

Decree of the Minister of the Environment of 22 June 1989, No. 295 (G.U. 194 of 21 August 1989; errata corrige in G.U. 213 of 12 September 1989). Financing of interventions against eutrophication.

Decree of the Minister of the Environment of 28 June 1989 (G.U. 159 of 10 July 1989). Additions to the Decree of 26 April 1989 on the guarantee of transfrontier shipment of wastes.

Decree of the Minister for the Coordination of Civil Protection of 8 July 1989

(G.U. 161 of 12 July 1989). Rules for the provisional stocking of the industrial wastes coming from Nigeria and carried by the ship "Deep Sea Carrier".

Ordinance of the Minister for the Coordination of Civil Protection of 8 July 1989 (G.U. 172 of 25 July 1989). Exceptional measures on the operations to be carried out in the harbour of Leghorn and in the Region of Tuscany for the analysis, provisional stocking and definitive disposal of industrial wastes coming from Nigeria and carried by the ship "Deep Sea Carrier".

Decree of the Ministers of Merchant Marine and for the Cultural Heritage of 12 July 1989 (G.U. 175 of 28 July 1989). Rules on the protection of marine areas of historical, artistic, and archaeological interest.

Decree of the Minister of the Environment of 14 July 1989 (G.U. 295 of 19 December 1989). Creation of a marine natural reserve in the Archipelago of Tremiti in the Adriatic Sea. The reserve includes the waters delimited by the 70-meter isobath around the islands of San Domino, San Nicola, Caprara, and Pianosa.

Decree of the Minister of Merchant Marine of 20 July 1989 (G.U. 181 of 4 August 1989). Interdiction of the granting of new permits for fishing with driftnets and interdiction of the use of such nets for fishing of swordfish and albacore tuna during the month of October.

Decree of the Minister of Merchant Marine of 20 July 1989 (G.U. 181 of 4 August 1989). Provisions on fishing of clams.

Decree of the Minister of the Environment of 21 July 1989 (G.U. 177 of 31 July 1989). Provisional bordering and provisional measures for the protection of the national park of the Archipelago of Tuscany in the Tyrrhenian Sea. The provisional bordering of the park includes some of the islands of the archipelago, namely Montecristo, Capraia, Gorgona, Giannutri, and the surrounding waters as delimited by the 100-meter isobath.

Decree of the Minister of Merchant Marine of 25 July 1989 (G.U. 175 of 28 July 1989). Anticipation of the temporary stop of fishing vessels in the Adriatic Sea.

Decree of the President of the Council of Ministers of 27 July 1989 (G.U. 257 of 3 November 1989). Establishment of a network of monitoring and of elaboration of environmental data for the Adriatic Sea.

Decree of the Minister of Merchant Marine of 31 July 1989 (G.U. 181 of 4 August 1989). Anticipation of the temporary stop of fishing vessels for the maritime departments of Manfredonia, Molfetta, Bari and Brindisi.

Decree of the Minister of Merchant Marine of 4 August 1989 (G.U. 200 of 28 August 1989). Provisions on fishing of molluscs in the Tyrrhenian Sea.

Ordinance of the Minister of the Environment of 11 August 1989 (G.U. 193 of 19 August 1989). Urgent measures against mucilages along the Adriatic Coast.

Ordinance of the Minister of the Environment of 11 August 1989 (G.U. 193 of 19 August 1989). Experimental interventions for limiting the effects of mucilages.

Ordinance of the Minister for the Coordination of Civil Protection of 11 August 1989 (G.U. 196 of 23 August 1989). Further measures for the definitive disposal in the Region Veneto of industrial wastes coming from Lebanon and carried on the ship "Jolly Rosso".

Law No. 305 of 28 August 1989 (G.U. 205 of 2 September 1989). Provisions on the triennial program for the protection of the environment.

Decree of the Minister of Merchant Marine of 6 September 1989 (G.U. 213 of

12 September 1989). Institution of a 1,500-meter zone of biological protection around the island of Pianosa in the Archipelago of Tuscany.

Ordinance of the Minister for the Coordination of Civil Protection of 13 September 1989 (G.U. 225 of 26 September 1989). Further measures for the definitive disposal of industrial wastes carried by the ship "Jolly Rosso".

Ordinance of the Minister of the Environment of 22 September 1989 (G.U. 250 of 25 October 1989). Experimental interventions for limiting the effects of mucilages.

Law No. 345 of 20 October 1989 (G.U. 248 of 23 October 1989). Italian financial contribution to the Mediterranean Action Plan for the years 1988–1989.

Decree of the Minister of Merchant Marine of 25 October 1989 (G.U. 255 of 31 October 1989). Interdiction of the use of driftnets for the fishing of swordfish and albacore tuna from 1 November 1989 to 31 March 1990.

Ordinance of the Minister for the Coordination of Civil Protection of 9 November 1989 (G.U. 271 of 20 November 1989). Further measures on the definitive disposal of the industrial wastes carried by the ship "Jolly Rosso".

Decree of the Minister of Merchant Marine of 16 November 1989 (G.U. 273 of 22 November 1989). Rules on fishing of bivalve molluscs.

Decree of the Minister of Merchant Marine of 1 December 1989 (G.U. 294 of 18 December 1989). Provisions on the fishing of fry within 3 miles from the coast.

Law No. 426 of 28 December 1989 (G.U. 10 of 13 January 1990). Financing of oceanographic research and studies for the implementation of the 1974 Agreement between Italy and Yugoslavia on the cooperation for the protection of the Adriatic Sea against pollution.

Decree of the Minister of Merchant Marine of 28 December 1989 (G.U. 12 of 16 January 1989). Interdiction of the granting of new permits for fishing with trawlnets.

Decree of the Minister of Merchant Marine of 28 December 1989 (G.U. 12 of 16 January 1989). Provisions on fishing of bivalve molluscs.

Law No. 424 of 30 December 1989 (G.U. 6 of 9 January 1990). Provisions for the granting of State contribution for the economic activities in the areas affected by the exceptional phenomena of eutrophication and mucilages occurred in 1989 in the Adriatic Sea.

Decree of the Minister of Merchant Marine of 16 January 1990 (G.U. 18 of 23 January 1990). Provisions on coastal fishing.

ITALIAN CASES

Patmos

On 30 March 1989 the Court of Appeal of Messina reformed the decision rendered by the Tribunal of Messina on 30 July 1986 on the Patmos case (*ILSNL*, no. 18, p. 10). The Court of Appeal acknowledges the right of the Italian State to be granted compensation for the ecological damage to the territorial waters and marine resources resulting from an oil spill.

The Court declares that the claim for ecological damage maintained by the State is consistent with the definition of "pollution damage" laid down in the 1969 Brussels Convention on civil liability for oil pollution damage, which includes any injury caused to the coastline and to the other connected interests of environmental character such as the conservation of marine fauna and flora. The Court maintains that the ecological damage is grounded neither on the costs incurred by the State in order to restore the injuries caused by pollution nor on any other alleged economic loss, but on the State function of protecting the interest of the national community in preserving the ecolog ical and biological balance of the territory, the territorial sea included. The concrete determination of the amount of the damage is left to a subsequent decision.

RECENT DEVELOPMENTS IN ITALIAN PRACTICE

A Forum on International Law of the Environment

In the Declaration adopted at the 1989 Paris Summit the seven most industrialized countries noted with interest "the initiative of the Italian Government to host in 1990 a Forum on international law of the environment with scholars, scientific experts and officials, to consider the need for a digest of existing rules and to give in-depth consideration to the legal aspects of environment at the international level" (para. 47). The Forum was held in Siena on 17–21 April 1990, and an introductory document has been prepared by the Italian Government.

Actions against Traffic in Drugs in the EEZ

In signing the Convention against illicit traffic in narcotic drugs (Vienna, 20 December 1988) Brazil declared that Art. 17, para. 11, of the Convention does not prevent any coastal State from requiring prior authorization for any action under this article by other States in its Exclusive Economic Zone.

On 27 December 1989 Italy made the following objection to the declaration of Brazil: "Italy, a member State of the European Community, being attached to the principle of freedom of navigation, in particular within the exclusive economic zone, regards the declaration of Brazil on Article 17, paragraph 11 of the Convention against illicit traffic in narcotic drugs, adopted in Vienna on December 20, 1988, as exceeding the rights granted to coastal States by international law".

Italian Statement at the UNGA on the Law of the Sea

On 20 November 1989 the delegate of Italy, Tullio Treves, made a statement on item "Law of the Sea" at the 44th session of the United Nations General Assembly.

As regards deep seabed mining, Italy pointed out that it wished to consolidate the atmosphere favorable to a dialogue for ensuring the universality of the LOS Convention inaugurated by the statements made at the Prepcom in September 1989 by the representatives of both the Group of 77 and the Group of 6 (Belgium, the Federal Republic of Germany, Italy, Japan, the Netherlands, and the United Kingdom):

> What should be clear from the outset is the objective for the dialogue, namely the creation of the conditions for making the Convention a universally accepted

instrument. Provided they help to reach this objective, all means the imagination of diplomats and lawyers can think of should be considered. The beginnings should be very cautious: a lot of mistrust has to be removed, a lot of ties have to be restored. Even though the willingness to engage in dialogue without preconditions is certainly a very positive element, it seems to us that, at least at the initial stages, to go too quickly would be even more dangerous than to go too slowly. We must, however, start moving. Time is not unlimited and the favourable atmosphere now prevailing, and reflected in the draft resolution, should be taken advantage of.

As regards other recent developments, Italy made comments on the difficulties that:

have emerged in two major international negotiations about striking the right balance between protection of the environmental or other interests of coastal States and the interest of navigation. These were the negotiations that led to the adoption, on 19 December 1988, of the U.N. Convention against illicit traffic in narcotic drugs and psychotropic substances and to the adoption, on 22 March 1989, of the Basel Convention on the control of transboundary movements of hazardous wastes and their disposal. In both negotiations certain States argued that particular powers should be recognized to the coastal States. As regards the negotiations on the drug convention, it was argued that permission should be requested not only to the flag State but also to the coastal State, in order to take measures regarding a vessel suspected of being engaged in illicit traffic of drugs and exercising freedom of navigation in the exclusive economic zone. In the hazardous wastes negotiation it was argued that the coastal State enjoys the right of giving permission for transit in its territorial waters to ships carrying hazardous wastes. In both cases these positions met with strong resistance: the right of innocent passage was invoked in the hazardous wastes negotiation and freedom of navigation in the economic zone was insisted upon in the drugs negotiation. In both cases the problem was solved with provisions that recall the relevant rules of international law, indirectly quoting the Law of the Convention. These episodes confirm once more the problem-solving function of the Convention. They indicate also, however, that the balance struck in the U.N. Law of the Sea Convention is under the pressure of new problems and concerns. This is once more an indication that the entry into force of the Convention for the widest and most representative group of states is of the utmost importance for preserving such balance. This is confirmed by the fact that certain reservations, which are highly questionable in the light of the Law of the Sea Convention, have been made to the above mentioned Conventions.

ITALIAN PARTICIPATION AT THE PREPARATORY COMMISSION

On 1 September 1989 the representative of Italy at the Prepcom, Minister Ramiro Ruggiero, speaking on behalf of the Group of Six (Belgium, the Federal Republic of Germany, Italy, Japan, the Netherlands, the United Kingdom), made the following statement:

Our negotiations are entering into a crucial stage. We are convinced that the United Nations Law of the Sea Convention constitutes a major achievement of the United Nations and of the process of codification and progressive development of international law. But the States belonging to the Group of Six hold the view that Part XI presents some serious problems which, if left unresolved, might jeopardize this achievement. We have, therefore, tirelessly worked in this forum to find appropriate solutions to the abovementioned difficulties, so as to pave the way for a universally acceptable Convention. We strongly believe that the achieving of this lofty objective might be greatly facilitated should all States agree to the launching of a dialogue, without pre-conditions and in the appropriate framework, aimed at achieving better understanding of those problems and solutions to them. We should, therefore, welcome developments in that direction and are ready to make our contribution. It is in this light . . . that we have heard with interest (and appreciation) the declaration just pronounced by the distinguished Chairman of the Group of 77, Ambassador Kapumpa.

Report of the Netherlands Institute for the Law of the Sea (NILOS) 1989*

INTRODUCTION

The Netherlands Institute for the Law of the Sea (NILOS) was established in September 1984 as part of the Faculty of Law of the University of Utrecht. NILOS now operates within the framework of the Netherlands Institute for Social and Economic Legal Research (NISER) of the Utrecht Law Faculty. During 1989, the fifth year of its existence, the NILOS staff consisted of:

> Prof. A. H. A. Soons, Director
> Dr. B. Kwiatkowska, Associate Director
> Mr. A. H. IJlstra, LL.M., Research Associate
> Mr. P. A. Nollkaemper, LL.M., M.A., Research Associate
> Mr. J. G. M. Peet, M. Sc., Research Associate
> Mr. A. G. Oude Elferink, M.A., Research Associate (from 1 July 1989)
> Mr. P. Ymkers, M.A., Research Assistant
> Ms. M. W. de Boer, Secretary

Additional support was provided by Mr. A. van Schalm, Administrative Manager of NISER, and Mrs. W. J. Vreekamp of NISER Secretariat, as well as by the Department of International, Social and Economic Law of the Faculty of Law.

Prof. A. W. Koers, until 1 July 1987 Director of NILOS and presently Director of the Program on the Use of Computers in Law of the Faculty of Law, University of Utrecht, as well as Dr. E. Hey, LL.M., M. Sc., formerly a staff member of NILOS and presently Senior Legal Officer in the Netherlands Ministry of Transport, Water Management and Public Works, continue to be involved in NILOS projects.

NILOS is one of the law of the sea institutes and programs existing in Europe. The basic aims of NILOS correspond to the current developments in the law of the sea, generated in particular by the 1982 United Nations Convention on the Law of the Sea. The Convention, now signed by 157 states as well as the UN Council for Namibia and the European Economic Community and ratified by 41 states and the UN Council for Namibia, is a forceful factor accelerating the development of state practice, and consequently of a new customary international law of the sea. For most states, particularly developing countries, this process results in a greatly expanded need for expertise in marine affairs and the law of the sea. Accordingly the two basic aims of NILOS are:

---

*EDITORS' NOTE.—This report is abridged from the *1989 Annual Report* of the Netherlands Institute for the Law of the Sea (NILOS) which was supplied by Dr. B. Kwiatkowska, Associate Director, NILOS, Faculty of Law, University of Utrecht, Janskerkhof 3, 3512 BK Utrecht, the Netherlands.

—to conduct research on all issues related to the law of the sea with special emphasis on questions relating to the conservation, utilization and management of marine natural resources, and

—to assist states, in particular the developing countries, in dealing with the law of the sea issues which confront them.

NILOS' program of work focuses on the international processes by which a new international law of the sea is formed, as well as the substantive outcome of such processes. Additionally, it examines the unprecedented problems of policy formulation and regulatory responsibilities in relation to the law of the sea at the national level. The aim of NILOS is to assist (developing) states in the formulation of national policies and legislation.

NILOS' general program of work is implemented first and foremost through specific research and training projects, both joint and individual. The results of those projects provide in most cases the basis for publications and papers presented at international conferences by NILOS staff members on various problems of the law of the sea that correspond to the fields of their respective interest.

RESEARCH PROJECTS

Joint Projects

International Organizations and the Law of the Sea—Documentary Yearbook (Martinus Nijhoff Publishers)
The International Organizations and the Law of the Sea Documentary Yearbook is an annual publication which reproduces systematically the most important law of the sea related documents issued each year by international organizations.

The *Yearbook* is edited by the staff of NILOS, with the following division of responsibilities: Dr. Kwiatkowska—the UN, IAEA, ICAO, UNEP, WMO, and regional organizations; Dr. Hey—FAO; Prof. Soons—IOC; Mr. IJlstra and Mr. Peet—IMO; and Mr. Nollkaemper—ILO and UNCTAD.

NILOS is guided and assisted in its editorial work by nine distinguished authorities on matters pertaining to the international law of the sea who are the members of the Editor's Advisory Board. These are: Thomas Clingan—Professor of Law, United States; Jens Evensen—Judge of the International Court of Justice; Guenther Jaenicke—Emeritus Professor of Law, Federal Republic of Germany; Douglas Johnston—Professor of Law, Canada; Albert Koers—Professor of Law, the Netherlands; Shigeru Oda—Judge of the International Court of Justice; Christopher Pinto—Secretary-General, Iran-United States Claims Tribunal, The Hague; Shabtai Rosenne–Professor of Law, Israel; and Tullio Treves—Professor of Law, Permanent Mission of Italy to the United Nations.

The fourth yearbook covering documents issued in 1988 is in print and the fifth one, covering the 1989 documents, is being prepared.

International Organization and Integration (I.O.I.): First Supplement—Law of the Sea (Martinus Nijhoff Publishers)
This project will result in the First Supplement (Law of the Sea) to Martinus Nijhoff's I.O.I. series. It contains:

—a reference guide to multilateral treaties, other instruments and cases related to the law of the sea which are systematically arranged on the basis of the 1982 Convention's provisions;

—a summary of the 1982 Convention by Prof. Koers; and

—the texts of the most important law of the sea instruments adopted after 1982.

The project is designed as a research tool for scholars and practitioners concerned with law of the sea and marine affairs. The Editor-in-Chief of the project is Prof. Soons, while Prof. B. Boczek (U. S. A.), Dr. Kwiatkowska and Mr. IJlstra are the co-editors.

Research for Fisheries Organizations
The purpose of this project, conducted for the Netherlands Organization for Fish and Fish Products, was to examine the legal implications of an exchange of fishery quotas (within the framework of the Common Fisheries Policy of the European Economic Community [EEC]), by the Netherlands Government with other EEC member states. The research was conducted by researchers of the Institute for European Law of the University of Utrecht and NILOS. Prof. Soons participated in the project on behalf of NILOS. The report was completed in October 1989.

NILOS Newsletter
1989 saw the birth of the NILOS *Newsletter*. The *Newsletter*, scheduled to be published twice a year, has two main objectives:

—to provide information on recent Netherlands' state practice,

—to provide information about NILOS activities on a more regular basis, and if appropriate also in a more extensive form, than is possible in the Annual Report. Editors of the Newsletter are Mr. IJlstra, Mr. Nollkaemper, Mr. Peet and Mr. Ymkers. The Newsletter can be obtained from NILOS free of charge.

In 1989 the *Newsletter* was published twice and sent to some 300 institutions and individuals with whom NILOS entertains regular contacts. It contained, inter alia, contributions on the possible establishment by the Netherlands of an exclusive economic zone and on recent Netherlands legislation. Summaries of papers presented by NILOS staff at various conferences were also included.

Individual Projects

In addition to joint projects NILOS staff members have been working on the following individual projects during 1989:

—Development of the Legal Regime of Marine Scientific Research since 1982: Experience with State Practice (Prof. Soons).

—Effects of Sea-Level Rise on Maritime Limits and Boundaries (Prof. Soons).

—Maritime Boundary Project of American Society of International Law (ASIL) directed by Prof. Jonathan J. Charney with the assistance of Prof. Lewis Alexander, and funded by the Ford and Mellon Foundations. Dr. Kwiatkowska participates in the project on behalf of NILOS as a subject expert for economic and environmental considerations. The publications of the project are scheduled for the second half of 1990 and will include the texts of over 120 maritime boundary delimitations studied, analyses of each delimitation, and analytical papers addressing issues common to the delimitations studied.

—Rocks Which Cannot Sustain Human Habitation or Economic Life of their Own (art. 121 para. 3 of the 1982 Law of the Sea Convention) (Dr. Kwiatkowska and Prof. Soons).

—Conflicting Uses of the North Sea, An International Legal Study (Mr. IJlstra—doctoral dissertation under supervision of Prof. Soons).

—Netherlands State Practice on the Law of the Sea (Mr. Nollkaemper and Mr. A. G. Oude Elferink).

—The International Regime for Transboundary Water Pollution. The Case of the Netherlands (Mr. Nollkaemper—doctoral dissertation, partly devoted to the regime for marine pollution through watercourses, under supervision of Prof. Soons).

—Integrated Management of Sea Areas (Mr. Peet—doctoral dissertation under the supervision of Prof. Wiggerts of the University of Technology in Delft and Prof. Soons).

—International Legal Aspects of Nature Conservation in Marine Areas (Mr. A. G. Oude Elferink).

COOPERATIVE RESEARCH AND TRAINING PROJECTS

ICLOS-NILOS Cooperation

The cooperative project with Indonesia is carried out within the framework of the Netherlands-Indonesian Cooperation on Legal Matters covering the period 1986–1990. The Project on the Law of the Sea (PLS) is formally a subproject of the Public International Law project within this Cooperation. The PLS is carried out between NILOS and the Indonesian Centre for the Law of the Sea (ICLOS), established as a result of PLS at the Faculty of Law, University of Padjadjaran (UNPAD) in Bandung. The basic objective of the PLS is to provide assistance to the Indonesian Centre in strengthening its programs of training, education, and research in the new law of the sea, as to enable the Indonesian Centre to play an important role in translating the new law of the sea regime into the framework of national law and policy.

In accordance with the Programme of Cooperation between ICLOS and NILOS which was signed at Bandung on 14 January 1989, the parties carried out the following activities in 1989:

—joint ICLOS-NILOS research projects on Fishery in the Indonesian Exclusive Economic Zone (1988–1990), and on Joint Development of the Resources of the Continental Shelf (1989–1990);

—cooperative activities aiming at library support for ICLOS and preparation of a Catalogue of Publications present at ICLOS;

—translation of a book by Prof. Koers, *The United Nations Convention on the Law of the Sea* into *Bahasa Indonesia* and other book development activities; and

—preparation of a second Penataran (an upgrading course) on the International Law of the Sea to be held at ICLOS, Bandung, on 8–20 January 1990.

The joint ICLOS-NILOS fisheries project referred to above covers rules of international law applicable to foreign fishing in coastal waters in general and an analysis of bilateral fisheries agreements and joint venture arrangements concluded for purposes of granting foreign vessels access to the waters of coastal states, as well as examination of Indonesian laws and policies applicable to domestic and foreign fishing and of the access arrangements concluded between Indonesia and other states or between Indonesian and foreign fishing companies. Publication of the final report of this research is scheduled for 1990.

During 1989, the parties reviewed their progress, discussed their various collaborative activities, and gave consideration to continuation of these undertakings beyond the year 1990, taking into account the important place ocean affairs and the law of the sea has in Indonesia's overall national development as reflected by the Action Plan for Sustainable Development of Marine and Coastal Resources. The Action Plan was prepared by the Indonesian National Development Planning Board (BAPPENAS) and includes medium-term (5 years) policies and programs to complement Indonesia's 1989–1994 five-year development plan (RAPELITA).

IOMAC-NILOS Cooperation

The collaborative relationship between NILOS and the Secretariat of the Indian Ocean Marine Affairs Cooperation Conference (IOMAC), Colombo, Sri Lanka, was established in 1987 out of the initiative of Dr. M. C. W. Pinto, the Secretary-General of the Iran–US Claims Tribunal in The Hague. Dr. Kwiatkowska participated in the IOMAC Meeting of Legal and Fisheries Experts and Second Meeting on IOMAC Statute held in Jakarta, Indonesia, on 20–24 January 1989, under the chairmanship of Ambassador Djalal. The meeting was devoted to an examination of tuna fisheries in the Indian Ocean in the context of the recent FAO's attempt to establish a new Indian Ocean Tuna Commission and the possible alternative solutions which would enable all Indian Ocean states to participate more actively in tuna fisheries. The draft of IOMAC's Statute was also given consideration with a view to completion of its final version for the Second IOMAC Conference in 1990.

IOMAC's documents being obtained by NILOS on a regular basis as well as experience resulting from the continuous IOMAC-NILOS collaborative relationship are used for the purposes of NILOS Documentary Yearbook and specialization of research on marine policy problems encountered by the Indian Ocean states. Further prospects for IOMAC-NILOS collaboration may be related to possible extension of the Netherlands development cooperation with Sri Lanka to marine sectors.

SEAPOL-NILOS Cooperation

The Southeast Asian Programme on Ocean Policy, Law and Management (SEAPOL) Phase I covered the years 1983–1987. Phase II commenced in 1989 under the continued directorship of Prof. Phiphat Tangsubkul. Following the National Seminar on

Implementation of the 1982 Law of the Sea Convention held in Bangkok on 19 April 1989, the first major activity of SEAPOL Phase II was the Workshop on Ocean Regime Building in Southeast Asia which was held in Phuket, Thailand, on 1–4 May 1989 with a view to further cultivating a cooperative research and training mechanism in the field of ocean policy and management within the region.

The first SEAPOL Phase II International Conference, "The Implementation of the Law of the Sea Convention in the 1990's," will be held in Bali, Indonesia, May/June 1990.

ESCAP-NILOS Cooperation

In connection with his participation in the May 1989 Phuket Workshop of SEAPOL, Mr. IJlstra established a contact with the representative of the United Nations Economic and Social Commission for Asia and the Pacific (ESCAP), Mrs. Alfsen, Marine Affairs Officer of ESCAP (Natural Resources Division), with whom he preliminarily discussed the possibility of organizing joint ESCAP-NILOS seminars in Southeast Asia. As a follow-up to the 1988 ESCAP seminar on the removal of offshore installations and structures, the first ESCAP-NILOS seminar would be devoted to disposal of such installations and structures which is becoming an important issue of ocean policy and management in the Southeast Asian Region.

International Policy Seminar on the Management of the Exclusive Economic Zone for Development

During 1989 NILOS further continued its efforts to organize an International Policy Seminar on the Management of the Exclusive Economic Zone for Development in close cooperation with the Institute for Social Studies in The Hague (the Netherlands). A first folder announcing the seminar has been printed and the first International Policy Seminar is planned for May/June 1991 and will be of five weeks' duration. The seminar aims primarily at training participants from developing countries in the management of Exclusive Economic Zone.

The Erasmus Program

The Erasmus program is a program of the European Communities providing scholarships to students from EC member states in order to follow courses in their discipline at other universities of the EC member states. NILOS has concluded a cooperation agreement with the universities of Brest and Parma in 1988. This agreement was joined in 1989 by the universities of Gent (Belgium) and Giessen (FRG). Mr. IJlstra is Erasmus-NILOS coordinator. In this network several students have benefited from attending courses abroad. The contracting partners are willing to extend their cooperation to universities in other EC member states, notably the United Kingdom. Preliminary discussions have taken place between Mr. IJlstra and Dr. Churchill (Cardiff), Dr. Barston (London School of Economics).

Dubrovnik Course

In cooperation with the universities of Zagreb (Prof. B. Vukas), Rome (Prof. U. Leanza), Parma (Prof. T. Scovazzi) and Brest (Prof. D. Le Morvan) NILOS co-organized the Annual Course on the Law of the Sea in the Inter-University Centre of Dubrovnik. Mr. IJlstra represented NILOS as a course-director. After the sessions in 1984, 1986, 1987, and 1988 this was the fifth time the annual course was organized. The 1989 theme was the "Law of the Sea: Recent Developments in State Practice." Lecturers came from France, Italy, Hungary, the Netherlands, Yugoslavia, and Austria. Special guests of the course-directors this year were Profs. J. Lebullenger (Rennes), Daniel Vignes (Brussels), I. Seidl-Hohenveldern (Vienna), and Ambassador F. Starcevic (Belgrad). The Dutch Friends of Dubrovnik organization (NFC Dubrovnik) provided scholarships so that three students from the Institute of Social Studies (The Hague) could come to Dubrovnik.

Other Activities

NILOS researchers use their expertise for the purpose of specialized teaching, both within and outside the University of Utrecht and consultancy for different governmental and non-governmental bodies and organizations. In 1989, the consulting activities included the preparation of a draft fisheries legislation concerning the 200-mile zone for the Netherlands Antilles by Prof. Soons, as well as a background paper on the implementation of marine scientific research provisions of the 1982 Law of the Sea Convention for the UN Office for Ocean Affairs and the Law of the Sea. The UN Marine Scientific Research Workshop was held in September 1989 and will result in a handbook on the practical implementation of the regime for marine research in areas under national jurisdiction. NILOS staff members are also active in several institutions and organizations with competences in marine affairs, both national and international. In order to be effective, NILOS activities are coordinated with those of similar institutions in other countries and implementation is facilitated by the participation of NILOS researchers at international conferences, by their visits abroad, and by the visits of staff members of other institutes to NILOS. In 1989 NILOS cosponsored the Twenty-third Annual Conference of the Law of the Sea Institute, which was held in Noordwijk aan Zee, the Netherlands, June 12–15, 1989 and in which NILOS, through special funds, ensured participation from the developing states.

Reports from Organizations

World Meteorological Organization (WMO) 1989–1990*

MARINE METEOROLOGY AND ASSOCIATED OCEANOGRAPHIC ACTIVITIES PROGRAMME OF WMO

The basic objectives of the Marine Meteorology and Associated Oceanographic Activities Programme are the provision of high-quality meteorological and oceanographic services in support of marine users and the maintenance of a comprehensive operational global marine meteorological and oceanographic observing system in support of all WMO programmes. Technical advice and support for the programme are provided principally through the Commission for Marine Meteorology (CMM) and the Joint IOC/WMO (IOC is the Intergovernmental Oceanographic Commission of Unesco) Committee for IGOSS (Integrated Global Ocean Services System).

The major event within the Programme for 1989 was the tenth session of CMM, which took place in Paris from 6 to 17 February 1989. The first two days were devoted to a highly successful Technical Conference on Ocean Waves, which is one of the main scientific-technical areas of concern to the Commission. Some 150 scientists participated, while over 100 delegates and observers from 44 Members and 10 international organizations took part in the Commission session proper. Major topics for discussion at the session included the reorganization of the existing system for the broadcast of high seas forecasts and warnings so as to be compatible with the telecommunication facilities of the Global Maritime Distress and Safety System (GMDSS) of the International Maritime Organization (IMO); the requirement for international coordination of meteorological input to marine pollution emergency response operations; the potential impact of data from the coming generation of oceanographic satellites on the provision of marine meteorological and oceanographic services; the requirement for enhanced long-term education and training facilities in marine meteorology and physical oceanography; and the organization of the Commission's substructure of working groups and rapporteurs.

On the basis of detailed guidelines provided by CMM-X, work began during 1989 on the restructuring of the coordinated high seas forecasts and warnings broadcast system, which, as noted above, will be required for the GMDSS. The GMDSS will enter into force on 1 February 1992 and it is important that at least a provisional system is agreed upon before that date. A first draft proposal to this effect has been prepared by the chairman of the CMM subgroup on observations and telecommunications, for the consideration of a session of the subgroup in early 1990. In addition, preliminary discussions have been held with the International Hydrographic Organi-

*Editor's Note.—This report was supplied by Mr. P. E. Dexter, Chief, Ocean Affairs Division, World Weather Watch Department, World Meteorology Organization, 41 Giuseppe-Motta, Case postale no. 2300, CH-1211, Geneva 2, Switzerland.

zation (IHO) and IMO on the coordination of the new system with the World-wide Navigational Warning Service operated by IHO.

In recognition of the urgent need to improve long-term education and training facilities in marine meteorology and physical oceanography, a detailed proposal (including syllabus) for a six-month certificate course at the RMTC Nairobi was prepared by two consultants from WMO and IOC, in consultation with the director and staff of the RMTC. Depending on the availability of funding and material support, it is hoped to implement the first (pilot) course in 1992.

Detailed, operational observations of certain ocean variables are essential for global climate monitoring, research and prediction, as well as in support of marine meteorological and oceanographic services. Existing operational ocean-observing system components are inadequate for this purpose, however, and both WMO's Executive Council at its forty-first session and the Fifteenth IOC Assembly adopted a proposal for the planning and implementation of a composite, operational, global ocean-observing system. The system's initial priority will be observations relating to short-term climate change. The first steps towards planning the system have already begun in a cooperative effort between IOC and WMO, and are directed principally through an Ocean Observing System Development Panel established jointly by the Joint Scientific Committee of WMO and ICSU for the WCRP and the Committee for Climate Changes and the Ocean (CCCO). It is intended that the system should be implemented as much as possible through a strengthening of the existing ocean-observing systems of WMO and IOC, including drifting buoys, ocean satellites, and new technologies as they become available.

MARINE METEOROLOGY AND ASSOCIATED OCEANOGRAPHIC ACTIVITIES MEETINGS FOR 1989

Date and Place	Title
6–7 February Paris, France	Technical Conference on Ocean Waves (in conjunction with CMM-X)
6–17 February Paris, France	Commission for Marine Meteorology—tenth session
15–17 May San Diego, USA	CLS/Service Argos Users' Conference (co-sponsored)
17–20 October Geneva	Drifting Buoy Co-operation Panel—fifth session
20–24 November Buenos Aires, Agentina	RAIII/RAIV Seminar on Marine Meteorological Services and Marine Forecast Techniques

Appendix B

Selected Documents and Proceedings

The documents and proceedings included in this appendix represent a selection of international agreements, legislation, proceedings of international conferences, and other documents bearing on important ocean-related developments. Commentaries on certain of these documents appear in the *Yearbook* and are referred to in the footnotes and noted in the index.

THE EDITORS

Selected Documents and Proceedings

Antarctic Treaty Special Consultative Meeting on Antarctic Mineral Resources: Final Act and Convention on the Regulation of Antarctic Mineral Resource Activities[1]

INTERNATIONAL LEGAL MATERIALS (*I.L.M.*) BACKGROUND

The Special Consultative Meeting on Antarctic Mineral Resources was convened in accordance with Recommendation XI-1 adopted by the Antarctic Treaty Consultative Parties at Buenos Aires in July 1981 (20 *I.L.M.* 1256 [1981]). It began its work in Wellington, June 14–25, 1982, and subsequent sessions were held in Wellington (January 17–28, 1983), Bonn (July 11–22, 1983), Washington, D.C. (January 18–27, 1984), Tokyo (May 23–31, 1984), Rio de Janeiro (February 26–March 12, 1985), Hobart (April 14–25, 1986), Montevideo (May 11–20, 1987), and Wellington (January 18–29, 1988).

The final session was held in Wellington, May 2–June 2, 1988. Representatives of the Antarctic Treaty Consultative Parties (Argentina, Belgium, Brazil, Chile, China, France, German Democratic Republic, Federal Republic of Germany, India, Italy, Japan, New Zealand, Norway, Poland, South Africa, Union of Soviet Socialist Republics, United Kingdom, United States, and Uruguay) participated in the session. At the invitation of the Consultative Parties, representatives of the thirteen Contracting Parties (Bulgaria, Canada, Czechoslovakia, Denmark, Ecuador, Finland, Greece, Republic of Korea, Netherlands, Papua New Guinea, Peru, Romania, and Sweden) also participated in the session. At the conclusion, the Consultative Parties adopted the Convention by consensus, and together with the Non-Consultative Parties participating in the final session, signed the Final Act to which the Convention is annexed. It was agreed that the Convention would be opened for signature at Wellington on November 25, 1988.

FINAL ACT ON ANTARCTIC MINERAL RESOURCES[2]

The final session of the Fourth Special Antarctic Treaty Consultative Meeting on Antarctic Mineral Resources was held at Wellington from 2 May to 2 June 1988. Representatives of the Consultative Parties to the Antarctic Treaty, namely Argentina,

---

1. Cite as 27 *International Legal Materials (I.L.M.)* (1988) 859.

2. The final Act of the Fourth Special Antarctic Treaty Consultative Meeting on Antarctic Mineral Resources is reprinted from 27 *International Legal Materials (I.L.M.)* (1988) 865–869. Note, the text of the Antarctic Treaty, done at Washington, December 1, 1959, and entered into force June 23, 1961, may be found at 12 *United States Treaty Series (UST)* 794; *U.S. Treaties and Other International Agreements (TIAS)* 4780; and 402 *United Nations Treaty Series (UNTS)* 71.

Australia, Belgium, Brazil, Chile, China, France, German Democratic Republic, Federal Republic of Germany, India, Italy, Japan, New Zealand, Norway, Poland, South Africa, Union of Soviet Socialist Republics, United Kingdom, United States, and Uruguay, participated in the Meeting. On the invitation of the Consultative Parties, Representatives of 13 Contracting Parties to the Antarctic Treaty that are not Consultative Parties, namely Bulgaria, Canada, Czechoslovakia, Denmark, Ecuador, Finland, Greece, Republic of Korea, Netherlands, Papua New Guinea, Peru, Romania, and Sweden, also participated in the Meeting.

As a result of their deliberations, the Consultative Parties adopted in the official languages of the Antarctic Treaty the "Convention on the Regulation of Antarctic Mineral Resource Activities," the text of which is annexed to this Final Act, and agreed that it would be opened for signature at Wellington on 25 November 1988.

Taking into account the decision reflected in Article 67 of the Convention that Chinese would be an authentic language, the Meeting agreed that the Drafting Committee would be reconvened by the Depositary, at a time and place to be agreed, for the purpose of bringing into concordance with the text of the Convention in the four official languages of the Antarctic Treaty, a Chinese text. To this end it was agreed that the Depositary would circulate in advance of such meeting a text of the Convention in the Chinese language.

The Meeting also agreed that the Drafting Committee should consider any questions of linguistic consistency, which might possibly be found to be necessary, in the authentic texts in the official languages of the Antarctic Treaty with a view to their rectification in accordance with the rules and procedures set forth in the Vienna Convention on the Law of Treaties 1969.

With respect to the decision reflected in Article 67(2) of the Convention the Meeting noted that at any time after the opening for signature of the Convention a Signatory or Acceding State could lodge with the Depositary an official translation of the Convention which would then be circulated in accordance with Article 67(2).

The Meeting also considered the question of continuing the restraint of Antarctic mineral resource activities agreed to in Recommendation IX-1 for the interim period before the entry into force of the Convention. Taking into account Recommendation IX-1 and the adoption by the Meeting of the Convention on the Regulation of Antarctic Mineral Resource Activities, the Meeting agreed that all States represented at the Meeting would urge their nationals and other States to refrain from Antarctic mineral resource activities as defined in the Convention pending its timely entry into force.

The Meeting recognized that unfair economic practices including certain forms of subsidies could cause adverse effects to the interests of Parties to the Convention and that such effects should be addressed in the context of the relevant multilateral agreements. To this end, the Meeting agreed that Parties to the Convention which are also Parties to such multilateral agreements will determine conditions of application of these agreements to Antarctic mineral resource activities.

The Meeting noted that mineral resources, as defined in Article 1(6) of the Convention, do not include ice and that if harvesting of ice, including icebergs, were to become a possibility in the future there could be impacts on the Antarctic environment and on dependent and associated ecosystems. The Meeting also noted that the harvesting of ice from the coastal region of Antarctica, more particularly if land-based facilities were required, could raise some of the environmental and other issues addressed in the Convention. The Meeting agreed that the question of harvesting Antarctic ice should be further considered by the Antarctic Treaty Consultative Parties at the next regular meeting.

The Meeting noted the requirement under Article 8 of the Convention for a separate Protocol on liability and agreed that it would be desirable to begin work on its elaboration at an early stage.

With respect to the financial obligations of Operators, the Meeting noted the importance for the operation of the Convention that an indication of the possible extent of the financial obligations of Operators should be available to them in reasonable time before applications for exploration permits are submitted.

The Meeting agreed that the area of regulation of Antarctic mineral resource activities defined in Article 5(2) of the Convention does not extend to any continental shelf appurtenant in accordance with international law to islands situated north of 60° south latitude.

The Meeting also agreed that the geographic extent of the continental shelf as referred to in Article 5(3) of the Convention would be determined by reference to all the criteria and the rules embodied in paragraphs 1 to 7 of Article 76 of the United Nations Convention on the Law of the Sea.

With respect to Articles 6 and 41(1)(d) of the Convention, the Meeting noted that the promotion and encouragement of international participation do not prejudice the right of any applicant to exercise freedom of choice over the partners in a joint venture, including the terms of their partnership, consistently with the Articles referred to above and any measures pursuant to them, in offering international participation in any proposed Antarctic mineral resource activity.

The Meeting agreed that Article 8(10) of the Convention was to be interpreted as excluding multiple judgments in respect of the same liability claim. Specifically, if a liability claim has been referred to adjudication in the courts of one Party, such claim would not be subject to additional adjudication while those proceedings are pending or after they have resulted in a final judgment. The Meeting also noted that Article 8(10) would apply in the period prior to entry into force of the Protocol referred to in Article 8(7) and it was understood that paragraph 10 should be interpreted in light of Article 37 of the Convention and that the Operators referred to in that paragraph were those defined in Article 1 of the Convention.

In relation to Article 29 of the Convention the Meeting agreed that the member or members of the Commission mentioned in Article 29(2)(a) are those identified by reference to Article IV(1)(a) of the Antarctic Treaty.

The members of the Commission mentioned in Article 29(2)(b) are those identified by reference to Article IV(1)(b) of the Antarctic Treaty.

The Meeting acknowledged that the specific formula in Article 29(3)(b) of the Convention ("at least three developing country members" of the Commission) accurately reflected the balance between developed and developing Consultative Parties as of the date of the adoption of the Convention. It was also recognized that, in the event of an increase of the size of the Commission in the future resulting in a significant alteration of this balance, there would be a case for considering, by way of an amendment in accordance with Article 64 of the Convention and by reference to paragraph 2(c)(ii) of that Article, the total membership of the Regulatory Committee.

The Meeting agreed that it was desirable that the decision-making process in the Regulatory Committee pursuant to Article 32 of the Convention should reflect all the interests represented in the Regulatory Committee. It was also agreed, in particular, that it was desirable that the two-thirds majority referred to in Article 32 should include at least one developing country.

With respect to Article 62 of the Convention, the Meeting agreed that all of the institutions of the Convention should not be established in respect of every area of Antarctica unless all the States referred to in Article IV(1)(a) and (b) of the Antarctic Treaty and at least four states referred to in paragraph 1(c) of that article were Parties to the Convention, and that these included at least three developing countries.

The Meeting agreed that the titles of Chapters and Articles in the Convention are indicative only and were included for the sole purpose of facilitating examination of the text and reference to different provisions of the Convention.

The Meeting also agreed that the contents of this Final Act are without prejudice to the legal position under the Antarctic Treaty of any Party.

Done at Wellington, this second day of June 1988, in a single original copy in the four official languages of the Antarctic Treaty to be deposited in the archives of the Government of New Zealand which will transmit a certified copy thereof to all Contracting Parties to the Antarctic Treaty.

CONVENTION ON THE REGULATION OF ANTARCTIC MINERAL RESOURCE ACTIVITIES[3]

Preamble (*I.L.M.* page 868)

The States Parties to this Convention, hereinafter referred to as the Parties,

Recalling the provisions of the Antarctic Treaty;

Convinced that the Antarctic Treaty system has proved effective in promoting international harmony in furtherance of the purposes and principles of the Charter of the United Nations, in ensuring the absence of any measures of a military nature and the protection of the Antarctic environment and in promoting freedom of scientific research in Antarctica;

Reaffirming that it is in the interest of all mankind that the Antarctic Treaty area shall continue forever to be used exclusively for peaceful purposes and shall not become the scene or object of international discord;

Noting the possibility that exploitable mineral resources may exist in Antarctica;

Bearing in mind the special legal and political status of Antarctica and the special responsibility of the Antarctic Treaty Consultative Parties to ensure that all activities in Antarctica are consistent with the purposes and principles of the Antarctic Treaty;

Bearing in mind also that a regime for Antarctic mineral resources must be consistent with Article IV of the Antarctic Treaty and in accordance therewith be without prejudice and acceptable to those States which assert rights of or claims to territorial sovereignty in Antarctica, and those States which neither recognize nor assert such rights or claims, including those States which assert a basis of claim to territorial sovereignty in Antarctica;

3. The Convention on the Regulation of Antarctic Mineral Resource Activities (CRAMRA) is reprinted from 27 *International Legal Materials (I.L.M.)* (1988) 859–900. Note, the Preamble is reprinted in its entirety. Due to space restrictions Articles 1–67 are as summarized by *I.L.M.*

Noting the unique ecological, scientific and wilderness value of Antarctica and the importance of Antarctica to the global environment;

Recognizing that Antarctic mineral resource activities could adversely affect the Antarctic environment or dependent or associated ecosystems;

Believing that the protection of the Antarctic environment and dependent and associated ecosystems must be a basic consideration in decisions taken on possible Antarctic mineral resource activities;

Concerned that participation in Antarctic mineral resource activities should be open to all States which have an interest in such activities and subscribe to a regime governing them and that the special situation of developing country Parties to the regime should be taken into account;

Believing that the effective regulation of Antarctic mineral resource activities is in the interest of the international community as a whole;

HAVE AGREED as follows: (summarized)

Chapter I: General Provisions (*I.L.M.* page 868)

Art. 1 Definitions [1–20]

Art. 2 Objectives and General Principles [Regarding mineral resource activities, to establish a means of assessing their environmental impact, determining their acceptability, governing their conduct if found acceptable, and ensuring their compliance with this Convention]

Art. 3 Prohibition of Antarctic Mineral Resource Activities Outside This Convention [In addition, exploration and development activities must operate pursuant to an approved Management Scheme (see Arts. 48 and 54).]

Art. 4 Principles Concerning Judgments on Antarctic Mineral Resource Activities [Activities must first be judged according to the criteria set forth in this article.]

Art. 5 Area of Application [The Antarctic Treaty area, including all ice shelves, south of the 60 degree south latitude, and in the seabed and subsoil of adjacent offshore areas up to the deep seabed (defined)]

Art. 6 Cooperation and International Participation [The participation of developing countries should be promoted.]

Art. 7 Compliance with this Convention [Each party shall take measures to ensure its compliance with this Convention and report such measures to the Executive Secretary. The Commission shall draw attention to non-compliance.]

Art. 8 Response Action and Liability [Response action must be timely to prevent, contain, clean up, and remove damaging conditions. The Executive Secretary must be notified of such actions. An operator causing damage shall be held strictly liable for damage it causes unless due to the intervention of an unforeseen natural disaster, armed conflict, or terrorism. Each party shall ensure that the Commission has standing to appear in its national courts to pursue liability claims.]

Art. 9 Protection of Legal Positions under the Antarctic Treaty [No activity may have any effect with respect to a claim to territorial sovereignty within the Treaty area.]

Art. 10 Consistency with the Other Components of the Antarctic Treaty System [The parties shall ensure such consistency.]

Art. 11 Inspection under the Antarctic Treaty [All areas of an activity shall be open to inspection by Antarctic Treaty observers.]

Art. 12 Inspection under this Convention [The Commission may designate observers; aerial inspections may be conducted; inspections shall not be unduly burdensome; prior arrangements must be made for inspections in Article 41 areas.]

Art. 13 Protected Areas [Activities are prohibited in Specially Protected Areas, Sites of Special Scientific Interest, areas designated pursuant to Art. IX(1) of the Treaty, and other protected areas designated by the Commission for reasons set forth.]

Art. 14 Non-Discrimination [Discrimination in implementing this Convention is prohibited.]

Art. 15 Respect for Other Uses of Antarctica [Other uses include stations, scientific research, conservation of marine living resources, tourism, historic monuments, navigation and aviation.]

Art. 16 Availability and Confidentiality of Data and Information [Data and information, unless of commercial value, shall be made freely available.]

Art. 17 Notifications and Provisional Exercise of Functions of the Executive Secretary [Through the Executive Secretary; alternatively through the depositary]

Chapter II: Institutions (*I.L.M.* page 876)

Art. 18 Commission [On membership and observer status]

    Art. 19 Commission Meetings

    Art. 20 Commission Procedure

    Art. 21 Functions of the Commission

    Art. 22 Decision Making in the Commission [By a three-quarters super-majority on matters of substance. By consensus (absence of formal objection) on budgetary matters and decisions under Arts. 21(1)(i) and 41(2). By simple majority on procedural matters]

    Art. 23 Advisory Committee [The Scientific, Technical and Environmental Advisory Committee. Membership is open to all parties. Provision for observer status]

    Art. 24 Advisory Committee Meetings

    Art. 25 Advisory Committee Procedure

    Art. 26 Functions of the Advisory Committee

    Art. 27 Reporting by the Advisory Committee [To the Commission and relevant Regulatory Committees (see Art. 29) on each of its meetings]

    Art. 28 Special Meeting of Parties

    Art. 29 Regulatory Committees [There shall be an Antarctic Mineral Resources Regulatory Committee for each of the areas identified by the Commission under Art. 41. Membership provisions]

    Art. 30 Regulatory Committee Procedure

    Art. 31 Functions of Regulatory Committees

    Art. 32 Decision Making in Regulatory Committee

    Art. 33 Secretariat [Secretariat of the Commission: powers, responsibilities, and functions]

    Art. 34 Cooperation with International Organisations [Specific organisations are named.]

Art. 35 Financial Provisions [Regarding a budget, financial regulations, financing by revenues and contributions by Commission members. Consequences of a member's failure to pay its contribution]

Art. 36 Official and Working Languages [English, French, Russian, and Spanish]

Chapter III: Prospecting (*I.L.M.* page 886)

Art. 37 Prospecting [Prospecting does not confer ownership rights on operators. Sponsoring states shall ensure that their operators respect the rights of other operators. Prospecting must be preceded by 9 months' notice that must disclose certain data and information unless it is of commercial value.]

Art. 38 Considerations of Prospecting by the Commission [The Commission may require modifications of prospecting plans to ensure compliance with this Convention.]

Chapter IV: Exploration (*I.L.M.* page 887)

Art. 39 Requests for Identification of an Area for Possible Exploration and Development [Any party may request designation by the Executive Secretary of an area for exploration and development. Certain specific information must be contained in the request.]

Art. 40 Action by the Advisory Committee and Special Meeting of Parties [The Advisory Committee shall advise the Commission on whether to approve the designation.]

Art. 41 Action by the Commission [Set forth are certain factors that the Commission shall consider in deciding whether to make the designation. A decision shall be reached by consensus.]

Art. 42 Revision in the Scope of an Identified Area [The Commission may amend the boundaries or designation of resources for any area it has identified, after seeking the views of the advisory Committee and the relating Regulatory Committee.]

Art. 43 Preparatory Work by Regulatory Committees [After identification of an area pursuant to Art. 41, a Regulatory Committee shall be established for the area. The Regulatory Committee adopts guidelines for activities in its area and carries out other duties and functions that are set forth.]

Art. 44 Application for an Exploration Permit [Application must be made for a permit to conduct activities in a designated area. Certain specified information and certifications must be included.]

Art. 45 Examination of Applications [The Regulatory Committee shall seek advice from the Advisory Committee in considering applications.]

Art. 46 Management Scheme

Art. 47 Scope of the Management Scheme [The scheme sets forth the terms and conditions for exploration and development within a specific area. A list of topics to be covered by a scheme are set forth.]

Art. 48 Approval of the Management Scheme [Approval shall be pursuant to Art. 32.]

Art. 49 Review [The Commission may be convened to review a decision of the

Regulatory Committee to approve a Management Scheme or issue a development permit.]

Art. 50 Rights of Authorised Operators [A Management Scheme may be changed or suspended only pursuant to Art. 51 or by consent of the Sponsoring State.]

Art. 51 Suspension, Modification or Cancellation of the Management Scheme and Monetary Penalties [The Regulatory Committee may modify a Management Scheme to avoid an unfavorable environmental impact. A scheme and related permit may be cancelled if an operator ceases to have a substantial and genuine link with the Sponsoring State.]

Art. 52 Monitoring in Relation to Management Schemes [The Regulatory Committee shall monitor compliance and assess the environmental impacts of activities in its area.]

Chapter V: Development (*I.L.M.* page 893)

Art. 53 Application for a Development Permit [The information to be contained in an application is set forth.]

Art. 54 Examination of Applications and Issue of Development Permits [The Regulatory Committee shall seek advice from the Advisory Committee in conducting its examination of applications.]

Chapter VI: Disputes Settlement (*I.L.M.* page 894)

Art. 55 Disputes Between Two or More Parties [Arts. 56–58 shall apply.]

Art. 56 Choice of Procedure [The ICJ or by arbitration. The choice may be made at the time of becoming a party to this Convention.]

Art. 57 Procedure for Dispute Settlement

Art. 58 Exclusion of Categories of Disputes [Declarations to the Convention may exclude any category except those listed in this article.]

Art. 59 Additional Dispute Settlement Procedures [To review an alleged improper decision to decline, suspend, modify, or cancel a Management Scheme or to decline the issuance of a development permit]

Chapter VII: Final Clauses (*I.L.M.* page 896)

Art. 60 Signature [Open for signature on 25 November 1988]

Art. 61 Ratification, Acceptance, Approval or Accession

Art. 62 Entry into Force [When 16 of the participants in the final session become parties]

Art. 63 Reservations, Declarations and Statements [No reservations are permitted. Declarations shall not limit the legal effect of this Convention, except as allowed in Art. 58.]

Art. 64 Amendment

Art. 65 Withdrawal [Two years after written notice]

Art. 66 Notifications by the Depositary

Art. 67 Authentic Texts, Certified Copies and Registration with the United Nations [Chinese, English, French, Russian, and Spanish] [Done at Wellington on 2 June 1988]

Annex for an Arbitral Tribunal (*I.L.M.* page 898)

Art. 1–12 [Designation of arbitrators; notification to commence proceedings; place of proceedings (at the headquarters of the Commission); counterclaims; provisional measures; intervention by third parties; evidence; proceeding in the absence of one party; submission of a dispute for a decision *ex aequo et bono*; finality and binding effect of awards; expenses (to be borne by the parties in equal shares); decisions shall be reached by a majority of the arbitrators.]

Selected Documents and Proceedings

IMO and the Peaceful Use of Nuclear Energy through the Last Thirty Years*

Emil Jansen

NUCLEAR POWER GENERATION ON LAND

Through the last 30 years, the lifetime of this Organization, we have seen a remarkable growth of the production of nuclear energy for peaceful purposes all over the world.

By the end of 1988, 428 nuclear power reactors were in operation world wide, delivering a total of 309,641 megawatts electric power (MWe). At the same time a further 111 units with a capacity of 88,648 MWe were under construction. At present nuclear power accounts for more than 16% of world electricity production. In France (70%), Belgium (65%), Hungary (49%), Republic of Korea (47%), and Sweden (47%) a substantial proportion of their total electricity during 1988 was produced in nuclear reactors (Table 1).

All nuclear electric power reactors in operation and under construction by the end of 1988 are shown in Table 2.

NUCLEAR PROPULSION OF SHIPS

Nuclear powered navy vessels were in operation before the birth of IMO. Unclassified published statistics show that in 1988, on a world basis, there were 376 active nuclear powered submarines and surface craft, while 47 were under construction or in the process of conversion. Table 3 gives information on types, number of units, and nationality of these nuclear powered navy vessels.

Parallel to this development, the Soviet Union has successfully built and put into operation a fleet of nuclear powered icebreakers, starting with the *Lenin,* commissioned in 1959. With her 44,000 SHP, the *Lenin* was superior to any conventionally powered icebreaker, for two main reasons. Firstly she would maintain a constant predetermined displacement, owing to the fact that the vessel would not have to carry large amounts of fuel oil, normally to be consumed with resulting reduced displacement and changing icebreaking properties.

Secondly, for all practical purposes, the nuclear powered icebreaker has an unlimited sailing independence. It has been claimed that *Lenin* could circumnavigate the world 7–10 times without refuelling. Further, *Lenin* could proceed through ice of 2 metres thickness at a continuous speed of about 3 knots.

*EDITORS' NOTE.—Mr. Emil Jansen of Norway was awarded the International Maritime Prize for 1988. He was Chairman of the Maritime Safety committee from 1984 until his retirement at the end of 1988, Chairman of the Marine Environment Protection Committee from 1978 to 1984, and has been actively involved in IMO's work since 1966. Mr. Jansen delivered this speech on World Maritime Day 1989, as reprinted in *IMO News* 1 (1990): 7–10.

TABLE 1.—COUNTRIES IN WHICH NUCLEAR ENERGY PROVIDES FOR MORE THAN 25% OF ELECTRIC POWER GENERATION

France	70%	Finland	36%
Belgium	65%	Spain	36%
Hungary	49%	Bulgaria	35%
Rep. of Korea	47%	Federal Rep. of Germany	33%
Sweden	47%	Japan	28%
Switzerland	37%	Czechoslovakia	27%

TABLE 2.—NUCLEAR POWER REACTORS IN OPERATION AND UNDER CONSTRUCTION BY THE END OF 1988

	In Operation		Under Construction	
	Units	MWe	Units	MWe
Canada	18	12,185	4	3,524
France	55	52,588	9	12,245
Federal Rep. of Germany	23	21,491	2	1,520
Japan	36	26,888	14	12,272
Spain	10	7,519		
Sweden	12	9,693		
United Kingdom	40	11,921	2	1,820
United States	108	95,273	7	7,689
USSR	57	34,020	25	20,800
Others	69	38,063	48	28,788
World	428	309,641	111	88,648

SOURCE.—IAEA Power Reactor Information System.

TABLE 3.—NUCLEAR POWERED NAVY VESSELS (1988)

Country	Vessel Type	Active	Building/ Conversions
USSR	Submarines	195	7
	Aircraft carriers	—	2
United States	Submarines	133	24
	Aircraft carriers	5	2
	Cruisers	9	
United Kingdom	Submarines	21	5
France	Submarines	13	7
Total		376	47

SOURCE.—*Jane's Fighting Ships, 1988.*

The later Soviet nuclear icebreakers are larger and more powerful than *Lenin*, with displacements up to 23,500 tons and 75,000 SHP.

The first nuclear powered merchant ship, the United States' *Savannah*, was launched on 21 July 1959, and put into service on 1 May 1962. The Federal Republic of Germany put their nuclear propelled merchant ship *Otto Hahn* into operation in October 1968, and in November the same year, the keel of the first Japanese nuclear ship, the *Mutsu*, was laid at the Ishikawajima Harima Heavy Industries in Tokyo.

IMO'S INITIAL NUCLEAR ENERGY ENGAGEMENT

By this time IMO had been in existence for nearly 10 years. Our Organization realized early that it could and would be influenced by the rapid development of the peaceful uses of nuclear energy on land and at sea, and assumed its responsibility in due course.

Assembly Resolution A.30 (II), adopted by the second session of the Assembly in April 1961, records the approval of the text of the Agreement between the International Atomic Energy Agency (IAEA) and the International Maritime Organization (IMO). The Agreement spells out in quite some detail the responsibilities of the two UN sister-organizations, outlines methods of cooperation and consultation, reciprocal representation and exchange of information and documents. The Agreement became a solid foundation for the close and efficient cooperation that appeared in the years to follow.

To underline IMO's adherence to activities solely concerned with the peaceful uses of nuclear energy, the Assembly at its third session in October 1963 adopted Resolution A.45 (III) giving its support to the conclusion and coming into force of the Treaty to Ban Nuclear Tests in the Atmosphere, in outer Space and under Water, signed in Moscow on 5 August 1963.

Even before the formal agreement between IMO and IAEA was implemented, the two organizations had sponsored a first Symposium on Nuclear Ship Propulsion which was held at Taormina in Italy, November 1960.

CHAPTER VIII OF THE 1960 SOLAS CONVENTION

The International Convention for the Safety of Life at Sea SOLAS, 1960, was signed in London in June 1960, only about 9 months before the IMO second Assembly session. At this second Assembly session, duties assigned to IMO under the 1960 Safety Convention were approved by the Assembly. Among these duties are also duties related to nuclear ships. Chapter VIII of the 1960 SOLAS Convention deals with nuclear ships, and so does Annex C to that Convention, "Recommendations applicable to nuclear ships." Chapter VIII is written in a general form, and indeed in few and brief Regulations, covering 3–4 pages only. The Nuclear Ship Recommendations at Annex C of the 1960 SOLAS Convention also appear in a brief form covering a modest 5 pages in the book.

The very remarkable feature of Chapter VIII is that through the 30 years since its conception, no amendment has ever been made to this chapter, while the Articles and all the other 7 chapters of the 1960 SOLAS have been through considerable and significant changes, amendments and even complete substitutions.

In the latest edition of the SOLAS Convention which incorporates the consolidated text of the 1974 SOLAS Convention, the 1978 SOLAS Protocol and the 1981 and 1983 SOLAS Amendments, there appear two footnotes in Chapter VIII. The first footnote is made at Regulation 2, and reads: "Reference is made to the Code of Safety for Nuclear Merchant Ships (resolution A.491 (XII)) which supplements the requirements of this chapter."

The second footnote appears at Regulation 11, and reads: "Reference is made to the IMO/IAEA Safety Recommendation on Use of Ports by Nuclear Merchant Ships."

These two footnotes are a reminder that even though the text of Chapter VIII has remained unchanged through all these years, IMO has been active and shown a degree of action and preparedness so characteristic for our Organization.

THE CODE OF SAFETY FOR NUCLEAR MERCHANT SHIPS

The International Conference on Safety of Life at Sea, 1974, in Resolution 1.7 called for a revision of relevant provisions of the 1960 SOLAS Convention with respect to nuclear merchant ships, having regard to the progress in nuclear engineering, experience gained by a number of countries in operating ships with nuclear propulsion units and the expected increase in the use of nuclear propulsion of ships.

Following this clear signal from the 1974 Conference, the Maritime Safety Committee decided that safety requirements for nuclear ships would be the prime responsibility of IMO and assigned to the Sub-Committee on Ship Design and Equipment the task in general and in particular instructed that sub-committee to coordinate work in this area within IMO.

At its 15th and 16th sessions in 1976 the DE-Sub-Committee initiated work on the Safety Code, and an ad hoc Working Group on the Safety of Nuclear Ships formulated a code outline and procedure and program for developing a safety code.

Under the auspices of OECD's Nuclear Energy Agency, an ad hoc Working Group on the Safety of Nuclear Ships had made available to IMO their consideration and findings. Likewise the Federal Republic of Germany, the United Kingdom, the United States, and the Soviet Union had submitted general papers on the subject matter.

A considerable effort was expanded in developing the Nuclear Ship Safety Code. Maritime and nuclear experts cooperated closely in many and long working group meetings—meetings hosted by the USSR in Leningrad, by Italy in Genoa, by Canada in Ottawa, and by the Federal Republic of Germany in Hamburg. The Safety Code was finally approved by the Maritime Safety Committee in April 1981, and adopted by the 12th session of the Assembly in November 1981 as Resolution A.491 (XII). In contrast to Chapter VIII and the original Nuclear Ship Recommendations which were brief and general in form and extent, the new Code of Safety for Nuclear Merchant Ships is extensive and quite detailed, covering 124 pages.

A nuclear powered lighter carrier, planned by the Soviet Union, for use in Arctic areas, designed according to the IMO—Nuclear Ship Safety Code, was reported on in 1982. It is not known to the author whether this project has been realized or not.

PROGRESS OF CIVILIAN NUCLEAR PROPULSION

In May 1971, a nuclear ship symposium, sponsored by IMO and IAEA, was held in Hamburg, Federal Republic of Germany. Since the preceding symposium sponsored by IMO and IAEA, referred to above, held in Taormina, Italy, in 1960, interest in nuclear ships had increased markedly. The Hamburg symposium in 1971 was attended by 470 participants from 30 countries and by representatives from 8 international organizations. The proceedings from the symposium cover more than 1,000 pages, and contain reports on all facets of nuclear shipping, ship design and construction, operation, maintenance, safety considerations and even economic and legal aspects of nuclear propelled merchant ships. The United States' *N.S. Savannah* had been in operation over 10 years, and the experience with the ship was reported to have been very good. Likewise, the Federal Republic of Germany's *N.S. Otto Hahn* had been put into service in 1968, and into regular freight service in the beginning of 1971. The reports from the operation of *Otto Hahn* were also promising. "The reactor plant has proven itself fully equivalent to a conventional installation," it was reported. The Symposium in Hamburg generated a mood of realistic optimism.

IMO's Secretary General at that time, Mr. Colin Goad, in his speech at the opening of the symposium stated:

I would like to express my conviction that this symposium will contribute significantly towards the promotion of nuclear shipping which may very well in the future replace ships driven by oil powered machinery, just as, some fifty years ago, the era of sailing ships was brought to an end by the introduction of the steamship.

The atmosphere of optimism and expectation during the 1970s had a noticeable bearing on IMO's endeavors to develop international safety regulations for nuclear ships, to meet the possible and anticipated nuclear ship building "explosion." It may be recalled, that the working group that developed the Safety Code was under a certain pressure to finalize its work.

THE USE OF PORTS BY NUCLEAR MERCHANT SHIPS

Opening the 1977 Hamburg Symposium on the Safety of Nuclear Ships, Mr. Däunert, Bundesministerium für Forschung und Technologie, stated:

A merchant vessel is designed to sail all over the world and it is of minor commercial value without the possibility of free port entrance anywhere at any time.

Mr. Däunert also explained that port entrances for nuclear ships had to be negotiated bilaterally with each individual country, sometimes resulting in comprehensive contracts, after lengthy negotiations.

On the 6th of August 1963, the United States' *N.S. Savannah* visited Oslo. The United States had applied for one visit, and permission was granted under a special agreement between the governments of the United States and Norway, and necessary, quite detailed preparation on land and at sea had been made. The visit of the *Savannah* was made without any problems.

Four years later, August 1967, a formal invitation was received in Norway from the Federal Republic of Germany, to initiate negotiation between the two countries, with the view of reaching an agreement covering regular routine operation of the *N.S. Otto Hahn* to the port of Narvik, situated way up north on the coast of Norway. The new aspects of regular operation, through our fishery zones, increased the complexity of the negotiations significantly, to the extent that *N.S. Otto Hahn* never visited Norway.

The need for international guidance and regulations had become more obvious than ever.

In 1968, safety standards in the use of ports and approaches by nuclear merchant ships were developed by consultants of IAEA together with representatives of a number of international organizations.

These safety standards were published in 1968 sponsored by both IMO and IAEA (No. 27 of IAEA Safety Series). The purpose of this publication was to provide guidance to governments and port authorities on the various procedures and precautionary measures that may be employed where nuclear merchant ships use ports and approaches.

During the development of IMO's Code of Safety for Nuclear Merchant Ships mentioned above, which had been initiated in 1976, reference was made to port entry procedures, which could either be an integral part of the Code or a separate publication. The Working Group proposed that the IMO/IAEA guidelines should be kept as a separate publication, improved and amended in step with IMO's rule-making for nuclear ships.

The Maritime Safety Committee in 1978 agreed, and steps were taken to improve and revise the port guidelines in cooperation between IMO and IAEA.

A joint IMO/IAEA Technical Committee on Port Entry Requirements for Nuclear Merchant Ships was held in May 1979 at Church House, Westminster, London, at the kind invitation of the United Kingdom.

In October the same year, the Maritime Safety Committee considered the draft developed by the joint Technical Committee, and approved the Safety Recommendations on the Use of Ports by Nuclear Merchant Ships. The Recommendations were similarly approved by IAEA, and issued on behalf of both organizations in 1980.

The world was really prepared for an international nuclear merchant fleet.

LIABILITY OF NUCLEAR SHIP OPERATORS

One particularly cumbersome problem had early been encountered during the negotiations preceding nuclear ship bilateral port visiting agreements, namely questions on liability of nuclear ship operators. For the purpose of resolving this problem, a diplomatic conference was convened in Brussels, Belgium.

The Convention on the Liability of Operators of Nuclear Ships was adopted and signed at the close of the Conference in May 1962 by 15 countries. The number of ratifications needed for the coming into force of this Convention is remarkably low, namely two only. An additional requirement is, however, that one of these must be from a Contracting State which operates or has authorized the operation of a nuclear ship under its flag. The Convention has so far been ratified by 6 Contracting States, none of which at any time in history has had a nuclear ship under their flag. Thus the Convention is still not in force.

Article 1 of the Convention defines the term "nuclear ship" as "any ship equipped with a nuclear power plant," and thus warships are not excluded. The inclusion of warships was a stumbling block for the Conference and a point of serious disagreement. It is no secret that the inclusion of warships was strongly objected to by governments operating nuclear powered navy vessels.

Another controversial issue was the determination of the liability of operators of nuclear ships. The Norwegian delegate, speaking on behalf of Denmark, Finland, Sweden and Norway, said in his statement relative to the specific amount of the maximum liability: "We believe that an excessive figure will hamper the development of nuclear ships and thus unduly retard an economic process which will benefit all nations."

The Conference settled for, however not unanimously, 1,500 million francs (100 million US dollars) in respect of any one nuclear incident.

It has been thought by some observers that the Convention on the Liability of Operators of Nuclear Ships was prematurely rushed forward to accommodate an anticipated rapid development of nuclear powered ships. Further, the Convention has probably been lacking the necessary support from international organizations. The IAEA did play a role in the preparation of the Convention while it is not known whether the IMO at any time had been consulted in the preparatory phases of the Convention.

IMO's Secretary General at that time, Mr. Ove Nielsen, participated as an observer in the first part of the Conference in April 1961. The second and concluding part of the Conference was held in May 1962, at which time the Secretary General of IMO was not present, but represented by an IMO officer.

REFLECTIONS

The three Symposia on nuclear ships that have been referred to throughout this paper, in Taormina, Italy, in 1960, and in Hamburg, Federal Republic of Germany, in 1971 and 1977, have been very useful for the industry and the shipping world. They have given the status of nuclear ship development and served as an important indicator of possibilities and probabilities over a period of close to 20 years.

At the time of the Taormina Symposium, the world anticipated that nuclear propulsion of merchant ships would soon be a reality. Only a few months before that Symposium the 1960 SOLAS Conference had adopted Chapter VIII of the Convention on nuclear ships. The message so to speak from Taormina was, go ahead and green lights.

The 1971 Hamburg Symposium was a manifestation of the belief and optimism through the 1960s. Comprehensive reports on the successful operation of *N.S. Savannah* and of *N.S. Otto Hahn* were presented, and reports on the preparation for the operation of the Japanese *N.S. Mutsu* were given. Many countries presented detailed studies of nuclear ship projects. Even economical comparisons between conventionally propelled and nuclear propelled oiltankers were presented, showing for certain sizes and speeds economical breakeven points.

And finally, at the 1977 Hamburg Symposium, one may sense a certain moderation with respect to the probability of a rapid nuclear ship development. At that time *N.S. Savannah* had been taken out of service and the only nuclear driven merchant ship still in operation was *N.S. Otto Hahn*. The representative of the Federal Republic of Germany's Bundesministerium für Forschung und Technologie stated at the Symposium: "In the present circumstances no one expects a spectacular commercial breakthrough of nuclear ship propulsion in the short term. By short term I mean the next 15 years."

How pathetically true this prediction would prove to be. Now at the end of the 1980s there are no signs of a reappearance of nuclear merchant shipping. The *N.S. Otto Hahn*'s operation has terminated, and thus no nuclear merchant ships are operating any longer. Further, there seem to be no immediate plans for design and construction of a new generation of commercial nuclear ships.

PREDICTION

With the comprehensive development and progress on land in the production of nuclear power for the generation of electricity, as referred to in the introductory paragraphs of this paper, and with the extensive use of nuclear propulsion for navy surface craft and submarines over the last three decades, it would seem strange and almost unbelievable that all this technical advancement could not be taken advantage of for the safe and economical operation of nuclear merchant ships in the relatively near future.

The arguments for nuclear propulsion for civilian ships are not merely those of technical advancement, safety and economy. Our generation's accelerated depletion of natural resources and environment, including the ozone layer, will not only make nuclear propulsion of civilian transportation acceptable, but also preferable. The correct and safe application of nuclear energy for peaceful purposes, also for merchant ships, is anticipated to become the most efficient and hygienic source of energy during the next century.

IMO has already on its program an item on the prevention of air pollution by exhaust gases from ships. The outcome of the deliberation of this item may well be one factor bringing the date of realization of nuclear propulsion of merchant ships closer. IMO is anticipated to be deeply involved in this development for years to come.

REFERENCES

1. Proceedings of Symposium on Nuclear Ship Propulsion, Taormina, Italy, 1960.
2. *N.S. Savannah* Safety Assessment, 1960.
3. IMO Assembly Resolution A.30 (II), Agreement between IAEA and IMO, 1961.
4. Proceedings of Conference on the Liability of Operators of Nuclear Ships, 1962.
5. IMO Assembly Resolution A.45 (III), Banning of Nuclear Tests in the Atmosphere, in Outer Space and Under Water, 1963.
6. Norwegian-Swedish Nuclear Marine Project Vol. 1, 1965.
7. Nuclear Research Ship *Otto Hahn*. Safety Assessment, 1968.
8. Safety Standards in Use of Ports and Approaches by Nuclear Merchant Ships. IMO/IAEA, 1968. (No. 27 of IAEA Safety Series.)

9. Proceedings of the Symposium on Nuclear Ships, Hamburg, 1971.
10. Proceedings of the Symposium on the Safety of Nuclear Ships, Hamburg, 1977.
11. Safety Recommendations on the Use of Ports by Nuclear Merchant Ships, IMO/IAEA, 1980.
12. IMO Assembly Resolution A.491 (XII), Code of Safety for Nuclear Merchant Ships, 1981.
13. Nuclear Powered Lighter Carrier, *Morskoy Flot no. 8*, 1982.
14. A. M. Petrosy'ants, *Problems of Nuclear Science and Technology*, Fourth Edition, 1984.
15. IAEA Power Reactor Information System, 1988.
16. *Jane's Fighting Ships, 1988.*

Selected Documents and Proceedings

Final Act of the Meeting on a Convention to Prohibit Driftnet Fishing in the South Pacific and Convention for the Prohibition of Fishing with Long Driftnets in the South Pacific*

Representatives from the Territory of American Samoa, Australia, the Cook Islands, Federated States of Micronesia, Fiji, France (French Polynesia, New Caledonia, and Wallis and Futuna), Republic of Kiribati, Republic of the Marshall Islands, Republic of Nauru, New Zealand, Niue, Republic of Palau, Pitcairn, Solomon Islands, Papua New Guinea, Tokelau, Kingdom of Tonga, Tuvalu, Republic of Vanuatu, and Western Samoa, and the Forum Fisheries Agency, South Pacific Commission, Forum Secretariat and South Pacific Regional Environment Programme/United Nations Environment Programme represented by observer delegations, accepted the invitation extended by the Government of New Zealand to participate in a Meeting on a Convention to Prohibit Driftnet Fishing in the South Pacific. The Representatives met in Wellington from 21 to 24 November 1989 under the Chairmanship of Mr. Christopher Beeby, leader of the delegation of New Zealand.

As a result of its deliberations the Meeting adopted the "Convention for the Prohibition of Fishing with Long Driftnets in the South Pacific", the text of which is annexed to this Final Act, and agreed that it should be opened for signature on 29 November 1989.

The Meeting also endorsed Protocols and an associated instrument for discussion with those States and entities or organisations eligible to sign. The texts of those instruments are annexed to this Final Act. The Meeting agreed that, upon completion of these discussions, the Protocols and associated instrument would be further considered for adoption by the participants in the Meeting.

The Meeting agreed that nothing in Article 8 of the Convention would prejudice the position of any Party on the content or scope of any conservation and management measures that might be adopted for South Pacific albacore tuna.

The Meeting further agreed that the words "adjacent to the Convention Area" in Article 7 of Protocol II relate to countries on or within the Pacific rim.

IN WITNESS WHEREOF the Representatives have signed this Final Act.

DONE at Wellington, this twenty-fourth day of November 1989, in a single original copy in the English and French langauges to be deposited in the archives of the Government of New Zealand, which will transmit a certified copy thereof to all the Representatives participating in the Meeting.

*EDITORS' NOTE.—This report provided by the New Zealand Ministry of Foreign Affairs. The text may also be found in *International Legal Materials* 29 (1990):1453–63.

CONVENTION FOR THE PROHIBITION OF FISHING WITH LONG
DRIFTNETS IN THE SOUTH PACIFIC

The Parties to this Convention,

RECOGNISING the importance of marine living resources to the people of the
South Pacific region;

PROFOUNDLY CONCERNED at the damage now being done by pelagic driftnet
fishing to the albacore tuna resource and to the environment and economy of the
South Pacific region;

CONCERNED ALSO for the navigational threat posed by driftnet fishing;

NOTING that the increasing fishing capacity induced by large scale driftnet fishing threatens the fish stocks in the South Pacific;

MINDFUL OF the relevant rules of international law, including the provisions
of the United Nations Convention on the Law of the Sea done at Montego Bay on 10
December 1982, in particular Parts V, VII and XVI;

RECALLING the Declaration of the South Pacific Forum at Tarawa, 11 July
1989[,] that a Convention should be adopted to ban the use of driftnets in the South
Pacific region;

RECALLING ALSO the Resolution of the 29th South Pacific Conference at
Guam, which called for an immediate ban on the practice of driftnet fishing in the
South Pacific Commision region;

HAVE AGREED as follows:

Article 1: Definitions

For the purposes of this Convention and its Protocols:

(a) the "Convention Area",

(i) subject to sub-paragraph (ii) of this paragraph, shall be the area lying within
10 degrees North latitude and 50 degrees South latitude and 130 degrees East longitude and 120 degrees West longitude, and shall also include all waters under the
fisheries jurisdiction of any Party to this Convention.

(ii) In the case of a State or Territory which is Party to the Convention by virtue
of paragraph 1(b) or 1(c) of Article 10, it shall include only waters under the fisheries
jurisdiction of that Party, adjacent to the Territory referred to in paragraph 1(b) or
1(c) of Article 10;

(b) "driftnet" means a gillnet or other net or a combination of nets which is more
than 2.5 kilometres in length the purpose of which is to enmesh, entrap or entangle
fish by drifting on the surface of or in the water;

(c) "driftnet fishing activities" means:

(i) catching, taking or harvesting fish with the use of a driftnet;

(ii) attempting to catch, take or harvest fish with the use of a driftnet;

(iii) engaging in any other activity which can reasonably be expected to result in
the catching, taking or harvesting of fish with the use of a driftnet, including searching
for and locating fish to be taken by that method;

(iv) any operations at sea in support of, or in preparation for any activity described in this paragraph, including operations of placing, searching for or recovering fish aggregating devices or associated electronic equipment such as radio beacons;

(v) aircraft use, relating to the activities described in this paragraph, except for flights in emergencies involving the health or safety of crew members or the safety of a vessel; or

(vi) transporting, transhipping and processing any driftnet catch, and cooperation in the provision of food, fuel and other supplies for vessels equipped for or engaged in driftnet fishing;

(d) the "FFA" means the South Pacific Forum Fisheries Agency; and

(e) "fishing vessel" means any vessel or boat equipped for or engaged in searching for, catching, processing or transporting fish or other marine organisms.

Article 2: Measures regarding Nationals and Vessels

Each Party undertakes to prohibit its nationals and vessels documented under its laws from engaging in driftnet fishing activities within the Convention Area.

Article 3: Measures against Driftnet Fishing Activities

(1) Each Party undertakes:

(a) not to assist or encourage the use of driftnets within the Convention Area; and

(b) to take measures consistent with international law to restrict driftnet fishing activities within the Convention Area, including but not limited to:

(i) prohibiting the use of driftnets within areas under its fisheries jurisdiction; and

(ii) prohibiting the transhipment of driftnet catches within areas under its jurisdiction.

(2) Each Party may also take measures consistent with international law to:

(a) prohibit the landing of driftnet catches within its territory;

(b) prohibit the processing of driftnet catches in facilities under its jurisdiction;

(c) prohibit the importation of any fish or fish product, whether processed or not, which was caught using a driftnet;

(d) restrict port access and port servicing facilities for driftnet fishing vessels; and

(e) prohibit the possession of driftnets on board any fishing vessel within areas under its fisheries jurisdiction.

(3) Nothing in this Convention shall prevent a Party from taking measures against driftnet fishing activities which are stricter than those required by the Convention.

Article 4: Enforcement

(1) Each Party shall take appropriate measures to ensure the application of the provisions of this Convention.

(2) The Parties undertake to collaborate to facilitate surveillance and enforcement of measures taken by Parties pursuant to this Convention.

(3) The Parties undertake to take measures leading to the withdrawal of good standing on the Regional Register of Foreign Fishing Vessels maintained by the FFA against any vessel engaging in driftnet fishing activities.

Article 5: Consultation with Nonparties

(1) The Parties shall seek to consult with any State which is eligible to become a Party to this Convention on any matter relating to driftnet fishing activities which appear to affect adversely the conservation of marine living resources within the Convention Area or the implementation of the Convention or its Protocols.

(2) The Parties shall seek to reach agreement with any State referred to in paragraph 1 of this Article, concerning the prohibitions established pursuant to Articles 2 and 3.

Article 6: Institutional Arrangements

(1) The FFA shall be responsible for carrying out the following functions:

(a) the collection, preparation and dissemination of information on driftnet fishing activities within the Convention Area;

(b) the facilitation of scientific analyses on the effects of driftnet fishing activities within the Convention Area, including consultations with appropriate regional and international organisations; and

(c) the preparation and transmission to the Parties of an annual report on any driftnet fishing activities within the Convention Area and the measures taken to implement this Convention or its Protocols.

(2) Each Party shall expeditiously convey to the FFA:

(a) information on the measures adopted by it pursuant to the implementation of the Convention; and

(b) information on, and scientific analyses on the effects of, driftnet fishing activities relevant to the Convention Area.

(3) All Parties, including States or Territories not members of the FFA and the FFA[,] shall cooperate to promote the effective implementation of this Article.

Article 7: Review and Consultation among Parties

(1) Without prejudice to the conduct of consultations among Parties by other means, the FFA, at the request of three Parties, shall convene meetings of the Parties to review the implementation of this Convention and its Protocols.

(2) Parties to the Protocols shall be invited to any such meeting and to participate in a manner to be determined by the Parties to the Convention.

Article 8: Conservation and Management Measures

Parties to this Convention shall cooperate with each other and with appropriate distant water fishing nations and other entities or organisations in the development of conservation and management measures for South Pacific albacore tuna within the Convention Area.

Article 9: Protocols

This Convention may be supplemented by Protocols or associated instruments to further its objectives.

Article 10: Signature, Ratification and Accession

(1) This Convention shall be open for signature by:
 (a) any member of the FFA; and
 (b) any State in respect of any Territory situated within the Convention Area for which it is internationally responsible; or
 (c) any Territory situated within the Convention Area which has been authorised to sign the Convention and to assume rights and obligations under it by the Government of the State which is internationally responsible for it.
 (2) This Convention is subject to ratification by members of the FFA and the other States and Territories referred to in paragraph 1 of this Article. The instruments of ratification shall be deposited with the Government of New Zealand which shall be the Depositary.
 (3) This Convention shall remain open for accession by the members of the FFA and the other States and Territories referred to in paragraph 1 of this Article. The instruments of accession shall be deposited with the Depositary.

Article 11: Reservations

This Convention shall not be subject to reservations.

Article 12: Amendments

(1) Any Party may propose amendments to this Convention.
 (2) Amendments shall be adopted by consensus among the Parties.
 (3) Any amendments adopted shall be submitted by the Depositary to all Parties for ratification, approval or acceptance.
 (4) An amendment shall enter into force thirty days after receipt by the Depositary of instruments of ratification, approval or acceptance from all Parties.

Article 13: Entry into Force

(1) This Convention shall enter into force on the date of deposit of the fourth instrument of ratification or accession.

(2) For any member of the FFA or a State or Territory which ratifies or accedes to this Convention after the date of deposit of the fourth instrument of ratification or accession, the Convention shall enter into force on the date of deposit of its instrument of ratification or accession.

Article 14: Certification and Registration

(1) The original of this Convention and its Protocols shall be deposited with the Depositary, which shall transmit certified copies to all States and Territories eligible to become Party to the Convention and to all States eligible to become Party to a Protocol to the Convention.

(2) The Depositary shall register this Convention and its Protocols in accordance with Article 102 of the Charter of the United Nations.

DONE at Wellington this twenty-fourth day of November 1989 in the English and French languages, each text being equally authentic.

IN WITNESS WHEREOF the undersigned, being duly authorised by their Governments, have signed this Convention.

PROTOCOL I

The Parties to this Protocol,

NOTING the provisions of the Convention for the Prohibition of Fishing with Long Driftnets in the South Pacific ("the Convention")

HAVE AGREED as follows:

Article 1: Application of the Convention

Nothing in this Protocol shall affect or prejudice the views or positions of any Party with respect to the law of the sea.

Article 2: Measures regarding Nationals and Vessels

Each Party undertakes to prohibit its nationals and fishing vessels documented under its laws from using driftnets within the Convention Area.

Article 3: Transmission of Information

Each Party shall expeditiously convey to the FFA:

(a) information on the measures adopted by it pursuant to the implementation of this Protocol; and

(b) information on, and scientific analyses on the effects of, driftnet fishing activities relevant to the Convention Area.

Article 4: Conservation and Management Measures

Parties to this Protocol shall cooperate with Parties to the Convention in the development of conservation and management measures for South Pacific albacore tuna within the Convention Area.

Article 5: Enforcement

Each Party shall take appropriate measures to ensure the application of the provisions of this Protocol.

Article 6: Withdrawal

At any time after three years from the date on which this Protocol has entered into force for a Party, that Party may withdraw from the Protocol by giving written notice to the Depositary. The Depositary shall immediately inform all Parties to the Convention or its Protocols of receipt of a withdrawal notice. Withdrawal shall take effect one year after receipt of such notice by the Depositary.

Article 7: Final Clauses

(1) This Protocol shall be open for signature by any State whose nationals or fishing vessels documented under its laws fish within the Convention Area or by any other State invited to sign by the Parties to the Convention.

(2) This Protocol shall be subject to ratification. Instruments of ratification shall be deposited with the Government of New Zealand, which shall be the Depositary.

(3) This Protocol shall enter into force for each State on the date of deposit of its instrument of ratification with the Depositary.

(4) This Protocol shall not be subject to reservations. DONE at Noumea this twentieth day of October 1990.

IN WITNESS WHEREOF the undersigned, being duly authorised by their Governments, have signed this Protocol.

PROTOCOL II

The Parties to this Protocol,

NOTING the provisions of the Convention for the Prohibition of Fishing with Long Driftnets in the South Pacific ("the Convention")

HAVE AGREED as follows:

Article 1: Application of the Convention

Nothing in this Protocol shall affect or prejudice the views or positions of any Party with respect to the law of the sea.

Article 2: Measures regarding Nationals and Vessels

Each Party undertakes to prohibit its nationals and fishing vessels documented under its laws from using driftnets within the Convention Area.

Article 3: Measures against Driftnet Fishing Activities

(1) Each Party undertakes:
(a) not to assist or encourage the use of driftnets within the Convention Area; and
(b) to take measures consistent with international law to restrict driftnet fishing activities, including but not limited to:
(i) prohibiting the use of driftnets within areas under its fisheries jurisdiction; and
(ii) prohibiting the transhipment of driftnet catches within areas under its jurisdiction.
(2) Each Party may also take measures consistent with international law to:
(a) prohibit the landing of driftnet catches within its territory;
(b) prohibit the processing of driftnet catches in facilities under its jurisdiction;
(c) prohibit the importation of any fish or fish product, whether processed or not, which was caught using a driftnet;
(d) restrict port access and port servicing facilities for driftnet fishing vessels; and
(e) prohibit the possession of driftnets on board any fishing vessel within areas under its fisheries jurisdiction.
(3) Nothing in this Protocol shall prevent a Party from taking measures consistent with international law against driftnet fishing activities which are stricter than those required by the Protocol.

Article 4: Transmission of Information

Each Party shall expeditiously convey to the FFA:
(a) information on the measures adopted by it pursuant to the implementation of this Protocol; and
(b) information on, and scientific analyses on the effects of, driftnet fishing activities relevant to the Convention Area.

Article 5: Enforcement

Each Party shall take appropriate measures to ensure the application of the provisions of this Protocol.

Article 6: Withdrawal

At any time after three years from the date on which this Protocol has entered into force for a Party, that Party may withdraw from the Protocol by giving written notice to the Depositary. The Depositary shall immediately inform all Parties to the Convention or its Protocols of receipt of a withdrawal notice. Withdrawal shall take effect one year after receipt of such notice by the Depositary.

Article 7: Final Clauses

(1) This Protocol shall be open for signature by any State the waters under the jurisdiction of which are contiguous with or adjacent to the Convention Area or by any other State invited to sign by the Parties to the Convention.

(2) This Protocol shall be subject to ratification. Instruments of ratification shall be deposited with the Government of New Zealand, which shall be the Depositary.

(3) This Protocol shall enter into force for each State on the date of deposit of its instruments or ratification with the Depositary.

(4) This Protocol shall not be subject to reservations.

DONE at Noumea this twentieth day of October 1990.

IN WITNESS WHEREOF the undersigned, being duly authorised by their Governments, have signed this Protocol.

Selected Documents and Proceedings

Preparatory Commission for the International Sea-Bed Authority and Sea-Law Tribunal Concludes Seventh Session, 14 August–1 September 1989*

"Developing Countries say they support efficient, cost-effective Sea-Law Convention, and would discuss any issues with signatories and non-signatories" (Press Release, United Nations SEA/ 1089, 1 September 1989).

Developing countries this morning called for the early establishment of institutions envisaged under the United Nations Convention on the Law of the Sea, urging the registered pioneer investors to meet their obligations under the Convention. They made the call as the Preparatory Commission for the International Sea-Bed Authority and for the International Tribunal for the Law of the Sea wound up its sessions for 1989.

The developing countries—the "Group of 77"—also reaffirmed their readiness for talks with signatories and non-signatories to the Convention on any issues related to the Convention, and their support for an efficient, cost-effective Authority.

The representative of **Denmark,** speaking on behalf of the Group of 11, known as "friends of the Convention", said the Convention on the Law of the Sea was a milestone in the history of international lawmaking which must not be allowed to fail. The matter of pioneer obligations was crucial and should be settled at the Commission's next session, so that other outstanding issues might be addressed.

The representative of **France,** speaking on behalf of the European Economic Community (EEC), said the Commission's work had been characterized by a spirit of openness which augured well for the achievement of universality. It was important to achieve that universality through dialogue. He spoke also on behalf of the registered pioneer investors, who, he said, favored the spirit of open-mindedness and would do everything possible to promote dialogue with a view to achieving universal acceptance of the Convention. The pioneer investors—France, India, Japan and the Soviet Union—would continue to demonstrate a spirit of compromise.

The representative of **Italy,** speaking for the Group of Six—Belgium, the Netherlands, the United Kingdom, Japan, the Federal Republic of Germany and Italy—said negotiations were entering a crucial stage and that the Convention constituted a major achievement of the United Nations and of the process of codification and progressive development of international law. His Group believed the achievement of universality of the Convention might be greatly facilitated if all States agreed to the launching of a dialogue, without pre-conditions and in the appropriate framework.

The representative of **Canada,** speaking for potential applicants for registration as pioneer investors, expressed regret that it had not been possible to resolve the issue of pioneer investor obligations during the current session. His group work towards its resolution and in support of the achievement of universality of the Convention.

*EDITORS' NOTE.—This report, Press Release SEA/1089 of 1 September 1989, was supplied by the United Nations, Department of Public Information, Press Section, New York.

The representative of **China** said the reports of the Chairmen of the various Special Commissions and that of the Plenary demonstrated that "smooth progress" had been achieved during the session. He expressed regret that the consultations on the obligations of pioneer investors had not proceeded satisfactorily and hoped that progress could be made in that area at the next session in Kingston.

The representative of **Bulgaria,** speaking for the Group of Socialist States of Eastern Europe, said they supported the position of the Group of 77 on the need for ensuring the universality of the Convention and would work towards that end.

The representative of **Oman** said his country supported international efforts to regulate passage between littoral States and others. It had made proposals during the work of the Commission and had, through the Group of 77, reached compromises that would satisfy all countries.

Statements were also made by the observer of the African National Congress of South Africa (ANC) and by a representative of the Asian-African Legal Consultative Committee.

Before adjourning, the Preparatory Commission considered the reports of the Chairmen of the Special Commissions and of the Plenary. The Preparatory Commission decided to hold its next session—its eighth—in Kingston, Jamaica, from 5 to 30 March 1990.

SPECIAL COMMISSION 1 REPORT (DEVELOPING LAND-BASED PRODUCER STATES)

The report of the Special Commission on Developing Land-Based Producer States, Special Commission 1, introduced by acting Chairman Luis Giotto Preval Paez (Cuba), said "significant results" had been achieved in deliberations in the Commission's working group which should lead to solutions to the issue of a system of compensation for those States likely to be seriously affected by future sea-bed production of cobalt, copper, nickel and manganese.

At its next session, the Commission would continue consideration of the provisional conclusions (document CRP.16)—starting with provisional conclusion 18—which might form the basis of its recommendations to the International Sea-Bed Authority. The remaining conclusions cover identification and measurement of effects of sea-bed production on developing land-based producer States, determination of the problems/difficulties that would be encountered by the affected land-based producer States, formulation of measures to minimize problems/difficulties of those States and background requirement: necessary data and information.

The report said new conclusions could be added to the list covering the following topics: bilateral trade in copper, nickel, cobalt and manganese, including barter trade in those metals; commodity agreements or commodity associations related to the four metals; projection of future supply-demand-price of the four metals; supplementary and updated information on existing international and multilateral economic measures which could be of relevance in alleviating the problems of the developing land-based producer States and updated information on various aspects of the four metals on a country-by-country basis, including responses to a second note verbale being transmitted to States.

SPECIAL COMMISSION 2 REPORT (THE ENTERPRISE)

Lennox Ballah (Trinidad and Tobago), Chairman of the Special Commission on the sea-bed mining Enterprise, Special Commission 2, said that during the session the Commission had completed an article-by-article reading of a working paper relating to the structure and organization of the Enterprise, concentrating primarily on the identification of those provisions of the Convention which called for annotations of various kinds (document LOS/PCN/SCN.2/WP.16). Annotations would serve several purposes such as allowing the Preparatory Commission to comment on provisions of the Convention in the interests of promoting reasonable interpretation and also to suggest additional draft rules on matters such as the confidentiality requirement, on which the Convention had no rule specific to the Enterprise.

The Commission had approved a paper on draft principles, policies, guidelines and procedures for a Preparatory Commission training program (document LOS/PCN/SCN.2/L.6), later also approved by the Plenary on 31 August.

Special Commission 2, which is charged with preparing for the operation of the Enterprise, also discussed the question of how to achieve the desired degree of autonomy for the Enterprise from organs of the Authority. It considered instances in which the Council, and the Assembly, on some matters, would exercise a significant degree of control. Discussion also centered on harmonization with Special Commission 3—which is drawing up a mining code for the deep sea-bed—on transfer of technology provisions relating to the powers of the Governing Board; and with the informal plenary, on the rights, privileges and immunities of the Enterprise at the seat of the Authority.

The Chairman's Advisory Group on Assumptions had examined current metal price movement, long-term projections and the establishment of a new set of assumptions. Data received from the expert of Australia and the Secretariat had shown that prices for nickel, copper and manganese had continued to rise, as much as twofold over prices in the sea-bed mining model submitted earlier by Australia. After updating the metal prices in that model, the Australian Bureau of Mineral Resources had again come to the conclusion that deep sea-bed mining was not yet economically viable. It was speculated whether, if the recent upward trend in metal prices were to continue, the aggregate metal price would reach a level at which sea-bed mining might be feasible with the current set of assumptions. At the same time, however, it was recognized that the long-term trend projections for the metal prices remain essentially the same.

In the absence of any substantial developments in sea-bed mining technology, the Group was unable, at present, to change the set of assumptions used in its model. The Secretariat was in the process of creating the necessary facilities for the establishment of a data bank to aid in the Group's work. The Group would continue to monitor metal price movements, long-term projections and technological developments relating to sea-bed mining.

The Chairman said that at its next session, Special Commission 2 would take up implementation of the training program; discuss "transitional arrangements" for the Enterprise; and continue its consideration of exploration. He said that in the event of his inability to attend the next session, arrangements had been made for any one of the three Vice-Chairmen of the Commission—Senegal, Yugoslavia and Canada, in that order—to take over the chairmanship in an acting capacity.

SPECIAL COMMISSION 3 REPORT (MINING CODE)

The Chairman of the Special Commission 3 on the mining code, Jaap Walkate (Netherlands), said the Commission had made "good progress" in its work by completing a preliminary first reading of draft regulations on production authorizations (document LOS/PCN/SCN.3/WP.6/Add.1). According to his report, delegations had been able to identify their positions on the issue of a production policy in general and the instrument of production limitation by means of a production ceiling as contained in the relevant article of the Convention (Article 151).

A seminar on issues relating to production policies with regard to the resources of the international sea-bed Area, such as the concerns of producers, consumers and potential sea-bed miners, was organized to assist the work of the Special Commission. The second reading of the regulations would take place on the basis of the revisions prepared by the Chairman in cooperation with the Secretariat. The Special Commission has responsibility for drafting the rules and regulations for production authorizations.

The Chairman said although progress had been made in his discussions with Secretariat officials on the revision of the working paper on transfer of technology, the revised document would not be out before the middle of 1990.

SPECIAL COMMISSION 4 REPORT (LAW OF THE SEA TRIBUNAL)

The Chairman of the Special Commission on the International Tribunal for the Law of the Sea, Special Commission 4, Gunter Goerner (German Democratic Republic), said in his report that the Commission continued to examine the draft Protocol on the Privileges and Immunities of the Tribunal and later requested the Secretariat to produce a revised draft in the light of suggestions presented. The Special Commission is to work on practical arrangements for the establishment of the Tribunal, which would be based in the Federal Republic of Germany.

The Commission also took up the question of relationship agreements between the Tribunal and other international institutions—such as the United Nations, the International Court of Justice and the projected International Sea-Bed Authority—examining the principles that would govern them and issues related to such agreements. On the basis of Secretariat working papers (documents CRP.36 and WP.7), the Commission identified the main substance of relationship agreements between the Tribunal and the three institutions. It was agreed that the main issues for the accords should be indicative in nature and that the possibility should be left open for additional subject matters to be added.

At its next session, the Special Commission would commence consideration of the institutional structure and initial staffing needs of the Tribunal, taking into account provisions of the Convention and the structure of existing international tribunals and courts and their experiences. It would also, in that context, take up the initial and subsequent financial implications arising from the envisaged institutional structure of the Tribunal. The Secretariat had been requested to prepare a relevant working paper on the subjects raised. The Commission would also continue the examination of the "main content of draft agreements/arrangements to be concluded between the Tribunal and the United Nations, the International Court of Justice and the International Sea-Bed Authority and, if necessary, with other international bodies having competence in the field of the law of the sea and relevant affairs." The Secretariat had been requested to prepare a working paper which would be the basis for a relationship agreement to be negotiated between the Tribunal and the United Nations.

PLENARY REPORT

The Chairman of the Preparatory Commission, Jose Luis Jesus (Cape Verde), said the three matters before the plenary at the current session had been consideration of the draft headquarters agreement between the Authority and the Government of Jamaica; the Finance Committee; and the General Protocol and Privileges and Immunities of the Authority.

On the headquarters agreement, great progress had been made, although some issues remained open pending further consultations. On the Finance Committee, the Chairman had undertaken informal consultations on problem areas, which he would continue at the Commission's next session. On the General Protocol on Privileges and Immunities, there had not been sufficient time to discuss the matter, but it would be taken up at the next session.

Referring to Resolution II of the Third United Nations Conference on the Law of the Sea which deals with the obligations of pioneer investors, he said a compromise had been reached on the mandate for the meeting of the Technical Group of Experts which took place during the meeting of the Preparatory Commission. The experts had prepared two reports, at the request of the Preparatory Commission, to assist it in discussing the register of investors. The General Committee of the Commission had considered those reports, and the Chairman had undertaken consultations with the delegations involved to narrow down identified differences. It was decided to review the paper circulated last March in Kingston, as revised by the recommendations of the Group of Experts, to help the delegations involved to come closer to an agreement. Consultations on the matter would be resumed from the first day of the next session.

On the long-term program of work, he said that four items should be dealt with during 1990, and that he would resume his consultations on certain hard-core issues. In 1991, two issues would remain on the Commission's agenda, and he would continue his consultations on hard-core issues, with a view to concluding them by the target date. He intended to begin those consultations at the next session, time permitting and under favorable conditions.

The representative of **Zambia,** Chairman of the Group of 77, speaking on the report of Special Commission 2, said that every registered pioneer was expected to provide training at all levels so that the Enterprise might be able to keep pace with the system of deep sea-bed mining. A training panel was to be set up to select nominees offered by various Governments. That training panel should be set up by the Preparatory Commission without delay, so that the common objective of the common heritage of mankind might be realized.

STATEMENTS

The representative of **Zambia,** speaking on behalf of the Group of 77, said the Group believed that the work of the Preparatory Commission should be organized in such a way as to ensure the early completion of the preparations for the entry into force of the Convention and the establishment of the Authority and its organs. While appreciating the need to be thorough, the Group believed that constant reviews and re-opening of issues previously agreed upon was retrogressive. It supported the proposal of the Chairman of the Preparatory Commission that 1991 should be the provisional target date for completion of the work of the Commission.

He said the developing countries continued to be ready to hold discussions, without any preconditions, with any delegation or group of delegations—whether signatories or non-signatories to the Convention—on any issues related to the Convention and work of the Preparatory Commission. Their willingness to do so was born out of a "genuine desire to ensure the universality of the Convention". The Group, contrary to false impressions created in some quarters, supported the establishment of an efficient, cost-effective Authority, the size of which would be no larger or smaller than was required to enable it to carry out its functions. A realistic size and budget of the Authority, upon entry into force of the Convention, would not impose great financial burdens. It was the hope of his Group that in view of their support for a cost-effective Authority, the question of burdensome costs could be laid to rest.

He said that although the first group of pioneer investors—France, India, Japan and the Soviet Union—had been registered in 1987, agreement had yet to be reached on an arrangement for the discharge of obligations. There seemed to be "an unfortunate deliberate delay" in resolving issues related to the subject. The registered pioneer investors should prove that they had always intended to discharge their obligations arising out of the special status created for them under the Convention and various understandings.

The representative of **Gabon,** referring to elements of the reports of Special Commissions 1 and 3, said he could not support development of a subsidies system regarding sea-bed mining. Such a system would deviate from the provisions of the Convention.

The representative of **Zaire** agreed with Gabon. The report of Special Commission 1 spoke of a system of compensation and of subsidization of sea-bed mining (paragraph 2 of document LOS/PCN/L.73). He saw no provision in the Convention empowering the Preparatory Commission to subsidize sea-bed mining, but only to study the effects of subsidization. The Special Commission's report should be revised to reflect that.

Referring to the report of Special Commission 3 (paragraph 14 of document LOS/PCN/L.74), he said his Government had never recognized subsidization of any type of production, and particularly of deep sea-bed mining, which had not yet even begun.

Finally, he proposed that **Zambia**'s statement on behalf of the Group of 77 be published as part of the historical record of work on the Law of the Sea.

The **Chairman** said he would call attention of the Special Commission Chairmen to the questions raised regarding subsidization. On the proposal to publish the statement of the Group of 77, he said an effort would be made to reproduce that statement.

The representative of **Gabon** said he insisted that the statements be published and distributed to delegations.

The Special Representative of the Secretary-General for the Law of the Sea, **Satya Nandan,** said the Secretariat was required to reproduce the decisions taken by the Commission, and, in that connection, also reproduced working documents. If it were to reproduce other statements, it would have to make provision for the recording, translation and reproduction of all statements in six languages. Several representatives had requested reproduction of specific statements. The Secretariat was willing to photocopy such statements for the benefit of delegations. If the Commission decided to request summary records, it would have to bring that decision to the General Assembly for its approval. But the current trend was to reduce summary records as far as possible because of the financial situation of the United Nations.

The Preparatory Commission took note of the reports of the Chairman of the Special Commissions and of the Preparatory Commission.

CLOSING STATEMENTS

The representative of **Denmark,** in his statement on behalf of the Group of 11, said the Convention on the Law of the Sea was a milestone in the history of international lawmaking which must not be allowed to fail. The convention expressed the aspirations of States, and his Group would work towards the achievement of its universal acceptance. He said that the matter of pioneer obligations was crucial and should be settled at the Commission's next session, so that other outstanding issues might be addressed.

The representative of **France,** on behalf of the European Community and its member States, said the Commission's work had been characterized by a spirit of openness which augured well for the achievement of universality. Although it had not yet entered into force, the Convention provided "an indispensable reference point" on law of the sea problems, and promoted a harmonization of international law. But lack of total acceptance entailed a risk that divergent practices might emerge. It was, therefore, important to achieve universality, through the opening of dialogue.

He then spoke on behalf of the pioneer investors, who, he said, favored the spirit of open-mindedness and would do everything possible to promote dialogue with a view to achieving universal acceptance of the Convention. The pioneer investors would continue to demonstrate a spirit of compromise. The Chairman's statement had indicated progress in two key areas: informal consultations on the technical experts' reports; and the existence of a favorable climate. He said the Chairman should begin negotiations on obligations, for the Commission's next session to have a positive outcome.

The representative of **Italy,** speaking for the Group of Six, said negotiations were entering a crucial stage, and the Convention constituted a major achievement of the United Nations and of the process of codification and progressive development of international law. But the Group held the view that provisions of the Convention covering the sea-bed mining Area and principles governing its exploitation (Part XI of Convention) presented some serious problems which, if left unresolved, might jeopardize that achievement. The members of the Group had worked tirelessly in the Preparatory Commission to find appropriate solutions that might pave the way for a universally acceptable Convention. They believed the achievement of that objective might be greatly facilitated should all States agree to the launching of a dialogue, without preconditions and in the appropriate framework. They would welcome developments in that direction.

The representative of **Canada,** speaking for potential applicants for registration as investors, expressed appreciation for the efforts of the Secretariat in attempting to resolve the issue of pioneer investor obligations. The potential applicants welcomed the reports of the Group of Experts and hoped the report, outlining plans for stages of exploration of a mine-site reserved for the Authority, would enable them to resolve some of the outstanding differences related to the obligations issue. They regretted, however, that it had not been possible to achieve that result during the session. They were ready to contribute actively and constructively to a dialogue, without preconditions, for a universally acceptable solution.

The representative of **China** said the reports of the Chairmen of the various Special Commissions and that of the Plenary demonstrated that "smooth progress" had been achieved during the session. He was confident that the goals of the Preparatory Commission could be achieved within the target date of 1991. He expressed regret that the consultations on the obligations of pioneer investors had not proceeded satisfactorily and hoped that progress could be made in that area at the next session in Kingston. He welcomed the positive attitude of the Group of 77 and joined it in inviting non-signatories to the Convention to participate in talks to resolve problems they might have. China would work for the attainment of the goals of the Commission.

The representative of **Bulgaria,** speaking for the Group of Socialist States of Eastern Europe, said they supported the position of the Group of 77 on the need for ensuring the universality of the Convention. They would work towards that end. He said the concept of the resources of the seas being the common heritage of mankind could not be achieved without the participation of all countries in realizing the objectives of the Convention. The Group was ready to support all positive steps aimed at surmounting difficulties involved in the attainment of the universality concept.

The representative of **Oman** said his country was a littoral State and had, therefore, supported international efforts to regulate passage between littoral States and others. His Government had made proposals and had, through the Group of 77, reached compromises that would satisfy all countries. It had ratified the Convention and had declared a 200-mile Exclusive Economic Zone. It would give due consideration to the rights of other States in conformity with provisions of the Convention.

The observer of the African National Congress of South Africa (ANC) said sea-bed resources formed part of the national resources of her country. But her people could not participate as a nation in the deliberations on that matter because of the system of *apartheid*. South Africa was in a political and economic crisis from which it could not extricate itself. The defiance campaign in her country was a non-violent, popular movement expressing the will of the majority of her people—both black and white—to eradicate *apartheid*. She repeated her call for creation of a non-racial democratic State in South Africa.

The Assistant Secretary-General of the Asian-African Legal Consultative Committee said the Law of the Sea Convention was unique. As a codifying Convention, it combined customary international law with new rules regarding contractual rights and obligation. As a constitutive instrument, it established such entities as the International Sea-Bed Authority and the Law of the Sea Tribunal. It envisaged creation of a commission to make recommendations to coastal States on the outer limits of their continental shelf. In addition to providing for settlement of disputes arising out of the Convention, it envisaged a significant role for international organizations.

He was concerned, however, about "furtive efforts" to amend the Convention before it had even entered into force. The developing countries had a feeling of frustration and betrayal at the industrialized countries' reluctance to ratify and become bound by the Convention. The principle of the common heritage of mankind was a vital element of the new legal order for the seas and oceans, and was part and parcel of contemporary international law. He, therefore, opposed "any premature amendments" aimed at removing the reservations of some industrialized countries, and suggested that there might be other solutions for those reservations.

The Commission had long waited for one of the major industrialized countries to participate in the negotiations, he said. It was time for all States to reaffirm their faith in negotiations and commence a dialogue with a view to finding generally acceptable solutions that would enable them to ratify the Convention.

The observer of the International Ocean Institute called attention to her Institute's training program on sea-bed mining, which had been restructured to meet the changing needs of the international community and the developing countries in particular. It was a demanding course on the high technologies related to sea-bed mining. Over the past two years, it had been held in Jamaica, Colombia and India, and arrangements were being made for it to be held alternately in Oslo and India. She said that the program made provision for 16 scholarships, and urged interested delegations to apply early, in view of the large number of applications received.

The **Chairman** said that, after consultation with the regional groups, he proposed that the next session of the commission be held at Kingston, Jamaica, from 5 to 30 March 1990.

The **Secretary** said the provision for the forthcoming eighth session had been made in the regular budget of the United Nations. The Commission decided to hold its next session from 5 to 30 March at Kingston.

The **Chairman** said he agreed with the need for universal acceptance of the Convention. Any action that the Preparatory Commission might take to strengthen the universality principle of the Convention was, therefore, welcome. He expressed the hope that, by the time of its entry into force, every nation would lend its support to the Convention.

MEMBERS OF COMMISSION

Under the provisions of Resolution I establishing the Preparatory Commission, the Commission consists of the representatives of States and Namibia (represented by the Council for Namibia), which have signed the Law of the Sea Convention or acceded to it. Representatives of signatories of the Conference's Final Act may participate fully in the deliberations of the Commission as observers, but are not entitled to participate in its decisions.

The following are the signatories to the Convention: Afghanistan, Algeria, Angola, Antigua and Barbuda, Argentina, Australia, Austria, Bahamas, Bahrain, Bangladesh, Barbados, Belgium, Belize, Benin, Bhutan, Bolivia, Botswana, Brazil, Brunei Darussalam, Bulgaria, Burkina Faso, Burma, Burundi, Byelorussia, Cameroon, Canada, Cape Verde, Central African Republic, Chad, Chile, China, Colombia, Comoros, Congo, Cook Islands, Costa Rica, Côte d'Ivoire, Cuba, Cyprus, Czechoslovakia, Democratic Kampuchea, Democratic People's Republic of Korea, Democratic Yemen, Denmark, Djibouti, Dominica, Dominican Republic, Egypt, El Salvador, Equatorial Guinea, Ethiopia, European Economic Community, Fiji, Finland, France, Gabon,

Gambia, German Democratic Republic, Ghana, Greece, Grenada, Guatemala, Guinea, Guinea-Bissau, Guyana, Haiti, Honduras, Hungary, Iceland, India, Indonesia, Iran, Iraq, Ireland, Italy, Jamaica, Japan, Kenya, Kuwait, Lao People's Democratic Republic, Lebanon, Lesotho, Liberia, Libya, Liechtenstein, Luxembourg, Madagascar, Malawi, Malaysia, Maldives, Mali, Malta, Mauritania, Mauritius, Mexico, Monaco, Mongolia, Morocco, Mozambique, Namibia (Council for Namibia), Nauru, Nepal, Netherlands, New Zealand, Nicaragua, Niger, Nigeria, Niue, Norway, Oman, Pakistan, Panama, Papua New Guinea, Paraguay, Philippines, Poland, Portugal, Qatar, Republic of Korea, Romania, Rwanda, Saint Kitts and Nevis, Saint Lucia, Saint Vincent and the Grenadines, Samoa, São Tomé and Principe, Saudi Arabia, Senegal, Seychelles, Sierra Leone, Singapore, Solomon Islands, Somalia, South Africa, Spain, Sri Lanka, Sudan, Suriname, Swaziland, Sweden, Switzerland, Thailand, Togo, Trinidad and Tobago, Tunisia, Tuvalu, Uganda, Ukraine, USSR, United Arab Emirates, United Republic of Tanzania, Uruguay, Vanuatu, Viet Nam, Yemen, Yugoslavia, Zaire, Zambia and Zimbabwe.

The Convention had been ratified by the following 41 countries: Antigua and Barbuda, Bahamas, Bahrain, Belize, Brazil, Cameroon, Cape Verde, Côte d'Ivoire, Cuba, Cyprus, Democratic Yemen, Egypt, Fiji, Gambia, Ghana, Guinea, Guinea-Bissau, Iceland, Indonesia, Iraq, Jamaica, Kenya, Kuwait, Mali, Mexico, Namibia (Council for Namibia), Nigeria, Paraguay, Philippines, Saint Lucia, São Tomé and Principe, Senegal, Somalia, Sudan, Togo, Trinidad and Tobago, Tunisia, United Republic of Tanzania, Yugoslavia, Zaire and Zambia.

OFFICERS

The Chairman of the Preparatory Commission is Jose Luis Jesus (Cape Verde). Its 14 Vice-Chairmen are the representatives of Algeria, Australia, Brazil, Cameroon, Chile, China, France, India, Iraq, Japan, Liberia, Nigeria, Sri Lanka and USSR.

The Rapporteur-General is Kenneth Rattray (Jamaica).

Hasjim Djalal (Indonesia) is Chairman of the Special Commission on Developing Land-Based Producer States (Special Commission 1); Lennox Ballah (Trinidad and Tobago) is Chairman of the Special Commission on the Enterprise (Special Commission 2); Jaap Walkate (Netherlands) is Chairman of the Special Commission on the Mining Code (Special Commission 3); and Theodore Chalkiopoulis (Greece) is Acting Chairman of the Special Commission on the International Tribunal for the Law of the Sea (Special Commission 4).

The Vice-Chairmen for Special Commission 1 are the representatives of Austria, Cuba, Romania and Zambia; for Special Commission 2, Canada, Mongolia, Senegal and Yugoslavia; for Special Commission 3, Gabon, Mexico, Pakistan and Poland; and for Special Commission 4, Colombia, Greece, Philippines and Sudan.

The Special Representative of the Secretary-General for the Law of the Sea is Satya N. Nandan, Under-Secretary-General for Ocean Affairs and Law of the Sea.

Selected Documents and Proceedings

Disarmament: The Sea-Bed Treaty and Its Third Review Conference in 1989*

BACKGROUND

The Treaty on the Prohibition of the Emplacement of Nuclear Weapons and Other Weapons of Mass Destruction on the Sea-Bed and the Ocean Floor and in the Subsoil Thereof, known as the Sea-Bed Treaty, represents an important step towards preventing an arms race in the vast area at the bottom of the seas and oceans that cover two-thirds of the surface of the globe. The Treaty was concluded in 1971. As the Secretary-General of the United Nations stated when the Treaty was opened for signature: "Fortunately, the world early recognized that the expansion of the arms race to the sea-bed and ocean floor would not only seriously interfere with the growing peaceful exploitation of the area, but would provide a new danger to international security and add a great and unnecessary burden to the already staggering world outlay for military purposes." The Treaty, the Secretary-General added, "may be regarded as the first step in the direction of barring any such undesirable development before it takes place."

The concern of the United Nations regarding the sea-bed—not only its military but also its economic potential—began to find concrete expression in 1967. At the session of the General Assembly that year, it was proposed that international action be taken to regulate the uses of the sea-bed and to ensure that the Area's exploitation would be for peaceful purposes only and for the benefit of all mankind.

In 1968, the Soviet Union proposed that the multilateral disarmament negotiating body in Geneva, the Eighteen-Nation Committee on Disarmament (predecessor of the present Conference on Disarmament), begin negotiations on the establishment of a regime to ensure the exclusively peaceful use of the sea-bed beyond territorial waters, in particular to prohibit the establishment of fixed military installations in that vast area. At the same time, the United States acknowledged the timeliness and relevance of dealing with the question and suggested that the Committee begin to define those factors vital to a workable, verifiable and effective international agreement which would prevent the sea-bed from being used for the emplacement of weapons of mass destruction.

It was then agreed between the Soviet Union and the United States that the purpose of the treaty under discussion would be to limit the military use of the sea-bed by banning from it nuclear and other weapons of mass destruction. To that end, the two States submitted a joint draft treaty to the Eighteen-Nation Committee on Disarmament in Geneva, which was extensively debated and subsequently revised a number of times. In the course of the debate various proposals were made, which

*EDITORS' NOTE.—This document has been supplied by the Department of Disarmament Affairs, United Nations, New York, N.Y. 10017 and has been published as *Disarmament Facts 69: The Sea-Bed Treaty and Its Third Review Conference in 1989* (December 1989).

concerned mainly the geographical area covered by the treaty; verification of compliance; the relationship of the obligations assumed under the treaty and other international obligations; the relationship of the treaty to international agreements concerning the establishment of nuclear-free zones; and the commitment of the parties to continue negotiations on further disarmament measures for the sea-bed and the ocean floor. In early September 1970, after intensive consultations, the final text of the draft treaty that incorporated the substance of most of the amendments and suggestions put forward by a number of States was approved and submitted to the General Assembly as part of the Committee's report.

On 7 December, the General Assembly commended the Treaty and requested its depositary Governments to open it for signature and ratification at the earliest possible date. In doing so, the Assembly expressed its conviction that the prevention of an arms race on the sea-bed and the ocean floor served the interests of maintaining world peace, and that it was in the common interest of mankind to reserve the sea-bed and the ocean floor exclusively for peaceful purposes.

The Treaty was opened for signature on 11 February 1971 and entered into force on 18 May 1972. Three nuclear-weapon States, the Soviet Union, the United Kingdom and the United States—which are the depositaries of the Treaty—and numerous other countries, in particular a number of important maritime Powers, are parties to it. By September 1989, when the States parties met (for the third time) to review the operation of the Treaty, 82 States had ratified it, while 23 States had signed but not yet ratified it.

The Treaty constitutes an arms limitation measure applicable to the sea-bed environment. As mentioned above, it was negotiated at a time of growing interest in the regulation of the use of the oceans and their resources. Efforts directed towards the broader objective of developing a comprehensive legal code to govern the use of the oceans culminated, in 1982, in the conclusion of the United Nations Convention on the Law of the Sea. Nothing contained in that Convention affects the rights and obligations assumed by States parties under the Sea-Bed Treaty.

MAIN PROVISIONS OF THE SEA-BED TREATY

In the Preamble, the States parties express their conviction that the Treaty constitutes a step towards the exclusion of the sea-bed, the ocean floor and its subsoil from the arms race.

All States parties undertake, in Article I, not to emplant or emplace nuclear and other weapons of mass destruction on the sea-bed beyond a 12-mile wide zone defined in Article II. In addition, no facilities specifically designed for storing, testing or using such weapons may be installed. The outer limit of the sea-bed zone is defined in Article II as being coterminous with the 12-mile outer limit of the zone referred to in an earlier international agreement, the 1958 Geneva Convention on the Territorial Sea and the Contiguous Zone.

Under Article III, each State party has the right to verify, through observation, other parties' activities on the sea-bed beyond the 12-mile zone, provided that such observations do not interfere with those activities. The Treaty also provides for the possibility of consultation and cooperation on such further verification procedures as may be agreed to, including appropriate inspection of objects, structures, installations or other facilities that may reasonably be expected to be of a kind prohibited by the Treaty. Verification may be undertaken by any State party using its own means or through appropriate international procedures within the framework of the United Nations. If, in spite of consultation and cooperation among the parties, there remains a serious question concerning fulfilment of the obligations under the Treaty, a State party may refer the matter to the Security Council.

As the Sea-Bed Treaty was negotiated at a time when the broader issues of international law applicable to the sea were being discussed in the preparatory phase for the United Nations Conference on the Law of the Sea, Article IV states that nothing in the Treaty shall be interpreted as supporting or prejudicing the position of any State party with respect to existing international conventions, including the 1958 Convention on the Territorial Sea and the Contiguous Zone, or with respect to any claim it may make related to waters off its coast, including territorial seas and contiguous zones, or to the sea-bed and the ocean floor, including continental shelves.

States parties undertake, in Article V, to continue negotiations in good faith concerning further measures in the field of disarmament for the prevention of an arms race on the sea-bed and the ocean floor and in its subsoil.

In Article VII the Treaty provides for review conferences in order to ensure that the purposes of the preamble and the provisions of the Treaty are being realized, taking into account any relevant technological developments.

As stated in Article IX, the Treaty in no way affects the obligations assumed by States parties under international instruments establishing nuclear-weapon-free zones.

FIRST AND SECOND REVIEW CONFERENCES

Prior to 1989, two review conferences were held in 1977 and 1983, respectively. In their Final Declaration of 1977, which they adopted by consensus, the participating States recognized the continuing importance of the Treaty and its objectives and affirmed their belief that universal adherence to it would enhance peace and security. They therefore called upon the States that had not yet become parties to the Treaty, particularly those possessing nuclear weapons or any other types of weapons of mass destruction, to do so at the earliest possible date. They emphasized that the Treaty had been faithfully observed and that it had demonstrated its effectiveness since its entry into force. They also reaffirmed their common interest in avoiding an arms race involving nuclear and other weapons of mass destruction on the sea-bed. The 1983 Final Declaration, also adopted by consensus, reached the same conclusions.

At the two Review Conferences, an examination of the various provisions of the Treaty was undertaken with a view to making recommendations regarding their further implementation.

Both Conferences affirmed that the zone covered by the Treaty reflected the right balance between the need to prevent an arms race on the sea-bed and the right of States to control verification activities close to their own coasts. Participants also noted that no verification procedures had been invoked under Article III, and that the provisions under that article included the right to agree to resort to various international procedures, such as ad hoc consultative groups of experts. In discussions at both Conferences, a number of countries pointed out that since most States parties did not possess adequate independent means of verification, the procedures provided for in Article III should be further elaborated.

The States parties reaffirmed their commitment to continue negotiations on further measures to prevent an arms race on the sea-bed. Since talks had not yet been held, the Geneva negotiating body was requested to proceed promptly with its consideration of such measures, in consultation with the States parties.

Even though no information was presented to the Review Conferences indicating that major technological developments affecting the operation of the Treaty had taken place since 1972, States parties recognized the need to keep such developments under continuous review. Certain parties expressed doubts about statements by other parties to the effect that no relevant military or peaceful technological developments had occurred.

The Conferences reaffirmed their conviction that nothing in the Treaty affected the obligations assumed by States parties to the Treaty under international instruments establishing zones free from nuclear weapons.

In 1983, the General Assembly welcomed with satisfaction the positive assessment of the Treaty made by the Second Review Conference and requested the Conference on Disarmament to proceed promptly with consideration of further disarmament measures for the prevention of an arms race on the sea-bed.

Subsequently, the Conference on Disarmament reported that, during consideration of the subject, the view was expressed that the scope of the Treaty should be broadened to allow for the fuller demilitarization of the sea-bed, that its provisions governing procedures for verification and compliance should be improved and that access to information on relevant technological developments should be facilitated. The Conference on Disarmament also noted that differences of opinion existed concerning the urgency of conducting negotiations on further measures.

THE THIRD REVIEW CONFERENCE

At the 1988 regular session of the United Nations General Assembly, States parties to the Treaty agreed that a further review conference should be held in 1989, and the General Assembly adopted a resolution to that effect. The Preparatory Committee for the Third Review Conference met in April 1989 and decided that the Conference would be held in September 1989 in Geneva. It also adopted a provisional agenda and draft rules of procedure for it.

The Conference was held from 19 to 28 September under the presidency of Ambassador Sergio de Queiroz Duarte of Brazil. Of the 82 States parties to the Treaty, 53 participated in the Review Conference, joined by 2 of the 23 signatory States. In addition, 13 non-signatory States were granted observer status at the Conference.

In a message addressed to the participants, the Secretary-General of the United Nations underscored that the Treaty was an important preventive measure in the field of arms limitation and disarmament. It reflected, he stated, the awareness of the international community that the extension of the arms race to two-thirds of the surface of our planet would only add new threats to international peace and security. The Treaty, he noted, called for continued negotiations on further measures for the prevention of an arms race on the sea-bed and the ocean floor and in the subsoil thereof. This commitment had been reaffirmed by the General Assembly in the Final

Document of its first special session devoted to disarmament as well as by the First and the Second Review Conferences. Furthermore, the Treaty recognized that scientific and technological advances could open possibilities for new military uses of the sea-bed. Consequently, the Secretary-General stated, one of the main tasks of the Conference was to conduct a thorough review of the situation, taking into account relevant technological developments.

The Review Conference held a general debate in which 28 speakers made statements. In the debate, the effectiveness of the Treaty in ensuring that no nuclear weapons or other weapons of mass destruction were emplaced in the sea-bed zone and on the ocean floor was stressed in spite of some long-standing differences regarding measures to improve the Treaty regime. It was also noted that no party had resorted to the verification arrangements provided for in Article III of the Treaty.

In addition to verification, the main subjects of discussion at the Third Review Conference were: the scope of the Treaty, both in terms of the geographical zone of application and in terms of extending its application to weapons other than those of mass destruction; the related question of the need for further measures to prevent an arms race on the sea-bed; technological developments relevant to the operation of the Treaty; the relationship between the Treaty and the 1982 United Nations Convention on the Law of the Sea; and the question of additional review conferences.

By far the most significant development that emerged from the Third Review Conference concerned the question of extending the geographical scope of the Treaty, a question that falls under Article II. In the general debate, the three depositaries of the Treaty declared, for the first time, that they "have not emplaced any nuclear weapons or other weapons of mass destruction on the sea-bed outside the zone of application of the treaty as defined by its Article II and have no intention to do so"—a statement generally understood to refer to territorial waters. The declarations were welcomed by many delegations. After discussion, the Conference confirmed in its Final Declaration that those statements held true for all States parties. At the final meeting of the Conference, the three nuclear-weapon States emphasized that this confirmation in the Declaration did not represent a legally-binding modification of the Treaty itself, but rather a statement of fact and of present intentions.

With respect to the extension of the scope of the Treaty to weapons other than those of mass destruction and the related question of further measures to prevent an arms race on the sea-bed, States reiterated views expressed during the previous Review Conferences. The non-aligned and socialist States favored the complete demilitarization of the sea-bed and urged further negotiations to that end. The United Kingdom stated once again that it had not identified any further measures that could be initiated in that regard. The United States, for its part, believed that no arms race on the sea-bed existed nor was one in the offing. However, all participating States agreed that if further measures were to be identified, that task would fall within the domain of the Conference on Disarmament. Ultimately, the Final Declaration, under Article V, repeated the request made at the Second Review Conference that the Conference on Disarmament, in consultation with the States parties to the Treaty and taking into account existing proposals and any relevant technological developments, proceed promptly with consideration of such measures. In addition, the Final Declaration stated that the parties recognized that "other arms limitation and disarmament negotiations on measures with wider application that will contribute to the general objectives of the Treaty have been completed, are under way or are contemplated, and will, when successfully implemented, contribute to the effectiveness of the Treaty." It was understood that this referred to the negotiations in connection with the USSR-US Treaty on the elimination of their intermediate and shorter-range missiles (INF Treaty), which have ended successfully, and the ongoing strategic arms reduction talks and the chemical weapons negotiations.

The question of technological developments relevant to the operation of the Treaty, which many parties consider as being the very *raison d'être* of review conferences, also received great attention. Several proposals were made on the monitoring of relevant technological developments. In general, the proposals which had aspects connected with the question of verification of compliance and had financial implications were not supported by Western States. In their view, the Treaty functioned well as it was, and proliferation of bodies and mechanisms should be avoided. Finally, the Review Conference decided to call on the Secretary-General of the United Nations to report by 1992, and every three years thereafter until the Fourth Review Conference is convened, on technological developments relevant to the Treaty and to the verification of compliance with the Treaty, including dual purpose technologies for peaceful and specified military ends. This is the first time that the Secretary-General has been requested to report on technological developments relevant to the verification of compliance with the Treaty. In carrying out this task, the Secretary-General should draw from official sources and from contributions by States parties, and could use the assistance of appropriate expertise. Parties were urged to assist him by providing information and drawing his attention to suitable sources.

The relationship between the Sea-Bed Treaty and the 1982 United Nations Convention on the Law of the Sea was referred to by many delegations, usually in the context of the need to avoid any weakening of the Sea-Bed Treaty's provisions. The view that nothing in the Law of the Sea Convention should affect the rights and obligations assumed by States parties under the Sea-Bed Treaty was reaffirmed.

All delegations expressed the view that the Conference had been a success and that this success was due not only to the fact that the Treaty's provisions had been effectively implemented by all States parties, but also to the prevailing relaxation of tension in international relations.

The Conference noted with concern that although the Treaty had demonstrated its effectiveness it did not yet enjoy universal adherence. The Conference called upon the States that had not yet become parties, particularly those possessing nuclear weapons or any other types of weapons of mass destruction, to do so at the earliest possible date. Such adherence would be a further significant contribution to international confidence.

Agreement was reached that a Fourth Review Conference would be held, in principle no earlier than 1996, if a majority of States parties to the Treaty so requested. If it was not convened in 1996, the depositary Governments would solicit the views of all States parties on it in 1997.

ACTION BY THE GENERAL ASSEMBLY, 1989

At its forty-fourth session, the General Assembly, *inter alia*, welcomed with satisfaction the Third Review Conference's positive assessment of the effectiveness of the Treaty since its entry into force, as reflected in its Final Declaration. The Assembly reiterated its expressed hope for the widest possible adherence to the Treaty, and invited all States that had not yet done so, particularly those possessing nuclear weapons or any other types of weapons of mass destruction, to ratify or accede to the Treaty as a significant contribution to international peace and security. Furthermore, it affirmed its strong interest in avoiding an arms race in nuclear weapons or any other types of weapons of mass destruction on the sea-bed and requested the Conference on Disarmament to proceed promptly with consideration of further measures to prevent an arms race in that environment. It also requested the Secretary-General to report by 1992 and at three year intervals thereafter on technological developments relevant to the Treaty and to the verification of compliance with it in accordance with the Final Declaration of the Conference.

FINAL DECLARATION OF THE THIRD REVIEW CONFERENCE

Preamble

The States Parties to the Treaty of the Prohibition of the Emplacement of Nuclear Weapons and Other Weapons of Mass Destruction on the Sea-Bed and the Ocean Floor and in the Subsoil Thereof which met in Geneva in September 1989 in accordance with the provisions of Article VII to review the operation of the Treaty with a view to assuring that the purposes of the Preamble and the provisions of the Treaty are being realized:

Recognizing the continuing importance of the Treaty and its objectives,

Recalling the Final Declaration of the First Review Conference of the Parties to the Treaty on the Prohibition of the Emplacement of Nuclear Weapons and Other Weapons of Mass Destruction on the Sea-Bed and the Ocean Floor and in the Subsoil Thereof held in Geneva from 20 June to 1 July 1977, as well as the Final Declaration of the Second Review Conference of the Parties to the Treaty held in Geneva from 12 to 23 September 1983,

Affirming their belief that universal adherence to the Treaty and particularly adherence by those States possessing nuclear weapons or any other weapons of mass destruction would enhance international peace and security,

Recognizing that an arms race in nuclear weapons or any other types of weapons of mass destruction on the sea-bed would present a grave threat to international security,

Recognizing also the importance of negotiations concerning further measures in the field of disarmament for the prevention of an arms race on the sea-bed, the ocean floor and the subsoil thereof,

Considering that a continuation of the trend towards a relaxation of tension and an increase of mutual trust in international relations would provide a favorable climate in which further progress can be made towards the cessation of the arms race towards disarmament,

Reaffirming their conviction that the Treaty constitutes a step towards the exclusion of the sea-bed, the ocean floor and the subsoil thereof from the arms race, and towards a treaty on general and complete disarmament under strict and effective international control,

Emphasizing the interest of all States, including specifically the interest of developing States, in the progress of the exploration and use of the sea-bed and the ocean floor and its resources for peaceful purposes,

Affirming that nothing contained in the Convention on the Law of the Sea of 10 December 1982 affects the rights and obligations assumed by States Parties under the Treaty,

Taking note of the information concerning the informal meeting held in 1989 under the auspices of the Conference on Disarmament as well as the communications from the Depositary Governments and other States,

Appealing to States to refrain from any action which might lead to the extension of the arms race to the sea-bed and ocean floor, and might impede the exploration and exploitation by States of the natural resources of the sea-bed and ocean floor for their economic development,

Declare as follows:

Purposes

The States Parties to the Treaty reaffirm their strong common interest in avoiding an arms race on the sea-bed in nuclear weapons or any other types of weapons of mass destruction. They reaffirm their strong support for the Treaty, their continued dedication to its principles and objectives and their commitment to implement effectively its provisions.

Article I
The review undertaken by the Conference confirms that the obligations asssumed under Article I of the Treaty have been faithfully observed by the States Parties. The Conference is convinced that the continued observance of this article remains essential to the objective which all States Parties share of avoiding an arms race in nuclear weapons or any other types of weapons of mass destruction on the sea-bed.

Article II
The Conference reaffirms its support for the provisions of Article II which define the zone covered by the Treaty. The Conference agrees that the zone covered by the Treaty reflects the right balance between the need to prevent an arms race in nuclear weapons and any other types of weapons of mass destruction on the sea-bed and the right of States to control verification activities close to their own coasts. All States Parties to the Treaty confirm that they have not emplaced any nuclear weapons or other weapons of mass destruction on the sea-bed outside the zone of application of the Treaty as defined by its Article II and have no intention to do so.

Article III
The Conference notes with satisfaction that no State Party has found it necessary to invoke the provisions of Article III, paragraphs 2, 3, 4 and 5, dealing with international complaints and verification procedures. The Conference considers that the provisions for consultation and cooperation contained in paragraphs 2, 3 and 5 include the right of interested States Parties to agree to resort to various international consultative procedures. These procedures could include ad hoc consultative groups

of experts in which all States Parties could participate, and other procedures. The Conference stresses the importance of cooperation between States Parties with a view to ensuring effective implementation of the international consultative procedures provided for in Article III of the Treaty, having regard also for the concerns expressed by some States Parties that they lack the technical means to carry out the verification procedures unaided.

The Conference reaffirms in the framework of Article III and Article IV that nothing in the verification provisions of this Treaty should be interpreted as affecting or limiting, and notes with satisfaction that nothing in these provisions has been identified as affecting or limiting, the rights of States Parties recognized under international law and consistent with their obligations under the Treaty, including the freedom of the high seas and the rights of coastal States.

The Conference reaffirms that States Parties should exercise their rights under Article III with due regard for the sovereign rights of coastal States as recognized under international law.

Article IV
The Conference notes the importance of Article IV which provides that nothing in this Treaty shall be interpreted as supporting or prejudicing the position of any State Party with respect to existing international conventions, including the 1958 Convention on the Territorial Sea and Contiguous Zone, or with respect to rights or claims which such State party may assert, or with respect to recognition or non-recognition of rights or claims asserted by any other State, related to waters off its coast, including, *inter alia,* territorial seas and contiguous zones, or to the sea-bed and the ocean floor, including continental shelves. The Conference also noted that obligations assumed by States Parties to the Treaty arising from other international instruments continue to apply.

Article V
The Conference reaffirms the commitment undertaken in Article V to continue negotiations in good faith concerning further measures in the field of disarmament for the prevention of an arms race on the sea-bed, the ocean floor and the subsoil thereof.

The Conference notes that negotiations aimed primarily at such measures have not yet taken place. Consequently, the Conference again requests that the Conference on Disarmament, in consultation with the States Parties to the Treaty, taking into account existing proposals and any relevant technological developments, proceed promptly with consideration of further measures in the field of disarmament for the prevention of an arms race on the sea-bed, the ocean floor and the subsoil thereof.

At the same time, the Conference notes that other arms limitation and disarmament negotiations on measures with wider application that will contribute to the general objectives of the Treaty have been completed, are under way or are contemplated, and will, when successfully implemented, contribute to the effectiveness of the Treaty.

Article VI
The Conference notes that over the 17 years of the operation of the Treaty no State Party proposed any amendments to this Treaty according to the procedure laid down in this article.

Article VII
The Conference notes with satisfaction the spirit of cooperation in which the Third Review Conference was held.

The Conference, recognizing the importance of the review mechanism provided in Article VII, and having considered the question of the timing of the next Review Conference and the necessary preparations thereto, decides that the Fourth Review Conference shall be convened in Geneva, in principle not earlier than 1996, at the request to the Depositary Governments of a majority of States Parties to the Treaty, if they consider that relevant developments make this advisable. If the Fourth Review Conference is not convened in 1996, the Depositary Governments shall solicit the views of all States Parties to this Treaty on the holding of the Conference in 1997. If 10 States Parties so request, the Depositary Governments shall take immediate steps to convene the Conference. If there is no such request, the Depositary Governments shall resolicit the views of States Parties at three-year intervals thereafter.

The Conference takes note of the fact that no information has been presented to it indicating that major technological developments have taken place since 1983 which affect the operation of the Treaty. The Conference, nevertheless, recognizes the need to keep such developments under continuing review, and the importance of relevant information in assisting States Parties to decide on the timing of the Fourth Review Conference.

To this end the Conference requests the Secretary-General of the United Nations to report by 1992, and every three years thereafter until the Fourth Review Conference is convened, on the technological developments relevant to the Treaty and to the verification of compliance with the Treaty, including dual purpose technologies for peaceful and specified military ends. In carrying out this task the Secretary-General should draw from official sources and from contributions by States Parties to the Sea-Bed Treaty, and could use the assistance of appropriate expertise. The Review Conference urges all States Parties to the Treaty to assist the Secretary-General by providing information and drawing his attention to suitable sources.

Article VIII
The Conference notes with satisfaction that no State Party has exercised its rights to withdraw from the Treaty under Article VIII.

Article IX
The Conference reaffirms its conviction that nothing in the Treaty affects the obligations assumed by States Parties to the Treaty under international instruments establishing zones free from nuclear weapons.

Article X
The Conference stresses that the 17 years that have elapsed since the date of entry of the Treaty into force have demonstrated its effectiveness. At the same time, the Conference notes with concern that the goal of the Parties that the Treaty should enjoy universal acceptance has not yet been achieved.

The Conference welcomes the adherence of 10 States to the Treaty since the Second Review Conference, thus bringing the total number of Parties to 82. The Conference calls upon the States that have not yet become Parties, particularly those possessing nuclear weapons or any other types of weapons of mass destruction, to do so at the earliest possible date. Such adherence would be a further significant contribution to international confidence.

PARTIES TO THE TREATY AT THE TIME OF THE THIRD REVIEW CONFERENCE

Afghanistan, Antigua and Barbuda, Argentina, Australia, Austria, Bahamas, Belgium, Benin, Botswana, Brazil, Bulgaria, Byelorussian Soviet Socialist Republic, Canada, Cape Verde, Central African Republic, Congo, Côte d'Ivoire, Cuba, Cyprus, Czechoslovakia, Democratic Yemen, Denmark, Dominican Republic, Ethiopia, Finland, German Democratic Republic, Germany, Federal Republic of, Ghana, Greece, Guinea-Bissau, Hungary, Iceland, India, Iran (Islamic Republic of), Iraq, Ireland, Italy, Jamaica, Japan, Jordan, Lao People's Democratic Republic, Lesotho, Luxembourg, Malaysia, Malta, Mauritius, Mexico, Mongolia, Morocco, Nepal, Netherlands, New Zealand, Nicaragua, Niger, Norway, Panama, Poland, Portugal, Qatar, Republic of Korea, Romania, Rwanda, São Tomé and Principe, Saudi Arabia, Seychelles, Singapore, Solomon Islands, South Africa, Spain, Swaziland, Sweden, Switzerland, Togo, Tunisia, Turkey, Ukrainian Soviet Socialist Republic, Union of Soviet Socialist Republics, United Kingdom of Great Britain and Northern Ireland, United States of America, Viet Nam, Yugoslavia and Zambia. (Total: 82)

SIGNATORIES TO THE TREATY AT THE TIME OF THE CONFERENCE

Bolivia, Burundi, Cameroon, Colombia, Costa Rica, Democratic Kampuchea, Equatorial Guinea, Gambia, Guatemala, Guinea, Honduras, Lebanon, Liberia, Madagascar, Mali, Myanmar, Paraguay, Senegal, Sierra Leone, Sudan, United Republic of Tanzania, Uruguay and Yemen. (Total: 23)

Selected Documents and Proceedings

Joint Contingency Plan against Pollution in the Bering and Chukchi Seas*

1. INTRODUCTION

1.1. Purpose

This Plan, including the operational appendix, is established under the Agreement Between the United States of America and Union of Soviet Socialist Republics Concerning Cooperation in Combatting Pollution in the Bering and Chukchi Seas in Emergency Situations, hereinafter referred to as "the Agreement", and provides for coordinated and combined responses to pollution incidents in the Bering and Chukchi Seas.

This Plan shall be implemented subject to the provisions of the Agreement. Nothing in this Plan shall affect in any way the rights and obligations of either Party resulting from other bilateral and multilateral international agreements.

This Plan primarily addresses international matters and as such is meant to augment pertinent national, state, republic, regional, and sub regional (local) plans of the Parties.

1.2. Objectives

The objectives of the Plan are:

a. To develop appropriate preparedness measures and systems for discovering and reporting the existence of a pollution incident within the areas of responsibility of each Party.

b. To provide the means to institute prompt measures to restrict the further spread of oil and other hazardous substances.

c. To provide the mechanism by which adequate resources may be employed to respond to a pollution incident.

1.3. Responsibility

The implementation of the Plan is the joint responsibility of the U.S. Coast Guard (Department of Transportation) and the USSR Marine Pollution Control and Salvage Administration, attached to the USSR Ministry of Merchant Marine. The two aforementioned agencies are the competent authorities and shall be assisted by other national agencies as appropriate and when required.

*EDITORS' NOTE.—This document was supplied by Mr. Roger Kohn, Information Officer, International Maritime Organization, 4 Albert Embankment, London SE1 7SR, United Kingdom.

1.4. Definitions

National Response Team (NRT)—The U.S. national planning, policy and coordinating body consisting of fourteen federal agencies with interests and expertise in various aspects of emergency response.

    Joint Response Team (JRT)—A joint U.S.-USSR planning, policy and coordinating body that provides guidance to the OSC prior to an incident and assistance as requested by the OSC during a pollution incident.

    On-Scene Commander/Coordinator (OSC)—The predesignated official that coordinates containment, removal and disposal efforts and resources during a pollution incident.

    Deputy On-Scene Commander/Coordinator (DOSC)—The designated official of the Party which is not providing the OSC who acts as the OSC's direct liaison with agencies of the Party which the DOSC represents.

    Joint Response Center (JRC)—The designated site of each Party where facilities are available to provide requirements to fulfill the provisions of the Plan.

    Situation Report (SITREP)—A report of the most current information relating to a pollution incident, including actions taken and progress made during the response.

    Pollutant—Oil or other hazardous substance.

2. PRINCIPLES AND PROCEDURES

2.1. The joint policy pursuant to the Plan is based on three fundamental aspects: planning, coordination of joint response and communications.

    2.2. The competent authorities of the U.S. and USSR will cooperate as fully as possible to respond expeditiously to a pollution incident that affects or threatens to affect both Parties. Actions taken pursuant to the Plan shall be consistent with the legal authorities, operational requirements and other obligations of each of the Parties.

    2.3. Any pollution incident that presents a potential threat to a Party will be reported promptly to the appropriate agency of that Party in accordance with the provisions of this Plan.

    2.4. In a response situation which falls within the scope of this Plan, the NRT member agencies in the U.S. and responsible agencies in the USSR will make available any resources they may have which could be used for joint response operations, subject to the availability of those resources.

2.5. Dispersants and Other Chemicals

The existing national decision making process of each Party will be followed to determine whether dispersants or other chemicals will be used to respond to a pollution incident. The use of dispersants or other chemicals in situations which can affect the interests of both Parties shall only be undertaken upon agreement between the Parties.

2.6. Mechanism for Invoking the Plan

2.6.1. The Plan may be invoked by agreement of the U.S. and USSR JRT Co-Chairmen in the event of a pollution incident which originated within the area of responsibility of one Party and which is accompanied by a threat of a pollutant spreading into the area of responsibility of the other Party, or where such spreading has already occurred.

2.6.2. The Plan may also be invoked by agreement of the U.S. and USSR JRT Co-Chairmen in the event of a pollution incident where no pollutants have spread or threaten to spread into both areas of responsibility but where the magnitude of the incident, or other factors, makes a joint response desirable.

2.6.3. The Plan may also be invoked by agreement of the U.S. and USSR JRT Co-Chairmen in the event of a pollution incident originating outside the areas of responsibility of both Parties, resulting in a threat of the spread of a pollutant into the area of responsibility of one or both Parties.

3. PLANNING AND RESPONSE ORGANIZATION

3.1. Joint Response Team

3.1.1. The Joint Response Team consists of representatives of responsible agencies of the U.S. and USSR. The JRT functions as an advisory team and will be activated by agreement in the event of a pollution incident occurring within the areas encompassed by this Plan.

3.1.2. The United States Coast Guard member to the JRT will be designated by the Commander Seventeenth Coast Guard District as U.S. Co-Chairman and will chair the JRT when a pollution incident occurs in the U.S. area of responsibility. The USSR member to the JRT will be designated by the President of the Far Eastern Shipping Company as USSR Co-Chairman and will chair the JRT when a pollution incident occurs in the USSR area of responsibility. When the Plan is invoked under section 2.6.3, the JRT Co-Chairman will determine by agreement which of them will chair the JRT.

3.1.3. U.S. members of the JRT shall consist of predesignated representatives of National Response Team agencies, state environmental response agencies and other agencies as stipulated by appropriate national and regional oil spill contingency plans. USSR members of the JRT shall consist of representatives of appropriate agencies in the USSR.

3.1.4. The JRT Co-Chairmen will compile a directory to be updated on 1 March and 1 September of each year, which will include data on names, positions, 24 hour telephone numbers, office addresses and when applicable, telex, TWX and facsimile numbers of all JRT members, OSCs and DOSCs. The directory shall be made part of the operational appendix and shall be distributed in timely fashion to all concerned.

3.1.5. The general functions of the JRT include planning and preparedness before a response action is taken and coordination and advice during joint response operations, as outlined below:

a. Provide advice and assistance to the OSC during pollution incidents (the JRT does not exercise operational control over the OSC).

b. Promote a coordinated response by all agencies to pollution incidents. This includes promotion of measures to implement agreements of the Parties relating to legal, financial, customs, and immigration matters.

c. Review post-incident reports from the OSC on the handling of pollution incidents for the purpose of analyzing response actions and recommending needed improvements in the contingency plans.

d. Forward to appropriate federal, state and regional authorities relevant reports and recommendations including OSC post-incident reports, JRT debriefing reports and recommendations concerning amendments to the Plan.

3.1.6. Some measure of response functions will be performed each time the Plan is invoked. The degree of response will be determined by the demands of each particular situation. The specific advisory and support functions of the JRT will include:

a. Monitoring incoming reports, evaluating the possible impact of reported pollution incidents and being at all times fully aware of the actions and plans of the OSC;

b. Coordinating the actions of the various agencies in supplying the necessary resources and assistance to the OSC;

c. Recruiting other agencies, industrial or scientific groups to participate as appropriate in support actions by acting through the JRT or OSC;

d. Determining a shift of OSC from one Party to the other as indicated by the circumstances of the spill;

e. Coordinating all reporting on the status of the pollution incident to the respective national authorities;

f. Ensuring that the OSC has adequate public information support.

g. Providing letter reports on JRT activities to the appropriate national authorities.

3.1.7. The JRT is responsible for scheduling periodic meetings and conducting and evaluating joint response exercises relating to the Plan. It is recognized that the continued viability of this Plan is primarily dependent on the development of working relationships through such periodic meetings and exercises. The expected frequency of these meetings and exercises is as follows:

a. JRT meetings—one meeting at least once every 18 months, alternating in each country.

b. JRT exercises—one exercise every two years, alternating in each country.

3.2. On-Scene Commander/Coordinator (OSC)

3.2.1. The coordination and direction of the joint pollution control efforts at the scene of a pollution incident shall be achieved through an official appointed as the OSC. The OSC is designated in the operational appendix to the Plan by the Commander Seventeenth Coast Guard District for the U.S. area of responsibility and by the President of the Far Eastern Shipping Company for the USSR area of responsibility. His or her responsibility will continue until a shift in OSCs between the Parties is agreed upon by the JRT or until a shift in OSCs within the jurisdiction of one Party is directed by his or her appropriate national authority. In the event of a pollution incident, the first official from an agency with responsibility under this Plan arriving at the site will assume coordination of activities under the Plan until the designated OSC becomes available to take charge of the operation.

3.2.2. The OSC will determine the pertinent facts concerning a particular pollution incident, including the nature, amount, the location of pollutant spilled, probable direction and time of travel of the pollutant resources available and needed, and the areas which may be affected. He or she will establish the priorities for protection.

3.2.3. The OSC will initiate and direct, as required, Phase II, Phase III and Phase IV operations as hereinafter described.

3.2.4. The OSC will obtain proper authorization, in accordance with applicable laws, to call upon and direct the deployment of available resources to initiate and continue containment, countermeasures, cleanup and disposal functions.

3.2.5. In carrying out this Plan the OSC will maintain an up-to-date and accurate information flow including submission of situation reports (SITREPS) to the JRT as significant developments occur to ensure the maximum effectiveness of the joint effort in protecting the natural resources and environment from pollution damage. The necessary direct liaison between personnel at all levels in the agencies of both countries is essential to both planning and operations.

3.2.6. At the conclusion of each joint response to a pollution incident the OSC will submit to the JRT a complete report on the response operations, actions taken, problems encountered and recommendations on new procedures and policy.

3.3. Deputy On-Scene Commander/Coordinator (DOSC)

The DOSC will be designated by the Party which is not providing the OSC. He or she will act as the OSC's direct liaison with the agencies of the Party which the DOSC represents. He or she will assist the OSC and will control his or her own Party's response resources to comply with the planned tactics of the OSC.

4. RESPONSE OPERATIONS

4.1. Response Actions

The actions which are taken to respond to a pollution incident separate into four relatively distinct phases. However, elements of a phase or an entire phase may take place concurrently with one or more other phases.

Phase I   Discovery and Alarm
Phase II   Evaluation and Plan Invocation
Phase III   Containment, Countermeasures, Cleanup and Disposal
Phase IV   Documentation and Cost Recovery

4.2. Phase I—Discovery and Alarm

4.2.1. The discovery of a pollution incident may be made through the normal planned surveillance activities, through the observations of agencies of the various levels of government, by those who caused the incident, or by the alertness and concern of the general public.

4.2.2. The severity of the incident which in itself is conditioned by the nature and the quantity of the pollutant and the locality, will determine the level of response required and whether or not there is a need to invoke the Plan.

4.2.3. The first agency, having a responsibility under the Plan, to be made aware of a pollution incident shall notify the appropriate designated OSC immediately. If the pollution incident threatens to affect the area of responsibility of the other Party, an immediate warning is to be given in accordance with the procedures established in Section 5.

4.3. Phase II—Evaluation and Plan Invocation

4.3.1. If it is the evaluation of the OSC receiving the first warning that the pollution incident will possibly affect the other Party, he or she will:

a. Notify the designated OSC of the other Party;

b. Make a recommendation to his or her own nation's JRT Co-Chairman on whether to invoke the Plan;

c. Formulate plans to deal with the incident; and

d. Initiate Phase III and IV actions as appropriate.

4.3.2. The Co-Chairmen may invoke the plan as provided in Section 2.6. The specific methods for warning the other Party and invoking the Plan are contained in Section 5.

4.4. Phase III—Containment, Countermeasures, Cleanup and Disposal

4.4.1. Containment is any measure, whether physical or chemical, which is taken to control or to restrict the spread of a pollutant.

4.4.2. Countermeasures embrace those activities, other than containment, which are implemented to reduce the impact and the effect of a pollutant on the public health, welfare, or environment.

4.4.3. Cleanup operations are directed towards reducing the impact of a pollution incident to the extent possible. It will include the removal of the pollutant from the water and shoreline using available technology.

4.4.4. Disposal of pollutants which are recovered as a result of cleanup actions will be in accordance with national procedures so as to preclude the possibility of further or continuing environmental damage.

4.5. Phase IV—Documentation and Cost Recovery

This phase is directed at the collection and maintenance of documentation to prove the source and circumstances of the incident, the responsible party or parties, the impact and potential impacts to the public health and welfare and the environment and recovery costs for response activities.

5. REPORTS AND COMMUNICATIONS

Joint Response Centers

Joint Response Centers are designated sites where facilities are available to provide the necessary requirements to fulfill the provisions of the Plan. The locations of the designated Joint Response Centers are contained in the operational Appendix. During an incident, the Joint Response Center (JRC) will be established in the designated facilities of the Party providing the OSC and will ordinarily be shifted to the facilities of the other Party if the OSC is shifted to that Party. Alternate sites closer to the scene of the incident may be specified, in lieu of the predesignated sites, at the discretion of the appropriate chairman of the JRT.

5.1. Rapid Alerting System

Any potential pollution threat to a Party will be reported to that Party without delay. The reporting points are:

U.S.	USSR
Seventeenth Coast Guard District	Far Eastern Shipping Company
Operations Center, Juneau, Alaska	Radio Center, Vladivostok

5.2. Warning Message

While it may take some assessment to decide whether or not to invoke the Plan, a warning that the Plan may be invoked should be given. This warning will not activate the Plan. It will, however, permit immediate preparation for the possibility of its invocation. The warning message shall be in the following format:

```
DTG
FM                          (sender)
TO                          (action addressee)
INFO                        (information addressees)
BT
UNCLAS
US/USSR POLLUTION INCIDENT (OR POTENTIAL POLLUTION
     INCIDENT) (identify the incident)
1. GEOGRAPHICAL POSITION
2. ANY OTHER DETAILS
3. ACKNOWLEDGE
BT
```

Such a message will normally be originated by the appropriate JRT Co-Chairman and will always be acknowledged by the action addressee.

5.3. Invocation

The Plan shall be activated only by formal invocation. This will normally be done by message from the appropriate JRT Co-Chairman. Telephonic invocation will be followed by an invocation message. This message should contain at least the following information:

```
DTG
FM                                  (sender)
TO                                  (action addressee)
INFO                                (information addressees)
BT
UNCLAS
CONTINGENCY PLAN INVOKED AT (time GMT)
OSC                                 (name)
JRT Co-Chairman                     (name)
JRC ESTABLISHED AT                  (location and telephone no.)
ACKNOWLEDGE
BT
```

If a warning message was not issued, the information that would have been contained in that message should be added to the invocation message. In the acknowledgment message to the above, the receiving party will report the name of the JRT Co-Chairman, the name of the DOSC and the DOSC's ETA at the locality of the headquarters established by the original message of invocation.

5.4. Situation Report Requirements (US/USSR SITREP)

5.4.1. Up-to-date information on a spill which has justified joint response activity is essential to the effective management and outcome of an incident. This information should be submitted by the designated OSC to the JRT in the format shown below. U.S./USSR SITREPS should be made as frequently as necessary to ensure that those who need to know have a full and timely appreciation of the incident and of actions taken and progress made during the response.
    5.4.2. Standard Message Format

```
DTG
FM                                  (sender)
TO                                  (action addressee)
INFO                                (information addressees)
BT
UNCLAS
US/USSR SITREP                      (report number)
POLLUTION INCIDENT                  (identify the case)
1. SITUATION:
2. ACTION TAKEN:
```

3. FUTURE PLANS/FURTHER ASSISTANCE REQUIRED:
4. CASE STATUS: (pends/closed)
BT

SITUATION: The situation section should provide the full details on the pollution incident including what happened, type and quantity of material, agencies involved, areas covered and threatened, success of control efforts, prognosis and any other pertinent data.

ACTION TAKEN: The action section should include a summary of all actions taken so far by the responsible party, local forces, government agencies and others.

FUTURE PLANS: The plans section should include all future action planned.

FURTHER ASSISTANCE REQUIRED: Any additional assistance required from the JRT by the OSC pertaining to the response will be included in this section.

CASE STATUS: The status section should indicate "case closed", "case pends", or "participation terminated", as appropriate.

5.5. Revocation

5.5.1. A recommendation to revoke the joint response to a particular incident shall be made by agreement of the OSC and DOSC. The JRT Chairman from the Party which originally invoked the joint response shall revoke it by message after consultation with the Chairman from the other Party. This message will clearly establish the date and time, in GMT, of the cessation of the joint response. The requirement to consult in no way diminishes the invoking Chairman's prerogative to decide upon revocation.

5.5.2. Standard Message Format for Revocation

DTG
FM   (sender)
TO   (action addressee)
INFO   (information addressees)
BT
UNCLAS
US/USSR CONTINGENCY PLAN REVOKED AT (date, time—in GMT at
    which joint operation will cease)
BT

5.6. Post-Incident Reports

The JRT may request the OSC and DOSC involved to submit reports and to prepare operational debriefings for the JRT, on the incident, the action taken and any observations or recommendations which need to be made.

6. ADMINISTRATION

6.1. Competent Authorities

The competent authorities for this Plan and any amendments thereto are: for the U.S., Commandant, United States Coast Guard; and for the USSR, the Head of the Marine Pollution Control and Salvage Administration, attached to the USSR Ministry of Merchant Marine.

6.2. Operational Appendix

The JRT Co-Chairmen will maintain the operational appendix to this Plan covering such regional topics as communications, reporting systems, designated and/or potential JRT members, useful points of contact and abbreviations.

6.3. Amendments

The Plan may be amended upon agreement between competent authorities by exchange of letters. The operational appendix may be amended upon agreement between the JRT Co-Chairmen. The JRT Co-Chairmen will notify their appropriate national authorities of all such amendments by letter.

The Plan shall become effective on the date of signature. It shall remain in force as long as the Agreement is in force.

DONE at London, in duplicate, this   day of October, 1989, in the Russian and English languages, both texts being equally authentic.

For the United States   For the Marine Pollution
Coast Guard   Control and Salvage Administration

Tables, Living Resources

The six tables of this appendix follow the typology of those in *Ocean Yearbook 6*. Data from previous years have been updated wherever possible. For additional information on living resources see the articles by John Bardach and Karen Eckert in this volume.

THE EDITORS

TABLE 1C.   --   WORLD NOMINAL MARINE CATCH BY CONTINENT*   (1,000 Metric Tons)

	1970	1980	1985	1986	1987	1986-87 (% change)
Africa	3,131	2,725.0	2,704.1	2,929.4	3,514.7	+20.0
America, N.	4,750	6,698.1	7,883.3	8,109.5	8,925.7	+10.1
America, S.	14,629	7,438.2	11,478.5	13,691.4	11,655.8	-14.9
Asia	19,453	26,258.0	30,606.0	32,984.3	33,320.5	+1.0
Europe	11,815	12,169.4	12,519.0	12,244.4	12,130.7	-0.9
Oceania	194	448.3	595.2	669.1	782.5	+16.9
USSR	6,399	8,722.6	9,617.2	10,333.0	10,171.2	-1.6
World Totals§	61,432	64,459.6	75,403.3	80,961.1	80,501.2	-0.6

SOURCE. -- FAO, <u>Yearbook of Fishery Statistics</u>.
*Continental Classification follows FAO usage.
§Exceeds the sum of the figures by continent due to the inclusion of catches not elsewhere included.

TABLE 2C. -- WORLD NOMINAL MARINE CATCH, BY MAJOR FISHING AREA

	Million Metric Tons				1986-87
	1970	1980	1986	1987	(% change)
Atlantic, N.W.	4.23	2.87	2.94	3.01	+2.38
Atlantic, N.E.	10.70	11.81	10.57	10.39	-1.70
Atlantic, N.	14.93	14.68	13.51	13.40	-0.81
Atlantic, W.C.	1.42	1.80	2.08	2.17	+4.33
Atlantic, E.C.	2.77	3.43	3.05	3.25	+6.56
Mediterranean & Black	1.15	1.64	2.01	1.91	-4.98
Atlantic, C.	5.34	6.87	7.14	7.33	+2.66
Atlantic, S.W.	1.10	1.27	1.74	2.23	+28.16
Atlantic, S.E.	2.52	2.17	2.12	2.69	+26.89
Atlantic, S.	3.62	3.90	4.32	5.36	+24.07
Atlantic	23.53	25.45	24.97	26.09	+4.49
Indian, W.	1.72	2.09	2.61	2.59	-0.77
Indian, E.	.81	1.46	2.32	2.38	+2.59
Indian*	2.53	3.55	4.93	4.97	+0.81
Pacific, N.W.	13.01	18.76	25.87	25.91	+0.15
Pacific, N.E.	2.65	1.97	3.21	3.38	+5.30
Pacific, N.	15.60	20.73	29.08	29.29	+0.72
Pacific, W.C.	4.22	5.49	6.57	6.50	-1.07
Pacific, E.C.	.91	2.42	2.63	2.43	-7.60
Pacific C.	5.12	7.91	9.20	8.93	-2.93
Pacific, S.W.	0.17	0.38	0.75	0.90	+20.00
Pacific, S.E.	13.76	6.23	11.98	10.27	-14.27
Pacific, S.	13.93	6.62	12.73	11.17	-12.25
Pacific	34.71	35.26	51.01	49.39	-3.18
World Total	61.43	64.39	80.96	80.50	-0.57
Northern Regions	30.59	35.41	42.59	42.69	+0.23
Central Regions	13.00	18.33	21.27	21.23	-0.19
Southern Regions	17.54	10.65	17.09	16.57	-3.04

SOURCE. -- FAO.
NOTE. -- Totals and subtotals may differ from the sum of included figures
due to rounding and the inclusion of catches not elsewhere included.
  *Temperate and tropical.

TABLE 3C. -- WORLD NOMINAL FISH CATCH, DISPOSITION*   (Million Metric Tons)

	1970	1980	1985	1986	1987	1986-87 (% change)
Human Consumption	44.6 (100.0)	53.3 (100.0)	60.7 (100.0)	65.5 (100.0)	67.1 (100.0)	+2.4
Marketing Fresh	18.6 (41.7)	14.9 (27.9)	16.1 (26.5)	19.2 (29.4)	20.2 (30.1)	+5.2
Freezing	9.8 (22.0)	16.3 (30.6)	20.4 (33.6)	21.8 (33.3)	22.1 (32.9)	+1.4
Curing	8.1 (18.1)	11.3 (21.3)	12.7 (21.1)	13.0 (19.9)	13.2 (19.7)	+1.5
Canning	8.1 (18.2)	10.8 (20.2)	11.5 (18.9)	11.5 (17.4)	11.6 (17.3)	+0.9
Other Purposes	26.0 (100.0)	18.9 (100.0)	25.3 (100.0)	26.8 (100.0)	25.6 (100.0)	-4.5
Reduction	25.0 (96.2)	18.1 (96.0)	24.3 (96.0)	25.8 (96.3)	24.6 (96.1)	-4.7
Miscellaneous	1.0 (3.8)	0.8 (4.0)	1.0 (4.0)	1.0 (3.7)	1.0 (3.9)	0.0
World Total	70.6	72.2	86.0	92.3	92.7	+0.4

SOURCE. -- FAO.
NOTE. -- Percentages of subtotals shown in parentheses.
*The figures for disposition are based on "live weight" and include freshwater catches.

TABLE 4C. -- WORLD NOMINAL MARINE CATCH, BY COUNTRY* (1,000 Metric Tons)

	1970	1980	1986	1987	1986-87 (% change)
Anglo-America:					
Canada	1,290.1	1,292.7	1,463.7	1,410.5	-3.6
Greenland	39.8	103.6	101.0	101.0#	0.0
U.S.A.	2,729.3	3,565.0	4,871.2	5,661.3	+16.2
Other	0.9	20.4	8.3	8.5	+2.5
Total	4,060.1	4,981.7	6,444.2	7,181.3	+11.4
Latin America:					
Argentina	186.1	376.9	411.8	551.6	+33.9
Brazil	432.7	548.5	617.5#	578.1#	-6.4
Chile	1,200.3	2,816.7	5,570.6	4,813.4	-13.6
Colombia	21.3	29.3	27.4	23.4	-14.6
Costa Rica	7.0	19.7	20.6	19.7	-4.4
Cuba	105.3	180.1	227.0	197.6	-13.0
Dominican Rep.	5.0	8.2	16.3	18.5	+13.5
Ecuador	91.4	639.0	1,002.5	678.1	-32.4
Guyana	17.4	30.8	39.4	40.8	+3.6
Mexico	344.1	1,212.6	1,184.9	1,245.8	+5.1
Panama	52.2	216.4	129.4	167.9	+29.8
Peru	12,532.9	2,696.1	5,581.4	4,547.1	-18.5
Uruguay	13.2	120.1	140.1	136.9	-2.3
Venezuela	122.6	169.4	294.0	276.0	-6.1
Other	64.2	84.3	92.3	105.2	+13.9
Total	15,215.5	9,148.1	15,355.2	13,400.1	-12.7
Western Europe:					
Belgium	53.0	45.6	38.9	39.8	+2.3
Denmark	1,217.1	2,010.5	1,827.3	1,671.9	-8.5
Faeroe Is.	207.8	274.7	352.9	355.4	+0.7
Finland	62.6	111.0	121.8	126.8	+4.1
France	764.4	761.4	835.3#	803.8#	-3.8
Germany, F.R.	597.9	288.7	178.2	177.8	-0.2
Greece	91.5#	95.9	113.7	124.8	+9.8
Iceland	733.3	1,514.4	1,656.4	1,632.5	-1.4
Ireland	78.9	149.4	228.9	247.3	+8.0
Italy	379.3	467.3	509.7	499.3	-2.0
The Netherlands	298.8	338.4	450.5	430.0	-4.6
Norway	2,906.2	2,408.6	1,898.0	1,928.9	+1.6
Portugal	464.4	270.4	400.4	392.8	-1.9
Spain	1,517.0	1,281.8	1,407.2#	1,364.7#	-3.0
Sweden	284.2	230.7	211.9	210.7	-0.6
U.K.	1,091.3	844.2	844.9	947.8#	+12.2
Other	1.4	2.6	2.6	2.5	-4.1
Total	10,749.1	11,095.5	11,078.6	10,956.8	-1.1
Socialist Eastern Europe:					
Bulgaria	86.8	114.0	95.2	97.7	+2.6
German D.R.	308.2	223.1	189.3	173.5	-8.3
Poland	451.3	621.9	615.8	640.2	+4.0
Romania	24.8	120.9	205.3	197.5	-3.8
U.S.S.R.	6,386.5	8,722.6	10,333.0	10,171.2	-1.6
Yugoslavia	26.7	34.9	51.4	56.2	+9.3
Other	4.0	7.0	8.8	9.3	+5.7
Total	7,288.3	9,844.4	11,498.8	11,345.6	-1.3
Near East:					
Algeria	25.7	48.0#	70.0#	70.0#	0.0
Egypt	27.2	32.2	39.0	48.3#	+23.8
Iran	18.0#	40.0	121.8	120.0#	-1.5
Morocco	246.8	329.6	597.2	489.7#	-18.0
Oman	180.0#	79.0#	96.4	115.0	+19.3
Saudi Arabia	21.1	26.4	45.5	45.5#	0.0
Tunisia	24.0	60.1	92.6	99.2	+7.1
Turkey	165.3	394.6	539.6	580.9	+7.7
U.A. Emirates	40.0	64.4	79.5	85.4	+7.4
Yemen, A.R.	7.6#	17.0	22.2	22.3	+0.5
Yemen, Dem.	120.0	58.3	47.7	48.5	+1.7
Other	35.7	77.2	44.4	52.7	+18.8
Total	911.4	1,226.8	1,795.9	1,777.5	-1.0

TABLE 4C. -- (continued)

	1970	1980	1986	1987	1986-87 (% change)
Sub-Saharan Africa:					
Angola	368.2	77.6	50.5	73.3	+45.1
Cameroon	19.2	73.0#	64.0	62.5	-2.3
Cape Verde	5.1	8.8	6.5	6.9	+6.2
Congo	15.2	21.0	18.0	17.5	-2.8
Gabon	4.5#	18.0#	18.8	19.0#	+1.1
Ghana	141.5	191.9	267.7	317.8	+18.7
Guinea	5.6#	18.9#	28.0#	28.0#	0.0
Ivory Coast	66.5#	67.9	76.2	74.3	-2.5
Madagascar	13.1	12.2	17.6	17.6#	-0.1
Mauritania	47.6	15.6#	92.1#	93.3#	+1.3
Mozambique	7.6	30.4	31.2	35.9	+15.1
Namibia	711.2	252.6	201.2	519.4	+158.2
Nigeria	105.9	292.4	161.5	145.8	-9.7
Senegal	169.2	217.8	271.9	284.0#	+4.5
Sierra Leone	29.6	34.2	37.0#	37.0#	0.0
Somalia	30.0#	14.3	16.5	17.0	+3.0
South Africa	511.1	601.9	627.9	901.3	+43.5
Tanzania	18.3	38.0	44.1	47.8	+8.4
Togo	6.4	8.4	14.1	14.5	+2.8
Other	53.5	53.8	74.1	85.2	+14.9
Total	2,341.5	2,048.6	2,118.9	2,798.0	+32.1
South Asia:					
Bangladesh	90.0#	122.0	207.5	232.9	+12.2
India	1,085.6	1,554.7	1,716.9	1,681.5	-2.1
Maldives	34.5	34.6	45.8	46.9	+2.4
Pakistan	149.3	232.9	331.7	336.1	+1.3
Sri Lanka	89.8	165.2	142.9	153.5	+7.4
Total	1,449.2	2,109.4	2,444.8	2,450.9	+0.2
East Asia:					
China	2,192.5	2,995.4	4,636.6	5,408.4	+16.6
Hong Kong	133.3	187.5	207.8	221.6	+6.6
Japan	8,658.4	10,213.3	11,777.6	11,615.1	-1.4
Korea, D.P.R.	447.0	1,330.0#	1,600.0#	1,600.0#	0.0
Korea, Rep.	725.5	2,052.0	3,046.5	2,819.3	-7.5
Taiwan	540.4	761.3	828.5	930.7	+12.3
Other	9.6	6.9	8.0	3.5	-56.3
Total	12,706.7	17,546.4	22,105.0	22,598.6	+2.2
Southeast Asia:					
Burma	311.4	428.8	535.2	540.9	+1.1
Indonesia	804.0	1,387.0	1,850.0	1,967.9	+6.4
Malaysia	338.5	733.2	607.0	598.6	-1.4
Philippines	844.1	1,134.3	1,377.8	1,426.0	+3.5
Singapore	17.3	15.5	20.3	15.1	-25.6
Thailand	1,343.4	1,653.0	2,348.3	2,000.4	-14.8
Vietnam	668.3	398.7	582.1	620.4	+6.6
Other	21.7	45.2	19.6	19.6	0.0
Total	4,348.7	5,795.7	7,340.3	7,188.9	-2.1
Australasia:					
Australia	100.7	130.3	176.5#	197.5#	+11.9
Fiji	3.9	23.6	23.9	30.5	+27.6
Kiribati	9.7	18.9	33.6	43.9	+30.7
New Zealand	59.2	190.9	345.0	430.5	+24.8
Solomon Is.	1.0	34.8	55.4	44.5	-19.7
Other	20.7	15.9	23.3	24.0	+2.9
Total	195.2	414.4	657.7	770.9	+17.2
Other NEI§	240.6	327.9	119.1	34.1	-71.4
World total	59,495.0	64,539.5	80,961.1	80,501.2	-0.6

SOURCE. -- FAO, unless otherwise noted.
*Nominal marine catch = nominal catch in marine areas. Countries which reported marine catches of less than 10,000 metric tons in 1986 are included under "Other" for all years.
#FAO estimate.
§Not Elsewhere Included. These values differ from those found in the source due to the exclusion of Taiwan's catch from the FAO sum, the inclusion of the catches for the French Southern and Antarctic Territories in the FAO sum, and small discrepancies due to rounding.

TABLE 5C. -- TRADE OF FISHERIES COMMODITIES,* BY MAJOR IMPORTING AND
           EXPORTING COUNTRIES (million US $)

	1970	1980	1986	1987	1986-87 (% change)
Imports:					
Japan	291.9	3,158.7	6,593.4	8,308.1	+26.0
U.S.A.	835.8	2,633.2	4,748.9	5,662.3	+19.2
France	203.9	1,131.2	1,510.0	2,022.5	+33.9
Italy	159.6	831.7	1,264.5	1,738.2	+37.5
U.K.	294.0	1,033.7	1,216.6	1,386.8	+14.0
Spain	46.6	544.0	721.2	1,321.8	+83.3
Germany, F.R.	264.8	1,023.9	1,113.3	1,270.5	+14.1
Denmark	47.1	330.7	596.6	842.5	+41.2
Hong Kong	56.0	361.4	624.5	794.3	+27.2
Belgium	86.3	408.3	425.6	529.9	+24.5
Canada	51.0	301.6	433.3	511.9	+18.1
The Netherlands	93.2	389.4	387.8	509.4	+31.4
Portugal	32.6	100.0	256.6	424.7	+65.5
Sweden	98.8	325.2	333.4	404.9	+21.4
Switzerland	47.5	211.8	264.5	332.8	+25.8
Singapore	28.9	142.1	257.8	313.0	+21.4
Australia	41.6	178.5	226.7	299.6	+32.2
Thailand	3.9	23.4	283.4	267.1	-5.7
Korea, Rep.	5.0	35.7	117.7	212.5	+80.5
Nigeria	4.1	422.8	90.0	168.0	+86.6
Subtotal	2,692.6	13,587.2	21,465.7	27,320.9	+27.3
Other	582.4	2,370.2	2,728.7	3,188.1	+16.8
World Total	3,275.0	15,957.4	24,194.5	30,509.0	+26.1

TABLE 5C. -- (continued)

	1970	1980	1986	1987	1986-87 (% change)
Exports:					
Canada	257.3	1,094.5	1,751.8	2,092.2	+19.4
USA	111.9	1,001.7	1,481.0	1,836.5	+24.0
Denmark	165.6	1,000.0	1,381.5	1,750.7	+26.7
Korea, Rep.	42.0	681.8	1,171.1	1,505.7	+28.6
Norway	259.1	974.7	1,171.2	1,474.9	+25.9
Thailand	17.7	358.3	1,011.9	1,261.1	+24.6
Iceland	112.9	708.6	858.0	1,071.1	+24.8
The Netherlands	111.8	524.6	766.4	953.2	+24.4
China	NA£	348.4	645.8	912.5	+41.3
Japan	335.5	905.2	897.9	889.8	-0.9
U.K.	55.1	365.2	511.1	717.6	+40.4
France	37.0	320.3	501.2	654.5	+30.6
USSR	90.4	307.9	587.1	637.3	+8.5
Chile	27.4	323.0	516.0	635.6	+23.2
Mexico	71.5	580.0	457.4	569.9	+24.6
Hong Kong	18.4	165.3	417.9	501.8	+20.1
Ecuador	6.1	200.0	383.6	481.0	+25.4
Spain	95.5	344.4	398.7	474.8	+19.1
Indonesia	4.5	211.3	340.6	441.1	+29.5
Germany, F.R.	62.9	317.1	359.3	439.7	+22.4
Subtotal	1,882.6	10,732.3	15,609.6	19,300.7	+23.6
Other	(1,882.6)	4,568.2	7,447.0	8,775.0	+17.8
World Total		15,300.5	23,056.5	28,075.7	+21.8

SOURCE. -- FAO.
  *The term "Fishery Commodities" follows FAO usage and includes the seven principle fishery commodity groups. Countries are ranked according to 1987 figures.
  £1970 date for China not available.

TABLE 6C. -- FISHING FLEETS, BY COUNTRY, 1988*

| | Trawlers and Fishing Vessels | | | Factory Ships and Carriers | |
	grt	no.	over 500 grt (%)§	grt	no.
Albania	300	2	0.0		
Algeria	2,731	23	0.0		
Angola	17,334	73	6.1		
Antigua & Barbuda	754	2	0.0		
Argentina	90,741	199	63.0	4,264	2
Australia	48,118	267	3.2	291	1
Bahamas	1,667	11	0.0		
Bahrain	1,004	7	0.0		
Bangladesh	10,496	47	31.7	263	2
Barbados	3,368	27	0.0		
Belgium	21,851	117	2.5		
Benin	1,078	8	0.0		
Bermuda	1,481	3	77.4	706	1
Brazil	14,451	89	6.0		
Brunei	314	1	0.0		
Bulgaria	73,541	32	99.2	29,056	6
Burma	5,578	27	0.0		
Cameroon	5,470	35	0.0		
Canada	158,716	486	51.1	698	3
Cape Verde	1,851	8	0.0		
Cayman Islands	42,631	74	80.6		
Chile	80,237	176	51.8	1,257	1
China, P.R.	42,194	110	44.6	24,354	21
Colombia	2,258	17	0.0		
Congo	7,682	18	71.0		
Costa Rica	5,784	13	56.3		
Cuba	135,509	261	83.5		
Cyprus	5,549	9	79.3		
Denmark	201,630	563	49.5	312	1
Dominica	103	1	0.0		
Ecuador	24,750	84	24.5		
Egypt	1,790	5	69.2		
El Salvador	3,514	12	65.2		
Ethiopia	218	2	0.0		
Fiji	620	5	0.0		
Finland	2,371	17	0.0		
France	130,549	382	57.5	3,393	1
Gabon	2,108	13	23.7		
Gambia	256	2	0.0		
German D.R.	91,455	107	90.9	62,016	9
Germany, F.R.	38,532	93	64.1		
Ghana	55,737	101	33.6	6,069	4
Greece	39,844	115	47.0		
Guatemala	377	3	0.0		
Guinea	2,333	6	46.4		
Guinea-Bissau	1,941	6	66.8		
Guyana	4,894	46	0.0	100	1
Haiti	280	1	0.0		
Honduras	31,952	164	6.0	2,033	2
Hong Kong	2,104	7	61.0		
Iceland	107,482	342	30.1	2,830	1
India	24,387	146	10.9		
Indonesia	49,255	230	9.0	1,510	5
Iran	8,799	27	60.4		
Iraq	13,950	12	93.3	10,413	2
Ireland	28,388	75	49.7		
Israel	2,908	3	100.0		
Italy	68,478	243	45.8		
Ivory Coast	9,994	33	52.1	499	1
Jamaica	769	4	0.0		
Japan	912,547	2,647	32.1	159,936	146
Kenya	1,352	6	0.0		
Kiribati	121	1	0.0	608	1
Korea, D.R.	6,320	14	41.1	36,190	6
Korea, Rep.	396,216	1,041	36.0	67,366	43
Kuwait	10,329	67	6.6	788	1
Lebanon	560	4	0.0	738	1
Libya	5,322	32	0.0		

TABLE 6C. -- (continued)

| | Trawlers and Fishing Vessels | | | Factory Ships and Carriers | |
	grt	no.	over 500 grt (%)§	grt	no.
Madagascar	6,652	42	0.0		
Malaysia	3,472	15	0.0		
Maldive Islands	1,602	3	62.3	3,751	4
Malta	10,734	32	51.6	1,589	1
Mauritania	32,750	107	11.1	1,100	1
Mauritius	7,137	21	30.6		
Mexico	128,116	410	59.0		
Morocco	76,894	234	28.0		
Mozambique	16,778	76	26.1		
Nauru	948	1	100.0		
The Netherlands	143,624	435	35.2		
New Zealand	21,036	49	58.1		
Nicaragua	1,971	17	0.0		
Nigeria	24,271	111	33.1	4,923	3
Norway	263,201	596	55.1	5,682	13
Oman	945	4	0.0		
Pakistan	2,009	7	0.0		
Panama	154,223	387	44.6	23,164	18
Papua New Guinea	2,774	19	0.0		
Peru	146,974	542	31.0		
Philippines	70,175	287	6.9	1,374	6
Poland	216,839	303	89.5	95,241	14
Portugal	119,535	194	75.6		
Qatar	531	4	0.0		
Romania	139,497	52	99.4	94,350	12
St. Lucia	105	1	0.0		
St. Vincent	9,476	14	72.2	1,391	1
Sao Tome & Principe	993	2	53.4		
Saudi Arabia	3,146	15	0.0	354	1
Senegal	33,952	137	31.6		
Seychelles				1,827	1
Sierre Leone	7,561	27	0.0		
Singapore	3,875	15	0.0		
Solomon Islands	2,311	7	57.6		
Somalia	5,188	14	79.0		
South Africa	64,633	157	49.3	303	1
Spain	534,966	1,581	38.2	3,440	3
Sri Lanka	2,971	11	23.5		
Surinam	1,260	7	0.0		
Sweden	21,730	110	10.5		
Taiwan	86,318	278	13.4	2,165	3
Tanzania	911	4	0.0		
Thailand	6,147	12	57.2		
Togo	446	3	0.0		
Tonga	764	4	0.0		
Trinidad & Tobago	2,758	20	0.0		
Tunisia	2,685	16	0.0		
Turkey	3,639	9	47.8	668	2
Turks & Caicos Is.	124	1	0.0		
Tuvalu	173	1	0.0		
U.S.S.R.	3,671,809	2,801	95.5	3,171,500	517
U.A.E.	2,364	8	42.3		
U.K.	114,948	395	37.7		
U.S.A.	639,200	3,194	30.1	21,295	13
Uruguay	16,293	56	21.1		
Vanuatu	7,295	7	100.0		
Venezuela	33,707	96	51.8	199	1
Vietnam	7,626	30	36.8	2,494	3
Virgin Is. (Br.)	705	4	0.0		
Western Samoa	213	1	0.0		
Yemen, P.D.R.	3,366	5	82.8		
Yugoslavia	2,204	18	0.0	113	1
Zaire	4,793	14	53.5		
World Total	9,960,566	21,827	63.0	3,851,775	879

SOURCE. -- Lloyd's Register of Shipping Statistical Tables.
*Data exclude vessels of less than 100 grt. The smaller vessels used in artisanal fishing are therefore excluded.
§Percent of grt.

Tables, Nonliving Resources

The four tables of this appendix are modeled on those in *Ocean Yearbook 8* and have been updated wherever possible.

THE EDITORS

TABLE 1D.   --   WORLD PRODUCTION OF CRUDE OIL, TOTAL AND OFFSHORE
                 (Barrels per Day, in Thousands)

Year	World Production	Offshore Production	Offshore as % of World
1975	53,850.00	8,278.36	15.4
1976	57,210.00	9,431.91	16.5
1977	56,567.00	11,436.75	20.2
1978	60,337.00	11,480.75	19.0
1979	62,768.00	12,491.93	19.9
1980	59,812.00	13,587.49	22.7
1981	55,886.20	13,664.61	24.5
1982	53,191.00	13,541.25	25.5
1983	53,259.00	13,791.04	26.7
1984	54,090.00	15,311.50	28.3
1985	53,391.00	15,128.33	28.0
1986	55,864.00	13,923.39	24.9
1987	56,070.00	14,741.55	26.3
1988	58,009.00	14,402.61	24.8
1989	59,661.00	14,833.91	24.9

SOURCES. -- <u>Offshore</u> and <u>Oil and Gas Journal</u>.
     NOTE. -- 6.998 barrels of crude petroleum approximately equal 1.0
metric ton (ASTM-1P Petroleum Measurement Tables).

TABLE 2D. -- OFFSHORE CRUDE OIL PRODUCTION, BY REGION AND COUNTRY (Barrels per Day, in Thousands)

Area and Country	1970	1975	1980	1987	1988	1989	1988-89 (% change)
Total	7,532.0	8,278.4	13,587.5	14,741.6#	14,402.6#	14,833.9#	-6.7
Anglo-America:							
U.S.A.	1,557	909.6	1,038.1	1,212.3	1,129.0	1,053.0	-6.7
Latin America							
Argentina						2.8	N/A
Brazil	8.0	19.0	73.0	396.0	376.0	616.5	+64.0
Chile				23.0	20.9	18.2	-12.9
Mexico	35.0	45.0	500.2	1,671.4	1,666.3	1,523.2	-8.6
Peru		28.9	29.9	115.6	103.5	92.0	-11.1
Trinidad & Tobago	76	174.0	166.5	128.3	117.2	112.3	-4.2
Venezuela	2,460	1,737.1	1,095.6	1,061.6	989.0	932.0	-5.8
Subtotal	2,579	2,004.0	1,865.2	3,395.9	3,272.9	3,297.0	+0.7
Western Europe:							
Denmark		3.3	6.6	12.1	14.2	11.1	-21.8
Germany, F.R.					10.7	11.7	+9.3
Greece				26.2	24.2	18.2	-24.8
Italy	12	10.4	6.3	57.4	62.3	68.1	+9.3
The Netherlands				19.2	18.6	45.4	+59.0
Norway		189.6	628.8	864.9	907.0	1,516.5	+67.2
Spain		32.9	31.2	35.1	30.2	20.4	-32.5
U.K.		83.0	1,650.0	2,314.7	2,094.7	1,792.8	-14.4
Subtotal	12	319.2	2,322.9	3,329.6	3,162.0	3,484.2	+10.2
Socialist Eastern Europe:							
Romania				3.2	5.3	4.5	-15.1
U.S.S.R.	258	228.0	200.0	207.0	201.6	183.2	-9.1
Subtotal	258	228.0	200.0	210.2	206.9	187.7	-9.3
Near East:							
Divided Zone	257	315.1	403.0	252.0	192.0	188.0	-2.1
Egypt	322	165.0	390.3	602.1	591.3	531.0	-10.2
Iran		481.2	150.0	515.5	323.0	338.0	+4.6
Libya					20.0	47.0	+57.4
Qatar	172		247.6	191.3	153.2	191.0	+24.7
Saudi Arabia	1,251	1,385.8	2,958.0	1,510.3	1,481.3	1,381.1	-6.8
Tunisia		43.0	43.6	28.4	28.3	24.7	-12.7
UAE	339.0	750.4	1,676.9	591.7	590.8	625.4	+5.9
Subtotal	2,341	3,140.5	5,769.4	3,691.2	3,380.0	3,326.2	-1.6

TABLE 2D. -- (continued)

Area and Country	1970	1975	1980	1987	1988	1989	1988-89 (% change)
Sub-Saharan Africa:							
Angola/Cabinda	96	143.2	99.0	185.0	450.9	420.0	-6.8
Cameroon				201.3	187.1	155.2	-17.0
Congo		37.3	27.0	116.8	133.7	126.0	-5.7
Gabon	29	179.9	177.9	92.6	108.3	106.9	-1.3
Ghana			2.0	0.3	0.2		N/A
Ivory Coast			5.8	17.4	15.4	10.9	-29.2
Nigeria	275	431.3	579.1	456.2	471.6	481.2	+2.0
Zaire			21.6	17.5	13.5	11.9	-11.8
Subtotal	400	791.7	910.4	1,087.0	1,380.7	1,312.1	-5.0
South Asia:							
India			142.1	584.0	618.2	625.7	+1.2
East Asia:							
China	3		2.0	1.3	2.9	16.7	+82.6
Japan		0.9	1.6	1.3	1.2	3.0	+60.0
Taiwan				3.0	2.4	1.9	-20.8
Subtotal	3	0.9	3.6	2.6	4.1	21.6	+81.0
Southeast Asia:							
Brunei	146§	141.2	192.2	95.7	89.8	82.1	-8.6
Indonesia		246.4	533.0	451.5	443.6	413.9	-6.7
Malaysia		84.5	280.3	204.3	262.7	567.0	+53.7
Philippines			4.0	4.7	6.0	8.8	+46.7
Thailand				14.3	19.3	21.2	+9.8
Subtotal	146	472.1	1,009.5	770.5	821.3	1,093.0	+33.1
Australasia:							
Australia	216	412.5	323.2	434.6	404.0	405.4	+0.3
New Zealand			3.2	20.8	21.1	28.1	+33.2
Subtotal	216	412.5	326.4	455.3	425.1	433.5	+2.0

SOURCE. -- OFFSHORE.
§Brunei/Malaysia combined production.
#Subtotals do not equal total because of rounding.

TABLE 3D. -- WORLD PRODUCTION OF NATURAL GAS, TOTAL AND OFFSHORE
            (Cubic Feet, in Billions)

Year	World Production	Offshore Production	Offshore as % of World
1975	47,029.9	9,932.1	20.3
1976	50,407.5	10,847.0§	21.5
1977	53,883.7	6,663.3§	12.4
1978	53,859.5	9,509.0§	17.8
1979	57,194.6	9,369.0§	16.4
1980	58,636.4	10,160.9§	17.3
1981	57,816.0	10,085.1§	17.4
1982	55,893.9	10,326.7§	18.5
1983	55,066.5	10,360.1§	18.8
1984	59,932.3	12,196.8§	20.4
1985	62,721.4	12,451.4§	19.9
1986	63,683.2	11,643.3§	18.3
1987	68,168.0	11,299.6§	16.6
1988	70,978.7	10,948.1§	15.4
1989	74,224.8	10,869.1§	14.6

SOURCES. -- <u>Basic Petroleum Data Book</u> for 1973-75, <u>Oil and Gas Journal</u> and <u>Offshore</u> for 1976-89.
§Based on extrapolation from average daily rate.

TABLE 4D. -- OFFSHORE NATURAL GAS PRODUCTION BY REGION AND COUNTRY
(Cubic Feet per Day, in Millions)

Area and Country	1970	1975	1980	1987	1988	1989	1988-89 (% change)
Total	10,315.1	20,967.0	27,838.0	30,957.8	29,912.9	29,778.5	-0.4
Anglo-America:							
U.S.A.	8,591.8	11,664.3	14,703.7	11,363.7	10,444.4	9,280.3	-11.1
Latin America:							
Brazil		25.4	100.0	389.0	454.0	591.0	30.2
Chile				96.7	90.3	83.0	-8.1
Colombia			64.7				
Mexico			25.0*	212.8	245.0	255.0	4.1
Peru	64.7	77.0					
Trinidad & Tobago	11.0	123.0	420.0	386.8	374.6	366.6	-2.1
Venezuela				609.3	601.1	533.0	-11.3
Subtotal	75.7	225.0	609.7	1,694.6	1,765.0	1,828.6	3.6
Western Europe:							
Denmark				164.4	171.0	208.3	21.8
Greece				9.1	9.7	8.2	-15.5
Ireland			125.0	232.1	207.0	192.0	-7.2
Italy			22.0	420.7	365.3	440.0	20.4
The Netherlands		186.0	1,170.0	1,645.0	1,591.0	1,866.6	17.3
Norway		16.5	2,426.0	2,862.7	2,800.2	2,844.0	1.6
Spain				91.9	82.9	77.3	-6.7
U.K.	1,086.3	3,600.0	3,610.0	4,468.1	4,423.0	4,323.1	-2.3
Subtotal	1,086.3	3,802.5	7,353.0	9,894.0	9,650.1	9,959.5	3.2
Socialist Eastern Europe:							
U.S.S.R.		774.0	1,225.0	1,400.0	1,330.0	1,341.0	0.8
Near East:							
Egypt				120.9	115.0	126.0	9.6
Qatar				96.3	98.4	114.5	16.3
Saudi Arabia	494.8	3,825.4	0.0$	531.9	526.7	555.2	5.4
UAE			1,434.0	735.0	701.2	707.0	0.8
Subtotal	494.8	3,825.4	1,491.9	1,484.1	1,441.4	1,502.7	4.3

Sub-Saharan Africa:							
Angola/Cabinda			8.0	33.8	39.2	42.0	7.1
Ghana							-100.0
Ivory Coast		314.0		1.9	1.3		
Nigeria			500.0	96.9	114.1	101.0	-11.5
Subtotal		314.0	508.0	132.7	154.6	143.0	-7.5
East Asia:							
China			47.7	51.0	3.0	6.0	100.0
Japan				19.3	46.0	7.4	-83.9
Taiwan					20.2	16.0	-20.8
Subtotal			47.7	70.3	69.2	29.4	-57.5
South Asia:							
India			2.7	430.4	433.0	523.3	20.9
Southeast Asia:							
Brunei			984.8	822.0	814.0	788.0	-3.2
Indonesia		158.8	440.0	793.0	830.0	624.7	-24.7
Malaysia				1,412.0	1,320.0	1,810.0	37.1
Thailand				353.0	550.0	579.2	5.3
Subtotal		158.8	1,424.8	3,380.0	3,514.0	3,801.9	8.2
Australasia:							
Australia	60.5	203.0	388.0	821.0	840.2	1,098.7	30.8
New Zealand			84.0	287.0	271.0	270.0	-0.4
Subtotal	60.5	203.0	472.0	1,108.0	1,111.2	1,368.7	23.2

SOURCES. -- *Offshore* and *Oil and Gas Journal.*
*Estimate.
§All gas flared.

Tables, Transportation and Communication

The four tables of this appendix are continuations of the shipping tables that appeared in previous volumes of the *Yearbook*. For additional information on shipping see the articles by Hal Olson and Nancy Yamaguchi.

THE EDITORS

TABLE 1E. -- WORLD SHIPPING TONNAGE, BY TYPE OF VESSEL (Million grt as of July 1)

	1970	1980	1986	1987	1988	1987-88 (% change)
Oil tankers*	86.1	175.0	128.4	124.7	127.8	2.6
Liquified gas carriers#	1.4	7.4	9.8	9.8	9.8	-0.4
Chemical carriers	0.5	2.2	3.6	3.5	3.5	0.2
Miscellaneous tankers		0.2	0.3	0.3	0.2	-21.0
Bulk/oil carriers§	8.3	26.2	21.3	20.5	20.0	-2.2
Ore and bulk carriers	38.3	83.4	111.6	110.6	109.6	-0.9
General cargo£	72.4	82.6	73.2	72.2	71.9	-0.4
Miscellaneous cargo ships						0.0
Container ships (fully cellular)	1.9	11.3	19.6	21.1	22.1	4.8
Barge-carrying vessels						0.0
Vehicle carriers				4.5	3.9	-12.1
Fishing factories, carriers, and trawlers	7.8	12.8	13.4	13.5	13.8	2.3
Passenger liners	3.0					0.0
Ferries and other passenger vessels		7.6	8.8	9.2	9.7	4.6
All other vessels•	7.8	8.6	10.5	10.8	11.1	2.1

SOURCE. -- Lloyd's Register of Shipping Statistical Tables.
NOTE. -- grt = gross registered tons.
*Including oil/chemical tankers.
#I.e., ships capable of liquid natural gas or liquid petroleum gas or other similar hydrocarbon and chemical products that are all carried at pressures greater than atmosphere or at subambient temperature or a combination of both.
§Including ore/oil carriers.
£Including passenger/cargo.
•Including livestock carriers, supply ships and tenders, tugs, cable ships, dredgers, icebreakers, research ships, and others.

TABLE 2E. -- ESTIMATED AVERAGE SIZE OF SELECTED TYPES OF VESSELS: EXISTING WORLD FLEETS, MIDYEAR (grt)

	1970	1980	1986	1987	1988	1987-88 (% change)
Oil tankers (100 grt and above)	14,110	24,606	20,742	20,635	19,473	-5.6
Ore/bulk carriers (6,000 grt and above)*	18,450	23,408	25,201	25,697	26,031	1.3
Container ships (100 grt and above)	11,420	17,030	18,430	19,295	19,828	2.8
Liquified gas carriers (grt)	4,690	11,624	12,769	12,691	12,650	-0.3

SOURCE. -- Lloyd's Register of Shipping Statistical Tables.
NOTE. -- grt = gross registered tons.
   *Including bulk/oil carriers.

TABLE 3E. -- WORLD MERCHANT FLEETS, BY REGION AND COUNTRY (grt as of July 1)

	1980	1986	1987	1988	1987-88 (% change)
Anglo-America:					
Bermuda		1,208,276	1,925,297	3,774,298	96.0
Canada	3,180,126	3,160,043	2,971,155	2,902,394	-2.3
U.S.A.	18,464,271	19,900,843	20,178,236	20,832,137	3.2
Subtotal	21,644,297	24,269,162	25,074,688	27,508,829	9.7
Latin America:					
Anguilla	399	4,106	3,705	3,303	-10.9
Antigua & Barbuda	410	1,048	51,875	323,469	523.6
Argentina	2,546,305	2,117,017	1,901,026	1,876,673	-1.3
Bahamas	87,320	5,985,011	9,105,182	8,962,892	-1.6
Barbados	5,257	7,572	8,348	8,470	1.5
Belize	620	620	620	620	0.0
Bolivia	15,130	14,913	13,824	9,610	-30.5
Brazil	4,533,633	6,212,287	6,324,059	6,122,835	-3.2
Caymen Is.	256,715	1,389,903	706,160	476,505	-32.5
Chile	614,425	566,881	546,745	603,557	10.4
Colombia	283,457	380,074	423,631	412,321	-2.7
Costa Rica	20,333	13,325	14,781	15,080	2.0
Cuba	881,260	958,613	966,288	912,002	-5.6
Dominica		2,013	1,724	2,224	29.0
Dominican Rep.	37,659	42,241	43,560	48,306	10.9
Ecuador	275,142	437,682	421,361	428,066	1.6
El Salvador	501	3,819	3,819	3,819	0.0
Falkland Is.	7,907	6,907	6,907	6,907	0.0
Grenada	226	425	550	516	-6.2
Guatemala	13,626	9,432	4,694	4,694	0.0
Guyana	18,261	22,731	22,310	14,956	-33.0
Haiti	1,120	2,688	512	512	0.0
Honduras	213,421	555,202	506,374	582,170	15.0
Jamaica	13,307	9,419	13,118	14,433	10.0
Mexico	1,006,417	1,520,246	1,532,485	1,448,335	-5.5
Montserrat	1,010	711	711	711	0.0
Nicaragua	15,726	22,930	12,739	13,658	7.2

Panama	24,190,680	41,305,009	43,254,716	44,604,071	3.1
Paraguay	23,019	43,298	41,670	38,567	-7.4
Peru	740,510	754,179	788,171	674,952	-14.4
St. Kitts-Nevis	256	256	556	300	-46.0
St. Lucia	2,378	2,766	2,092	1,891	-9.6
St. Vincent	19,679	509,878	699,947	900,477	28.6
Suriname	14,921	12,655	11,457	11,457	0.0
Trinidad & Tobago	17,456	19,381	18,527	23,857	28.8
Turks & Caicos Is.	2,408	3,583	3,469	3,963	14.2
Uruguay	198,478	149,811	144,394	169,939	17.7
Venezuela	848,540	998,296	999,195	982,117	-1.7
Virgin Is. (U.K.)	5,826	8,077	8,179	6,806	-16.8
Subtotal	36,910,961	64,095,005	68,609,481	69,715,041	-1.6*
Western Europe:					
Austria	88,784	124,794	193,513	201,251	4.0
Belgium	1,809,829	2,419,661	2,268,383	2,118,422	-6.6
Denmark	5,390,365	4,651,224	4,754,837	4,371,796	-8.1
Faeroe Is.	66,085	115,394	118,628	129,931	9.5
Finland	2,530,091	1,469,927	1,122,249	837,952	-25.3
France	11,924,557	5,936,268	5,371,273	4,506,227	-16.1
Germany, F.R.	8,355,638	5,565,214	4,317,616	3,917,257	-9.3
Gibraltar	2,291	1,612,948	2,827,098	3,041,811	7.6
Greece	29,471,744	28,390,800	23,559,852	21,978,820	-6.7
Iceland	188,215	176,409	173,618	174,550	0.5
Ireland	208,926	149,308	153,637	172,768	12.5
Italy	11,095,694	7,896,569	7,817,353	7,794,247	-0.3
Luxembourg	132,861			1,731	
Malta	31,422	2,014,947	1,725,984	2,685,888	55.6
Monaco					
The Netherlands	5,723,845	4,324,135	3,908,231	3,726,464	-4.7
Norway	22,007,490	9,294,630	6,359,349	9,350,303	47.0
Portugal	1,355,989	1,114,444	1,048,197	988,844	-5.7
Spain	8,112,245	5,422,002	4,949,387	4,415,122	-10.8
Sweden	4,233,977	2,516,614	2,269,541	2,116,079	-6.8
Switzerland	310,775	346,220	354,614	259,427	-26.8
U.K.	27,135,155	11,567,117	8,504,605	8,260,431	-2.9
Subtotal	150,175,978	95,108,625	81,797,965	81,049,321	-0.9

TABLE 3E. -- (continued)

	1980	1986	1987	1988	1987-88 (% change)
Socialist Eastern Europe:					
Albania	56,127	56,133	56,133	56,133	0.0
Bulgaria	1,233,307	1,385,009	1,551,176	1,392,381	-10.2
Czechoslovakia	155,319	197,868	156,791	157,903	0.7
German D.R.	1,532,197	1,518,944	1,494,039	1,442,840	-3.4
Hungary	74,997	86,395	77,377	76,121	-1.6
Poland	3,639,078	3,457,242	3,469,670	3,489,449	0.6
Romania	1,856,292	3,233,906	3,263,823	3,560,736	9.1
U.S.S.R.	23,433,534	24,960,888	25,232,091	25,783,969	2.2
Yugoslavia	2,466,574	2,872,613	3,164,893	3,476,354	9.8
Subtotal	34,475,421	37,768,998	38,465,993	39,435,886	2.5
Sub-Saharan Africa:					
Angola	65,310	92,285	91,712	91,038	-0.7
Benin	4,557	4,887	4,665	4,665	0.0
Cameroon	62,080	76,660	57,871	57,348	-0.9
Cape Verde	11,426	14,095	14,579	17,148	17.6
Comoros	1,116	1,261	1,795	1,187	-33.9
Congo	6,784	8,458	8,458	8,458	0.0
Djibouti	3,135	3,051	3,051	3,051	0.0
Equatorial Guinea	6,412	6,412	6,412	6,412	0.0
Ethiopia	23,811	66,926	73,456	74,143	0.9
Gabon	77,095	97,967	23,843	24,843	4.2
Gambia	3,907	2,588	3,878	3,529	-9.0
Ghana	250,428	165,644	142,421	125,679	-11.8
Guinea	5,648	7,179	7,179	7,179	0.0
Guinea-Bissau	577	4,070	4,070	4,070	0.0
Ivory Coast	186,127	120,679	118,952	118,952	0.0
Kenya	17,371	9,040	7,872	7,872	0.0
Liberia	80,285,176	52,649,444	51,412,029	49,733,615	-3.3
Madagascar	91,211	73,715	64,162	91,549	42.7
Malawi	200	424	424	424	0.0
Mali					
Mauritania	874	22,752	29,644	36,914	24.5
Mauritius	37,675	151,978	162,749	156,698	-3.7

Mozambique	37,887	42,801	35,957	36,006	0.1
Nigeria	498,202	563,912	593,582	586,868	-1.1
St. Helena	3,150	3,640	3,640	3,640	0.0
Sao Tome & Principe		1,488	1,488	1,488	0.0
Senegal	34,499	50,429	46,448	49,113	5.7
Seychelles	4,602	3,813	3,233	3,233	0.0
Sierra Leone	3,738	6,979	8,756	13,716	56.6
Somalia	45,553	15,719	17,896	12,785	-28.6
South Africa	728,926	599,509	533,092	485,526	-8.9
Sudan	104,803	95,742	96,699	96,699	0.0
Tanzania	55,916	50,726	31,551	32,123	1.8
Togo	14,886	54,882	59,690	47,772	-20.0
Uganda	5,510	5,091	5,091	5,091	0.0
Zaire	91,894	65,833	56,393	56,393	0.0
Zambia					
Subtotal	82,772,023	55,140,079	53,732,738	52,005,227	-3.2#
Near East:					
Algeria	1,218,621	881,670	892,553	896,691	0.5
Bahrain	10,248	51,713	43,833	54,417	24.1
Cyprus	2,091,089	10,616,809	15,650,207	18,390,642	17.5
Egypt	555,786	1,063,020	1,074,192	1,226,725	14.2
Iran	1,283,629	2,911,359	3,976,873	4,336,609	9.0
Iraq	1,465,949	1,016,343	1,002,236	953,069	-4.9
Israel	450,216	556,628	514,815	545,642	6.0
Jordan	496	42,365	32,884	32,198	-2.1
Kuwait	2,529,491	2,580,924	2,087,856	735,318	-64.8
Lebanon	267,787	484,624	460,876	405,311	-12.1
Libya	889,908	825,231	816,570	830,172	1.7
Morocco	359,552	416,482	418,451	436,997	4.4
Oman	6,954	14,793	25,321	25,470	0.6
Qatar	91,934	306,673	306,443	308,668	0.7
Saudi Arabia	1,589,668	2,978,016	2,692,044	2,269,398	-15.7
Syria	39,255	63,142	63,077	64,101	1.6
Tunisia	131,079	285,535	285,483	281,456	-1.4

TABLE 3E. -- (continued)

	1980	1986	1987	1988	1987-88 (% change)
Turkey	1,454,838	3,423,745	3,336,093	3,281,153	-1.6
United Arab Emirates	158,210	653,525	732,013	824,990	12.7
Yemen, A.R.	2,979	7,115	199,909	195,876	-2.0
Yemen, P.D.R.	12,230	12,543	12,278	11,177	-9.0
Subtotal	14,609,918	29,192,255	34,624,007	36,106,080	4.3
South Asia:					
Bangladesh	353,586	378,563	410,721	431,831	5.1
India	5,911,367	6,540,121	6,725,776	6,160,773	-8.4
Maldives	136,037	84,808	100,200	104,424	4.2
Pakistan	478,019	434,079	394,407	366,059	-7.2
Sri Lanka	93,471	622,226	594,491	410,381	-31.0
Subtotal	6,972,480	8,059,797	8,225,595	7,473,468	-9.1
East Asia:					
China	6,873,608	11,566,974	12,341,477	12,919,876	4.7
Hong Kong	1,717,230	8,179,670	8,034,668	7,328,984	-8.8
Japan	40,959,683	38,487,773	35,932,177	32,074,417	-10.7
Korea, D.P.R.	230,695	407,253	406,647	405,777	-0.2
Korea, Rep.	4,344,114	7,183,617	7,214,070	7,333,704	1.7
Taiwan	2,039,123	4,272,795	4,512,749	4,631,474	2.6
Subtotal	56,164,453	70,098,082	68,441,788	64,694,232	-5.5
Southeast Asia:					
Brunei	899	1,973	352,276	354,313	0.6
Burma	78,519	125,524	239,261	272,665	14.0
Indonesia	1,411,688	2,085,635	2,120,531	2,126,016	0.3
Kampuchea	3,558	3,558	3,558	3,558	0.0
Malaysia	702,145	1,743,629	1,688,523	1,608,155	-4.8
Philippines	1,927,869	6,922,499	8,681,227	9,311,555	7.3
Singapore	7,664,229	6,267,627	7,098,116	7,208,974	1.6
Thailand	391,456	533,138	510,991	515,314	0.8
Vietnam	240,895	338,668	360,470	337,875	-6.3
Subtotal	12,430,258	18,022,251	21,054,953	21,738,425	3.2

Australasia:

Australia	1,642,594	2,368,462	2,404,559	2,365,923	-1.6
Fiji	14,733	29,954	35,324	37,162	5.2
Kiribati	980	3,197	3,332	3,538	6.2
Nauru	54,004	66,725	65,777	60,109	-8.6
New Zealand	263,543	314,206	334,193	336,808	0.8
Papua New Guinea	24,904	30,922	36,346	37,678	3.7
Solomon Islands		6,002	6,387	8,647	35.4
Tongo	25,395	16,349	18,295	13,585	-25.7
Tuvalu		526	526	526	0.0
Vanuatu	12,541	164,953	540,088	789,506	46.2
Western Samoa	4,765	26,087	26,087	26,087	0.0
Subtotal	2,046,167	3,027,403	3,470,914	3,679,569	6.0

SOURCE. -- Lloyd's Register of Shipping Statistical Tables.
NOTE. -- grt = gross registered tons.
*Excluding Panama, the 1987-88 change for Latin America is -1.0%.
#Excluding Liberia, the 1987-88 change for Sub-Saharan Africa is -2.1%.

TABLE 4E. -- VESSELS LOST, BY COUNTRY (grt)

Country	1982	1983	1984	1985	1986	1987
Argentina	3,122 (3)		442 (1)	2,122 (1)	132 (1)	
Australia	1,265 (2)		138 (1)	124 (1)		
Belgium	3,729 (4)	141 (1)				
Brazil						
Canada	11,513 (9)	2,437 (3)	6,466 (2)	8,395 (2)	11,372 (1)	499 (2)
Cyprus	77,903 (18)	39,614 (9)	1,541 (4)	726 (2)	807 (2)	193,165 (12)
Denmark	279 (2)	2,002 (7)	243,285 (10)	177,804 (13)	361,760 (21)	4,080 (5)
Finland			3,355 (6)	857 (2)	896 (5)	487 (1)
France		447 (3)	1,323 (1)	499 (1)	2,326 (1)	3,859 (2)
German D.R.			5,644 (4)	458 (3)	641 (1)	
Germany, F.R.	119 (1)	6,324 (5)	27,389 (2)	1,743 (1)	8,227 (1)	1,881 (3)
Greece	352,822 (49)	418,173 (38)	2,179 (5)	5,020 (3)	1,286 (3)	155,503 (6)
Hong Kong	139 (1)		480,115 (21)	342,828 (20)	217,055 (13)	21,384 (4)
India	10,050 (2)	2,939 (1)	41,238 (2)	299 (1)	92,630 (13)	67,371 (4)
Italy	43,401 (11)	3,607 (6)	39,741 (2)	33,969 (4)	299 (1)	6,152 (4)
Japan	16,990 (42)	9,872 (26)	21,196 (8)	6,994 (7)	4,705 (6)	12,413 (24)
Korea, Rep.	27,644 (20)	43,377 (14)	13,890 (47)	19,181 (46)	10,006 (27)	147,711 (14)
Liberia	212,493 (6)	174,458 (7)	20,965 (11)	89,699 (17)	127,851 (13)	108,053 (3)
The Netherlands	4,297 (5)		424,300 (10)	107,251 (6)	849,663 (7)	122 (1)
Norway	12,983 (7)	16,259 (10)	5,146 (1)	125 (1)	5,945 (4)	3,673 (6)
Panama	292,855 (70)	341,746 (63)	115,549 (6)	829 (4)	178,309 (38)	141,634 (23)
Philippines	4,737 (6)	22,041 (8)	231,123 (49)	295,895 (46)	4,189 (5)	75,726 (7)
Poland	927 (1)	1,991 (1)	8,318 (8)	13,307 (7)	3,008 (1)	10,970 (1)
Portugal	2,881 (3)		8,750 (2)	1,974 (1)	320 (1)	
Singapore	17,371 (5)	11,836 (3)	803 (1)	3,575 (2)	19,593 (7)	43,346 (3)
Spain	6,142 (9)	151,010 (14)	10,867 (3)	13,032 (7)	113,845 (15)	1,895 (5)
Sweden		500 (1)	11,687 (18)	500 (5)	1,949 (3)	
Turkey	14,796 (4)	775 (2)	349 (1)	118,211 (5)	80,580 (3)	45,006 (6)
U.S.S.R.	8,069 (3)	6,912 (3)	43,623 (3)	10,944 (3)	47,317 (3)	19,071 (3)
U.K.	23,655 (10)	10,292 (9)	28,907 (7)	3,818 (8)	20,220 (7)	10,060 (8)
U.S.A.	72,245 (32)	34,753 (24)	64,510 (18)	10,979 (14)	16,578 (8)	77,034 (13)
Yugoslavia	1,445 (2)		36,034 (5)	12,162 (1)	2,175 (1)	
Others*	408,058 (75)	171,105 (82)	455,068 (68)	368,144 (79)	425,051 (67)	133,066 (65)
Total	1,631,930(402)	1,472,611(340)	2,353,941(327)	1,651,210(307)	2,608,735(265)	1,284,161(219)

SOURCE. -- Lloyd's Register of Shipping Statistical Tables.
NOTE. -- grt = gross registered tons; no. of vessels shown in parentheses.
*Others includes, inter alia, China, P.R.; Indonesia; Kuwait; Mexico; Saudi Arabia; and Taiwan.

Tables, Military Activities

The seven tables of this appendix were begun by Andrzej Karboszka in *Ocean Yearbook 2* and have been updated in each of the subsequent volumes of the *Yearbook.* For additional information on military activities the reader is referred to the articles in this volume by Joseph Morgan, Peter Haydon, Michael Morris, and Frank Barnaby.

THE EDITORS

TABLE 1F. -- WORLD STOCK OF AIRCRAFT CARRIERS, BY COUNTRY

	1950	1960	1970	1980	1987	1988	1989
World total:							
Attack	44	30	25	22	24	25	25
Other	75	42	30	21	26	28	27
U.S.A.:							
Attack	27	14	15	16	18	18	18
Other	75	40	24	14	12	13	13
U.S.S.R.:							
Attack					1	2	2
Other				3	6	6	6
U.K.:							
Attack	12	8	4	1			
Other		1	2	2	3	3	3
France:							
Attack	2	3	2	2	2	2	2
Other		1	1		1	1	1
Australia:							
Attack	1	2	1	1			
Other			1				
Canada:							
Attack	1	1					
Italy:							
Other					2	2	2
Netherlands:							
Attack	1	1					
Spain:							
Other			1	1	2	2	1
Argentina:							
Attack		1	2	1	1	1	1
Brazil:							
Other		1	1	1	1	1	1
India:							
Attack			1	1	2	2	2

SOURCE. -- J.E. Moore, ed. <u>Jane's Fighting Ships 1989/90</u> (London: MacDonald & Jane's, 1989).

NOTE. -- Other = antisubmarine, amphibious helicopter assault, and utility.

TABLE 2F. -- WORLD STOCK OF STRATEGIC SUBMARINES,* BY GROUPS OF COUNTRIES

	1960	1970	1980	1985	1987	1988	1989
World total:							
Nucl.	7	70	124	114	116	118	120
Conv.	10	26	30	17	16	16	18
U.S.A.:							
Nucl.	3	41	41	37	39	39	39
Conv.							
Other NATO:							
Nucl.		5	9	10	11	11	11
Conv.		1	1	1			
Total NATO:							
Nucl.	3	46	50	47	50	50	50
Conv.		1	1	1			
U.S.S.R.:							
Nucl.	4	24	74	65	62	63	65
Conv.	10	25	19	15	15	14	16
Other WTO:							
Nucl.							
Conv.							
Total WTO:							
Nucl.	4	24	74	65	62	63	65
Conv.	10	25	19	15	15	14	16
China:							
Nucl.				2	4	5	5
Conv.				1	1	2	2

SOURCE. -- See table 1F.
NOTE. -- Nucl. = nuclear powered; Conv. = conventionally powered;
WTO = Warsaw Treaty Organization, as of July 1989.
*Equipped with medium- or long-range ballistic missiles.

TABLE 3F. -- WORLD STOCK OF NUCLEAR-POWERED ATTACK SUBMARINES,* BY COUNTRY

	1955	1960	1965	1970	1976	1980	1987	1988	1989
World total	1	15	48	108	157	189	250	256	263
U.S.A.	1	11	22	46	68	81	97	99	100
France							4	5	5
U.K.				4	9	12	15	16	16
U.S.S.R.		4	26	58	80	96	128	132	137
China							4	4	4
India									1

SOURCE. -- See table 1F.
*Includes nuclear-powered cruise missile submarines.

TABLE 4F. -- WORLD STOCK OF PATROL SUBMARINES,* BY GROUPS OF COUNTRIES

	1950	1960	1970	1980	1987	1988	1989
World total:							
Nucl.		15	108	189	250	257	263
Conv.	355	535	535	495	497	472	476
Developed:							
Nucl.		15	108	180	244	252	258
Conv.	351	502	459	326	294	283	272
U.S.A.:							
Nucl.		11	46	81	97	99	100
Conv.	194	158	52	6	4	4	3
Other NATO:							
Nucl.			4	12	19	20	21
Conv.	105	89	75	82	82	80	77
Total NATO:							
Nucl.		11	50	93	116	120	121
Conv.	299	247	127	88	86	84	80
U.S.S.R.:							
Nucl.		4	58	96	128	132	137
Conv.	46	238	283	155	151	140	136
Other WTO:							
Nucl.							
Conv.			7	6	8	9	7
Total WTO:							
Nucl.		4	58	96	128	132	137
Conv.	46	238	290	161	159	149	143
Other Europe:§	3	12	27	32	21	22	22
Other developed:§	3	5	15	24	28	28	27
Total developing countries:							
Nucl.					4	5	5
Conv.	4	33	76	169	203	189	204
Middle East§		10	16	12	15	15	15
South Asia							
Nucl.						1	1
Conv.			8	13	17	19	20
China#							
Nucl.					4	4	4
Conv.		12	27	83	102	86	98
Other East and Southeast Asia§		2	14	20	22	22	26

TABLE 4F. -- (continued)

	1950	1960	1970	1980	1987	1988	1989
Sub-Saharan Africa§							
North Africa§				6	8	10	10
Central America§				2	4	3	3
South America§	4	9	11	33	35	34	32

SOURCE. -- See table 1F.

NOTE. -- Nucl. = nuclear powered; Conv. = conventionally powered; WTO = Warsaw Treaty Organization; Other Europe = Albania, Finland, Ireland, Malta, Sweden, and Yugoslavia (Spain included in NATO beginning 1982); Other developed = Australia, Japan, New Zealand, South Africa, and Taiwan; Middle East = Bahrain, Cyprus, Egypt, Iran, Iraq, Israel, Jordon, Kuwait, Lebanon, Oman, Qatar, Saudi Arabia, Syria, and the Yemens; South Asia = Bangladesh, India, Maldives, Pakistan, and Sri Lanka; Other East & S.E. Asia = Burma, Indonesia, Kampuchea, Malaysia, North Korea, Singapore, South Korea, Thailand, the Philippines, Vietnam, and the Pacific island nations; Sub-Saharan Africa = Angola, Cameroon, Cape Verde, Comoros, Congo, Ethiopia, Gabon, Ghana, Guinea, Ivory Coast, Kenya, Madagascar, Mauritania, Mozambique, Nigeria, Senegambia, Seychelles, Sierra Leone, Somalia, Sudan, Tanzania, Togo, and Zaire; North Africa = Algeria, Libya, Morocco, and Tunisia; Central America includes the Caribbean island nations; South America = Argentina, Brazil, Chile, Colombia, Ecuador, Guyana, Peru, Suriname, Uruguay, and Venezuela.

*Post-World War II submarines displacing 700 tons or more.

§All conventionally powered.

#Figures for China may vary considerably between years because of changes in the quality of available data.

TABLE 5F. -- WORLD STOCK OF COASTAL SUBMARINES,* BY GROUPS OF COUNTRIES

	1950	1960	1970	1980	1985	1987	1988	1989
World total	313	179	72	76	58	61	51	63
Developed	299	162	64	60	46	47	38	45
Developing	14	17	8	16	12	14	13	18
U.S.A.								
Other NATO		20	34	37	45	43	34	41
Total NATO		20	34	37	45	43	34	41
U.S.S.R.	273	127	22	20				
Other WTO								
Total WTO	273	127	22	20				
Other Europe	26	15	8	3	1	4	4	4
Other developed								

SOURCE. -- See table 1F.
NOTE. -- WTO = Warsaw Treaty Organization.
*Submarines displacing less than 700 tons; all conventionally powered.

TABLE 6F. -- WORLD STOCK OF MAJOR SURFACE WARSHIPS,* BY GROUPS OF COUNTRIES

	1950	1960	1970	1980	1987	1988	1989
World total:							
Miss.		18	191	560	812	880	901
Conv.	1,783	1,789	1,414	501	326	259	202
Developed:							
Miss.		18	189	480	641	697	707
Conv.	1,600	1,520	1,128	335	214	160	119
U.S.A.							
Miss.		15	77	148	210	233	230
Conv.	817	704	478	67	24	5	0
Other NATO:							
Miss.			57	160	214	220	215
Conv.	520	402	294	105	73	51	41
Total NATO:							
Miss.		15	134	308	428	453	445
Conv.	1,337	1,106	772	172	97	56	41
U.S.S.R.:							
Miss.		1	39	107	134	140	142
Conv.	150	260	206	86	74	74	70
Other WTO:							
Miss.			1	3	4	6	9
Conv.	5	14	9	2	6	7	4
Total WTO:							
Miss.		1	40	110	138	146	151
Conv.	155	274	215	88	80	81	74
Other Europe:							
Miss.		2	6	16	3	4	4
Conv.	67	75	71	23			
Other developed:							
Miss.			9	46	72	94	107
Conv.	41	63	64	52	42	23	4
Total developing:							
Miss.			2	80	171	183	194
Conv.	183	269	286	166	112	99	83
Middle East:							
Miss.			1	8	22	21	18
Conv.	18	15	21	9	3	5	5

TABLE 6F. -- (continued)

	1950	1960	1970	1980	1987	1988	1989
South Asia:							
Miss.				8	22	23	31
Conv.	16	26	32	27	16	14	13
China:							
Miss.				22	46	46	46
Conv.	4	23	21	6	10	10	5
Other East and S.E. Asia:§							
Miss.		1	6	25	29	31	32
Conv.	50	71	109	48	37	31	24
Sub-Saharan Africa:							
Miss.				1	1	1	1
Conv.			1	2	3	3	3
North Africa:							
Miss.				1	4	7	7
Conv.			2	1	1	1	1
Central America:							
Miss.					2	4	5
Conv.	30	29	22	31	10	12	8
South America:							
Miss.				34	45	50	54
Conv.	65	105	78	42	32	23	24

SOURCE. -- See table 1F.
NOTE. -- Miss. = missile armed (both SSMs and SAMs); Conv. = conventionally armed. See table 4F for definitions of groups of countries.
*Cruisers, destroyers, frigates, and escorts (over 1,000 tons displacement).
§Taiwan included under "other developed" beginning 1980.

TABLE 7F. -- WORLD STOCK OF LIGHT NAVAL FORCES,* BY GROUPS OF COUNTRIES

	1950	1960	1970	1980	1987	1988	1989
World total:							
Miss.		5	281	822	1,134	1,130	1,123
Conv.	987	1,849	2,457	3,696	3,653	3,529	3,421
Developed:							
Miss.		5	188	368	481	483	478
Conv.	822	1,422	1,290	1,145	1,073	1,001	963
U.S.A.:							
Miss.			1	1	6	6	6
Conv.	147	35	35	21	25	24	24
Other NATO:							
Miss.			1	106	142	136	142
Conv.	190	230	241	249	278	262	286
Total NATO:							
Miss.			2	107	148	142	148
Conv.	337	265	276	270	303	286	310
U.S.S.R.:							
Miss.		5	150	183	148	156	161
Conv.	395	769	600	422	327	309	300
Other WTO:							
Miss.			28	37	59	59	54
Conv.	16	141	164	183	195	163	130
Total WTO:							
Miss.		5	178	220	207	215	215
Conv.	411	910	764	605	522	472	430
Other Europe:							
Miss.			8	34	55	55	54
Conv.	60	180	198	212	154	154	128
Other developed:							
Miss.				7	71	71	61
Conv.	14	67	54	58	94	89	95
Total developing:							
Miss.			93	454	653	647	645
Conv.	156	427	1,141	2,556	2,580	2,528	2,458
Middle East:							
Miss.			32	105	163	160	158
Conv.	11	77	140	325	346	338	299

TABLE 7F. -- (continued)

	1950	1960	1970	1980	1987	1988	1989
South Asia:							
Miss.				20	30	31	32
Conv.		1	7	52	94	106	97
China:							
Miss.			15	181	237	228	220
Conv.		150	408	762	754	690	697
Other East &							
S.E. Asia:							
Miss.			12	61	90	98	106
Conv.	55	149	403	717	729	752	741
Sub-Saharan							
Africa:							
Miss.				15	29	37	33
Conv.		5	54	228	259	262	251
North Africa:							
Miss.			16	31	52	49	54
Conv.		2	45	67	70	68	88
Central America:							
Miss.			18	26	23	18	18
Conv.	16	18	72	191	251	243	215
South America:							
Miss.				15	29	26	24
Conv.	42	26	38	109	77	69	70

SOURCE. -- See table 1F.
NOTE. -- Miss. = missile armed; Conv. = conventionally armed. WTO = Warsaw Treaty Organization. See table 4F for definitions of groups of countries.
   *This category includes corvettes, fast patrol boats, torpedo boats, and large and coastal patrol crafts (gun armed). Riverine craft are excluded. Corvettes are not included for the years 1950-1976.

Appendix G

Tables, General Information

Since the signing and general acceptance of the 1982 United Nations Law of the Sea Convention many states have brought their maritime claims into line with the Convention. These tables provide general information about coastal states and their maritime claims and the status of the Convention.

THE EDITORS

Appendix 4G: a)   UNEP REGIONAL SEA AREAS AND ACTION PLANS

	Regional Sea Area	Action Plan Adopted		Published*
A	Mediterranean	February	1975	No. 34 (1983, rev. 1985)
B	Gulf	April	1978	No. 35 (1983)
C	West/Central Africa	March	1981	No. 27 (1983)
D	Southeast Pacific	November	1981	No. 20 (1983)
E	Red Sea	February	1982	No. 81 (1986)
F	Caribbean	April	1981	No. 26 (1983)
G	Eastern Africa	June	1985	No. 61 (1985)
H	South Pacific	March	1982	No. 29 (1983)
I	East Asia	October	1981	No. 24 (1983)
K	South Asia	in preparation	(expected completion date 1991)	
L	Black Sea	in preparation		
M	Northwest Pacific	in preparation		

*Published in UNEP Regional Seas Reports and Studies

For an earlier map showing the action plan regions as initially planned, see E. Mann
Borgese, "The Law of the Sea," Scientific American 248, no. 3(1983):35.

Appendix 4G: b) UNEP Regional Seas Programmes Agreements

1. MEDITERRANEAN

	Barcelona Convention[a]		Dumping Protocol[b]		Emergency Protocol[c]	
	signed/acceded	in force	signed/acceded	in force	signed/acceded	in force
Albania	30 May 1990	29 Jun 1990	30 May 1990	29 Jun 1990	30 May 1990	29 Jun 1990
Algeria	16 Feb 1981	18 Mar 1981	16 Mar 1981	15 Apr 1981	16 Mar 1981	15 Apr 1981
Cyprus	16 Feb 1976	19 Dec 1979	16 Feb 1976	19 Dec 1979	16 Feb 1976	19 Dec 1979
EEC	13 Sep 1976	15 Apr 1978	13 Sep 1976	15 Apr 1978	13 Sep 1976	11 Sep 1981
Egypt	16 Feb 1976	23 Sep 1978	16 Feb 1976	24 Aug 1978	16 Feb 1976	23 Sep 1978
France	16 Feb 1976	10 Apr 1978*	16 Feb 1976	10 Apr 1978*	16 Feb 1976	10 Apr 1978*
Greece	16 Feb 1976	2 Feb 1979	11 Feb 1977	2 Feb 1979	16 Feb 1976	2 Feb 1979
Israel	16 Feb 1976	2 Apr 1978*	16 Feb 1976	31 Mar 1984	16 Feb 1976	2 Apr 1978
Italy	16 Feb 1976	5 Mar 1979	16 Feb 1976	5 Mar 1979	16 Feb 1976	5 Mar 1979
Lebanon	16 Feb 1976	12 Feb 1978	16 Feb 1976	12 Feb 1978	16 Feb 1976	12 Feb 1978
Libya	31 Jan 1977	2 Mar 1979	31 Jan 1977	2 Mar 1979	31 Jan 1977	2 Mar 1979
Malta	16 Feb 1976	12 Feb 1978	16 Feb 1976	12 Feb 1978	16 Feb 1976	12 Feb 1978
Monaco	16 Feb 1976	12 Feb 1978	16 Feb 1976	12 Feb 1978	16 Feb 1976	12 Feb 1978
Morocco	16 Feb 1976	15 Feb 1980	16 Feb 1976	15 Feb 1980	16 Feb 1976	15 Feb 1980
Spain	16 Feb 1976	12 Feb 1978	16 Feb 1976	12 Feb 1980	16 Feb 1976	12 Feb 1978
Syria	26 Dec 1978	25 Jan 1979*	26 Dec 1978	25 Jan 1979	26 Dec 1978	25 Jan 1979
Tunisia	25 May 1976	12 Feb 1978	25 May 1976	12 Feb 1978	25 May 1976	12 Feb 1978
Turkey	16 Feb 1976	6 May 1981	16 Feb 1976	6 May 1981	16 Apr 1976	6 May 1981
Yugoslavia	15 Sep 1976	12 Feb 1978	15 Sep 1976	13 Jan 1978	15 Sep 1976	12 Feb 1978

[a] Convention for the Protection of the Mediterranean Sea Against Pollution, adopted at Barcelona on 16 February 1976; entered into force 12 February 1978.

[b] Protocol for the Prevention of Pollution of the Mediterranean Sea by Dumping from Ships and Aircraft, adopted at Barcelona on 16 February 1976; entered into force 12 February 1978.

[c] Protocol Concerning Co-operation in Combating Pollution of the Mediterranean Sea by Oil and Other Harmful Substance in Cases of Emergency, adopted at Barcelona on 16 February 1976; entered into force 12 February 1978.

* with reservation

1. MEDITERRANEAN cont'd

	Land-Based Sources Protocol[d]		Protected Areas Protocol[e]	
	signed/acceded	in force	signed/acceded	in force
Albania	30 May 1990	29 Jun 1990	30 May 1990	29 Jun 1990
Algeria	2 May 1983	17 Jun 1983	16 May 1985	23 Mar 1986
Cyprus	17 May 1980	28 Jul 1988	28 Jun 1988	28 Jul 1988
EEC	17 May 1980	6 Nov 1983	30 Mar 1983	23 Mar 1986
Egypt	18 May 1983	17 Jun 1983	16 Feb 1983	23 Mar 1986
France	17 May 1980	17 Jun 1983	3 Apr 1982*	2 Oct 1986*
Greece	17 May 1980	25 Feb 1987	3 Apr 1982	25 Feb 1987
Israel	18 May 1980		4 Apr 1982	27 Nov 1987
Italy	17 May 1980	3 Aug 1985	3 Apr 1982	23 Mar 1986
Lebanon	17 May 1980			
Libya	17 May 1980	5 Jul 1989	6 Jun 1989	5 Jul 1989
Malta	17 May 1980	31 Mar 1989	3 Apr 1982	10 Feb 1988
Monaco	17 May 1980	17 Jun 1983	3 Apr 1982	28 Jun 1989
Morocco	17 May 1980	11 Mar 1987	3 Apr 1982	23 Mar 1986
Spain	17 May 1980	5 Jul 1984	3 Apr 1982	21 Jan 1988
Syria				
Tunisia	17 May 1980	17 Jun 1983	3 Apr 1982	23 Mar 1986*
Turkey	21 Feb 1983	17 Jun 1983	6 Nov 1986	6 Dec 1986
Yugoslavia				

[d]Protocol for the Protection of the Mediterranean Sea Against Pollution from Land-Based Sources, adopted at Athens on 17 May 1980; entered into force 17 June 1983.

[e]Protocol Concerning Mediterranean Specially Protected Areas, adopted at Geneva on 3 April 1982.

[f]Note.--According to the information available at UNEP, yet to be confirmed by the Depositary, Yugoslavia has acceded to the Protocol, and Israel has ratified it.

*with reservation

2. PERSIAN/ARABIAN GULF

	Kuwait Convention[a]		Emergency Protocol[b]	
	signed	in force	signed	in force
Bahrain	24 Apr 1978	1 Jul 1979	24 Apr 1978	1 Jul 1979
Iran	24 Apr 1978	1 Jun 1980	24 Apr 1978	1 Jun 1980
Iraq	24 Apr 1978	1 Jul 1979	24 Apr 1978	1 Jul 1979
Kuwait	24 Apr 1978	1 Jul 1979	24 Apr 1978	1 Jul 1979
Oman	24 Apr 1978	1 Jul 1979	24 Apr 1978	1 Jul 1979
Qatar	24 Apr 1978	1 Jul 1979	24 Apr 1978	1 Jul 1979
Saudi Arabia	24 Apr 1978	26 Mar 1982	24 Apr 1978	26 Mar 1982
United Arab Emirates	24 Apr 1978	1 Mar 1980	24 Apr 1978	1 Mar 1980

[a] Kuwait Regional Convention for Co-operation on the Protection of the Marine Environment from Pollution, adopted on 23 April 1978; entered into force 1 July 1979.

[b] Protocol Concerning Regional Co-operation in Combating Pollution by Oil and Other Harmful Substances in Cases of Emergency, adopted on 23 April 1978; entered into force 1 July 1979.

3. GULF OF GUINEA

	Abidjan Convention[a]		Emergency Protocol[b]	
	signed/acceded	in force	signed/acceded	in force
Angola				
Benin	23 Mar 1981		23 Mar 1981	
Cameroon	1 Mar 1983	5 Aug 1984	1 Mar 1983	5 Aug 1984
Cape Verde				
Congo	23 Mar 1981	19 Feb 1988	23 Mar 1981	19 Feb 1988
Cote d'Ivoire	23 Mar 1981	5 Aug 1984	23 Mar 1981	5 Aug 1984
Equatorial Guinea				
Gabon	23 Mar 1981	11 Feb 1989	23 Mar 1981	11 Feb 1989
Gambia	23 Mar 1981	5 Feb 1985	23 Mar 1981	5 Feb 1985
Ghana	23 Mar 1981	18 Sep 1989	23 Mar 1981	18 Sep 1989
Guinea	23 Mar 1981	5 Aug 1984	23 Mar 1981	5 Aug 1984
Guinea-Bissau				
Liberia	23 Mar 1981		23 Mar 1981	
Mauritania	22 Jun 1981		22 Jun 1981	
Namibia				
Nigeria	23 Mar 1981	5 Aug 1984	23 Mar 1981	5 Aug 1984
Sao Tome & Principe				
Senegal	23 Mar 1981	5 Aug 1984	23 Mar 1981	5 Aug 1984
Sierra Leone				
Togo	23 Mar 1981	5 Aug 1984	23 Mar 1981	5 Aug 1984
Zaire				

[a]Convention for Co-operation in the Protection and Development of the Marine and Coastal Environment of the West and Central African Region, adopted at Abidjan on 23 March 1981; entered into force 5 August 1984.

[b]Protocol Concerning Co-operation in Combating Pollution in Cases of Emergency in the West and Central African Region, adopted at Abidjan on 23 March 1981; entered into force 5 August 1984.

4. SOUTHEAST PACIFIC

	Lima Convention[a]		Emergency Agreement[b]		Supplementary Protocol[c]		Land-Based Sources Protocol[d]	
	signed	in force	signed	in force	signed	in force	signed	in force
Chile	12 Nov 1981	19 May 1986	12 Nov 1981	14 Jul 1986	22 Jul 1983	20 May 1987	22 Jul 1983	23 Sep 1986
Colombia	12 Nov 1981	19 May 1986	12 Nov 1981	14 Jul 1986	22 Jul 1983	20 May 1987	22 Jul 1983	23 Sep 1986
Ecuador	12 Nov 1981	19 May 1986	12 Nov 1981	14 Jul 1986	22 Jul 1983	11 Jan 1988	22 Jul 1983	11 Jan 1988
Panama	12 Nov 1981	21 Sep 1986	12 Nov 1981	21 Sep 1986	22 Jul 1983	20 May 1987	22 Jul 1983	23 Sep 1986
Peru	12 Nov 1981	25 Feb 1989	12 Nov 1981	18 Apr 1989	22 Jul 1983	18 Apr 1989	22 Jul 1983	25 Feb 1989

[a]Convention for the Protection of the Marine Environment and Coastal Area of the Southeast Pacific, adopted at Lima on 12 November 1981; entered into force 19 May 1986.

[b]Agreement on Regional Co-operation in Combating Pollution of the Southeast Pacific by Hydrocarbons or Other Harmful Substances in Cases of Emergency, adopted at Lima on 12 November 1981; entered into force 14 July 1986.

[c]Supplementary Protocol to the Agreement on Regional Co-operation in Combating Pollution of the Southeast Pacific by Hydrocarbons or Other Harmful Substances in Cases of Emergency, adopted at Quito on 22 July 1983; entered into force 20 May 1987.

[d]Protocol for the Protection of the Southeast Pacific Against Pollution from Land-Based Sources, adopted at Quito on 22 July 1983; entered into force 23 September 1986.

4. SOUTHEAST PACIFIC cont'd

	Protected Area Protocol[e]		Radioactive Contamination Protocol[f]	
	signed	in force	signed	in force
Chile	21 Sep 1989		21 Sep 1989	
Colombia	21 Sep 1989		21 Sep 1989	
Ecuador	21 Sep 1989		21 Sep 1989	
Panama	21 Sep 1989		21 Sep 1989	
Peru	21 Sep 1989		21 Sep 1989	

[e]Protocol for the Conservation and Management of Protected Marine and Coastal Areas of the Southeast Pacific, adopted at Paipa on 21 September 1989; not yet in force.

[f]Protocol for the Protection of the Southeast Pacific Against Radioactive Contamination, adopted at Paipa on 21 September 1989; not yet in force.

5.   RED SEA

	Jeddah Convention[a]		Emergency Protocol[b]	
	signed/acceded	in force	signed	in force
Democratic Yemen	14 Feb 1982		14 Feb 1982	
Egypt	21 May 1990	20 Aug 1990	21 May 1990	20 Aug 1990
Jordan	14 Feb 1982	7 Feb 1989	14 Feb 1982	7 Feb 1989
Palestine*	14 Feb 1982	20 Aug 1985	14 Feb 1982	20 Aug 1985
Saudi Arabia	14 Feb 1982	20 Aug 1985	14 Feb 1982	20 Aug 1985
Somalia	14 Feb 1982	30 May 1988	14 Feb 1982	30 May 1988
Sudan	14 Feb 1982	20 Aug 1985	14 Feb 1982	20 Aug 1985
Yemen Arab Republic	14 Feb 1982	20 Aug 1985	14 Feb 1982	20 Aug 1985

[a]Regional Convention for the Conservation of the Red Sea and Gulf of Aden Environment, adopted at Jeddah on 14 February 1982; entered into force 20 August 1985.

[b]Protocol Concerning Regional Co-operation in Combating Pollution by Oil and Other Harmful Substances in Cases of Emergency, adopted at Jeddah on 14 February 1982; entered into force 20 August 1985.

6. CARIBBEAN

	Cartagena Convention[a]		Oil Spills Protocol[b]		Protected Areas Protocol[c]	
	signed/acceded	in force	signed/acceded	in force	signed/acceded	in force
Antigua & Barbuda	11 Sep 1986	11 Oct 1986	11 Sep 1986	11 Oct 1986	18 Jan 1990	
Bahamas						
Barbados	5 Mar 1984	11 Oct 1986	5 Mar 1984	11 Oct 1986		
Colombia	24 Mar 1983	2 Apr 1988	24 Mar 1983	2 Apr 1988	18 Jan 1990	
Costa Rica						
Cuba	15 Sep 1988	15 Oct 1988	15 Sep 1988	15 Oct 1988	18 Jan 1990	
Dominica						
Dominican Republic						
EEC	24 Mar 1983		24 Mar 1983			
France	24 Mar 1983	11 Oct 1986*	24 Mar 1983	11 Oct 1986*	18 Jan 1990	
Grenada	24 Mar 1983	16 Sep 1987	24 Mar 1983	16 Sep 1987		
Guatemala	5 Jul 1983	17 Jan 1990	5 Jul 1983	17 Jan 1990	18 Jan 1990	
Guyana						
Haiti						
Honduras	24 Mar 1983	1 May 1987	24 Mar 1983	1 May 1987	18 Jan 1990	
Jamaica	24 Mar 1983	11 Oct 1986	24 Mar 1983	11 Oct 1986	18 Jan 1990	
Mexico	24 Mar 1983	11 Oct 1986	24 Mar 1983	11 Oct 1986	18 Jan 1990	
Netherlands[d]	24 Mar 1983		24 Mar 1983			
Nicaragua	24 Mar 1983		24 Mar 1983			
Panama	24 Mar 1983	6 Nov 1987	24 Mar 1983	6 Nov 1987		
St. Christopher and Nevis						
St. Lucia	24 Mar 1983	11 Oct 1986	24 Mar 1983	11 Oct 1986	18 Jan 1990	
St. Vincent and Grenadines						
Suriname						
Trinidad & Tobago	24 Jan 1986	11 Oct 1986	24 Jan 1986	11 Oct 1986	18 Jan 1990	
United Kingdom[e]	24 Mar 1983	11 Oct 1986*	24 Mar 1983	11 Oct 1986*	18 Jan 1990	
USA	24 Mar 1983	11 Oct 1986	24 Mar 1983	11 Oct 1986	18 Jan 1990	
Venezuela	24 Mar 1983	17 Jan 1987	24 Mar 1983	17 Jan 1987	18 Jan 1990	

[a] Convention for the Protection and Development of the Marine Environment of the Wider Caribbean Region, adopted at Cartagena on 24 March 1983; entered into force 11 October 1986.

[b] Protocol Concerning Co-operation in Combating Oil Spills in the Wider Caribbean Region, adopted at Cartagena on 24 March 1983; entered into force 11 October 1986.

[c] Protocol concerning specially protected areas and wildlife to the Convention for the Protection and Development of the Marine Environment of the Wide Caribbean Region, adopted at Kingston 18 January 1990; not yet in force.

[d] On behalf of Aruba and the Netherlands Antilles Federation.

[e] On behalf of the Cayman Islands and the Turks and Caicos Island, reserving the right to extend it at a future date to include the other territories of the United Kingdom particpating in the Caribbean Action Plan (Anguilla, British Virgin Islands, and Montserrat)

*with reservation

7. INDIAN OCEAN

	Nairobi Convention[a]		Protected Areas Protocol[b]		Emergency Protocol[c]	
	signed	in force	signed	in force	signed	in force
Comoros						
EEC	19 Jun 1986				19 Jun 1986	
France[d]	21 Jun 1985	18 Aug 1989	21 Jun 1985	18 Aug 1989	21 Jun 1985	18 Aug 1989
Kenya						
Madagascar	21 Jun 1985		21 Jun 1985		21 Jun 1985	
Mauritius						
Mozambique						
Seychelles	21 Jun 1985*		21 Jun 1985*		21 Jun 1985*	
Somalia	21 Jun 1985*		21 Jun 1985*		21 Jun 1985*	
United Republic of Tanzania						

[a]Convention for the Protection, Management and Development of the Marine and Coastal Environment of the Eastern African Region, adopted at Nairobi on 21 June 1985; not yet in force.

[b]Protocol Concerning Protected Areas and Wild Fauna and Flora in the Eastern African Region, adopted at Nairobi on 21 June 1985; not yet in force.

[c]Protocol Concerning Co-operation in Combating Marine Pollution in Cases of Emergency in the Eastern African Region, adopted at Nairobi on 21 June 1985; not yet in force.

[d]On behalf of Reunion

*According to the information available at UNEP, yet to be confirmed by the Depositary, Seychelles and Somalia have ratified the Convention and both protocols.

8. SOUTHWEST PACIFIC

	Noumea Convention[a]		Emergency Protocol[b]		Dumping Protocol[c]	
	signed/acceded	in force	signed/acceded	in force	signed/acceded	in force
Australia	24 Nov 1987		24 Nov 1987		24 Nov 1987	
Cook Islands	25 Nov 1986		25 Nov 1986		25 Nov 1986	
Federated States of Micronesia	9 Apr 1987		9 Apr 1987		9 Apr 1987	
Fiji	18 Sep 1989		18 Sep 1989		18 Sep 1989	
France	25 Nov 1986		25 Nov 1986		25 Nov 1986	
Kiribati						
Marshall Islands	25 Nov 1986		25 Nov 1986		25 Nov 1986	
Nauru	15 Apr 1987		15 Apr 1987		15 Apr 1987	
New Zealand	25 Nov 1986		25 Nov 1986		25 Nov 1986	
Palau	25 Nov 1986		25 Nov 1986		25 Nov 1986	
Papua New Guinea	3 Nov 1987		3 Nov 1987		3 Nov 1987	
Solomon Islands	10 Aug 1989		10 Aug 1989		10 Aug 1989	
Tonga						
Tuvalu	14 Aug 1987		14 Aug 1987		14 Aug 1987	
United Kingdom	16 Jul 1987		16 Jul 1987		16 Jul 1987	
USA	25 Nov 1986		25 Nov 1986		25 Nov 1986	
Vanuatu						
Western Samoa	25 Nov 1986+		25 Nov 1986+		25 Nov 1986+	

[a]Convention for the Protection of the Natural Resources and Environment of the South Pacific Region, adopted at Noumea on 25 November 1986; not yet in force.

[b]Protocol Concerning Co-operation in Combating Pollution Emergencies in the South Pacific Region, adopted at Noumea on 25 November 1986; not yet in force.

[c]Protocol for the Prevention of Pollution of the South Pacific Region by Dumping, adopted at Noumea on 25 November 1986; not yet in force.

+Note.--According to the information available at UNEP, yet to be confirmed by the Depositary, Western Samoa has ratified the Convention. With ratification of the Convention by Western Samoa, the number of Parties would increase to ten, i.e., the requisite number for the Convention to enter into force.

Appendix 4G:　　　UNEP Global Marine Environment Guidelines

c.　　Conclusions of the study of legal aspects concerning the environment related to
　　　offshore mining and drilling within the limits of national jurisdiction

Endorsed by decision 10/14(VI) of the Governing Council of UNEP on 31 May 1982, with the
participation of the following Member States:

Argentina*	Mexico
Bangladesh	Morocco*
Belgium	Netherlands*
Botswana	New Zealand
Brazil*	Oman
Bulgaria	Pakistan
Burundi	Peru
Byelorussian Soviet Socialist	Poland*
Republic	Saudi Arabia
Canada*	Senegal
Chile	Spain
China	Sri Lanka
Colombia*	Sudan*
Egypt	Sweden*
Ethiopia	Switzerland*
France*	Thailand
Gabon*	Ukrainian Soviet Socialist
Germany, Federal Republic of*	Republic
Ghana*	Union of Soviet Socialist
Greece*	Republics*
Guinea	United Arab Emirates
Iceland	United Kingdom of Great Britain and
India*	Northern Ireland*
Indonesia	United Republic of Tanzania
Jamaica*	United States of America*
Japan*	Uruguay
Kenya	Venezuela*
Libyan Arab Jamahiriya	Yugoslavia
Malaysia	Zaire

*Members of the Working Group of Experts on Environmental Law, which drafted the
conclusions and adopted them at its eighth session in Geneva on 13 February 1981
(UNEP/WG.54/4).

In addition, the following States which were not members of the UNEP Governing Council
at the time of endorsement of the conclusions also particpated in the Working Group:
Australia, Austria, Finland, Iraq, Nigeria, Norway, and Tunisia.

Appendix 4G: UNEP Global Marine Environment Guidelines

d: Montreal guidelines for the protection of the marine environment
            against pollution from land-based sources

Endorsed by decision 13/18(II) of the Governing Council of UNEP on 24 May
1985, with the participation of the following Member States:

Algeria	Malaysia
Argentina*	Malta
Australia*	Mexico*
Austria	Nepal
Belgium*	Nigeria*
Botswana	Norway
Brazil*	Oman
Bulgaria*	Panama
Canada*	Papua New Guinea
Chile*	Peru*
China	Philippines*
Colombia*	Poland
Finland*	Rwanda
France*	Saudi Arabia
Germany, Federal Republic of*	Sri Lanka*
Ghana*	Sudan
Hungary	Tunisia
India*	Turkey
Indonesia*	Uganda
Italy*	Ukrainian Soviet Socialist
Ivory Coast	Republic
Jamaica	Union of Soviet Socialist
Japan	Republics*
Jordan	United Kingdom of Great Britain and
Kenya*	Northern Ireland*
Kuwait*	United States of America*
Lesotho	Venezuela*
Libyan Arab Jamahiriya*	Yugoslavia*
	Zaire

*Participated in the <u>Ad Hoc</u> Working Group of Experts on the Protection of the
Marine Environment Against Pollution from Land-Based Sources, which drafted
the guidelines and adopted them at its third session in Montreal on 19 April
1985 (UNEP/WG.120/3).

In addition, the following States which were not members of the UNEP Governing
Council at the time of adoption of the guidelines also participated in the
Working Group: Bolivia, Cameroon, Dominican Republic, Egypt, Gabon, German
Democratic Republic, Greece, Honduras, Malawi, The Netherlands, Pakistan,
Portugal, Sweden, Tanzania, and Uruguay.

Appendix 5G: UNEP-administered Cites and Bonn Conventions

    a.   GLOBALLY PROTECTED MARINE SPECIES (CITES and CMS)

Listed on appendices I, II, and III of the UNEP-administered 1973 Washington Convention on International Trade in Endangered Species of Wild Fauna and Flora (CITES), in force since 1 July 1975 (<u>United Nations Treaty Series</u> 993/243), as amended at the 6th meeting of the Conference of the Parties (Ottawa 1987), in force 22 October 1987:

SCIENTIFIC NAME	COMMON NAME
<u>Cetacea</u> spp. II (Bonn I/II)*	Whales, dolphins, and porpoises
<u>Physeter macrocephalus (catadon)</u> I	Sperm whale
<u>Berardius</u> spp. I, <u>Hyperoodon</u> spp. I	Bottlenose whales
<u>Sotalia</u> spp. I, <u>Sousa</u> spp. I	Humpback dolphins
<u>Neophocoena phoncoenoides</u> I	Indian finless porpoise
<u>Phocoena sinus</u> I	Gulf of California harbor porpoise
<u>Eschrichtius robustus</u> I	Grey whale
<u>Balaenoptera acutorostrata</u> I	
(West Greenland population II)	Minke whale
<u>Balaenoptera borealis</u> I	Sei whale
<u>Balaenoptera edeni</u> I	Bryde's whale
<u>Balaenoptera musculus</u> I (Bonn I)	Blue whale
<u>Megaptera novaeangliae</u> I (Bonn I)	Humpback whale
<u>Balaenoptera physalus</u> I	Fin whale
<u>Balaena</u> spp. I (CMS I)	Right whales
<u>Enhydra lutris nereis</u> I	Southern sea otter
<u>Lutra felina</u> I	Marine otter
<u>Arctocephalus</u> spp. II	Southern fur seals
<u>Arctocephalus townsendi</u> I	Guadalupe fur seal
<u>Mirounga</u> spp. II	Elephant seals
<u>Monachus</u> spp. I (Bonn I/II)	Monk seals
<u>Odobenus rosmarus</u> III (Canada)	Walrus
<u>Dugong dugon</u> I	
(Australian population II) (Bonn II)	Dugong
<u>Trichechus inunguis</u> I	South American manatee
<u>Trichechus manatus</u> I	Caribbean manatee
<u>Trichechus senegalensis</u> II	West African manatee
<u>Cheloniidae</u> spp. I (Bonn I/II)	Sea turtles
<u>Dermochelys coriacea</u> I (Bonn I/II)	Leatherback turtle
<u>Crocodylus porosus</u> I (Australian,	
Indonesian and New Guinean populations	
II) (Bonn II)	Saltwater crocodile
<u>Amblyrhynchus cristatus</u> II	Galapagos marine iguana
<u>Latimeria chalumnae</u> II	Coelacanth fish
<u>Acipenser brevirostrum</u> I	Shortnose sturgeon
<u>Acipenser sturio</u> I	Baltic sturgeon
<u>Acipenser oxyrhynchus</u> II	Atlantic sturgeon
<u>Tridacnidae</u> spp. II	Giant clams

a.   GLOBALLY PROTECTED MARINE SPECIES (CITES and CMS)   cont'd

<u>Antipatharia</u> spp. II	Black corals
<u>Seriatopora</u> spp. II	Bush corals
<u>Pocillopora</u> spp. II	Cauliflower corals
<u>Stylopora</u> spp. II	Hood corals
<u>Acropora</u> spp. II	Flower corals
<u>Pavona</u> spp. II	Leat corals
<u>Fungia</u> spp. II	Mushroom corals
<u>Halomitra</u> spp. II	Bowl corals
<u>Polyphyllia</u> spp. II	Boomerang corals
<u>Favia</u> spp. II	Knob corals
<u>Platygyra</u> spp. II	Valley corals
<u>Merulina</u> spp. II	Crispy crust corals
<u>Lobophyllia</u> spp. II	Lobed cup corals
<u>Pectinia</u> spp. II	Lettuce corals
<u>Euphyllia</u> spp. II	Bear corals
<u>Millepora</u> spp. II	Fire corals
<u>Heliopora</u> spp. II	Blue corals
<u>Tubipora</u> spp. II	Organ pipe corals

---

*Bonn = Also listed, wholly or partly, in appendices I and/or II of the UNEP-administered 1979 Bonn Convention on the Conservation of Migratory Species of Wild Animals (Bonn) in force since 1 November 1983 (<u>International Legal Materials</u> 19/15), as amended at the first meeting of the Conference of the Parties (Bonn 1985), in force 24 January 1986. Appendix II of Bonn, in addition, lists the following marine species not currently listed under CITES: <u>Phoca vitulina</u> (Harbor seal, Baltic and Wadden Sea population) and <u>Halichoerus grypus</u> (Grey seal, Baltic Sea population).

NOTE that the present list is arbitrarily restrictive, as it only contains aquatic fauna and does not include other terrestrial species (e.g., sea birds) also listed under CITES and/or Bonn, which because of their predominantly maritime/coastal habitat range may equally be considered as 'marine species' in a wider sense.

Appendix 5G: UNEP-administered CITES and Bonn Conventions

b.   List of Parties to the Convention on International Trade in Endangered
         Species of Wild Fauna and Flora (CITES)*

Afghanistan	Guatemala	Peru*
Algeria	Guinea	Philippines
Argentina	Guyana	Portugal (Bonn)
Australia	Honduras	Rwanda
Austria*	Hungary (Bonn)	Saint Lucia
Bahamas	India (Bonn)	Senegal (Bonn)
Bangladesh	Indonesia	Seychelles
Belgium	Iran, Islamic Rep.	Singapore*
Belize	Israel (Bonn)	Somalia (Bonn)
Benin (Bonn)	Italy (Bonn)	South Africa
Bolivia	Japan*	Spain
Botswana	Jordan	Sri Lanka
Brazil*	Kenya	Sudan
Cameroon (Bonn)	Liberia	Suriname*
Canada	Liechtenstein	Sweden (Bonn)
Central African Republic	Luxembourg (Bonn)	Switzerland
Chile (Bonn)	Madagascar	Tanzania, United Rep.
China	Malawi	Thailand
Colombia	Malaysia	Togo
Congo	Mauritius	Trinidad and Tobago
Costa Rica	Monaco	Tunisia (Bonn)
Cyprus	Morocco	Union of Soviet
Denmark (Bonn)	Mozambique	Socialist Republics
Dominican Republic	Nepal	United Kingdom
Ecuador	Netherlands (Bonn)	of Great Britain and
Egypt (Bonn)	Nicaragua	Northern Ireland (Bonn)
El Salvador	Niger (Bonn)	United States of America
Finland	Nigeria (Bonn)	Uruguay
France	Norway* (Bonn)	Venezuela
Gambia	Pakistain (Bonn)	Zaire
German Democratic Rep.	Panama	Zambia
Germany, Federal Rep. (Bonn)	Papua New Guinea	Zimbabwe
Ghana (Bonn)	Paraguay	

* = With reservations concerning marine species.

Bonn = Also Party to the UNEP-administered 1979 Bonn Convention on the Conservation of
Migratory Species of Wild Animals (Bonn), together with the following Bonn Parties that
are not Parties to CITES: Ireland, Mali, and the European Economic Community (however,
the EEC already implements CITES on the basis of Council Regulation EEC/3626/82).

Table 6G — MARITIME CLAIMS OF COUNTRIES, BY JURISDICTION

Country or Territory	Area (km2)	Coastline (km)	Ports (no.) Major	Ports (no.) Minor	Territorial Sea	Contiguous Zone	Jurisdictional Claims (nm) Exclusive Economic Zone	Exclusive Fishing Zone	Continental Shelf	Security Zone	Pollution Zone
Albania	28,750	362	1	3	15	-	-	-	-	15	-
Algeria	2,381,740	998	12	11	12	-	-	-	-	-	-
Angola	1,246,700	1,600	3	5	20	-	-	200	-	-	-
Antigua & Barbuda	440	153	1	1	12	24	200	-	-	-	-
Argentina	2,766,890	4,989	7	30	200/a	-	-	-	b	-	c
Australia	7,686,850	25,760	12	-	3	-	-	200	b	-	c
Bahamas, The	13,940	3,542	2	9	3	-	-	200	-	-	-
Bahrain	620	161	1	1	3	-	-	-	d	-	-
Bangladesh	144,000	580	2	7	12	18	200	-	e	f	200c
Barbados	430	97	1	2	12	-	200	-	-	-	-
Belgium	30,510	64	6	1	12	-	-	200	d	-	c
Belize	22,960	386	2	6	3	-	-	-	-	-	-
Benin	112,620	121	1	-	200	-	-	-	-	-	-
Bermuda	50	103	3	-	3	-	-	200	b	-	-
Brazil	8,511,970	7,491	8	23	200	-	-	-	b	-	c
Brunei Darussalam	5,770	161	1	4	12	-	-	200	-	-	-
Bulgaria	110,910	354	3	6	12	-	200	-	b	-	-
Burma	676,550	3,060	4	6	12	24	200	-	g	f	200c
Cameroon	475,440	402	1	3	50	-	-	-	g	-	-
Canada	9,976,140	243,791	25	225	12	-	-	200	b	-	c(1)

Cape Verde	4,030	965	2	2	12/h	-	200	-	-	-	-	-
Chile	756,950	6,435	10	13	12	24	200	-	i/j	-	-	-
China	9,596,960	14,500	15	180	12	-	-	-	d	-	-	-
Colombia	1,138,910	3,208	5	5	12	-	200	-	b	-	-	c
Comoros	2,170	340	1	2	12/h	-	200	-	-	-	-	-
Congo	342,000	169	2	-	200	-	-	-	-	-	-	-
Cook Islands	240	120	-	2	12	-	200	-	g	-	-	-
Costa Rica	50,900	1,290	1	4	12	-	200	-	i	-	-	-
Cuba	110,860	3,735	20	50	12	-	200	-	i	-	-	-
Cyprus	9,250	648	3	11	12	-	-	-	k	-	-	-
Dem. Kampuchea	181,040	443	2	5	12	24	200	-	i	-	-	-
Denmark	43,070	3,379	19	41	3	4	-	200	1	-	-	-
Djibouti	22,000	314	1	-	12	24	200	-	-	-	-	-
Dominica	750	148	1	1	12	24	200	-	-	-	-	-
Dominican Rep.	48,730	1,288	4	17	6	24	200	-	g	-	-	-
Ecuador	283,560	2,237	4	6	200	-	-	-	i/m(2)	-	-	-
Egypt	1,001,450	2,450	5	15	12	18	200	-	b	-	f	-
El Salvador	21,040	307	2	1	200/a	-	-	-	-	-	-	-
Eq. Guinea	28,050	296	1	3	12	-	200	-	-	-	-	-
Ethiopia	1,221,900	1,094	2	-	12	-	-	-	-	-	-	-
Falkland Is.	12,170	1,288	1	4	3	-	-	200/n	b	-	-	-
Faero Is.	1,400	764	2	8	3	4	-	200	b	-	-	-
Fiji	18,270	1,129	1	6	12	-	200	-	b	-	-	-
Finland	337,030	1,126	11	34	4	6	-	12	b	-	-	-

TABLE 6G — (continued)

Country or Territory	Area (km2)	Coastline (km)	Ports (no.) Major	Ports (no.) Minor	Territorial Sea	Contiguous Zone	Jurisdictional Claims (nm) Exclusive Economic Zone	Exclusive Fishing Zone	Continental Shelf	Security Zone	Pollution Zone
France	547,030	3,427	26	6	12	24	200	-	b	-	-
French Guiana	91,000	378	1	7	12	-	200	-	b	-	-
Fr. Polynesia	3,941	2,525	1	6	12	-	200	-	b	-	-
Gabon	267,670	885	2	3	12	24	200	150	-	-	-
Gambia, The	11,300	80	1	-	12	18	-	200	i	-	-
German Dem. Rep.	108,330	901	4	13	12	-	-	200	b	-	-
Germany, F. Rep.	248,580	1,488	10	11	3/p	-	-	200	b	-	-
Ghana	238,540	539	2	-	12	24	200	-	q	-	-
Gibraltar	7	12	1	-	3	-	-	-	b	-	-
Greece	131,940	13,676	15	42	6	-	-	-	b	-	-
Greenland	2,175,600	44,087	8	14	3	4	-	200	b	f	-
Grenada	340	121	1	1	12	-	200	-	-	-	-
Guadeloupe	1,780	306	1	3	12	-	-	-	b	-	-
Guatemala	108,890	400	2	3	12	-	200	-	b	-	-
Guinea	245,860	320	1	2	12	-	200	-	-	-	-
Guinea-Bissau	36,120	350	1	-	12	-	200	-	-	-	-
Guyana	214,970	459	1	6	12	-	-	200	g	-	-
Haiti	27,750	1,771	2	12	12	24	200	-	b	-	-
Honduras	112,090	820	1	4	12	24	200	-	b	-	-
Hong Kong	1,040	733	1	-	3	-	-	-	b	-	-

Iceland	103,000	4,988	4	50	12	–	200	–	g/a	.	–
India	3,287,590	7,000	9	79	12	24	200	–	g/a	f	–
Indonesia	1,919,440	54,716	15	70	12/h	–	200	–	b	–	–
Iran	1,648,000	3,180	6	12	12	–	–	50/r	f	–	–
Iraq	434,920	58	3	–	12	–	–	–	d	–	o
Ireland	70,280	1,448	8	38	12	–	–	200	d	–	–
Israel	20,770	273	3	5	6	–	–	–	b/s	–	–
Italy	301,230	4,996	20	40	12	–	–	–	b	–	o
Ivory Coast	322,460	515	2	1	12	–	200	–	i	–	–
Jamaica	10,990	1,022	2	10	12	–	–	–	b	–	–
Japan	372,310	13,685	132	2,000	12/t	–	–	200	–	–	–
Jordan	91,880	26	1	–	3	–	200	–	b	–	–
Kenya	582,650	536	1	–	12	–	200	–	–	–	–
Kiribati	717	1,143	2	–	12	–	200	–	–	50	–
Korea D.P.R.	120,540	2,495	6	26	12	–	200	200	d	–	o
Korea, Rep.	98,480	2,413	11	32	12/u	–	–	–	d	–	–
Kuwait	17,820	499	3	6	12	–	–	–	–	f	–
Lebanon	10,400	225	2	6	12	–	–	–	b	–	–
Liberia	111,370	579	1	6	200	–	–	–	–	(4)	–
Libyan Arab Jamahiriya	1,759,540	1,770	6	21	12	–	–	–	–	–	–
Macau	16	40	1	–	6	–	12	–	–	–	–
Madagascar	587,040	4,828	4	–	12	–	200	–	i/m	–	–
Malaysia	329,750	4,675	6*	26	12	–	–	200	b	–	–

TABLE 6G — (continued)

Country or Territory	Area (km2)	Coastline (km)	Ports (no.) Major	Ports (no.) Minor	Territorial Sea	Contiguous Zone	Jurisdictional Claims (nm) Exclusive Economic Zone	Exclusive Fishing Zone	Continental Shelf	Security Zone	Pollution Zone
Maldives	300	644	-	2	12	-	-	-	v	-	-
Malta	320	140	3	1	12	24	-	25	b	-	-
Martinique	1,100	290	1	5	12	-	-	-	b	-	-
Mauritania	1,030,700	754	2	-	12	-	200	-	g	-	-
Mauritius	1,860	177	1	-	12	-	200	-	g	-	-
Mexico	1,972,550	9,330	11	20	12	-	200	-	b	-	-
Monaco	2	4	-	1	12	-	-	-	-	-	c
Morocco	446,550	1,835	9	15	12	24	200	-	b	-	c
Mozambique	801,590	2,470	3	2	12	-	200	-	-	-	-
Namibia	824,290	1,489	1	-	6	-	-	12	-	-	-
Nauru	20	24	-	1	12	-	-	200	-	-	-
Netherlands	37,310	451	10	2	12	12	-	200	b	-	-
Neth. Antilles	960	364	3	6	12	-	-	-	-	-	-
New Caledonia	19,060	2,254	1	21	12	-	200	-	b	-	-
New Zealand	268,680	15,134	3	3	12	-	200	-	g	-	c
Niue	260	64	-	-	12	-	200	-	-	-	-
Nicaragua	129,494	910	6	11	200	-	-	-	i	-	-
Nigeria	923,770	853	6	-	30	-	200	-	b	-	-
Norway	324,220	21,925	20	58	4	10	200	-	b	-	-
Oman	212,460	2,092	2	5	12	-	200	-	b	-	-

Pakistan	803,940	1,046	2	4	12	-	200	-	g/q	-	-
Panama	78,200	2,490	2	8	200	-	-	-	-	-	-
Papua New Guinea	461,690	5,152	5	9	12/h	-	200	200	b	-	c
Peru	1,285,220	2,414	7	25	200	-	-	-	g	-	-
Philippines	300,000	36,289	10	-	w/h	-	200	200	d/s	-	c
Poland	312,680	491	4	12	12	-	-	200	b	-	-
Portugal	92,080	1,793	7	34	12	-	200	-	b	f	-
Qatar	11,000	563	2	1	3	-	x	200	d	-	-
Reunion	2,510	201	1	-	12	-	200	-	b	-	-
Romania	237,500	225	4	7	12	-	200	-	b	-	-
St. Lucia	620	158	1	1	3	-	200	12	a/g	-	-
St. Vincent and the Grenadines	340	84	1	1	3	24	200	-	-	-	c
São Tomé & Príncipe	960	209	1	-	12/h	-	200	-	-	-	-
Saudi Arabia	2,149,690	2,510	7	17	12	18	-	-	d	-	-
Senegal	196,190	531	1	2	12	24	200	-	g	-	-
Seychelles	455	491	1	-	12	-	200	-	g	-	c
Sierra Leone	71,740	402	1	2	200	-	-	-	b	-	-
Singapore	580	193	3	-	3	-	-	12	-	-	c
Solomon Islands	28,450	5,313	-	5	12/h	-	200	-	-	-	-
Somalia	637,660	3,025	3	-	200	-	-	-	-	-	-
South Africa	2,881	2,881	8	-	12	-	-	200	b	-	-
Spain	504,750	4,964	23	175	12	-	200	-	b	-	-
Sri Lanka	65,610	1,340	3	9	12	24	200	-	g	f	-

TABLE 6G — (continued)

Country or Territory	Area (km2)	Coastline (km)	Ports (no.) Major	Ports (no.) Minor	Territorial Sea	Contiguous Zone	Jurisdictional Claims (nm) Exclusive Economic Zone	Exclusive Fishing Zone	Continental Shelf	Security Zone	Pollution Zone
Sudan	2,505,810	853	1	-	12	18	-	-	b	f	-
Suriname	163,270	386	1	6	12	-	200	-	-	-	-
Sweden	449,960	3,218	17	30	12	-	-	200	b	-	-
Syrian Arab Rep.	185,180	193	3	2	35	-	-	-	b	-	-
Taiwan	35,980	1,448	5	4	12	-	200	-	-	-	-
Tanzania, United Rep.	945,090	1,424	3	-	12	-	-	-	b	-	-
Thailand	514,000	3,219	2	16	12	-	200	-	b	-	-
Togo	56,790	56	1	1	30	-	200	-	-	-	-
Tonga	700	419	-	2	12	-	200	-	b	-	-
Trinidad & Tobago	5,130	362	1	8	12/h	-	200	-	b	-	-
Tunisia	163,610	1,148	7	14	12	-	-	-	-	-	-
Turkey	780,580	7,200	14	18	6/x	-	200/x	-	-	-	-
Tuvalu	26	24	-	2	12	-	200	-	-	-	-
USSR	22,402,200	42,777	53	180	12	-	200	-	b	-	c
United Arab Emirates	83,600	1,448	7	25	3(4)	-	200	-	-	-	-
United Kingdom	244,820	12,429	25	180	12/y	-	-	200	b	-	-
USA	9,372,610	19,924	44	-	12	12	200	-	b	-	c
Uruguay	176,220	660	1	9	200/a	-	-	-	b	-	c
Vanuatu	14,760	2,528	-	3	12	24	200	-	g	-	c
Venezuela	912,050	2,800	6	17	12	15	200	-	b	-	-

Vietnam	329,560	9	23	12	24	200	-	g	f	-
Wallis & Futuna	274	-	2	12	-	200	-	b	-	-
Western Sahara	266,000	-	2	z	-	-	-	-	-	-
Western Samoa	2,860	1	1	12	-	200	-	-	-	-
Yemen, Arab Rep. (North Yemen/Sanaa)	195,000	1	3	12	18	-	-	i	-	-
Yemen, Dem. Rep. (South Yemen/Aden)	332,970	1	5	12	24	200	-	g	f	c
Yugoslavia	255,800	9	24	12	-	-	-	b	-	c
Zaire	2,345,410	2	1	12	-	-	200	-	-	-

Sources:

Central Intelligence Agency, *The World Factbook 1988* (Washington, D.C.: US Government Printing Office, 1988), CPAS WF 88-001, May 1988.

U.S. Department of State, Bureau of Intelligence and Research, Office of the Geographer, National Claims to Maritime Jurisdiction, 5th Revision, <u>Limits of the Seas</u>, no. 36. edited by Robert W. Smith (Washington, D.C.: Department of State, May 1985).

Law of the Sea Bulletin (New York: United Nations, Office for Ocean Affairs and the Law of the Sea, 1988), Nos. 11-13.

U.S. Department of State, Bureau of Oceans and International Environmental and Scientific Affairs, "Notice to Research Vessel Operators No. 61 (Rev. 6): Claimed Maritime Jurisdictions" (Washington, D.C.: Department of State, June 14, 1989).

NOTES:

a Overflight and navigation permitted beyond 12 nm
b Claims continental shelf to 200 meters (m) or depth of exploitability
c Claims pollution zone
d Claims continental shelf without specific limits
e Claims continental shelf to the outer limits of the continental margin
f Claims security zone (breadth given if known)
g Claims Continental Shelf to the edge of the continental margin or 200 nm
h Claims archipelagic waters
i Claims continental shelf to 200 nm
j Continental shelf of 350 nm applies to Sala y Gomez and Easter Island
k Continental shelf beyond 200 m if part of the natural prolongation of the land territory
l 1958 Continental Shelf Convention definition
m Continental shelf to 200 nm or 100 nm from the 2,500 m isobath
n Exclusive Fishing Zone of 200 nm enforced to only 150 nm (February 1, 1987)
p Territorial Sea of 3 nm but extends to 16 nm at one point in the Helgolander Bucht (by decree of November 12, 1984 for the prevention of tanker casualties)
q Claims continental shelf to 100 fathoms or exploitability
r Exclusive Fishing Zone of 50 nm in the Sea of Oman and to median-line boundaries in the Persian Gulf
s Continental Shelf claimed to depth of exploitability
t Territorial Sea of 12 nm, but 3 nm in the international straits - La Perouse or Soya, Tsugaru, Osumi, and Eastern and Western Channels of Tsushima or Korea Strait
u Territorial Sea of 12 nm, and 3 nm in the Korea Strait
v Exclusive Economic Zone of from 37 nm to 310 m
w Territorial Sea of an irregular polygon extending up to 100 nm from the coastline as defined by 1898 treaty; since late 1970s has also claimed polygonal-shaped area in South China Sea up to 285 nm in breadth
x Territorial Sea of 6 nm (12 nm in the Black Sea and Mediterranean Sea); Exclusive Economic Zone of 200 nm in the Black Sea only
y Territorial Sea delimited by loxodromes between England and France (November 2, 1988)
z Maritime claims contingent upon resolution of sovereignty issue

(1) Claims a pollution zone in the Arctic region
(2) Continental shelf of 100 nm from the 2,500 meter isobath applies to the Galapagos Islands
(3) Claims non-specific security zone
(4) Claims Gulf of Sidra as Libyan sovereign territory; Gulf of Sidra closing line at 32° 30' N
(5) Territorial Sea of 12 nm claimed by Sharjah

TABLE 7G -- Large Oil Spills: A List of 66 Spills Greater Than 2 Million Gallons, 1967 to Present

No.	Date	Spill	Location	Volume (millions of gallons)	Ref(s)
1	1979-1980	Ixtoc 1, Well Blowout	Mexico	139-428*	abgh
2	1983	Nowruz Oil Field, Well Blowout(s)	Persian Gulf	80-185	ab
3	1983	Castillo de Bellver/Broke, Fire	South Africa	50-80*	abe
4	1978	Amoco Cadiz/Grounding	France	67-76	abfhm
5	1979	Aegean Captain/Atlantic Empress	off Tobago	49*	abl
6	1980-1981	D-103 Libya, Well Blowout	Libya	42	a
7	1979	Atlantic Empress/Fire	Barbados	41.5*	abl
8	1967	Torrey Canyon/Grounding	England	35.7-38.6*	bcf
9	1980	Irenes Serenade/Fire	Greece	12.3-36.6*	am
10	1972	Sea Star/Collision, Fire	Gulf of Oman	35.3*	bf
11	1981	Kuwait Nat'l Petroleum Tank	Kuwait	31.2	a
12	1976	Urquiola/Grounding	Spain	27-30.7*	bf
13	1970	Othello/Collision	Sweden	18.4-30.7	bcf
14	1977	Hawaiian Patriot/Fire	N Pacific	30.4*	bf
15	1979	Independenta	Turkey	28.9	a
16	1978	No. 126 Well/Pipe	Iran	28	a
17	1975	Jakob Maersk	Portugal	25*	f
18	1985	BP Storage Tank	Nigeria	23.9	a
19	1985	Nova/Collision	Iran	21.4	a
20	1978	BP, Shell Fuel Dept.	Zimbabwe	20	a
21	1971	Wafra	South Africa	19.6*	cf
22	1989	Kharg 5, Explosion	Morocco	19	q
23	1974	Metula/Grounding	Chile	16	cf
24	1983	Assimi/Fire	off Oman	15.8*	c
25	1970	Polycommander	Spain	3-15.3	c
26	1978	Tohoku Storage Tanks, Earthquake	Japan	15	a
27	1978	Andros Patria	Spain	14.6	a
28	1983	Pericles GC	Qatar	14	a
29	1985	Ranger, TX, Well Blowout	Texas	6.3-13.7	bk
30	1968	World Glory/Hull Failure	South Africa	13.5	bcf
31	1970	Ennerdale/Struck Granite	Seychelles	12.6	cf
32	1974	Mizushima Refinery, Tank Rupture	Japan	11.3	cdf
33	1973	Napier	SE Pacific	11*	f
34	1980	Juan A. Lavalleja	Algeria	11	a
35	1989	Exxon Valdez/Grounding	Alaska	10.8	i
36	1978	Turkish Petroleum Corporation	Turkey	10.7	a
37	1979	Burmah Agate/Collision, Fire	Texas	1.3-10.7*	abo
38	1971	Texaco Oklahoma, 120 mi. offshore	North Carolina	9.2-10.7	cf
39	1972	Trader	Mediterranean	10.4	f
40	1976	St. Peter	SE Pacific	10.4	f
41	1977	Irene's Challenge	Pacific	10.4	f

TABLE 7G -- Large Oil Spills   cont'd

No.	Date	Spill	Location	Volume (millions of gallons)	Ref(s)
42	1972	Golden Drake	NW Atlantic	9.5	f
43	1970	Chryssi	NW Atlantic	9.5	f
44	1969	Pacocean/Broke in two	NW Pacific	9.2	f
45	1977	Caribbean Sea	E Pacific	9.2	f
46	1976	Grand Zenith/Disappearance	NW Atlantic	8.9	f
47	1976	Cretan Star	Indian Ocean	8.9	bf
48	1969	Keo/Hull failure	Massachusetts	8.8	b
49	1969	Storage Tank	New Jersey	8.4	bf
50	1977	Ekofisk Bravo, Well Blowout	North Sea	4.6-8.2	f
51	1972	Giuseppi Guilietti	NE Atlantic	8	ef
52	1977	Venpet and Venoil/Collision	South Africa	7.4-8	bfh
53	1976	Argo Merchant/Grounding	Massachusetts	7.7	n
54	1967	Humble Oil Pipeline, Offshore Leak	Louisiana	6.7	c
55	1973	Jawacta	Baltic Sea	6.1	c
56	1967	R.C. Stoner	Wake Island	6	c
57	1970	Marlena	Sicily	4.3	c
58	1970	Pipeline	Saudi Arabia	4.2	c
59	1971	Oil Well	Persian Gulf	4.2	j
60	1980	Tanio/Broke amidships	France	4.2	b
61	1988	Ashland Storage Tank, Rupture	Pennsylvania	3.8	dfp
62	1969	Santa Barbara Channel, Well Blowout	California	1.4-3.4	ch
63	1970	Arrow/Grounding	Nova Scotia	1.5-3.1	c
64	1970	Storage Tank	Pennsylvania	3	b
65	1984	Alvenus/Grounding	Louisiana	2.8	c
66	1970	Offshore Platform, Well Blowout	Louisiana	2.7	

a. A list of the 20...., 1989.
b. Reuters, 1989.
c. Van Gelder-Ottway..., 1976.
d. A Basic Spill...., 1981.
e. Lord et al., 1987.
f. Butler, 1978.

g. Woods and Hannah, 1981.
h. Teal and Howarth, 1984.
i. Caleb Brett, 1989.
j. Ganten, 1985.
k. Quina et al., 1987.
l. Horn and Neil, 1981.

m. Bao-Kang, 1987.
n. Tracey, 1988.
o. Ocean Industry, 1980.
p. NRC, 1975.
q. Journal of Commerce, 1/4/90.

Tanker spills from the Iran/Iraq war were not generally available

*Fire burned part of spill

SOURCE: Congress of the United States, Office of Technology Assessment, "Coping With An Oiled Sea: An Analysis of Oil Spill Response Technologies," OTA-BP-O-63 (Washington, D.C.: U.S. Government Printing Office, March 1990), 70p.

Tsuneo Akaha, a native of Japan, received his M.A. and Ph.D. in international relations from the University of Southern California. He is currently an associate professor of international policy studies and Asian studies at the Monterey Institute of International Studies, Monterey, California. He previously taught at the University of Southern California, Kansas State University, and Bowling Green State University. He is author of *Japan in Global Ocean Politics* (Honolulu: University of Hawaii Press and Law of the Sea Institute, 1985) and of numerous articles in such journals as *Asian Survey, Pacific Affairs, Coastal Zone Management Journal, Ocean Development and International Law, Ecology Law Quarterly, Millennium: Journal of International Studies,* and *Peace and Change.* He is also editor of the *International Handbook of Transportation Policy* (New York: Greenwood Press, 1990) and coeditor of the forthcoming *International Political Economy: A Reader* (New York: HarperCollins, 1991).

Francis Auburn is an associate professor in law at the University of Western Australia. He is author of a major work on Antarctica and of numerous articles on Antarctica, the Law of the Sea, constitutional law, conflict of laws, and computers and the law for journals in the United States, England, Canada, West Germany, France, Australia, New Zealand, Denmark, Argentina, Japan, and Ghana.

John E. Bardach is interim director of the Environment and Policy Institute of the East-West Center and adjunct professor of both oceanography and geography at the University of Hawaii. He is known for his work in fisheries, aquaculture, coastal zone management, and climatic change, especially in Asia and the Pacific. He was director of the Hawaii Institute of Marine Biology, University of Hawaii, and board member of the Law of the Sea Institute. He taught for 17 years at the University of Michigan. He received his B.A. from Queens University, Kingston, Ontario (1946), and his M.S. and Ph.D. in zoology from the University of Wisconsin (1948, 1949).

Frank Barnaby, a nuclear physicist by training, worked at the Atomic Weapons Research Establishment, Aldermaston (1951–57), and at University College, London (1957–67). He has been executive secretary of the Pugwash Conferences on Science and World Affairs (1967–70) and director of the Stockholm International Peace Research Institute (1971–81). He was guest professor at the Free University, Amsterdam (1981–85), and is currently a consultant and writer on military technology. His books include *The Invisible Bomb, The Automated Battlefield, The Gaia Peace Atlas* (editor), *Prospects for Peace, Star Wars Brought Down to Earth, Future Warfare* (editor), *Verification Technologies* (coauthor), *Man and the Atom,* and *Nuclear Energy.* Other publications include articles on military technology and defense and security issues.

Patricia W. Birnie, director of the IMO International Maritime Law Institute in Malta (IMLI), is an Oxford, United Kingdom, law graduate, holds a Ph.D. from the University of Edinburgh in Scotland, and is also a nonpracticing barrister. She has specialized for the past 20 years in public international law, teaching this subject at the University of Edinburgh from 1968 to 1983. She taught until 1989, when she became the first director of IMLI at the London School of Economics. She has a special interest in all aspects of the Law of the Sea and has been widely published in these fields. Her publications include *The Maritime Dimension* (London: Allen and Unwin, 1980), coedited with R. Barson, and *The International Regulations of Whaling,* 2 vols. (Dobbs Ferry, New York: Oceana Publications, 1984). She is currently working with A. Boyle on a book on international environmental law, to be published by Oxford University Press in 1992.

Elisabeth Mann Borgese is professor of political science at Dalhousie University, chairman of the Planning Council of the International Ocean Institute, Malta, and chairman of the board of directors of the International Centre for Ocean Development (ICOD), Canada, and was an adviser to the delegation of Austria at UNCLOS III. For some years she was a senior fellow at the Center for the Study of Democratic Institutions, Santa Barbara. She has written numerous books, monographs, and essays on international ocean affairs and marine resource management, including *The Ocean Regime* (1968), *The Drama of the Oceans* (1976), *The New International Economic Order and the Law of the Sea* (with Arvid Pardo, 1976), *Seafarm* (1980), *The Mines of Neptune* (1984), and *The Future of the Oceans* (1986).

Chia Lin Sien, B.S. (Sydney), M.S. (McGill), and Ph.D. (Singapore), is acting head of the geography department at the National University of Singapore. His training is in climatology, and he specializes in maritime geography, which includes maritime traditions, seafarers, marine transportation, Law of the Sea, marine resource and coastal zone management, and pollution of the sea with special reference to Singapore and Southeast Asia. He has numerous publications, among them the jointly authored/edited books *Southeast Asian Transport: Issues in Development, Southeast Asian Seas: Frontiers of Development, Environmental Management in Southeast Asia, Handmaiden of Trade: A Student on*

ASEAN-Australia Shipping, and *The Coastal Environmental Profile of Singapore.* He is a member of the editorial board of the *Singapore Journal of Tropical Geography.*

Daniel J. Dzurek is a research associate in the Environment and Policy Institute of the East-West Center. He studied physics at the Illinois Institute of Technology (B.S.) and the Technische Universität München, under a Fulbright-Hays fellowship, and studied international relations at the University of Chicago (A.M.). Previously, he served in the U.S. Department of State as a senior analyst for nuclear proliferation in the Office of Strategic Forces Analysis and as chief of the Spatial, Environmental, and Boundary Analysis Division in the Office of the Geographer.

Karen L. Eckert received her Ph.D. in zoology and a certificate in global policy studies from the University of Georgia. She has been active for more than a decade in the fields of marine turtle research and international conservation policy. In 1984 she received the National Marine Fisheries Service (Department of Commerce) Recognition Award for "outstanding efforts in sea turtle conservation." She is a member of the U.S. Recovery Team for Marine Turtles and the Marine Turtle Specialist Group of the IUCN Species Survival Commission. Presently she serves as a team member and director of the Secretariat for the Wider Caribbean Sea Turtle Recovery Team and Conservation Network (WIDECAST), a United Nations–sponsored program with the task of developing sea turtle recovery action plans for Caribbean governments. She is particularly interested in the multilateral conservation of depleted migratory species, and her research has taken her to the Caribbean and the South China seas. She has served as a consultant to various governments, as well as to intergovernmental and nongovernment organizations, and has published many scientific and general-interest articles.

Peter M. Haas is a professor of political science at the University of Massachusetts at Amherst, where he researches and teaches on international environmental issues, international cooperation, and north-south politics. He is the author of *Saving the Mediterranean: The Politics of International Environmental Protection* (New York: Columbia University Press, 1990) and a number of articles and chapters on international environmental policy and cooperation. He has served as a consultant for the United Nations Environmental Programme, U.S. National Academy of Sciences, and the World Resources Institute.

Peter Haydon retired from the Canadian navy in 1988 after a 30-year career during which he served in submarines and destroyers, as well as in a wide range of staff positions. The last 9 years of his navy career were spent as a strategic analyst for the Department of National Defense in Ottawa, where he headed a strategic analysis team, and at the headquarters of the Supreme Allied Commander Atlantic in Norfolk, Virginia. Since retiring he has divided his time between graduate studies in international relations at Dalhousie University, Halifax, and acting as a consultant on all aspects of naval strategy to several nongovernment organizations, including the Centre for Foreign Policy Analysis at Dalhousie University, the Canadian Institute for Strategic Studies, and the Canadian Centre for Arms Control and Disarmament. He is a published author and has been a frequent commentator on naval strategic issues. He now makes his home in Bedford, Nova Scotia.

Mochtar Kusuma-Atmadja, of Djakarta, Indonesia, is a professor of law at Padjadjaran University Law School and University of Indonesia Law School in Djakarta, as well as an advisor to the Minister of Justice, Minister of Mines and Energy, and Minister for Foreign Affairs of the Republic of Indonesia. He received his first law degree from the University of Indonesia Law School in September 1955, having received his master of laws (LL.M.) from Yale Law School in June 1955. He obtained his doctor of laws (Ph.D.) from Padjadjaran University Law School in 1962. He was a member of the Indonesian delegation to the UN Conferences on the Law of the Sea in Geneva in 1958 and 1960 and was leader of the Indonesian delegation to the Third UN Conference on the Law of the Sea from 1973 until 1982. In 1970 he became a senior partner of the Mochtar, Karuwin, and Komar law firm in Djakarta and resigned in January 1974 when he became a member of the Indonesian government as Minister of Justice, where he remained until 1978. From 1978 to 1988 he was Minister of Foreign Affairs of the Republic of Indonesia. He rejoined the law firm of Mochtar, Karuwin, and Komar after resigning from the government in February 1988. In 1990 he was appointed to the Panel of Conciliators and Arbitrators of the ICSID, World Bank, Washington, D.C. (He was a member of the same panel from 1970 to 1979.) He is also a member of the *Ocean Development and International Law Journal* editorial board in Washington, D.C., and a member of the board of directors, Law of the Sea Institute, University of Hawaii. He is the author of various books and articles written in Indonesian and English.

Joseph R. Morgan spent 25 years in the U.S. Navy, retiring with the rank of captain. He then obtained his M.A. and Ph.D. degrees in geography from the University of Hawaii. He is currently associate professor of geography at the University of Hawaii, a research associate in the Environment and Policy

Institute at the East-West Center, and coeditor of *Ocean Yearbook*. His primary academic interest is marine political geography.

Michael A. Morris received his Ph.D. in international relations from the School of Advanced International Studies of the Johns Hopkins University. He is a professor in the political science department at Clemson University, South Carolina. While affiliated with Clemson University, he has held several visiting research positions, including one as a fellow with the Stockholm International Peace Research Institute (SIPRI) from 1979 to 1981, and as a senior fellow in the Marine Policy Program at the Woods Hole Oceanographic Institution from 1984 to 1985. During 1987–88 he also was a Fulbright Exchange Professor in Britain. Among his publications are *International Politics and the Sea: The Case of Brazil* (1979), *Expansion of Third-World Navies* (1987), and *The Strait of Magellan* (1989).

Hal F. Olson holds a joint appointment as research associate at the East-West Center, University of Hawaii, and as instructor in geography at the Leeward Community College campus of the University of Hawaii. He is a graduate of the U.S. Coast Guard Academy (B.S., engineering) and has a master's degree in geography from the University of Hawaii at Manoa, as well as a master's license in the U.S. Merchant Marine. Retired as a captain after 24 years' service in the Coast Guard, he has been a consultant to the International Maritime Organization (IMO) and the U.S. Agency for International Development (AID) and has served as coordinator for the Marine Technology Program at Leeward Community College.

Harvey A. Shapiro is a professor in Japan's first Department of Environmental Planning (established in 1971) at Osaka Geijutsu University in western Japan. He did his graduate work at the University of Pennsylvania under Professor Ian L. McHarg, ecological planner and author of the famous book *Design with Nature*. In 1969 Dr. Shapiro conducted the first ecological planning study of a Japanese region. That study provided the basis for his teaching, research, and other professional activities in Japan, where he has lived since July 1970. In addition to his position at Osaka Geijutsu University, he has also taught at Japan's second Department of Environmental Planning at Kobe National University in Kobe City from 1976 to 1984, and he has been teaching in the geography department at Nara University in Nara City since then. He has been a consultant to Japan's first ecological planning firm, R.P.T. Associates, Inc., in Tokyo, since its founding in 1974. His interest in Japan's coastal environment and coastal area management dates from the mid-1970s. Since then he has done extensive research and has published numerous professional papers in Japanese and English on these and related subjects. Among his publications are papers such as "Coastal Area Management in Japan: An Overview" (*Coastal Zone Management Journal* 12 [1984]: 21–56) and "Japan's Coastal Environment and Responses to Its Recent Changes" (in *The Coastal Zone: Man's Response to Change* [Geneva: Harwood Academic Publishers, 1988], pp. 491–520). He is a member of the International Union for the Conservation of Nature's (IUCN) Environmental Planning Commission and has been working in that capacity for the past several years to introduce and apply the concept of "sustainable development" to Japan and its neighbors in East Asia, on some of which he is now doing research and ecological planning studies.

Paul G. Sneed is a Ph.D. candidate in the geography department at the University of Hawaii, Honolulu. He holds a B.A. in anthropology from the University of California at Santa Barbara, and an M.A. in environmental planning from the University of British Columbia. He has had a varied career as an administrator and instructor in several community colleges, as a cultural and natural heritage resource manager for the government of British Columbia, as an environmental consultant, and as an ecotourism entrepreneur in northern Canada. He is currently doing research on ways to apply ecosystem management and landscape-planning approaches to the conservation of biodiversity.

Nancy Yamaguchi is a research fellow with the energy program of the Resource Systems Institute, East-West Center, Honolulu. She has been affiliated with the energy program for over 7 years, and her research efforts have focused on a variety of energy issues in the South Pacific, Asia, and the U.S. West Coast, with emphasis on computer modelling of oil refining, demand, transport, trade, and environmental concerns. She has a B.A. in communication and environmental policy from the University of California at Davis, where she received the Gilhooly Award for Most Outstanding Senior Woman in Service and Scholarship, class of 1981. She did her graduate studies in the geography department of the University of Hawaii and received her M.A. in 1985 and Ph.D. in 1988. Prior to joining the East-West Center, she was employed by the California State Energy Commission in the Conservation Division. Her work has been published in a variety of journals, and she has served as a consultant to a number of agencies and firms in the region, including the United Nations, the government of Indonesia, the Pacific International Center for High Technology Research, and many of the key energy firms operating in the Asia-Pacific region.